AN INTRODUCTION TO DIGITAL COMMUNICATIONS

AN INTRODUCTION TO DIGITAL COMMUNICATIONS

JACK KURZWEIL
San Jose State University

JOHN WILEY & SONS, INC.
New York / Chichester / Weinheim / Brisbane / Singapore / Toronto

Acquistions Editor   *Bill Zobrist*
Marketing Manager   *Katherine Hepburn*
Senior Production Editor   *Robin Factor*
Illustration Editor   *Sigmund Malinowski*
Senior Designer   *Harry Nolan*

This book was set in *New Times Roman* by *TechBooks* and printed and bound by *Hamilton Printing*. The cover was printed by *Phoenix Color.*

The book is printed on acid-free paper.

Library of Congress Cataloging in Publication Data:

Kurzweil, Jack.
    An introduction to digital communications / Jack Kurzweil.
        p.   cm.
    ISBN 0-471-15772-4 (alk. paper)
    1. Digital communications.   I. Title.
  TK5103.7.K87   1999
  621.382–dc21   99-27789
                                                        CIP

Printed in the United States of America

10 9 8 7 6 5 4 3 2 1

to my parents
Sam and Rose Kurzweil
who arrived in this country as immigrants
worked all of their lives
and saw to the education of their children

CONTENTS

PREFACE

This textbook is being published at a moment when the 56 Kbps modem has reached the limit of data transmission on the Plain Old Telephone System (POTS) and the attention of the industry and academic researchers has long since turned to the high-speed digital subscriber loop, coaxial cable, and wireless transmission. This is accompanied by technological changes that make it possible to incorporate the digital signal processing necessary to support high-speed data transmission onto special-purpose chips. Nevertheless, the new technology relies on theory that has been accumulated over decades; indeed, in some instances, it requires the rediscovery of theory that has lain dormant for decades.

Every textbook tells a story that comes from the accumulation of the author's experience. My experience started with some years at Racal-Vadic in Sunnyvale, California, working on the V.22 (bis) modem. This book evolved from developing a two-semester graduate-level course in digital communications at San Jose State University. Since the book comes from my syllabus, it includes only material that I have actually taught. This choice of material is consequently incomplete but does represent one viewpoint of what should go into such a course. This book deals with adaptive equalization, carrier recovery, and timing recovery in greater depth than most other digital communication texts. A central portion of the presentation is the development of simulation programs for testing the performance of different modulation techniques and equalization strategies. On the other hand, there is hardly any discussion of channel coding except in relation to understanding the meaning of channel capacity. There is considerable discussion of the time-domain characterization of channels, because that is the basis for adaptive equalization, but little discussion of continuous-phase modulation. There is no discussion of spread spectrum, satellite communications, or communications protocols. This book asks students to understand the code of the simulation programs and even write some code for the implementation of particular algorithms rather than using communications simulation packages, because it is at the code level that the student is required really to understand how the algorithm works.

In short, I have written this as a relatively focused book that is both theoretical and design-oriented. It is aimed toward graduate students or those advanced undergraduates who are prepared to engage in a combination of somewhat involved mathematics and computer simulation. All students taking the course are assumed to have a background in spectral analysis, z-transforms, Laplace transforms, and probability theory.

I have typically covered the first five to six chapters in the fall semester and the remainder in the spring. It is reasonable to tread lightly over Chapter 1, Fourier Series and Transforms, if the students are well prepared. The review of probability theory in Chapter 2, Spectral Analysis of Data Signals and Noise, is very modest and the discussion of random processes centers on communications applications. Of particular interest is the use of the Matlab Symbolic Toolbox for analyzing Markov processes. Chapter 3, Baseband Data Transmission, and Chapter 4, Bandpass Data Transmission, should be covered in their entirety because they form the basis for understanding most of the modulation techniques currently in use. There is an emphasis on the development of a simulation model for a communication system. An instructor may choose to be selective with Chapter 5, Maximum Likelihood Signal Detection and Some Applications, but I would urge careful attention to the development of the Viterbi Algorithm. This might allow time to begin Chapter 6, Carrier Phase and Timing Recovery. I would urge that this chapter be covered in some detail in the spring semester. The major purpose of Chapter 7, Channel Models for Communication Systems, is to establish the sampled impulse response model for a communication channel and give a brief view of how different channels can be modeled in this way. Extensions and variations of the material in this chapter are an excellent source of ideas for term papers and projects. In Chapter 8, Channel Capacity and coding, I have attempted to provide some kind of intuitive understanding of what the issue of capacity is about. I have also pointed to derivations of the capacity of some channels such as the Digital Subscriber Loop, in which the design of modems to come near achieving capacity continues to be an active project. Chapter 9, Trellis Coding and Multidimensional Signaling, develops the basic ideas of combined modulation and coding and examines a few codes in detail. A deeper look at these codes might very well be part of a separate course on algebraic coding theory. Chapter 10, Equalization of Distorted Channels, deals in detail with baud-rate, fractionally spaced, and decision-feedback equalizers. The performance of these equalizers is derived mathematically and simulated through Matlab scripts. Chapter 11, Adaptive Equalization and Echo Cancellation, examines the LMS algorithm for equalizer convergence as well as algorithms for rapid convergence. A program that simulates convergence for various equalizer structures and convergence algorithms is included.

All scripts and programs are available on the website:

http://www.wiley.com/college/kurzweil

Anyone who has designed modems owes an immense debt of gratitude to the extraordinary collection of researchers at Bell Labs who essentially invented the bulk of the theory. An examination of the references at the ends of the chapters shows these names repeated many times over. In particular, I have been heavily influenced by the work that culminated in *Data Communication Principles*, by Richard Gitlin, Jeremiah Hayes, and Stephen Weinstein, Plenum, 1992. As my book was taking shape, theirs was the assigned text for my course.

Since no single textbook can adequately address all of the issues of digital communications and each of them has something unique to offer, the reader will not be surprised to find references to the many fine books in this area by E. Lee and D. Messerschmitt; J. Proakis; B. Sklar; D. Smith; I. Korn; S. Benedetto, E. Biglieri, and V. Castellani; R. Zeimer and R. Peterson; A.P. Clark; and J. Bergmans.

I want to thank Kim Maxwell and John A.C. Bingham for taking the risk of bringing a somewhat theoretical and abstract academic into the R&D process at Racal-Vadic. John Bingham has, over the years, been a technical sounding board as well as a friend. His two books on modem design as well as his papers are prominent among the references. Every fumbling engineer needs a great technician and I was particularly blessed with Duane Marcroft, who guided me in the translation of ideas into reality.

My friends Steve Willet and Bill Swanson were always there to help with programming problems. Carlos Puig, lent a hand with problem solutions and many excellent ideas and insights. John Cavalli produced many of the simulations. Jack Stotes Berry kept my computer running.

I did not actually have the idea of translating my notes into a book until a Wiley representative wandered into my office with the innocent-sounding question, "Are you working on anything?" The Wiley staff have been most helpful and, when appropriate, firm in bringing this project to fruition. I would like to thank Bill Zobrist, Penny Perrota, Jennifer Welter, Robin Factor, and Andrew Wilson. I also want to thank the reviewers for their very helpful suggestions and insights: S. Hossein Mousavinezhad, Western Michigan

University; James S. Kang, California State Polytechnic University; John R. Barry, Georgia Institute of Technology; Haniph A. Latchman, University of Florida, Rich S. Blum, Lehigh University; Vijay K. Bhargava, University of Victoria; and Upamanyu Madhow, University of Illinois. San Jose State University allowed me to make a major leap in transforming a set of notes into a book by granting me a sabbatical leave.

I hope that the errors in this book are few and not very serious. They are, of course, entirely my responsibility.

On a more personal note, thanks go to Jennifer, Joshua, and Shoshana, who seemed to grow from adolescence to adulthood as this project proceeded, for their wonderful support. Finally, I want to pay tribute to Dr. Claudette Hoover, Professor of Literature and Women's Studies, my wife and strongest advocate, who passed away on December 21, 1997.

AN INTRODUCTION TO DIGITAL COMMUNICATIONS

FOURIER SERIES AND TRANSFORMS

FOURIER SERIES and transforms form the mathematical basis for a discussion of both analog and digital communications. This chapter presents a review of this material, one that is specifically directed toward digital communications and that contains a number of examples that will be developed in subsequent chapters.

1.1 THE DEFINITION OF THE FOURIER SERIES

Let us begin the definition by considering the real function $f(t)$ shown in fig. 1.1-1, defined on the finite time interval $[t_1, t_1 + T)$ or, equivalently, $t_1 \leq t < (t_1 + T)$. We want to represent this function on that time interval by an infinite series having the form

$$\hat{f}(t) = f_0 + 2 \sum_{n=1}^{\infty} |f_n| \cos(n\omega_0 t + \theta_n)$$

$$= \sum_{n=-\infty}^{\infty} f_n e^{jn\omega_0 t} \tag{1.1-1a}$$

where $\omega_0 = 2\pi / T$ is called the *fundamental frequency*. The frequencies contained in series are all integer multiples of the fundamental and $n\omega_0$ is referred to as the *nth harmonic*. Note that $\hat{f}(t)$ has been used in eq. 1.1-1a in order to make a temporary distinction between $f(t)$ and its series representation. The first form of the series is the trigonometric form and the second is the exponential form. For these two representations to be equivalent it is necessary that $f_n = f_{-n}^*$, which means that

$$f_n = |f_n| e^{j\theta_n} \qquad \text{and} \qquad f_{-n} = |f_n| e^{j\theta_{-n}} \tag{1.1-1b}$$

or

$$|f_n| = |f_{-n}| \qquad \text{and} \qquad \theta_n = -\theta_{-n} \tag{1.1-1c}$$

A graphical representation of the f_n coefficients in the trigonometric representation therefore always looks like fig. 1.1-2. The graph of $|f_n|$ vs. ω in fig. 1.1-2a, known as the *amplitude spectrum*, is always an even function of ω and the graph of θ_n vs. ω in fig. 1.1-2b, the *phase spectrum*, is always odd in ω. Since the f_n coefficients exist only for discrete values of $\omega = n\omega_0$, both the amplitude and phase spectra are characterized as discrete or line spectra. The coefficient f_0 is the average value or dc component of $f(t)$. These spectra are understood to be the *frequency-domain* representation of $f(t)$.

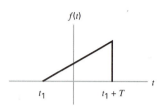

FIGURE 1.1-1

The average power of $f(t)$ over the defined time interval is

$$P_{av} = \frac{1}{T} \int_{t_1}^{t_1+T} f^2(t)\,dt = \frac{1}{T} \int_{t_1}^{t_1+T} \sum_{n=-\infty}^{\infty} f_n e^{jn\omega_0 t} \sum_{m=-\infty}^{\infty} f_m^* e^{-jm\omega_0 t}\,dt$$

$$= \frac{1}{T} \sum_{n=-\infty}^{\infty} f_n \sum_{m=-\infty}^{\infty} f_m^* \int_{t_1}^{t_1+T} e^{j(n-m)\omega_0 t}\,dt$$

$$= \sum_{n=-\infty}^{\infty} |f_n|^2 \tag{1.1-2a}$$

since

$$\int_{t_1}^{t_1+T} e^{j(n-m)\omega_0 t}\,dt = \begin{cases} T; & m = n \\ 0; & m \neq n \end{cases} \tag{1.1-2b}$$

Equation 1.1-2a is an expression of *Parseval's Theorem*, which relates average power computed in the time domain to average power computed in the frequency domain. The simplicity of the frequency-domain expression for average power is a consequence of eq. 1.1-2b. As we shall see in Chapter 5, this is an example of the orthogonality of functions over a given time interval.

We now turn to the problem of finding the set of Fourier Series coefficents f_n that will make the representation of eq. 1.1-1a as good as possible. First define the finite series

$$f_N(t) = \sum_{n=-N}^{N} f_n e^{jn\omega_0 t} \tag{1.1-3a}$$

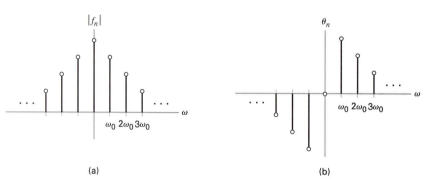

(a)   (b)

FIGURE 1.1-2

The error between $f(t)$ and $f_N(t)$ is defined as

$$e_N(t) = f(t) - f_N(t) \tag{1.1-3b}$$

for $t \in [t_1, t_1 + T)$. We want to choose the f_n coefficients in order to minimize the energy in $e_N(t)$. This energy is given by

$$\varepsilon_N = \int_{t_1}^{t_1+T} |e_N(t)|^2 \, dt = \int_{t_1}^{t_1+T} (f(t) - f_N(t))(f(t) - f_N^*(t)) \, dt$$

$$= \int_{t_1}^{t_1+T} \left(f(t) - \sum_{n=-N}^{N} f_n e^{jn\omega_0 t} \right) \left(f(t) - \sum_{m=-N}^{N} f_m^* e^{-jm\omega_0 t} \right) dt \tag{1.1-3c}$$

We will achieve the minimization by taking the partial derivative of ε_N with respect to each of the f_n coefficients and setting each result to zero. According to the rules of partial differentiation,

$$\frac{\partial \varepsilon_N}{\partial f_n} = \int_{t_1}^{t_1+T} \left[-f(t) e^{jn\omega_0 t} + e^{jn\omega_0 t} \sum_{m=-N}^{N} f_m^* e^{-jm\omega_0 t} \right] dt = 0 \tag{1.1-3d}$$

or, using eq. 1.1-2b,

$$\int_{t_1}^{t_1+T} \left[f(t) e^{jn\omega_0 t} \right] dt = \sum_{m=-N}^{N} f_m^* \int_{t_1}^{t_1+T} e^{-j(n-m)\omega_0 t} \, dt = T f_n^* \tag{1.1-3e}$$

so that:

$$f_n = \frac{1}{T} \int_{t_1}^{t_1+T} \left[f(t) e^{-jn\omega_0 t} \right] dt \tag{1.1-3f}$$

This is the required expression for the Fourier Series coefficient.
    Substituting eq. 1.1-3f into eq. 1.1-3c, we get

$$\varepsilon_N = \int_{t_1}^{t_1+T} f^2(t) \, dt - T \sum_{n=-N}^{N} |f_n|^2 = E - E_N \tag{1.1-3g}$$

where E_N is the energy in $f_N(t)$. Since $|f_n|^2$ is a positive quantity, it follows that ε_N will get smaller as N increases, going to zero asymptotically. Consequently eq. 1.1-1a leads to the conclusion that in the time interval $t \in [t_1, t_1 + T)$

$$f_N(t) \xrightarrow[N \to \infty]{} \hat{f}(t) = f(t) \tag{1.1-3h}$$

on an energy basis. This means that the series and the function $f(t)$ will be equal except at some isolated points that contribute zero energy. We state without proof that, at points of discontinuity the series will equal the average value of $f(t)$ at the discontinuity.
    We may now note that $\hat{f}(t)$, as defined in eq. 1.1-1a, is periodic with period T. Consequently if $f(t)$ is also periodic with period T, then $f(t)$ and the series are equal everywhere except at discontinuities.

1.2   EXAMPLES

EXAMPLE 1   *Periodic Rectangular Pulses*

The object is to find the f_n coefficients for the periodic rectangular pulses shown in fig. 1.2-1. A single rectangular pulse is defined as

$$\text{Pa}(t) = \begin{cases} 1; & -\frac{a}{2} \le t \le \frac{a}{2} \\ 0; & \text{otherwise} \end{cases} \tag{1.2-1a}$$

and the periodic version of this pulse is defined as

$$\text{Pa}(t, T) = \sum_{n=-\infty}^{\infty} \text{Pa}(t - nT) \tag{1.2-1b}$$

The *duty cycle* $= a/T$ of $f(t) = A\,\text{Pa}(t, T)$ is defined as the ratio of the time on to the period. According to eq. 1.1-3f, the Fourier Series coefficients for this function are given by

$$f_n = \frac{1}{T} \int_{t_1}^{t_1+T} \left[f(t)e^{-jn\omega_0 t} \right] dt \tag{1.1-3f}$$

where t_1 is arbitrary. We have deliberately chosen $f(t)$ to be centered at the origin and consequently chosen $t_1 = -T/2$. A Time-Shifting Theorem, which will enable us easily to find the Fourier coefficients of a shifted waveform from those of the symmetrical one, will be developed subsequently. Consequently:

$$f_n = \frac{1}{T} \int_{-T/2}^{T/2} f(t)e^{-jn\omega_0 t}\, dt = \frac{A}{T} \int_{-a/2}^{a/2} e^{-jn\omega_0 t}\, dt = \frac{A}{T} \frac{e^{-jn\omega_0 t}}{jn\omega_0} \Big|_{-a/2}^{a/2}$$

$$= A\frac{a}{T} \frac{\text{Sin}(n\omega_0\, a/2)}{(n\omega_0\, a/2)} = A\frac{a}{T} \frac{\text{Sin}(n\pi a/T)}{(n\pi a/T)} \tag{1.2-1c}$$

These coefficients can also be expressed in terms of the function

$$\text{Sinc}\,(x) = \frac{\text{Sin}(x)}{x} \tag{1.2-2a}$$

which is shown in fig. 1.2-2. The apparent discontinuity at $x = 0$ is resolved by using L'Hospital's Rule to show that

$$\text{Sinc}\,(0) = \lim_{x \to 0} \frac{\frac{d}{dx}\text{Sin}(x)}{\frac{d}{dx}x} = \lim_{x \to 0} \frac{\text{Cos}(x)}{1} = 1 \tag{1.2-2b}$$

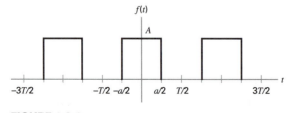

FIGURE 1.2-1

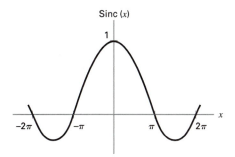

FIGURE 1.2-2

It has been shown that in general $f_n = f^*_{-n}$ but in this case f_n is real so that $f_n = f_{-n}$. As we shall subsequently see, when f_n is negative it will be interpreted as having an angle of an odd multiple of π.

We will now examine some specific examples in order to get a better idea of how this expression for f_n is actually used.

$a = T/4$

This waveform is shown in fig. 1.2-3a. According to eq. 1.2-1c,

$$f_n = A\frac{a}{T}\frac{\mathrm{Sin}\left(n\pi a/T\right)}{\left(n\pi a/T\right)} = \frac{A}{4}\frac{\mathrm{Sin}(n\pi/4)}{(n\pi/4)} \tag{1.2-3a}$$

$$f_0 = \frac{A}{4}; \quad \text{the dc component}$$

$$f_n = \frac{A}{n\pi}\mathrm{Sin}(n\pi/4); \quad \text{for } n \neq 0$$

$$f_1 = f_{-1} = \frac{A}{\pi}\mathrm{Sin}(\pi/4) = \frac{A}{\sqrt{2}\pi}$$

$$f_2 = f_{-2} = \frac{A}{2\pi}\mathrm{Sin}(\pi/2) = \frac{A}{2\pi}$$

$$f_3 = f_{-3} = \frac{A}{3\pi}\mathrm{Sin}(3\pi/4) = \frac{A}{3\sqrt{2}\pi}$$

$$f_4 = f_{-4} = \frac{A}{4\pi}\mathrm{Sin}(\pi) = 0$$

$$f_5 = f_{-5} = \frac{A}{5\pi}\mathrm{Sin}(5\pi/4) = -\frac{A}{5\sqrt{2}\pi}$$

and so forth. The spectrum is shown in fig. 1.2-3b. Note that the components corresponding to $n = \pm4, \pm8, \pm12, \ldots$ are all zero. Recalling that $f_n = |f_n|e^{j\theta_n}$, a real and negative value of f_n corresponds to a value of $\theta_n = \pm n\pi$. With this, the Fourier Series expression for this waveform is

$$f(t) \approx f_5(t) = f_0 + 2\sum_{n=1}^{5}|f_n|\mathrm{Cos}(n\omega_0 t + \theta_n)$$

$$= \frac{A}{4} + \frac{A}{\pi\sqrt{2}}\left[\mathrm{Cos}(\omega_0 t) + \frac{1}{\sqrt{2}}\mathrm{Cos}(2\omega_0 t) + \frac{1}{3}\mathrm{Cos}(3\omega_0 t) - \frac{1}{5}\mathrm{Cos}(5\omega_0 t)\right]$$

$$\tag{1.2-3b}$$

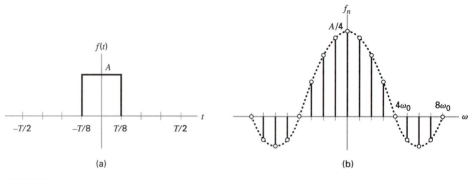

FIGURE 1.2-3

We also know from eq. 1.1-2a that the average power in the waveform can be computed both in the time domain and in the frequency domain.

$$P_{\mathrm{av}} = \frac{1}{T} \int\limits_{-T/2}^{T/2} f^2(t)\, dt = f_0^2 + 2 \sum_{n=1}^{\infty} |f_n|^2 \tag{1.1-2a}$$

A simple time-domain computation shows that $P_{\mathrm{av}} = A^2/4$ so that the ratio of the power, P_n, in the first n components to the average power is, for $N = 3$

$$\frac{P_3}{P_{\mathrm{av}}} = \frac{\frac{A^2}{16} + 2\frac{A^2}{2\pi^2}\left[1 + \frac{1}{2} + \frac{1}{9}\right]}{A^2/4} = \frac{1}{4} + \frac{4}{\pi^2}\left[1 + \frac{1}{2} + \frac{1}{9}\right] = .902 \tag{1.2-3c}$$

It is a simple matter to extend this computation to any value of N.

This computation is significant because it is a measure of the effective bandwidth of the waveform. Analytically, the spectrum would appear to be infinite in extent but it is clear that the bulk of the power is confined to a relatively small band. The effective bandwidth may then be defined as the band containing a specified fraction of the total average power.

$$a = T/2$$

This waveform is shown in fig. 1.2-4a. Since it is regarded as a square wave version of the cosine function, it is often referred to as "cos". The computation of the f_n coefficients

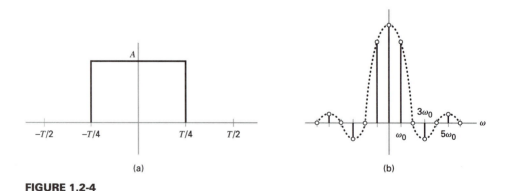

FIGURE 1.2-4

follows the previous example:

$$f_n = A\frac{a}{T}\frac{Sin(n\pi a/T)}{(n\pi a/T)} = \frac{A}{2}\frac{Sin(n\pi/2)}{(n\pi/2)} \tag{1.2-4a}$$

$$f_0 = \frac{A}{2}; \quad \text{the dc component}$$

$$f_n = \frac{A}{n\pi}Sin(n\pi/2); \quad \text{for } n \neq 0$$

$$f_1 = f_{-1} = \frac{A}{\pi}Sin(\pi/2) = \frac{A}{\pi}$$

$$f_2 = f_{-2} = \frac{A}{2\pi}Sin(\pi) = 0$$

$$f_3 = f_{-3} = \frac{A}{3\pi}Sin(3\pi/2) = -\frac{A}{3\pi}$$

$$f_4 = f_{-4} = \frac{A}{4\pi}Sin(\pi) = 0$$

$$f_5 = f_{-5} = \frac{A}{5\pi}Sin(5\pi/2) = \frac{A}{5\pi}$$

A plot of these coefficients is shown in fig. 1.2-4b. The spectrum of this "cos" square wave is quite important in a variety of engineering applications and is one of those things that a student should simply know.

The trigonometric series is given by

$$\text{"Cos"}(t) \approx \frac{A}{2} + \frac{2A}{\pi}\left[Cos(\omega_0 t) - \frac{1}{3}Cos(3\omega_0 t) + \frac{1}{5}Cos(5\omega_0 t) + \cdots\right] \tag{1.2-4b}$$

By delaying the "cos" square wave by one quarter of a period, $t = T/4$ or $90°$, we get the closely related waveform "sin". The Fourier Series for this waveform is given by:

$$\begin{aligned}\text{"sin"}(t) &\approx \frac{A}{2} + \frac{2A}{\pi}\left[Cos\left[\omega_0\left(t - \frac{T}{4}\right)\right] - \frac{1}{3}Cos\left[3\omega_0\left(t - \frac{T}{4}\right)\right]\right.\\ &\quad \left. + \frac{1}{5}Cos\left[5\omega_0\left(t - \frac{T}{4}\right)\right] + \cdots\right]\\ &= \frac{A}{2} + \frac{2A}{\pi}\left[Sin(\omega_0 t) + \frac{1}{3}Sin(3\omega_0 t) + \frac{1}{5}Sin(5\omega_0 t) + \cdots\right]\end{aligned} \tag{1.2-4c}$$

In drawing these spectra we have made use of the continuous version of the Fourier Series coefficient

$$A\frac{a}{T}\frac{Sin(n\omega_0 a/2)}{(n\omega_0 a/2)} \xrightarrow{n\omega_0 \to \omega} A\frac{a}{T}\frac{Sin(\omega a/2)}{(\omega a/2)} \tag{1.2-5}$$

in order to bound the spectrum. This continuous curve is called the "envelope" of the spectrum. Note that the first zero of this function occurs at $\omega = \pm 2\pi/a$. This interval is called the *main lobe*. The intervals between successive zeros are called the sidelobes. ∎

EXAMPLE 2 *Periodic Cosine Pulses*

Following the logic of the previous example in which we analyzed periodic rectangular pulses, we now want to find the f_n coefficients for the periodic cosine pulses shown in fig. 1.2-5a. A single cosine pulse is defined as

$$\text{Ca}(t) = \begin{cases} \text{Cos}(\frac{\pi}{a}t); & -\frac{a}{2} \leq t \leq \frac{a}{2} \\ 0; & \text{otherwise} \end{cases} \tag{1.2-6a}$$

and the periodic version of this pulse is defined as

$$\text{Ca}(t, T) = \sum_{n=-\infty}^{\infty} \text{Ca}(t - nT). \tag{1.2-6b}$$

Using eq. (1.1-3f),

$$f_n = A\frac{\pi}{2}\frac{a}{T}\frac{\text{Cos}(n\omega_0 a/2)}{(\pi/2)^2 - (n\omega_0 a/2)^2} \tag{1.2-6c}$$

Note that this expression is strictly real so that, as in the previous example, the angle of f_n will be a multiple of π. In order to find the main lobe of the envelope of this spectrum we may replace $n\omega_0 \rightarrow \omega$, a continuous variable.

$$f_n \rightarrow A\frac{\pi}{2}\frac{a}{T}\frac{\text{Cos}(\omega a/2)}{(\pi/2)^2 - (\omega a/2)^2} \tag{1.2-6d}$$

The spectrum is shown in fig. 1.2-5b. L'Hospital's rule can be used to show that there is no discontinuity at $\omega = \pm\pi/a$. Consequently, the main lobe extends to $\omega = \pm 3\pi/a$. ∎

EXAMPLE 3 *Periodic Triangular Pulses*

We now want to find the f_n coefficients for the periodic triangular pulses shown in fig. 1.2-6a. A single triangular pulse is defined as

$$\text{Ta}(t) = \begin{cases} 1 + \dfrac{2}{a}t; & -\frac{a}{2} \leq t < 0 \\ 1 - \dfrac{2}{a}t; & 0 \leq t \leq \frac{a}{2} \\ 0; & \text{otherwise} \end{cases} \tag{1.2-7a}$$

and the periodic version of this pulse is defined as

$$\text{Ta}(t, T) = \sum_{n=-\infty}^{\infty} \text{Ta}(t - nT). \tag{1.2-7b}$$

The Fourier Series coefficients of this waveform are:

$$f_n = \frac{A}{2}\frac{a}{T}\frac{\text{Sin}^2(n\omega_0 a/4)}{(n\omega_0 a/4)^2} = \frac{A}{2}\frac{a}{T}\text{Sinc}^2(n\omega_0 a/4) \tag{1.1-3c}$$

Note that in this example the main lobe extends to $\omega = \pm 4\pi/a$, as shown in fig. 1.2-6b.

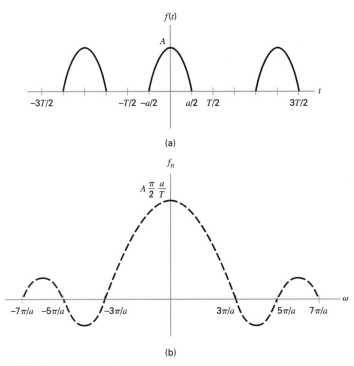

(a)

(b)

FIGURE 1.2-5

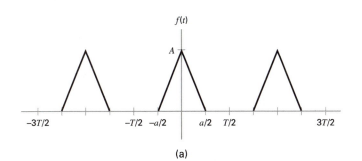

(a)

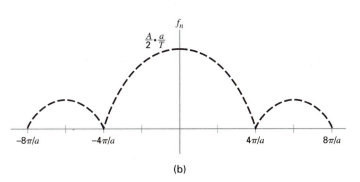

(b)

FIGURE 1.2-6

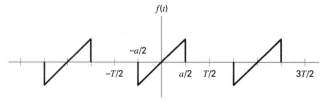

FIGURE 1.2-7

EXAMPLE 4   *Periodic Ramp Functions*

In the previous examples the functions $f(t)$ were even with respect to the origin, $f(t) = f(-t)$, and consequently the f_n coefficients were strictly real. By way of contrast in this example, as shown in fig. 1.2-7, the function that we are analyzing is odd, $f(t) = -f(-t)$, and the f_n coefficients will be strictly imaginary.

A single ramp pulse is defined as

$$\text{Ra}(t) = \begin{cases} \dfrac{t}{a}; & -\dfrac{a}{2} \leq t < \dfrac{a}{2} \\ 0; & \text{otherwise} \end{cases} \qquad (1.2\text{-}8a)$$

and the periodic version of this pulse is defined as

$$\text{Ra}(t, T) = \sum_{n=-\infty}^{\infty} \text{Ra}(t - nT). \qquad (1.2\text{-}8b)$$

The Fourier Series coefficients for this waveform are:

$$f_n = j\frac{A}{2}\frac{a}{T}\left[\frac{\text{Cos}(n\omega_0 a/2)}{(n\omega_0 a/2)} - \frac{\text{Sin}(n\omega_0 a/2)}{(n\omega_0 a/2)^2}\right] \qquad (1.2\text{-}8c)$$

The interpretation of f_n as imaginary is actually quite straightforward. Recall that in general the Fourier Series coefficients are considered to be complex, i.e.,

$$f_n = |f_n|e^{j\theta_n} \Rightarrow 2|f_n|\,\text{Cos}(n\omega_0 t + \theta_n) \qquad (1.1\text{-}1)$$

and that $j = e^{j\pi/2}$. In effect, an imaginary f_n coefficient turns a Cos term into a Sin term. ∎

EXAMPLE 5   *Periodic Impulse Functions*

The periodic impulse functions shown in fig. 1.2-8 have been displaced from the origin by an amount t_0 in order to give a somewhat more general result and provide a preview of the Time-Shifting Theorem, which will be discussed in the next section.

The periodic impulse train has a special name $\delta_T(t)$ where

$$\delta_T(t - t_0) = \sum_{n=-\infty}^{\infty} \delta(t - t_0 - nT) \qquad (1.2\text{-}9a)$$

To find the Fourier Series coefficients:

$$f_n = \frac{1}{T}\int_{-T/2}^{T/2} f(t)e^{-jn\omega_0 t}\,dt = \frac{A}{T}\int_{-T/2}^{T/2} \delta(t - t_0)e^{-jn\omega_0 t}\,dt = \frac{A}{T}e^{-jn\omega_0 t_0} \qquad (1.2\text{-}9b)$$

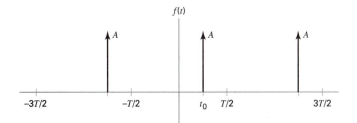

FIGURE 1.2-8

where $|f_n| = A/T$ and $\theta_n = -n\omega_0 t_0$. This means that all of the f_n coefficients are equal in magnitude for periodic impulses. The phase of the coefficents depends upon the time displacement from the origin.

All of the following representations are equivalent:

$$f(t) = \delta_T(t - t_0) = \sum_{n=-\infty}^{\infty} \delta(t - t_0 - nT)$$

$$= \sum_{n=-\infty}^{\infty} \frac{1}{T} e^{-jn\omega_0 t_0} = \frac{1}{T} \left[1 + 2 \sum_{n=1}^{\infty} \cos(n\omega_0(t - t_0)) \right] \qquad (1.2\text{-}9c)$$

and all will be used at some point in the text.

1.3 PROPERTIES OF THE FOURIER SERIES

The Fourier Series has a collection of mathematical properties, some of which are given below. These properties have physical significance and are directly related to specific applications.

THEOREM 1 *Linearity and Superposition*

If two signals $f_1(t)$ and $f_2(t)$ are each periodic with the same period T and their respective Fourier coefficients are f_{n_1} and f_{n_2}, then the Fourier coefficients of $f(t) = af_1(t) + bf_2(t)$ are $f_n = af_{n_1} + bf_{n_2}$ where a, b are any real numbers.

Proof:

$$f_n = \frac{1}{T} \int_{t_1}^{t_1+T} f(t)e^{-jn\omega_0 t}\, dt = \frac{1}{T} \int_{t_1}^{t_1+T} [af_1(t) + bf_2(t)]e^{-jn\omega_0 t}\, dt$$

$$= a \left[\frac{1}{T} \int_{t_1}^{t_1+T} f_1(t)e^{-jn\omega_0 t}\, dt \right] + b \left[\frac{1}{T} \int_{t_1}^{t_1+T} f_2(t)e^{-jn\omega_0 t}\, dt \right]$$

$$= af_{n_1} + bf_{n_2} \qquad (1.3\text{-}1)$$

end of proof

THEOREM 2   *Time Shifting*

Let $f(t)$ be a periodic function with period T having Fourier coefficients f_n. Then the Fourier coefficients of $f(t - t_0)$, the function shifted to the right by an amount t_0, are

$$f(t - t_0) \Leftrightarrow f_n e^{-jn\omega_0 t_0} \tag{1.3-2a}$$

Similarly, the Fourier coefficents of the function shifted to the left by t_0, $f(t + t_0)$, are given by

$$f(t + t_0) \Leftrightarrow f_n e^{jn\omega_0 t_0} \tag{1.3-2b}$$

Proof:

$$\tilde{f}_n = \frac{1}{T} \int_{t_1}^{t_1+T} f(t - t_0) e^{-jn\omega_0 t}\, dt = \frac{1}{T} \int_{t_1+t_0}^{t_1+t_0+T} f(t') e^{-jn\omega_0 (t'+t_0)}\, dt'$$

$$= \left[\frac{1}{T} \int_{t_1'}^{t_1'+T} f(t') e^{-jn\omega_0 t'}\, dt' \right] e^{-jn\omega_0 t_0} = f_n e^{-jn\omega_0 t_0}$$

end of proof

Another way of looking at the Time-Shifting Theorem is to consider the idea of distortionless transmission through a system consisting of only a time delay and no amplitude distortion. Such a system has a transfer function

$$H(\omega) = 1 \cdot e^{-j\omega t_0} \tag{1.3-2c}$$

with unit magnitude and a linear phase shift. If the input to this system is

$$f(t) = \sum f_n e^{-jn\omega_0 t_0} \tag{1.3-2d}$$

then the output is

$$g(t) = \sum_n f_n H(n\omega_0) e^{-jn\omega_0 t} = \sum_n f_n e^{-jn\omega_0 (t-t_0)} = f(t - t_0) \tag{1.3-2d}$$

∎

EXAMPLE 6   *Using Linearity, Superposition, and Time Shifting*

Recalling that $\mathrm{Pa}(t)$ is a rectangular pulse, having width equal to a, centered at the origin, we can express the function $f(t)$ shown in fig. 1.3-1a as

$$f(t) = \mathrm{Pa}(t - t_0) + \mathrm{Pa}(t + t_0) \tag{1.3-3a}$$

Recalling the expression for the Fourier coefficients for $\mathrm{Pa}(t)$ and using both theorems developed above, Time Shifting as well as Linearity and Superposition, the Fourier coefficients for $f(t)$ are

$$f_n = A\frac{a}{T}\mathrm{Sinc}(n\omega_0 a/2)e^{-jn\omega_0 t_0} + A\frac{a}{T}\mathrm{Sinc}(n\omega_0 a/2)e^{+jn\omega_0 t_0}$$

$$= A\frac{a}{T}\mathrm{Sinc}(n\omega_0 a/2)\left[e^{-jn\omega_0 t_0} + e^{+jn\omega_0 t_0}\right]$$

$$= 2A\frac{a}{T}\mathrm{Sa}(n\omega_0 a/2)\mathrm{Cos}(n\omega_0 t_0) \tag{1.3-3b}$$

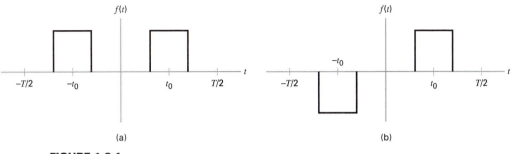

FIGURE 1.3-1

On the other hand, we may define a composite waveform as

$$g(t) = \text{Pa}(t - t_0) - \text{Pa}(t + t_0) \qquad (1.3\text{-}3c)$$

as shown in fig. 1.3-1b.

$$g_n = A\frac{a}{T}\text{Sinc}\,(n\omega_a/2)e^{-jn\omega_0 t_0} - A\frac{a}{T}\text{Sinc}\,(n\omega_0 a/2)e^{+jn\omega_0 t_0}$$

$$= A\frac{a}{T}\text{Sinc}\,(n\omega_0 a/2)\left[e^{-jn\omega_0 t_0} - e^{+jn\omega_0 t_0}\right]$$

$$= -2jA\frac{a}{T}\text{Sinc}\,(n\omega_0 a/2)\,\text{Sin}(n\omega_0 t_0) \qquad (1.3\text{-}3d)$$

∎

The next operation that we shall examine, frequency shifting, is central to understanding a variety of modulation techniques that are considered in this text. Before making a formal statement of the theorem, we will see what happens when two sinusoids are multiplied together. For puposes of illustration, let ω_1 be a low frequency and let ω_c be a relatively higher one. Then

$$[2|f_1|\cos(\omega_1 t + \theta_1)][2\cos(\omega_c t)]$$

$$= \left[f_1 e^{j\omega_1 t} + f_{-1}e^{-j\omega_1 t}\right]\left(e^{j\omega_c t} + e^{-j\omega_c t}\right)$$

$$= \left[f_1 e^{j(\omega_c + \omega_1)t} + f_{-1}e^{j(\omega_c - \omega_1)t}\right] + \left[f_1 e^{-j(\omega_c + \omega_1)t} + f_{-1}e^{-j(\omega_c - \omega_1)t}\right] \qquad (1.3\text{-}4a)$$

According to fig. 1.3-2, the two spectral lines representing the sinusoid $[2|f_1|\cos(\omega_1 t + \theta_1)]$ are shifted to the right and the left to be centered around the spectral lines representing $[2\cos(\omega_c t)]$. As we can see from the mathematics, multiplication by $e^{j\omega_c t}$ causes a right shift and $e^{-j\omega_c t}$ causes a left shift. Gathering terms, we get that

$$[2|f_1|\cos(\omega_1 t + \theta_1)][2\cos(\omega_c t)]$$

$$= 2|f_1|\cos[(\omega_c + \omega_1)t + \theta_1] + 2|f_1|\cos[(\omega_c - \omega_1)t + \theta_1] \qquad (1.3\text{-}4b)$$

More generally, if a periodic function $f(t)$, represented by its Fourier Series, is multiplied by $2\cos(\omega_c t)$, the result is:

$$2f(t)\cos(\omega_c t) = \sum_{n=-\infty}^{\infty} f_n e^{j(\omega_c - n\omega_0)t} + \sum_{n=-\infty}^{\infty} f_n e^{j(\omega_c + n\omega_0)t} \qquad (1.3\text{-}4c)$$

shifting the spectrum to the right and to the left. Since $\text{Sin}(\omega_c t) = (e^{j\omega_c t} - e^{-j\omega_c t})/2j$, it

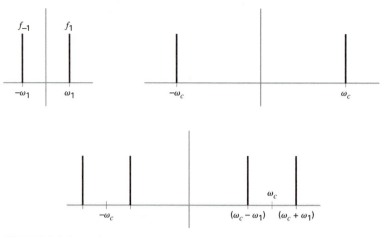

FIGURE 1.3-2

also follows that

$$2f(t)\operatorname{Sin}(\omega_c t) = \sum_{n=-\infty}^{\infty} j f_n e^{j(\omega_c - n\omega_0)t} - \sum_{n=-\infty}^{\infty} j f_n e^{j(\omega_c + n\omega_0)t} \qquad (1.3\text{-}4d)$$

We are now able to make a formal statement of the Frequency-Shifting Theorem.

THEOREM 3   *Frequency Shifting*

We denote the Fourier Series coefficient of a periodic signal $f(t)$ as

$$f_n[n\omega_0] = \frac{1}{T} \int_{t_1}^{t_1+T} f(t)e^{jn\omega_0 t}\, dt \qquad (1.3\text{-}5a)$$

where the $[\ldots]$ signifies that f_n is a function of the argument $n\omega_0$. The Fourier Series coefficients of $f(t)e^{j\omega_c t}$ is $f_n[n\omega_0 - \omega_c]$.

Proof:

The Fourier Series coefficient of $f(t)e^{j\omega_c t}$ is given by

$$\frac{1}{T} \int_{t_1}^{t_1+T} \left[f(t)e^{j\omega_c t} \right] e^{-j\omega_0 t}\, dt = \frac{1}{T} \int_{t_1}^{t_1+T} f(t)e^{-j(n\omega_0 - \omega_c)t}\, dt = f_n[n\omega_0 - \omega_c] \qquad (1.3\text{-}5b)$$

indicating a right shift.

end of proof

∎

1.4   APPLICATIONS OF FOURIER SERIES

1.4.1 Application 1 Double-Sideband Suppressed Carrier (DSB-SC) Amplitude Modulation

Double-Sideband Suppressed Carrier (DSB-SC) Amplitude Modulation is a very direct application of the frequency-shifting theorem. As shown in fig. 1.4-1a, a message signal

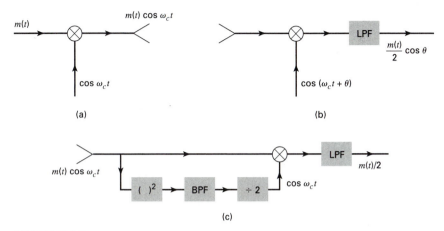

FIGURE 1.4-1

$m(t)$ is multiplied by a carrier $\mathrm{Cos}(\omega_c t)$ where the carrier frequency is typically greater than twice the highest frequency in $m(t)$. For purposes of illustration, let $m(t) = \mathrm{Cos}(\omega_1 t)$ where $\omega_c > 2\omega_1$. The frequency-shifting theorem then operates on the signal

$$m(t)\,\mathrm{Cos}(\omega_c t) = \mathrm{Cos}(\omega_1 t)\,\mathrm{Cos}(\omega_c t) \qquad (1.4\text{-}1a)$$

as shown in fig. 1.3-2. Observe that frequency shifting involves shifting the message (or baseband) signal to the right and to the left, around the carrier spectral line. In the result, the spectral line of the carrier no longer exists: hence, a suppressed carrier signal.

The DSB-SC receiver is structurally very straightforward. If the demodulating carrier has the same frequency and phase as the transmitter carrier, then, as shown in fig. 1.4-1b, the incoming signal $m(t)\,\mathrm{Cos}(\omega_c t)$ is multiplied by $\mathrm{Cos}(\omega_c t)$, forming

$$m(t)\,\mathrm{Cos}^2(\omega_c t) = \frac{m(t)}{2}[1 + \mathrm{Cos}(2\omega_c t)]. \qquad (1.4\text{-}1b)$$

The LPF (low-pass filter) is designed to pass $m(t)$ and reject the $2\omega_c$ term, so that the output is $m(t)/2$. The average power in the output is $\frac{1}{4}m^2(t)$.

However, the system has a major problem. If the demodulating carrier at the receiver is not exactly of the same frequency and phase, fig. 1.4-1b, the output will be $(m(t)/2)\,\mathrm{Cos}[\theta(t)]$, where $\theta(t)$ represents both a fixed phase difference θ_0 and a frequency offset ω_0, so that $\theta(t) = \theta_0 + \omega_0 t$. Clearly this prevents any kind of reasonable receiver functioning. Consequently, any DSB-SC receiver must incorporate means of synchronizing the demodulating carrier to the transmitted carrier, even though there is no explicit carrier component in the transmitted signal. One way of accomplishing this is shown in fig. 1.4-1c. The received signal is squared, yielding

$$m^2(t)\,\mathrm{Cos}^2(\omega_c t) = \frac{m^2(t)}{2}[1 + \mathrm{Cos}(2\omega_c t)] \qquad (1.4\text{-}1c)$$

Since $m^2(t)$ will always have a dc component, it follows that the squarer output will have a component at $2\omega_c$. This frequency can be extracted by a very narrow BPF (bandpass filter). Using a combination of analog and digital techniques, the signal frequency can be divided by two, yielding a frequency and phase replica of the transmitter carrier. Often the combined functions of filtering and frequency division are realized by using a PLL (phase-locked loop). These topics will be discussed in detail in Chapter 6.

This complexity involved in the demodulation process for DSB-SC is a direct consequence of the absence of an explicit carrier component in the transmitted signal. On the other hand, the carrier signal contains none of the information that we wish to transmit. From that point of view, any power devoted to transmitting the carrier is wasted. DSB-SC is typically used for digital communications in a power- and bandwidth-limited environment.

1.4.2 Application 2 Double-Sideband Large Carrier (DSB-LC) Amplitude Modulation

The definition of a Double-Sideband Large Carrier (DSB-LC) Amplitude-Modulated signal is

$$f(t) = A(1 + \alpha m(t)) \cos(\omega_c t) \tag{1.4-2a}$$

where $m(t)$ is the message to be transmitted and α is the index of modulation. This signal differs from DSB-SC in that the carrier is clearly explicitly present in the transmitted signal. If $m(t) = \cos(\omega_1 t)$, the resulting spectrum is shown in fig. 1.4-2a. Note that the carrier component is at ω_c and the message components are sidebands at $\omega_c \pm \omega_m$.

The average power in the signal is

$$\overline{f^2(t)} = A^2 \overline{(1 + \alpha m(t))^2 \cos^2(\omega_c t)}$$

$$= \frac{A^2}{2} \overline{(1 + 2\alpha m(t) + \alpha^2 m^2(t))(1 + \cos(2\omega_c t))}$$

$$= \frac{A^2}{2} \overline{(1 + \alpha^2 m^2(t))} = \frac{A^2}{2}(1 + \alpha^2 \overline{m^2(t)}) \tag{1.4-2b}$$

since $m(t)$ has no dc component. Clearly the transmitted power has separate carrier and message components. When $m(t) = \cos(\omega_1 t)$,

$$\overline{f^2(t)} = \frac{A^2}{2}\left(1 + \frac{\alpha^2}{2}\right) \tag{1.4-2c}$$

and the ratio of message or sideband power to total power in the transmitted signal is

$$\frac{\text{sideband power}}{\text{total power}} = \frac{P_S}{P_T} = \left(\frac{\alpha^2}{2 + \alpha^2}\right). \tag{1.4-2d}$$

There are two basic ways of demodulating a DSB-LC signal at the receiver. The first, synchronous detection, is the same as the approach used for DSB-SC. The received signal is multiplied by a replica of the carrier:

$$f(t)\cos(\omega_c t) = A(1 + \alpha m(t))\cos^2(\omega_c t)$$

$$= \frac{A}{2}(1 + \alpha m(t))(1 + \cos(2\omega_c t)) \tag{1.4-2e}$$

and then low-pass filtered to extract $m(t)$. As with DSB-SC, it is necessary to guarantee that the demodulating carrier is of the proper frequency and phase.

The second way of demodulating a DSB-LC signal, envelope detection, does not involve a demodulating carrier at all but does require that $0 < \alpha < 1$. This issue can be most easily illustrated if $m(t)$ is sinusoidal and

$$f(t) = A(1 + \alpha \cos(\omega_1 t))\cos(\omega_c t). \tag{1.4-2f}$$

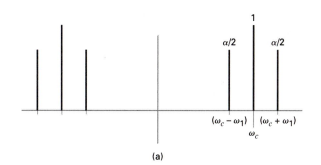

(a)

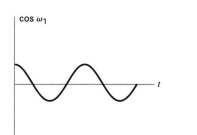

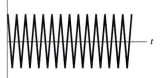

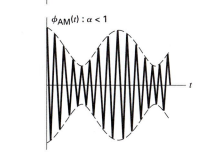

(b)

FIGURE 1.4-2 ([Str 90] fig. 5.18, fig. 5.22 reprinted with permission)

As shown in fig. 1.4-2b, if $\alpha < 1$ the actual shape of $m(t)$ is clearly visible as a slowly varying envelope, multiplying the carrier. But if $\alpha > 1$, the envelope is clearly distorted, no longer being a picture of $m(t)$. An envelope detector, suitable for $\alpha < 1$, is shown in fig. 1.4-2c as a diode followed by a simple lowpass RC filter. Although envelope detection is both crude and limiting, restricting $P_S/P_T < 1/3$, it is very simple to implement and has been historically used in AM radio.

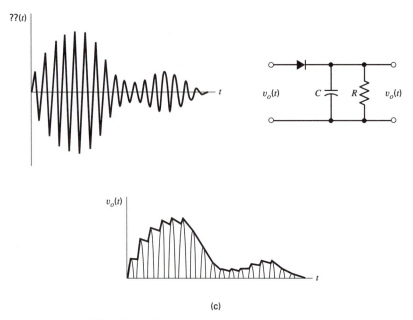

(c)

FIGURE 1.4-2   (*Continued.*)

1.4.3 Application 3 Frequency-Shift Keying (FSK)

Frequency-Shift Keying, or FSK, is a very direct way of transmitting digital data on a bandpass channel. The unique properties of the spectrum of FSK may be revealed through an application of the theorems on linearity and superposition and time shifting, as well as frequency shifting.

As shown in fig. 1.4-3a, digital data appears at a rate of f_b bits per second. The time interval between bits is $T_b = 1/f_b$ seconds. A data bit 0 generates a burst of cosine having frequency f_0 and T_b seconds long, and a data bit 1 generates a similar burst of cosine having frequency f_1 and T_b seconds long. Although it is not necessary in practice, it is crucial to our analysis that f_0 and f_1 be integer multiples of f_b in order to guarantee that the waveform at the boundary of two frequency bursts will be continuous. Although it certainly is possible to generate FSK without requiring phase continuity, CPFSK (Continuous-Phase FSK) has the property of being very bandwidth-efficient.

Since we are able at this point only to examine the spectra of periodic functions, we will assume that the data pattern is not random; rather, it is a repeating sequence 01010101 . . . . This sequence, known as a 'dotting pattern', is a standard test sequence in digital communications that will be made use of, as well, in subsequent chapters.

In our example we shall make

$$f_0 = 11 f_b \qquad \text{and} \qquad f_1 = 13 f_b \qquad (1.4\text{-}3a)$$

In order to find the Fourier Series coefficients for this function $f(t)$, we will start with the rectangular pulse $P_{T_p/2}(t)$, as shown in fig. 1.4-3b. Note that $T_p = 2T_b$. Accordingly, the function $f(t)$ may be written as

$$f(t) = P_{T_p/2}(t - T_p/4)\operatorname{Cos}(2\pi f_0 t) + P_{T_p/2}(t - 3T_p/4)\operatorname{Cos}(2\pi f_1 t) \qquad (1.4\text{-}3b)$$

where $T_p = 2T_b$ is the period of the 01 data pattern and $f_p = 1/T_p$. Note that $f(t)$ written in this way is phase-continuous. Since the Fourier Series of the basic pulse shape is given

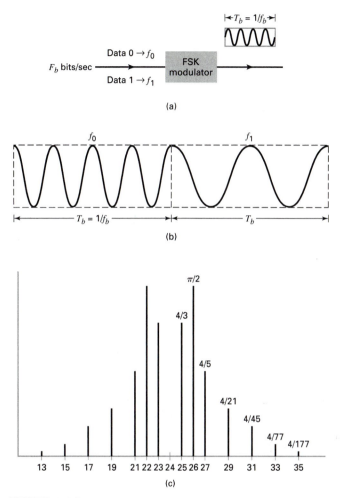

FIGURE 1.4-3

by

$$P_{T_p/2}(t) = \sum_{n=-\infty}^{\infty} \frac{1}{2} \operatorname{Sinc}\left(\frac{n\pi}{2}\right) \operatorname{Cos}(2\pi n f_p t) \qquad (1.4\text{-}3c)$$

it follows, using linearity, superposition, and time shifting, that

$$f(t) = \left\{ \sum_{n=-\infty}^{\infty} \frac{1}{2} \operatorname{Sinc}\left(\frac{n\pi}{2}\right) e^{-jn2\pi f_p T_p/4} \operatorname{Cos}(2\pi n f_p t) \right\} \operatorname{Cos}(2\pi \cdot 22 f_p t)$$

$$+ \left\{ \sum_{n=-\infty}^{\infty} \frac{1}{2} \operatorname{Sinc}\left(\frac{n\pi}{2}\right) e^{-jn2\pi f_p 3T_p/4} \operatorname{Cos}(2\pi n f_p t) \right\} \operatorname{Cos}(2\pi \cdot 26 f_p t) \quad (1.4\text{-}3d)$$

or

$$f(t) = \left\{ \sum_{n=-\infty}^{\infty} f_{n_1} \operatorname{Cos}(2\pi n f_p t) \right\} \operatorname{Cos}(2\pi \cdot 22 f_p t)$$

$$+ \left\{ \sum_{n=-\infty}^{\infty} f_{n_2} \operatorname{Cos}(2\pi n f_p t) \right\} \operatorname{Cos}(2\pi \cdot 26 f_p t) \qquad (1.4\text{-}3e)$$

The first step is to list both f_{n_1} and f_{n_2}.

n	$f_{n_1} = \frac{1}{2}\text{Sinc}\left(\frac{n\pi}{2}\right)e^{-jn\pi/2}$	$f_{n_2} = \frac{1}{2}\text{Sinc}\left(\frac{n\pi}{2}\right)e^{-jn3\pi/2}$	
-10	0	0	
-9	$j/9\pi$	$-j/9\pi$	
-8	0	0	
-7	$j/7\pi$	$-j/7\pi$	
-6	0	0	
-5	$j/5\pi$	$-j/5\pi$	
-4	0	0	
-3	$j/3\pi$	$-j/3\pi$	
-2	0	0	
-1	j/π	$-j/\pi$	
0	1/2	1/2	(1.4-3f)
1	$-j/\pi$	j/π	
2	0	0	
3	$-j/3\pi$	$j/3\pi$	
4	0	0	
5	$-j/5\pi$	$j/5\pi$	
6	0	0	
7	$-j/7\pi$	$j/7\pi$	
8	0	0	
9	$-j/9\pi$	$j/9\pi$	

In order to find the Fourier Series coefficients for the composite waveform, we must shift the f_{n_1} coefficients to be centered around $n = \pm22$ and multiply them by 1/2, and we must similarly shift the f_{n_2} coefficients to be centered around $n = \pm26$ and multiply them by 1/2 as well. We then add them up. This is detailed in eq. 1.4-3g below, where only positive values of n have been listed and all of the terms have been factured by $j/2\pi$.

n	$\frac{2\pi}{T} fn_1[n-22]$	$\frac{2\pi}{T} fn_2[n-26]$	$\left(\frac{2\pi}{T}\right)\cdot$ sum	
13	1/9	$-1/13$	4/177	
14	0	0	0	
15	1/7	$-1/11$	4/77	
16	0	0	0	
17	1/5	$-1/9$	4/25	
18	0	0	0	
19	1/3	$-1/7$	4/21	
20	0	0	0	
21	1	$-1/5$	4/5	
22	$-j\pi/2$	0	$-j\pi/2$	
23	-1	$-1/3$	$-4/3$	(1.4-3g)
24	0	0	0	
25	$-1/3$	-1	$-4/3$	
26	0	$-j\pi/2$	$-j\pi/2$	
27	$-1/5$	1	4/5	
28	0	0	0	
29	$-1/7$	1/3	4/21	
30	0	0	0	
31	$-1/9$	1/5	4/45	
32	0	0	0	
33	$-1/11$	1/7	4/77	
34	0	0	0	
35	$-1/13$	1/9	4/117	

Note that in the range $n = 22$ to $n = 26$, the magnitudes of the two components add, whereas outside that range the components subtract. This gives the overall signal a considerably narrower effective bandwidth than either of the two rectangular pulses that compose it. Since the overall signal $f(t)$ is a sinusoid of one frequency or another and having amplitude equal to unity, it has an average power $P_{av} = 1/2$. This amplitude spectrum is shown in fig. 1.4-3c.

Taking advantage of the symmetry and restoring the scaling factor, the fraction of the average power contained in the frequencies corresponding to $n = 21$ through $n = 27$ is

$$\text{fraction} = \frac{4\,|(j/2\pi)|^2\,[(4/3)^2 + |(\pi/2j)|^2 + (4/5)^2]}{1/2} = .98994 \qquad (1.4\text{-}3h)$$

This number should give the reader an appreciation for the spectral compactness of CPFSK.

Even though it is never generated, in this system the carrier frequency is defined as the average of the two frequencies f_0 and f_1 so that $f_c = 12 f_b$. One of the important measures in FSK systems is the ratio of the frequency deviation from the carrier to the bit frequency. This quantity is referred to as β, the index of modulation. In this system described above, $\beta = \nabla f / f_b = 1$. If we make $f_0 = 9 f_b$ and $f_1 = 15 f_b$ then we have the same carrier frequency but the new value of $\beta = \nabla f / f_b = 3$. It has been shown that the effective bandwidth of the transmitted signal, in terms of average power, is closely approximated by Carson's Rule:

$$\text{BW} = 2(1 + \beta) f_b. \qquad (1.4\text{-}4)$$

In our detailed example, where $\beta = 1$, we saw that a bandwidth of $4 f_b$ included 99% of the transmitted power.

1.4.4 Application 4 Frequency Modulation

A frequency-modulated (FM) signal has the form

$$f(t) = A\,\text{Cos}(\theta(t)) = A\,\text{Cos}\left(\omega_c t + k_f \int_{-\infty}^{t} m(\lambda)\,d\lambda\right) \qquad (1.4\text{-}5a)$$

where the instantaneous frequency is defined as

$$\omega_i = d\theta(t)/dt = \omega_c + k_f m(t) \qquad (1.4\text{-}5b)$$

and the information in this signal is carried in the instantaneous frequency.

Unlike AM, where the sidebands are linearly dependent on the message signal $m(t)$, the sidebands in FM have a nonlinear relation to $m(t)$. This can be seen by a closer examination of the signal

$$f(t) = A\,\text{Cos}\left(\omega_c t + k_f \int_{-\infty}^{t} m(\lambda)\,d\lambda\right)$$

$$= A\,\text{Cos}(\omega_c t + g(t)) = \text{Re}\left[A e^{j(\omega_c t + g(t))}\right]$$

$$= \text{Re}\left[A[\text{Cos}\,\omega_c t + j\,\text{Sin}\,\omega_c t]\left[1 + jg(t) - \frac{1}{2}g^2(t) - j\frac{1}{3\cdot 2}g^3(t) + \cdots\right]\right]$$

$$(1.4\text{-}5c)$$

The nonlinear relationship comes from the multiplication of the carrier by higher powers of $g(t)$.

FSK, which was examined above, is a special case of FM corresponding to $m(t)$ equal to a square wave. The other analytically tractable example of FM corresponds to a sinusoidal modulating signal

$$m(t) = a \, \text{Cos} \, \omega_m t \qquad (1.4\text{-}5d)$$

In this instance, the instantaneous frequency is

$$\omega_i = d\theta(t)/dt = \omega_c + ak_f \, \text{Cos}(\omega_m t) = \omega_c + \nabla\omega \, \text{Cos}(\omega_m t) \qquad (1.4\text{-}5e)$$

where $\nabla\omega$ is called the peak frequency deviation. Under these circumstances

$$\theta(t) = \omega_c t + ak_f \int^t \text{Cos}(\omega_m \lambda) \, d\lambda = \omega_c t + \frac{\nabla\omega}{\omega_m} \, \text{Sin}(\omega_m t) \qquad (1.4\text{-}5f)$$

where $\nabla\omega/\omega_m = \beta$, the index of modulation. This definition of β is essentially the same as given in the example of FSK.

In order to examine the spectrum of this sinusoidally modulated FM signal, we can revisit the transmitted signal as

$$f(t) = A \, \text{Cos}(\omega_c t + \beta \, \text{Sin}(\omega_m t)) = \text{Re}\left[A e^{j\omega_c t} e^{j\beta \, \text{Sin}(\omega_m t)}\right]. \qquad (1.4\text{-}6a)$$

The essential point is that the quantity $e^{j\beta \, \text{Sin}(\omega_m t)}$ is periodic with period $T_m = 2\pi/\omega_m$ so that it can be represented as a Fourier series:

$$e^{j\beta \, \text{Sin}(\omega_m t)} = \sum_{n=-\infty}^{\infty} f_n e^{jn\omega_m t} \qquad (1.4\text{-}6b)$$

where

$$f_n = \frac{1}{T_m} \int_{-T_m/2}^{T_m/2} e^{j\beta \, \text{Sin}(\omega_m t)} e^{-jn\omega_m t} \, dt = \frac{1}{2\pi} \int_{-\pi}^{\pi} e^{j(\beta \, \text{Sin}\delta - n\delta)} \, d\delta = J_n(\beta) \qquad (1.4\text{-}6c)$$

and where $J_n(\beta)$ is the Bessel Function of the first kind and nth order. A graph of these functions is given in fig. 1.4-4a.

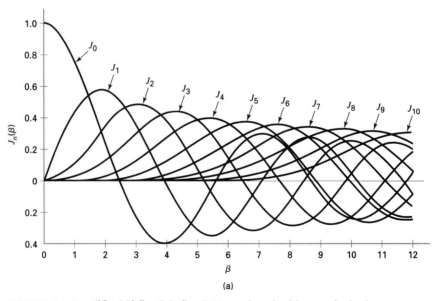

(a)

FIGURE 1.4-4a   ([Str90] fig. 6.9, fig. 6.10 reprinted with permission)

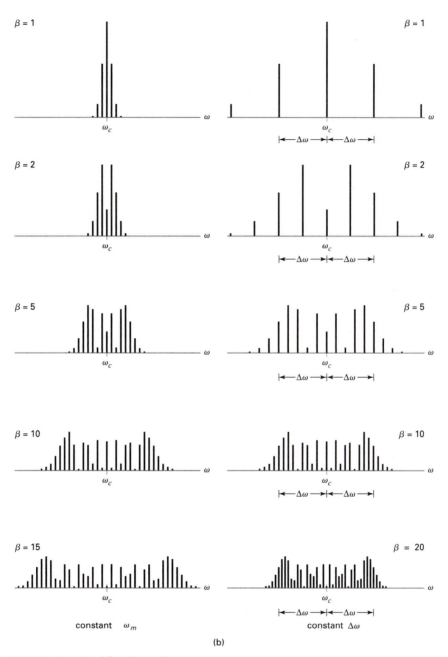

FIGURE 1.4-4b   (*Continued.*)

It follows that

$$e^{j\beta \, \text{Sin}(\omega_m t)} = \sum_{n=-\infty}^{\infty} J_n(\beta) e^{jn\omega_m t} \tag{1.4-6d}$$

and consequently

$$f(t) = A \, \text{Cos}(\omega_c t + \beta \, \text{Sin}(\omega_m t)) = \text{Re}\left[A e^{j\omega_c t} e^{j\beta \, \text{Sin}(\omega_m t)} \right]$$

$$= A \sum_{n=-\infty}^{\infty} J_n(\beta) \, \text{Cos}(\omega_c + n\omega_m)t \tag{1.4-6e}$$

This means that even with a single sinusoid as the modulating signal, the spectrum of the resulting FM signal, centered around the carrier frequency ω_c, contains, in theory, an infinite number of sidebands at all frequencies $\omega_c \pm n\omega_m$. We can expect that not all of these sidebands will contribute significant amounts of power to the overall signal.

Some properties of $J_n(\beta)$ are worth noting:

1. $$J_n(\beta) \text{ is real}$$

2. $$J_n(\beta) = J_{-n}(\beta) \quad \text{for } n \text{ even}$$

3. $$J_n(\beta) = -J_{-n}(\beta) \quad \text{for } n \text{ odd} \qquad (1.4\text{-}6f)$$

4. $$\sum_{n=-\infty}^{\infty} J_n^2(\beta) = 1$$

A sinusoidally modulated FM signal can therefore be represented as:

$$
\begin{aligned}
f(t) = A\,\mathrm{Cos}(\omega_c t + \beta\,\mathrm{Sin}(\omega_m t)) &= A \sum_{n=-\infty}^{\infty} J_n(\beta)\,\mathrm{Cos}\,(\omega_c + n\omega_m)t \\
&= A J_0(\beta)\,\mathrm{Cos}(\omega_c t) \\
&\quad + A J_1(\beta)[\mathrm{Cos}(\omega_c + \omega_m)t - \mathrm{Cos}(\omega_c - \omega_m)t] \\
&\quad + A J_2(\beta)[\mathrm{Cos}(\omega_c + 2\omega_m)t + \mathrm{Cos}(\omega_c - 2\omega_m)t] \\
&\quad + A J_3(\beta)[\mathrm{Cos}(\omega_c + 3\omega_m)t - \mathrm{Cos}(\omega_c - 3\omega_m)t] + \cdots \qquad (1.4\text{-}6g)
\end{aligned}
$$

Some examples of FM spectra for differing values of β are shown in fig. 1.4-4b.

As an example, consider the circumstance of $\beta = 5$. With $A = 1$, the average power in $f(t) = A\,\mathrm{Cos}(\omega_c t + \beta\,\mathrm{Sin}(\omega_m t))$ is $P_{av} = 1/2$, since $f(t)$ is simply a sinusoid having a varying frequency. According to Carson's Rule (eq. 1.4-3), the effective bandwidth should extend from $(\omega_c - 6\omega_m)$ to $(\omega_c + 6\omega_m)$. The average power in the signal within this frequency range divided by the total average power is:

$$
\begin{aligned}
\frac{P_{bw}}{P_{av}} &= \left[J_0^2(5) + 2\sum_{n=1}^{6} J_n^2(5) \right] \\
&= (-.18)^2 + 2[(-.33)^2 + (.05)^2 + (.36)^2 + (.39)^2 + (.26)^2 + (.13)^2] \\
&= .9876 \qquad (1.4\text{-}6h)
\end{aligned}
$$

which certainly ought to be a persuasive affirmation of Carson's Rule.

1.5 FROM THE FOURIER SERIES TO THE FOURIER TRANSFORM

The starting point for the development of Fourier Series was the series representation of some function $f(t)$ in a finite time interval T. As long as the interval T was finite, the series was composed of a countably infinite number of sinusoids having frequencies $n\omega_0$; $n = 0, \pm1, 2, \pm3 \ldots$, where $\omega_0 = 2\pi/T$. It was established that the series was equal, on an energy basis, to the function within the time interval and was equal to the periodic expansion of the function outside that interval. If the function is itself periodic then the two are equal, again on an energy basis, everywhere.

Of course, $f(t)$ cannot really be periodic because that would mean infinite duration and infinite energy, neither of which is possible in a real physical waveform. Nevertheless, periodicity is such a useful concept in describing certain kinds of signals that have existed for a long time (for example, the output of a square wave generator that has been on for more than a few seconds) that the idea of a real signal actually being periodic has crept into the thinking of spectral analysis. For this reason it is necessary to treat the transition from the Fourier Series to the Fourier Transform with some care.

The Fourier Transform arises as the limit of the f_n coefficients in the Fourier Series under the following conditions:

1. the time interval T in which the signal is being approximated by the series grows large, making the fundamental frequency $\omega_0 = 2\pi/T$ grow small, and

2. the signal energy contained in that time interval is finite.

The process of arriving at the Fourier Transform involves maintaining the amplitude and duration of the rectangular pulse of fig. 1.5-1a but allowing the interval $[-T/2,\ T/2)$ to get very large by allowing T to get very large. The energy in $f(t)$ remains the same, $E = A^2 a$, but the average power becomes small. Turning to the spectrum, fig. 1.5-1b, as T gets large, the fundamental frequency $\omega_0 = 2\pi/T$ gets small so that more frequency components fit into the main lobe, which has width $2\pi/a$. Note that the width of the main lobe is fixed at $2\pi/a$.

Therefore as T increases and ω_0 decreases, the number of harmonics in the main lobe increases, all with the total signal energy fixed. Consequently the amount of energy in each harmonic decreases, tending toward zero, as the frequency components become very dense. Along with this, the average power decreases toward zero.

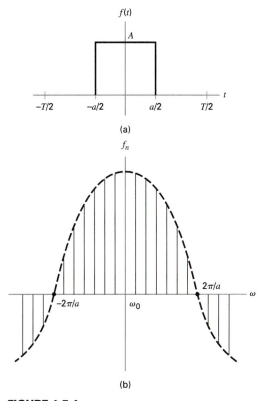

(a)

(b)

FIGURE 1.5-1

This limiting process is described mathematically in the following way. The series representation of $f(t)$ in the interval $[-T/2, T/2]$ is

$$f(t) = \sum_{n=-\infty}^{\infty} f_n e^{jn\omega_0 t} \tag{1.1-1a}$$

where

$$f_n = \frac{1}{T} \int_{-T/2}^{T/2} f(\lambda) e^{-jn\omega_0 \lambda} \, d\lambda \tag{1.1-3f}$$

so that

$$f(t) = \sum_{n=-\infty}^{\infty} \frac{1}{T} \int_{-T/2}^{T/2} f(\lambda) e^{-jn\omega_0 \lambda} \, d\lambda \, e^{jn\omega_0 t}$$

$$= \frac{1}{2\pi} \sum_{n=-\infty}^{\infty} \int_{-T/2}^{T/2} f(\lambda) e^{-jn\omega_0 \lambda} \, d\lambda \, e^{jn\omega_0 t} \omega_0 \tag{1.5-1a}$$

Referring to the spectrum of fig. 1.5-1b, it is clear that ω_0 is the difference between successive values of $n\omega_0$ (i.e., $-\omega_0 = (n+1)\omega_0 - n\omega_0 = \Delta\omega_n$). Consequently, in the limit as T becomes large, $n\omega_0 \to \omega$, a continuous variable, and $\Delta\omega_n \to d\omega$, a differential displacement, so that the summation of eq. 1.5-1a becomes

$$f(t) = \frac{1}{2\pi} \int_{-\infty}^{\infty} \left\{ \int_{-\infty}^{\infty} f(\lambda) e^{-jn\omega_0 \lambda} \, d\lambda \right\} e^{jn\omega_0 t} \, d\omega \tag{1.5-1b}$$

where the Fourier Transform pair is

$$F(\omega) = \int_{-\infty}^{\infty} f(t) e^{-j\omega t} \, dt \tag{1.5-2a}$$

$$f(t) = \frac{1}{2\pi} \int_{-\infty}^{\infty} F(\omega) e^{j\omega t} \, d\omega \tag{1.5-2b}$$

There is a clear relationship between the expression for the Fourier Series coefficients, eq. 1.1-3f, and the Fourier Transform, eq. 1.5-2a. The Fourier Series f_n coefficients for periodic rectangular pulses are:

$$f_n = A\frac{a}{T} \text{Sinc}(n\omega_0 a/2) \tag{1.2-1c}$$

and it is easy to show that the Fourier Transform of a single rectangular pulse is:

$$F(\omega) = Aa \, \text{Sinc}(\omega a/2) \tag{1.5-3}$$

In general, if:

1. the Fourier Series coefficient f_n represents a discrete spectrum of a power signal that is the periodic repetition of a time-limited pulse; and

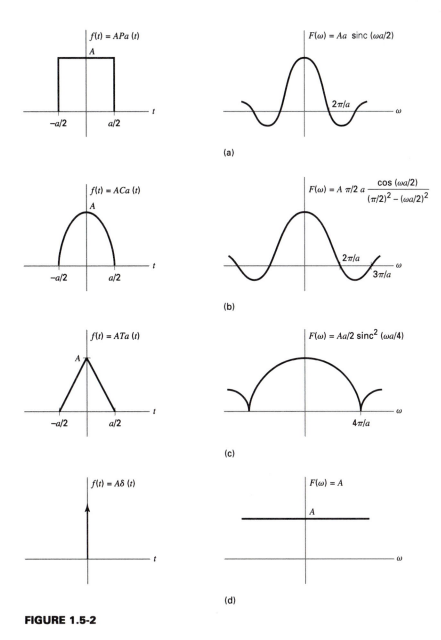

FIGURE 1.5-2

2. the Fourier Transform $F(\omega)$ is the continuous spectrum of an energy signal which is the same time-limited pulse; then

the relationship between f_n and $F(\omega)$ is:

$$\lim(T f_n) \xrightarrow{n\omega_0 \to \omega} F(\omega) \qquad (1.5\text{-}4)$$

Consequently, we can easily find the Fourier Transforms of the non-periodic versions of each of the finite time pulses that were examined in the chapter on Fourier Series. These are shown in fig. 1.5-2.

We can also use eq. 1.5-2a to find the Fourier Transform of any finite-energy signal. For example, if

$$f(t) = e^{-\alpha t} u(t) \tag{1.5-5a}$$

then

$$F(\omega) = \int_{-\infty}^{\infty} f(t) e^{-j\omega t}\, dt = \int_{0}^{\infty} e^{-\alpha t} e^{-j\omega t}\, dt = \frac{1}{\alpha + j\omega} \tag{1.5-5b}$$

With this, it is easy to show that

$$f(t) = e^{\alpha t} u(-t) \Leftrightarrow F(\omega) = \frac{1}{\alpha - j\omega} \tag{1.5-5c}$$

and, using linearity and superposition,

$$f(t) = e^{\alpha |t|} \Leftrightarrow F(\omega) = \frac{2\alpha}{\alpha^2 + \omega^2} \tag{1.5-5d}$$

This transform is shown in fig. 1.5-3a.

We shall also evaluate the transform of the Gaussian pulse. Let

$$f(t) = e^{-t^2/2a^2} \tag{1.5-6a}$$

then:

$$F(\omega) = \int_{-\infty}^{\infty} e^{-t^2/2a^2} e^{-j\omega t}\, dt = \int_{-\infty}^{\infty} e^{-(t/a\sqrt{2}+j\omega a/\sqrt{2})^2} e^{-(\omega a/\sqrt{2})^2}\, dt$$

$$= a\sqrt{2} e^{-(\omega a/\sqrt{2})^2} \int_{-\infty}^{\infty} e^{-x^2}\, dx = [a\sqrt{2\pi}] e^{-(\omega a/\sqrt{2})^2} \tag{1.5-6b}$$

so the Fourier Transform of a Gaussian pulse is Gaussian. This is shown in fig. 1.5-3b.

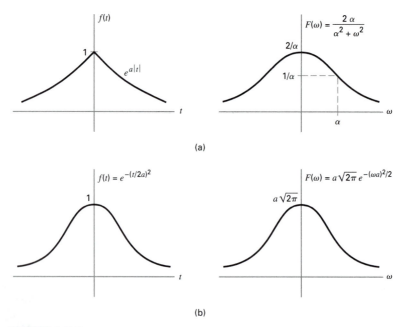

(a)

(b)

FIGURE 1.5-3

1.6   THEOREMS OF THE FOURIER TRANSFORM

The observant reader will already have noted that the Fourier Transform integral is very much like the Laplace Transform integral, as well as the integral that defines the Fourier Series coefficient. Consequently, the theorems describing the properties of the Transform will be reminiscent of previously encountered theorems. We shall therefore limit the formal proofs given below to those where there is additional information to be added.

Restating the Fourier Transform pair:

$$F(\omega) = \int_{-\infty}^{\infty} f(t)e^{-j\omega t}\,dt \tag{1.5-2a}$$

$$f(t) = \frac{1}{2\pi}\int_{-\infty}^{\infty} F(\omega)e^{j\omega t}\,d\omega \tag{1.5-2b}$$

THEOREM 1   *Linearity and Superposition*

$$\text{If } f(t) = af_1(t) + bf_2(t) \quad \text{then } F(\omega) = aF_1(\omega) + bF_2(\omega) \tag{1.6-1}$$

■

THEOREM 2   *Time Shifting*

$$\text{If } f(t) \Leftrightarrow F(\omega) \quad \text{then } f(t \pm t_0) \Leftrightarrow F(\omega)e^{\pm j\omega t_0} \tag{1.6-2}$$

■

THEOREM 3   *Multiplication by an Exponential*

a. $\quad$ If $f(t) \Leftrightarrow F(\omega)$ $\quad$ then $f(t)e^{-\alpha t} \Leftrightarrow F(\omega + \alpha/j)$ $\tag{1.6-3a}$

Example

$$f(t) = e^{-bt}u(t) \Leftrightarrow F(\omega) = \frac{1}{b+j\omega} \quad \text{then}$$

$$f(t)e^{-\alpha t} \Leftrightarrow F(\omega) = \frac{1}{b+j\,(\omega + \alpha/j)} = \frac{1}{(\alpha + b) + j\omega}$$

b. $\quad$ If $f(t) \Leftrightarrow F(\omega)$ $\quad$ then $f(t)e^{\pm j\beta t} \Leftrightarrow F(\omega \mp \beta)$ $\tag{1.6-3b}$

Example

$$f(t) = e^{-at}u(t) \Leftrightarrow F(\omega) = \frac{1}{a+j\omega} \quad \text{then}$$

$$f(t)e^{j\beta t} \Leftrightarrow \frac{1}{a+j(\omega - \beta)} \quad \text{and}$$

$$f(t)e^{-j\beta t} \Leftrightarrow \frac{1}{a+j(\omega + \beta)}$$

■

THEOREM 4   *Time Differentiation*

$$\text{If } f(t) \Leftrightarrow F(\omega) \quad \text{then } \frac{d^n f(t)}{dt^n} \Leftrightarrow (j\omega)^n F(\omega) = F_1(\omega) \qquad (1.6\text{-}4)$$

Proof:

$$F_1(\omega) = \int_{-\infty}^{\infty} \frac{df(t)}{dt} e^{-j\omega t}\, dt = f(t)e^{-j\omega t}\Big|_{-\infty}^{\infty} - \int_{-\infty}^{\infty} f(t)(-j\omega)e^{-j\omega t}\, dt$$

$$= (j\omega) \int_{-\infty}^{\infty} f(t)e^{-j\omega t}\, dt$$

where the fact that $f(t)$ is a finite-energy signal causes the first part of the integral to disappear. ■

THEOREM 5   *Time Integration*

$$\text{If } f(t) \Leftrightarrow F(\omega) \quad \text{then } \int_{-\infty}^{t} f(\lambda)\, d\lambda \Leftrightarrow \frac{1}{j\omega} F(\omega) + \pi F(0)\delta(\omega). \qquad (1.6\text{-}5)$$

Partial Proof:

$$F_1(\omega) = \int_{-\infty}^{\infty} \left[\int_{-\infty}^{t} f(\lambda)\, d\lambda \right] e^{-j\omega t}\, dt = \frac{e^{-j\omega t}}{-j\omega} \left[\int_{-\infty}^{t} f(\lambda)\, d\lambda \right]\Bigg|_{-\infty}^{\infty} - \int_{-\infty}^{\infty} f(t)\frac{e^{-j\omega t}}{-j\omega}\, dt$$

The impulse function arises if $f(t)$ has non-zero area. The integral will then contain that area so that it will not be a finite-energy signal. When we subsequently deal with the Fourier Transforms of power signals, we will return to the proof of this theorem. ■

THEOREM 6   *Frequency Differentiation*

$$\text{If } f(t) \Leftrightarrow F(\omega) \quad \text{then } t^n f(t) \Leftrightarrow (j)^n \frac{d^n F(\omega)}{d\omega^n} \qquad (1.6\text{-}6)$$

Proof:

Left to the problems. ■

THEOREM 7   *Scaling (Time-Bandwidth)*

$$\text{If } f(t) \Leftrightarrow F(\omega) \quad \text{then } f(at) \Leftrightarrow \frac{1}{|a|} F\left(\frac{\omega}{a}\right) \qquad (1.6\text{-}7a)$$

Proof:

Let $a > 0$. Then

$$\int_{-\infty}^{\infty} f(at)e^{-j\omega t}\, dt = \int_{-\infty}^{\infty} f(x)e^{-j\omega x/a}\frac{dx}{a} = \frac{1}{a} F\left(\frac{\omega}{a}\right)$$

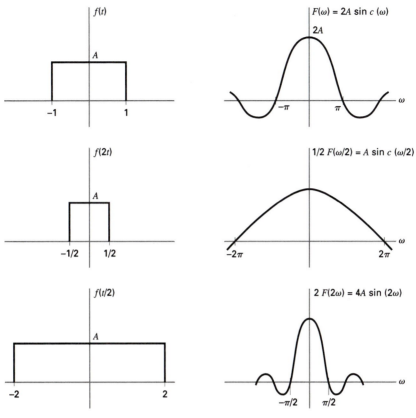

FIGURE 1.6-1

and

$$\int_{-\infty}^{\infty} f(-at)e^{-j\omega t}\, dt = \int_{-\infty}^{\infty} f(x)e^{j\omega x/a}\frac{dx}{(a)} = \frac{1}{a}F\left(-\frac{\omega}{a}\right)$$

The implications of this theorem are shown in fig. 1.6-1. The basic point is that as the time-domain signal gets narrower, the spectrum gets broader. The limit of this is the time impulse function, which has a uniform and infinite spectrum.

Special Case: Time Reversal

$$\text{If } f(t) \Leftrightarrow F(\omega) \quad \text{then } f(-t) \Leftrightarrow F(-\omega) \tag{1.6-7b}$$

Example:

$$\left\{e^{-at}u(t) \Leftrightarrow \frac{1}{a+j\omega}\right\} \Leftrightarrow \left\{e^{at}u(-t) \Leftrightarrow \frac{1}{a-j\omega}\right\}$$

∎

THEOREM 8 *Duality or Symmetry*

$$\text{If } f(t) \Leftrightarrow F(\omega) \quad \text{then } F(t) \Leftrightarrow 2\pi f(-\omega) \tag{1.6-8}$$

Proof:

From eq. 1.24b,

$$f(t) = \frac{1}{2\pi} \int_{-\infty}^{\infty} F(\omega)e^{j\omega t}\, d\omega$$

Now let $t \rightarrow -t$ so that

$$f(-t) = \frac{1}{2\pi} \int_{-\infty}^{\infty} F(\omega)e^{-j\omega t}\, d\omega$$

and exchanging t and ω

$$2\pi f(-\omega) = \int_{-\infty}^{\infty} F(t)e^{-j\omega t}\, dt$$

Some examples are given in Figs. 1.6-2 and 1.6-3.

The Symmetry Theorem also provides a simple way of evaluating some important integrals of infinite time functions. Since

$$F(\omega) = \int_{-\infty}^{\infty} f(t)e^{-j\omega t}\, dt \Rightarrow F(0) = \int_{-\infty}^{\infty} f(t)\, dt \tag{1.6-9a}$$

so that in fig. 1.6-3

$$A = \int_{-\infty}^{\infty} \frac{A\omega_c}{2\pi} \text{Sa}\,(\omega_c t/2)\, dt \Rightarrow \frac{\omega_c}{\pi}\text{Sa}(\omega_c x) \xrightarrow{\omega_c \to \infty} \delta(t) \tag{1.6-9b}$$

and

$$A = \int_{-\infty}^{\infty} \frac{A\omega_c}{4\pi} \text{Sa}^2\,(\omega_c t/4)\, dt \Rightarrow \frac{\omega_c}{\pi}\text{Sa}^2(\omega_c x) \xrightarrow{\omega_c \to \infty} \delta(t) \tag{1.6-9c}$$

∎

$f(t) \Leftrightarrow F(\omega)$		$F(t) \Leftrightarrow f(-\omega)$	
(a)	$Ae^{-at}u(t) \Leftrightarrow \dfrac{A}{a+j\omega}$	$\dfrac{A}{a+jt} \Leftrightarrow 2\pi A e^{a\omega}u(-\omega)$	
(b)	$Ae^{at}u(-t) \Leftrightarrow \dfrac{A}{a-j\omega}$	$\dfrac{A}{a-jt} \Leftrightarrow 2\pi A e^{-a\omega}u(\omega)$	
(c)	$Ae^{-a\lvert t\rvert} \Leftrightarrow \dfrac{2aA}{a^2+\omega^2}$	$\dfrac{2aA}{a^2+t^2} \Leftrightarrow 2\pi A e^{-a\lvert\omega\rvert}$	

FIGURE 1.6-2

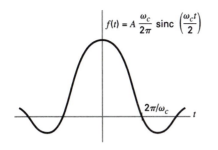

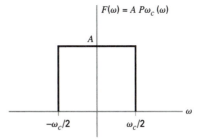

(a)

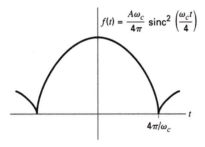

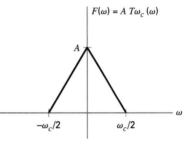

(b)

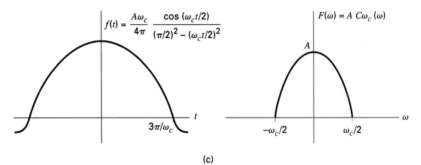

(c)

FIGURE 1.6-3

THEOREM 9   *Time Convolution*

If

$$f(t) \Leftrightarrow F(\omega), \quad h(t) \Leftrightarrow H(\omega), \quad g(t) \Leftrightarrow G(\omega)$$

then

$$G(\omega) = F(\omega)H(\omega) \Leftrightarrow g(t)$$

$$= \int\limits_{-\infty}^{\infty} f(\lambda)h(t-\lambda)\,d\lambda = \int\limits_{-\infty}^{\infty} h(\lambda)f(t-\lambda)\,d\lambda \qquad (1.6\text{-}10)$$

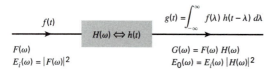

FIGURE 1.6-4

Proof:

$$G(\omega) = \int\limits_{-\infty}^{\infty} g(t)e^{-j\omega t}\,dt = \int\limits_{-\infty}^{\infty}\left[\int\limits_{-\infty}^{\infty} f(\lambda)h(t-\lambda)\,d\lambda\right]e^{-j\omega t}\,dt$$

$$= \int\limits_{-\infty}^{\infty} f(\lambda)e^{-j\omega\lambda}\left[\int\limits_{-\infty}^{\infty} h(t-\lambda)e^{-j\omega(t-\lambda)}\,dt\right]d\lambda$$

$$= H(\omega)\int\limits_{-\infty}^{\infty} f(\lambda)e^{-j\omega\lambda}\,d\lambda = F(\omega)H(\omega)$$

∎

The Time Convolution Theorem for Fourier Transforms is yet another version of the basic input-output relationship for linear systems. In this relationship $h(t)$ is the impulse response of the system, $H(\omega)$ is the transfer function, and convolution in time is the frequency domain equivalent of multiplication. This relationship is summed up in fig. 1.6-4.

THEOREM 10 *Frequency Convolution*

Let $g(t) = f_1(t)f_2(t)$, where

$$f_1(t) \Leftrightarrow F_1(\omega), \quad f_2(t) \Leftrightarrow F_2(\omega) \quad \text{and} \quad g(t) \Leftrightarrow G(\omega)$$

then

$$G(\omega) = \frac{1}{2\pi}\int\limits_{-\infty}^{\infty} F_1(\beta)F_2(\omega-\beta)\,d\beta = \frac{1}{2\pi}F_1(\omega)\otimes F_2(\omega) \qquad (1.6\text{-}11)$$

Proof:

$$G(\omega) = \int\limits_{-\infty}^{\infty} f_1(t)f_2(t)e^{-j\omega t}\,dt = \int\limits_{-\infty}^{\infty}\left[\frac{1}{2\pi}\int\limits_{-\infty}^{\infty} F_1(\beta)e^{j\beta t}\,d\beta\right]f_2(t)e^{-j\omega t}\,dt$$

$$= \frac{1}{2\pi}\int\limits_{-\infty}^{\infty} F_1(\beta)\left[\int\limits_{-\infty}^{\infty} f_2(t)e^{j(\omega-\beta)t}\,dt\right]d\beta = \frac{1}{2\pi}\int\limits_{-\infty}^{\infty} F_1(\beta)F_2(\omega-\beta)\,d\beta$$

The Frequency Convolution Theorem will be central to proving the Sampling Theorem, which forms the basis of both digital communications and digital signal processing. ∎

THEOREM 11 *Parseval's Theorem*

We have already seen a version of Parseval's Theorem for Fourier Series in eq. 1.1-2a. This is a more general version.

Let $f_1(t) \Leftrightarrow F_1(\omega)$ and $f_2(t) \Leftrightarrow F_2(\omega)$ where both time functions are real. Then

$$\int_{-\infty-}^{\infty} f_1(t)f_2(t)\,dt = \frac{1}{2\pi}\int_{-\infty-}^{\infty} F_1(\omega)F_2^*(\omega)\,d\omega. \tag{1.6-12a}$$

In particular, if $f_1(t) = f_2(t) = f(t)$, then

$$E = \int_{-\infty-}^{\infty} f^2(t)\,dt = \frac{1}{2\pi}\int_{-\infty-}^{\infty} |F(\omega)|^2\,d\omega. = \frac{1}{2\pi}\int_{-\infty-}^{\infty} E(\omega)\,d\omega \tag{1.6-12b}$$

where E is the energy contained in the signal and $E(\omega)$ is called the *Energy Spectral Density*.

Proof:

$$\int_{-\infty-}^{\infty} f_1(t)f_2(t)\,dt = \int_{-\infty-}^{\infty}\left[\frac{1}{2\pi}\int_{-\infty-}^{\infty} F_1(\omega)e^{j\omega t}\,d\omega.\right]f_2(t)\,dt$$

$$= \frac{1}{2\pi}\int_{-\infty-}^{\infty} F_1(\omega)\left[\int_{-\infty-}^{\infty} f_2(t)e^{j\omega t}\,dt.\right]d\omega$$

$$= \frac{1}{2\pi}\int_{-\infty-}^{\infty} F_1(\omega)F_2(-\omega)\,d\omega$$

where $F_2(-\omega) = F_2^*(\omega)$ since $f_2(t)$ is real.

end of proof

The Energy Spectral Density $E(\omega)$ is used to define another input-output relationship for a linear system. Consider that in fig. 1.6-4 the input energy spectral density is $E_i(\omega) = |F(\omega)|^2$ and that $E_0(\omega) = |G(\omega)|^2$ where $G(\omega) = H(\omega)F(\omega)$. It follows that

$$E_0(\omega) = |H(\omega)|^2 E_i(\omega) \tag{1.6-12c}$$

■

THEOREM 12 *Correlation Functions and the Weiner-Khinchin Theorem*

Let $f(t)$ be an energy signal having a Fourier Transform $F(\omega)$ and an energy spectral density $E(\omega) = |F(\omega)|^2$. The *autocorrelation function* of $f(t)$, $R_f(\tau)$, is defined as the inverse Fourier Transform of $E(\omega)$:

$$R_f(\tau) \Leftrightarrow E(\omega) \quad \text{or} \quad R_f(\tau) = \frac{1}{2\pi}\int_{-\infty}^{\infty} E(\omega)e^{j\omega\tau}\,d\omega \tag{1.6-13a}$$

Then

$$R_f(\tau) = \int\limits_{-\infty}^{\infty} f(t)f^*(t-\tau)\,dt. \tag{1.6-13b}$$

This Fourier Transform pair is known as the Weiner-Khinchin Theorem.

Proof:

$$R_f(\tau) = \frac{1}{2\pi}\int\limits_{-\infty}^{\infty} E(\omega)e^{j\omega\tau}\,d\omega = \frac{1}{2\pi}\int\limits_{-\infty}^{\infty} F(\omega)F^*(\omega)e^{j\omega\tau}\,d\omega$$

$$= \frac{1}{2\pi}\int\limits_{-\infty}^{\infty}\left[\int\limits_{-\infty}^{\infty} f(t)e^{-j\omega t}\,dt\right]F^*(\omega)e^{j\omega\tau}\,d\omega$$

$$= \int\limits_{-\infty}^{\infty} f(t)\left[\frac{1}{2\pi}\int\limits_{-\infty}^{\infty} F^*(\omega)e^{-j\omega(t-\tau)}\,d\omega\right]dt$$

$$= \int\limits_{-\infty}^{\infty} f(t)\left[\frac{1}{2\pi}\int\limits_{-\infty}^{\infty} F(\omega)e^{j\omega(t-\tau)}\,d\omega\right]^*dt$$

$$= \int\limits_{-\infty}^{\infty} f(t)f^*(t-\tau)\,dt$$

end of proof

∎

EXAMPLE:

Find the autocorrelation function of $f(t) = e^{-at}u(t)$. According to fig. 1.6-5,

For $\tau > 0$

$$R_f(\tau) = \int\limits_{-\infty}^{\infty} f(t)f(t-\tau)\,dt = \int\limits_{\tau}^{\infty} e^{-at}e^{-a(t-\tau)}\,dt$$

$$= e^{a\tau}\int_{\tau}^{\infty} e^{-2at}\,dt = e^{a\tau}\frac{e^{-2at}}{-2a}\bigg|_{\tau}^{\infty} = \frac{e^{-a\tau}}{2a}$$

For $\tau < 0$

$$R_f(\tau) = \int\limits_{-\infty}^{\infty} f(t)f(t-\tau)\,dt = \int\limits_{0}^{\infty} e^{-at}e^{-a(t-\tau)}\,dt$$

$$= e^{a\tau}\int\limits_{0}^{\infty} e^{-2at}\,dt = e^{a\tau}\frac{e^{-2at}}{-2a}\bigg|_{0}^{\infty} = \frac{e^{a\tau}}{2a}$$

This function is denoted as $R_f(\tau) = e^{-a|\tau|}/(2a)$. We have already seen in fig 1.6-2 that the Fourier Transform of this two-sided decaying exponential is $E(\omega) = 1/(a^2 + \omega^2)$. ∎

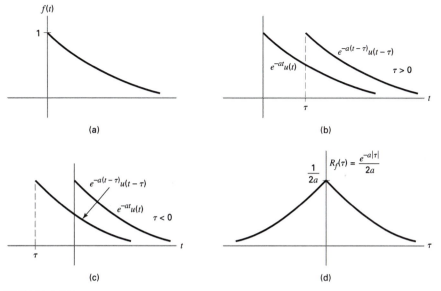

FIGURE 1.6-5

Properties of Autocorrelation Functions

1. $R(\tau) = R^*(-\tau)$ (1.6-14a)

Proof:

$$R(\tau) = \int_{-\infty}^{\infty} f(t)f^*(t - \tau)\,dt = \int_{-\infty}^{\infty} f(t' + \tau)f^*(t')\,dt' = R^*(-\tau)$$

2. $R(0) = $ energy contained in $f(t)$ (1.6-14b)

Proof:

$$R(0) = \int_{-\infty}^{\infty} f(t)f^*(t)\,dt = \int_{-\infty}^{\infty} |f(t)|^2\,dt = E$$

3. $R(0) \geq \pm R(\tau)$ for $\tau \neq 0$ (1.6-14c)

Proof:

$$\int_{-\infty}^{\infty} [f(t) \pm f(t - \tau)]][f(t) \pm f(t - \tau)]^*\,dt$$

$$= \int_{-\infty}^{\infty} |f(t)|^2\,dt + \int_{-\infty}^{\infty} |f(t + \tau)|^2\,dt \pm \int_{-\infty}^{\infty} f(t)f^*(t - \tau)\,dt \pm \int_{-\infty}^{\infty} f^*(t)f(t - \tau)\,dt$$

$$= 2\int_{-\infty}^{\infty} |f(t)|^2 dt \pm 2\,\mathrm{Re}\left[\int_{-\infty}^{\infty} f(t)f^*(t - \tau)\,dt\right] \geq 0$$

or

$$\int_{-\infty}^{\infty} |f(t)|^2 dt \geq \text{Re}\left[\int_{-\infty}^{\infty} f(t)f^*(t-\tau)\,dt\right]$$

or

$$R(0) \geq \pm R(\tau) \quad \text{for } \tau \neq 0$$

The *cross-correlation function* of two functions $f_1(t)$ and $f_2(t)$ is defined as

$$R_{12}(\tau) = \int_{-\infty}^{\infty} f_1(t)f_2^*(t-\tau)\,dt \tag{1.6-15a}$$

and

$$R_{21}(\tau) = \int_{-\infty}^{\infty} f_2(t)f_1^*(t-\tau)\,dt$$

$$R_{21}(\tau) = \int_{-\infty}^{\infty} f_2(t)f_1^*(t-\tau)\,dt \tag{1.6-15b}$$

$$\text{Clearly} \qquad R_{12}(\tau) = R_{21}^*(-\tau). \tag{1.6-15c}$$

EXAMPLE

Consider the cross-correlation of $f_1(t)$ and $f_2(t)$ as shown in fig. 1.6-6. Clearly R_{12} is not an even function, nor is it odd. R_{21} is the mirror image of R_{12}.

It is useful to make a clear distinction between the two operations of correlation and convolution. At first glance these two integrals appear quite similar. In fact, they are

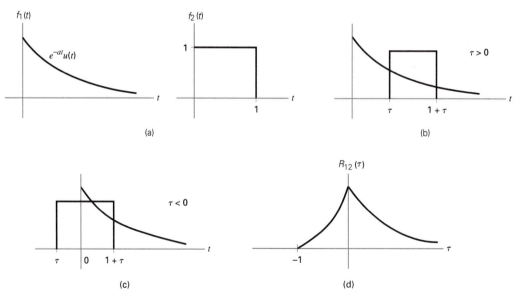

FIGURE 1.6-6

quite different.

$$\text{convolution:} \quad f_3(t) = \int_{-\infty}^{\infty} f_1(\lambda) f_2(t - \lambda) \, d\lambda \qquad (1.6\text{-}16\text{a})$$

$$\text{correlation:} \quad R_{12}(\tau) = \int_{-\infty}^{\infty} f_1(t) f_2(t - \tau) \, dt \qquad (1.6\text{-}16\text{b})$$

In terms of the operation of integration, note that in the convolution integral that the variable of integration is λ and that $f_2(t - \lambda)$ is a *time reversal* and shifting by t. In the correlation integral, the variable of integration is t, $f_2(t - \tau)$ is a shift by τ, and there is *no time reversal*.

To find the Fourier Transform of the cross-correlation function, note that

$$R_{12}(\tau) = \int_{-\infty}^{\infty} f_1(t) f_2^*(t - \tau) \, dt = \int_{-\infty}^{\infty} f_1(t) f_2^*[-(\tau - t)] \, dt \qquad (1.6\text{-}17\text{a})$$

which is the convolution of $f_1(t)$ and $f_2^*(-t)$. Since $f_2^*(-t) \Leftrightarrow F_2^*(\omega)$, it follows that

$$R_{12}(\tau) \Leftrightarrow F_1(\omega) F_2^*(\omega) = S_{12}(\omega) \qquad (1.6\text{-}17\text{b})$$

and similarly

$$R_{21}(\tau) = R_{12}^*(-\tau) \Leftrightarrow F_2(\omega) F_1^*(\omega) = S_{21}(\omega) = S_{12}^*(\omega) \qquad (1.6\text{-}16\text{c})$$

where $S_{21}(\omega)$ and $S_{12}(\omega)$ are the *cross-energy-spectral densities*. ∎

1.7 THE FOURIER TRANSFORM OF POWER SIGNALS

Until this point we have been taking Fourier Transforms of signals having finite energy. These energy signals may be limited in time (e.g., a rectangular pulse) or infinite in duration (e.g., a decaying exponential). In order to round out the theory of Fourier Transforms, it is necessary to extend them to periodic signals. Periodic signals are one form of *power signals*, signals that have infinite energy but finite average power. The two other forms of power signals that are of interest are data signals and noise. These, however, are statistical rather than deterministic signals so that their analysis will be delayed until the next chapter.

The extension of Fourier Transforms to power signals will also lead to the Sampling Theorem, an essential piece of the analytic framework of both digital communications and digital signal processing.

We shall proceed through a series of examples.

EXAMPLE 1 *The Fourier Transform of* sgn(t).

The function sgn(t), pronounced 'signum', is defined as

$$\text{sgn}(t) = \begin{cases} -1, & t < 0 \\ +1, & t > 0 \end{cases} \qquad (1.7\text{-}1\text{a})$$

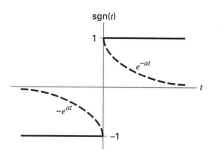

FIGURE 1.7-1

and is shown in fig. 1.7-1. Although we cannot find the Fourier Transform of sgn(t) directly, we can find the transform of

$$\{f(t) = e^{-at}u(t) - e^{at}u(-t)\} \Leftrightarrow \left\{ F(\omega) = \frac{1}{a + j\omega} - \frac{1}{a - j\omega} \right\} \qquad (1.7\text{-}1b)$$

In the limit as $a \to 0$, the two exponentials approach sgn(t) and

$$\text{sgn}(t) \Leftrightarrow \frac{2}{j\omega} \qquad (1.7\text{-}1c)$$

Using the symmetry theorem , we can also determine that

$$\frac{1}{\pi t} \Leftrightarrow -j\,\text{sgn}(\omega) = \begin{cases} j, & \omega < 0 \\ 0, & \omega = 0 \\ -j, & \omega > 0 \end{cases} \qquad (1.7\text{-}1d)$$

The significance of this transform is that it is the transfer function of an ideal 90° phase shifter, an operation that will be key to understanding the Hilbert Transform. These transforms will be examined in Chapter 4 in connection with bandpass data transmission. To illustrate the 90° phase shift property, consider that

$$A\,\text{Cos}(\omega_c t) = \frac{A}{2}e^{j\omega_c t} + \frac{A}{2}e^{-j\omega_c t} \qquad (1.7.1e)$$

and

$$A\,\text{Cos}\left(\omega_c t - \frac{\pi}{2}\right) = \frac{A}{2}e^{j\left(\omega_c t - \frac{\pi}{2}\right)} + \frac{A}{2}e^{-j\left(\omega_c t - \frac{\pi}{2}\right)}$$

$$= \frac{A}{2}e^{-j\frac{\pi}{2}}e^{j\omega_c t} + \frac{A}{2}e^{j\frac{\pi}{2}}e^{-j\omega_c t}$$

$$= -j\frac{A}{2}e^{j\omega_c t} + j\frac{A}{2}e^{-j\omega_c t} \qquad (1.7.1f)$$

It follows that $H(\omega) = -j\,\text{sgn}(\omega)$ is the transfer function of this 90° phase shift filter. Note that $|H(\omega)| = 1$, indicating that the amplitudes of the component signals are unaffected by the filter. ∎

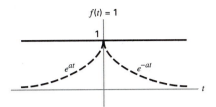

$f(t) = 1$

1

e^{at} e^{-at}

t

FIGURE 1.7-2

EXAMPLE 2   *The Fourier Transform of a Constant*

Again, we can find the Fourier Transform of a constant as a result of a limiting process as shown in fig. 1.7-2. We have already found that

$$\left\{ f(t) = e^{-a|t|} \right\} \Leftrightarrow \left\{ F(\omega) = \frac{2a}{a^2 + \omega^2} \right\} \tag{1.7-2a}$$

In the limit, as $a \to 0$, the two-sided exponential becomes a constant. Clearly, the Fourier Transform gets large at $\omega = 0$ and also becomes very narrow. We shall now demonstrate that this Fourier Transform becomes an impulse function. Note that the area under the transform is

$$\int_{-\infty}^{\infty} \frac{2a}{a^2 + \omega^2} \, d\omega = 2 \int_{-\infty}^{\infty} \frac{dx}{1 + x^2} = 2 \operatorname{Tan}^{-1}(x)|_{-\infty}^{\infty} = 2 \left[\frac{\pi}{2} - \left(-\frac{\pi}{2} \right) \right] = 2\pi \tag{1.7-2b}$$

independent of the value of a. This is an impulse function having area 2π. Consequently we have the transform pair

$$\{ f(t) = 1 \} \Leftrightarrow \{ F(\omega) = 2\pi \delta(\omega) \} \tag{1.7-2c}$$

■

EXAMPLE 3   *The Fourier Transform of a Unit Step Function*

Using the results for $\operatorname{sgn}(t)$ and for the constant, we can find the Fourier Transform of the unit step function $u(t)$. Invoking linearity and superposition yields

$$\left\{ f(t) = u(t) = \frac{1}{2} + \frac{1}{2} \operatorname{sgn}(t) \right\} \Leftrightarrow \left\{ F(\omega) = \pi \delta(\omega) + \frac{1}{j\omega} \right\} \tag{1.7-3}$$

We can now go back and complete the proof of the time-integration theorem.   ■

THEOREM 5   *Time Integration*

$$\text{If } f(t) \Leftrightarrow F(\omega) \quad \text{then } \int_{-\infty}^{t} f(\lambda) \, d\lambda \Leftrightarrow \frac{1}{j\omega} F(\omega) + \pi F(0)\delta(\omega). \tag{1.6-5}$$

Proof:
Define

$$f_1(t) = \int_{-\infty}^{t} f(\lambda) \, d\lambda = \int_{-\infty}^{\infty} f(\lambda) u(t - \lambda) \, d\lambda$$

so that $f_1(t)$ is the convolution of $f(t)$ and $u(t)$. Consequently, $F_1(\omega)$ is the product of the two transforms:

$$F_1(\omega) = F(\omega) \left[\frac{1}{j\omega} + \pi\delta(\omega) \right] = \frac{1}{j\omega} F(\omega) + \pi F(0)\delta(\omega)$$

end of proof

∎

THEOREM 13 *The Fourier Transform of a Periodic Signal*

We are now in a position to find the Fourier Transform of a periodic signal. We have seen that a periodic signal can be represented by a Fourier Series

$$f(t) = \sum_{n=-\infty}^{\infty} f_n e^{jn\omega_0 t} = f_0 + 2 \sum_{n=1}^{\infty} |f_n| \mathrm{Cos}(n\omega_0 t + \theta_n) \qquad (1.1\text{-}1a)$$

which is, after all, a time function.

The key to being able to find this Fourier Transform is finding the transform of $e^{jn\omega_0 t}$. This can be accomplished by starting with the Fourier Transform of a constant

$$\{f(t) = 1\} \Leftrightarrow \{F(\omega) = 2\pi\delta(\omega)\} \qquad (1.7\text{-}2c)$$

and then invoking the frequency-shifting theorem

$$\{f(t)e^{jn\omega_0 t} = 1 \cdot e^{jn\omega_0 t}\} \Leftrightarrow \{F(\omega - n\omega_0) = 2\pi\delta(\omega - n\omega_0)\}. \qquad (1.7\text{-}4a)$$

As a special case of eq. 1.7-4a, we can find the Fourier Transforms of Cos and Sin.

$$\left\{ \mathrm{Cos}(n\omega_0 t) = \frac{1}{2} e^{jn\omega_0 t} + \frac{1}{2} e^{-jn\omega_0 t} \right\} \Leftrightarrow \{\pi\delta(\omega - n\omega_0) + \pi\delta(\omega + n\omega_0)\} \quad (1.7\text{-}4b)$$

and

$$\left\{ \mathrm{Sin}(n\omega_0 t) = \frac{1}{2j} e^{jn\omega_0 t} - \frac{1}{2j} e^{-jn\omega_0 t} \right\}$$

$$\Leftrightarrow \left\{ \frac{\pi}{j}\delta(\omega - n\omega_0) - \frac{\pi}{j}\delta(\omega + n\omega_0) \right\} \qquad (1.7\text{-}4c)$$

as shown in fig. 1.7-3.

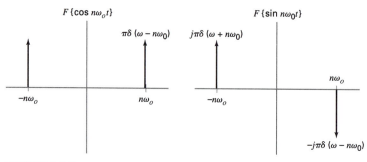

FIGURE 1.7-3

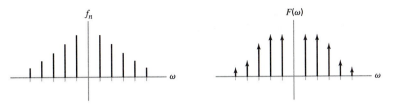

FIGURE 1.7-4

In general,

$$\left\{ f(t) = \sum_{n=-\infty}^{\infty} f_n e^{jn\omega_0 t} \right\} \Leftrightarrow \left\{ F(\omega) = 2\pi \sum_{n=-\infty}^{\infty} f_n \delta(\omega - n\omega_0) \right\} \qquad (1.7\text{-}4\text{d})$$

which is the Fourier Transform of any periodic signal.

The spectrum of a periodic signal is shown in fig. 1.7-4. ∎

This presents the reader with an interesting conceptual problem with which to struggle. Shown in fig. 1.7-4 are the spectra of both the Fourier Series and the Fourier Transform of that series. Both pictures consist of lines at the harmonic frequencies $n\omega_0$. In the spectrum of the series these line are simply representational, i.e., a way of graphically depicting a set of numbers; in the transform, the lines are impulse functions and the transform is therefore well defined mathematically. Clearly, however, the visual relationship between these two objects is quite close. To get the transform of a periodic function, simply take the spectral lines of the series, multiply them by 2π, and put arrows on them.

In particular, we are interested in the Fourier Transform of periodic time domain impulse functions, fig. 1.7-5. We know that for these periodic impulses, the Fourier Series coefficent is $f_n = 1/T$ so that

$$\left\{ \delta_T(t) = \sum_{n=-\infty}^{\infty} \delta(t - nT) = \frac{1}{T} \sum_{n=-\infty}^{\infty} e^{jn\omega_0 t} \right\}$$

$$\Leftrightarrow \left\{ F(\omega) = \frac{2\pi}{T} \sum_{n=-\infty}^{\infty} \delta(\omega - n\omega_0) = \omega_0 \delta_{\omega_0}(\omega) \right\} \qquad (1.7\text{-}5)$$

Therefore uniform impulses in the time domain give us uniform impulses in the frequency domain. The general point is that the Fourier Transform of a periodic function is a collection of impulse functions at the harmonic frequencies $n\omega_0$ and that the areas of these impulses are equal to the corresponding Fourier Series coefficient multiplied by 2π.

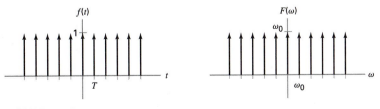

FIGURE 1.7-5

1.8   THE POISSON SUM FORMULAS

In subsequent discussions involving infinite sums of random data we will find the Poisson Sum formulas to be quite useful. They can all be derived from a single theorem.

THEOREM 14

Let

$$x(t) = \sum_{n=-\infty}^{\infty} g(t)h(t - nT_s) \quad \text{where } \omega_s = 2\pi/T_s \qquad (1.8\text{-}1a)$$

Then

$$X(\omega) = \frac{1}{T_s} \sum_{n=-\infty}^{\infty} H(n\omega_s)G(\omega - n\omega_s) \qquad (1.8\text{-}1b)$$

∎

Proof:

Since $\sum_{n=-\infty}^{\infty} h(t - nT_s)$ is a periodic function, it can be written as a Fourier Series so that

$$\sum_{n=-\infty}^{\infty} h(t - nT_s) = \sum_{m=-\infty}^{\infty} h_m e^{jm\omega_s t} \quad \text{where } h_m = \frac{H(m\omega_s)}{T_s}$$

Consequently:

$$x(t) = \sum_{n=-\infty}^{\infty} g(t)h(t - nT_s) = g(t)\sum_{n=-\infty}^{\infty} h(t - nT_s) = g(t)\sum_{m=-\infty}^{\infty} \frac{H(m\omega_s)}{T_s} e^{jm\omega_s t}$$

Taking the Fourier Transform, we get:

$$X(\omega) = \frac{1}{T_s} \sum_{m=-\infty}^{\infty} H(m\omega_s) \int_{-\infty}^{\infty} g(t)e^{jm\omega_s t}e^{-j\omega t}\, dt = \frac{1}{T_s} \sum_{m=-\infty}^{\infty} H(m\omega_s)G(\omega - m\omega_s)$$

end of proof

We can now use eq. 1.8-1a,b to prove the following identities:

1. $$x(t) = \sum_{n=-\infty}^{\infty} \delta(t - nT_s) \Rightarrow X(\omega) = \frac{1}{T_s} \sum_{m=-\infty}^{\infty} \delta(\omega - m\omega_s) \qquad (1.8\text{-}2a)$$

Proof:
In eq. 1.8-1a, let

$$\sum_{n=-\infty}^{\infty} h(t - nT_s) = \sum_{m=-\infty}^{\infty} h_m e^{jm\omega_s t} \quad \text{where } h_m = \frac{H(m\omega_s)}{T_s}$$

and let $h(t) = \delta(t)$ so that $H(\omega) = 1$. It follows that

$$x(t) = \sum_{n=-\infty}^{\infty} \delta(t - nT_s) = \frac{1}{T_s} \sum_{m=-\infty}^{\infty} e^{jm\omega_s t}$$

and the Fourier Transform is

$$X(\omega) = \frac{2\pi}{T_s} \sum_{m=-\infty}^{\infty} \delta(\omega - m\omega_s) = \omega_s \sum_{m=-\infty}^{\infty} \delta(\omega - m\omega_s)$$

end of proof

2. $x(t) = \sum_{n=-\infty}^{\infty} (-1)^n \delta(t - nT_s) \Rightarrow X(\omega) = \frac{1}{T_s} \sum_{m=-\infty}^{\infty} \delta(\omega - m\omega_s - \omega_{s/2})$

(1.8-2b)

Proof:
In eq. 1.8-1a,

$$\sum_{n=-\infty}^{\infty} h(t - nT_s) = \sum_{m=-\infty}^{\infty} c_m e^{jm\omega_s t} \quad \text{where } c_m = \frac{H(m\omega_s)}{T_s}$$

and let $h(t) = \delta(t) - \delta(t - T_s)$ so that

$$H(\omega) = 1 - e^{j\omega T_s} = \begin{cases} 0; & \text{if } m \text{ is even} \\ 2; & \text{if } m \text{ is odd} \end{cases}.$$

Since $h(t)$ has a period equal to $2T_s$ and, therefore, a fundamental frequency equal to $\omega_s/2$, it follows that

$$x(t) = \frac{2}{2T_s} \sum_{m=-\infty}^{\infty} e^{j(2m+1)\omega_s/2t} = \frac{1}{T_s} \sum_{m=-\infty}^{\infty} e^{j(m\omega_s + \omega_s/2)t}$$

so that the Fourier Transform is

$$X(\omega) = \frac{1}{T_s} \sum_{m=-\infty}^{\infty} \delta(\omega - m\omega_s - \omega_s/2)$$

end of proof

There are three more identities of interest whose proofs will be left to the Problems.

3. $$\sum_{n=-\infty}^{\infty} h(nT_s) = \frac{1}{T_s} \sum_{m=-\infty}^{\infty} H(m\omega_s)$$ (1.8-2c)

4. $$\sum_{n=-\infty}^{\infty} h(t - nT_s) = \frac{1}{T_s} \sum_{m=-\infty}^{\infty} H(m\omega_s) e^{jm\omega_s t}$$ (1.8-2d)

5. $$\sum_{n=-\infty}^{\infty} h(nT_s) e^{-jm\omega_s t} = \frac{1}{T_s} \sum_{m=-\infty}^{\infty} H(\omega - m\omega_s)$$ (1.8-2e)

1.9 TIME-AVERAGED AUTOCORRELATION FUNCTIONS AND POWER SPECTRAL DENSITY FOR PERIODIC SIGNALS

In a previous section we developed the twin ideas of energy spectral density and autocorrelation function for signals that have finite energy. We now want to extend that idea to periodic signals that have infinite energy but finite average power. The results that are obtained here will be extended to data signals and to random noise in the next chapter.

It is useful to show where we want to go by restating the salient facts about autocorrelation and energy spectral density for energy signals. Recall that for the energy function $f(t)$

1. The autocorrelation function of $f(t)$ is

$$R_f(\tau) = \int_{-\infty}^{\infty} f(t)f^*(t - \tau)\,dt \qquad (1.6\text{-}13b)$$

2. The energy spectral density is

$$E(\omega) = |F(\omega)|^2 \qquad (1.6\text{-}12b)$$

3. The autocorrelation function and the energy spectral density are a Fourier Transform pair

$$R_f(\tau) \Leftrightarrow E(\omega) \quad \text{or} \quad R_f(\tau) = \frac{1}{2\pi} \int_{-\infty}^{\infty} E(\omega)e^{j\omega\tau}\,d\omega \qquad (1.6\text{-}13a)$$

4. The energy contained in $f(t)$ is given by

$$E = \int_{-\infty}^{\infty} |f(t)|^2\,dt = R_f(0) = \frac{1}{2\pi} \int_{-\infty}^{\infty} E(\omega)\,d\omega \qquad (1.6\text{-}12b)$$

We want to construct an analagous set of relations for periodic signals based on the average power of such a signal. In general, the average power is given by

$$P_{\text{av}} = \lim_{T\to\infty} \frac{1}{T} \int_{-T/2}^{T/2} |f(t)|^2\,dt \qquad (1.9\text{-}1a)$$

and for periodic functions having period T

$$P_{\text{av}} = \frac{1}{T} \int_{-T/2}^{T/2} |f(t)|^2\,dt. \qquad (1.9\text{-}1b)$$

Similarly, we define the *time-average autocorrelation function* $\bar{R}_f(\tau)$ of a power signal as

$$\bar{R}_f(\tau) = \lim_{T\to\infty} \frac{1}{T} \int_{-T/2}^{T/2} f(t)f^*(t - \tau)\,dt \qquad (1.9\text{-}2a)$$

and for periodic functions having period T as

$$\bar{R}_f(\tau) = \frac{1}{T} \int\limits_{-T/2}^{T/2} f(t) f^*(t-\tau)\, dt \tag{1.9-2b}$$

We want to have a *power spectral density* $P(\omega)$ such that

$$\bar{R}_f(\tau) \Leftrightarrow P(\omega) \tag{1.9-3a}$$

and

$$P_{av} = \bar{R}_f(0) = \frac{1}{2\pi} \int\limits_{-\infty}^{\infty} P(\omega)\, d\omega \tag{1.9-3b}$$

and we want to have the input-output relationship for power spectral density

$$P_{out}(\omega) = P_{in}(\omega) |H(\omega)|^2 \tag{1.9-4}$$

when the signal is passed through a linear system having transfer function $H(\omega)$.

For a periodic signal $f(t) = \sum_{n=-\infty}^{\infty} f_n e^{jn\omega_0 t}$, the time-average autocorrelation function according to eq. 1.9-2b is

$$\bar{R}_f(\tau) = \frac{1}{T} \int\limits_{-T/2}^{T/2} \left[\sum_{n=-\infty}^{\infty} f_n e^{jn\omega_0 t} \sum_{m=-\infty}^{\infty} f_m^* e^{-jm\omega_0(t-\tau)} \right] dt$$

$$= \frac{1}{T} \sum_{n=-\infty}^{\infty} \sum_{m=-\infty}^{\infty} f_n f_m^* e^{jm\omega_0\tau} \int\limits_{-T/2}^{T/2} e^{j(n-m)\omega_0 t}\, dt$$

$$= \begin{cases} 0; & n \neq m \\ \sum_{n=-\infty}^{\infty} |f_n|^2 e^{jn\omega_0\tau}; & n = m \end{cases}$$

or

$$\bar{R}_f(\tau) = \sum_{n=-\infty}^{\infty} |f_n|^2 e^{jn\omega_0\tau} \tag{1.9-5a}$$

Clearly, this time-average autocorrelation function is periodic with period $T = 2\pi/\omega_0$. According to our theory, the Fourier Transform of $\bar{R}_f(\tau)$ should be the power spectral density $P(\omega)$. Directly taking this transform, we get

$$P(\omega) = 2\pi \sum_{n=-\infty}^{\infty} |f_n|^2 \delta(\omega - n\omega_0) \tag{1.9-5b}$$

and integrating to find the average power:

$$P_{av} = \bar{R}_f(0) = \frac{1}{2\pi} \int\limits_{-\infty}^{\infty} P(\omega)\, d\omega = \sum_{n=-\infty}^{\infty} |f_n|^2 \tag{1.9-5c}$$

which is the same expression for the average power in a periodic signal as eq. 1.1-2a. If this periodic signal $f(t)$ is applied to the system having transfer function $H(\omega)$, the output is

$$g(t) = \sum_{n=-\infty}^{\infty} g_n e^{jn\omega_0 t} = \sum_{n=-\infty}^{\infty} f_n H(n\omega_0) e^{jn\omega_0 t} \tag{1.9-6a}$$

and the output power is

$$P_0 = \sum_{n=-\infty}^{\infty} |f_n|^2 |H(n\omega_0)|^2 = \frac{1}{2\pi} \int_{-\infty}^{\infty} P_i(\omega)|H(\omega)|^2 \, d\omega = \frac{1}{2\pi} \int_{-\infty}^{\infty} P_0(\omega) \, d\omega \qquad (1.9\text{-}6b)$$

We also want to see the relationship between the time-average autocorrelation function of a periodic signal and the autocorrelation function of the energy function that forms a single period. Let

$$f_T(t) = \begin{cases} f(t); & -T/2 \leq t \leq T/2 \\ 0; & \text{otherwise} \end{cases} \qquad (1.9\text{-}7a)$$

so that

$$\begin{aligned}
\bar{R}_f(\tau) &= \frac{1}{T} \int_{-T/2}^{T/2} \sum_{n=-\infty}^{\infty} f_T(t - nT) f_T^*(t - nT - \tau) \, dt \\
&= \frac{1}{T} \sum_{n=-\infty}^{\infty} \int_{-\infty}^{\infty} f_T(t - nT) f_T^*(t - nT - \tau) \, dt \\
&= \frac{1}{T} \sum_{n=-\infty}^{\infty} R_{f_T}(\tau - nT) \qquad (1.9\text{-}7b)
\end{aligned}$$

In short, the time-average autocorrelation function is the periodic extension of the autocorrelation function of the basic waveform $f_T(t)$, as shown in fig. 1.9-1a. This periodic extension may involve the overlapping of adjacent basic functions, as shown in fig. 1.9-1b.

Finally, the inverse Fourier Transform of the power spectral density may be derived from eq. 1.9-5b as:

$$F^{-1}[P_f(\omega)] = \sum_{n=-\infty}^{\infty} |f_n|^2 e^{jn\omega_0 \tau} \qquad (1.9\text{-}8a)$$

and we now want to show that this is in fact equal to the time-average autocorrelation function. Since $\bar{R}_f(\tau)$ is periodic (eq. 1.53b), it can also be expanded into a Fourier Series

$$\bar{R}_f(\tau) = \frac{1}{T_s} \sum_{n=-\infty}^{\infty} b_n e^{jn\omega_0 \tau} \qquad (1.9\text{-}8b)$$

where

$$b_n = \frac{E_{f_T}(n\omega_0)}{T_s} = \frac{|F_T(n\omega_0)|^2}{T_s} \qquad (1.9\text{-}8c)$$

so that

$$\bar{R}_f(\tau) = \sum_{n=-\infty}^{\infty} \left| \frac{F_1(n\omega_0)}{T_s} \right|^2 e^{jn\omega_0 \tau} = \sum_{n=-\infty}^{\infty} |f_n|^2 e^{jn\omega_0 \tau} \qquad (1.9\text{-}8d)$$

We have therefore shown that for a periodic function $\bar{R}_f(\tau) \Leftrightarrow P_f(\omega)$, which is what we set out to do. This concept will be extended to data signals and to noise in the next chapter. It forms a basic analytic tool in the understanding of random processes.

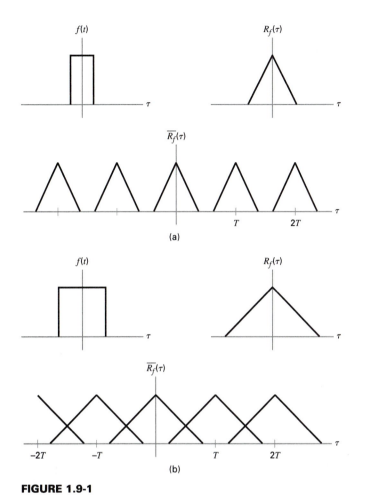

FIGURE 1.9-1

1.10   THE SAMPLING THEOREM AND APPLICATIONS

The process of sampling signals, transmitting the samples, and then recovering the signal from its samples is central to digital communication and to digital signal processing. The development of the Sampling Theorem requires that we use many of the ideas that have already developed. To begin, consider the following propositions:

Proposition 1
The convolution of two impulse functions is an impulse function.

$$\int_{-\infty}^{\infty} \delta(x - a)\delta(y - x - b)\,dx = \delta(y - (a + b)) \qquad (1.10\text{-}1a)$$

Proof:
A rectangular pulse having width ε and height $1/\varepsilon$ becomes an impulse function as $\varepsilon \to 0$. The convolution of this rectangle with itself yields a triangle of width 2ε and height $1/\varepsilon$. Since the triangle also becomes an impulse function in the limit as $\varepsilon \to 0$, the point is made.

Proposition 2

According to the frequency convolution theorem, the product of two functions in the time domain is equivalent to the convolution of their Fourier Transforms in the frequency domain.

Let $g(t) = f_1(t) f_2(t)$ where $f_1(t) \Leftrightarrow F_1(\omega)$, $f_2(t) \Leftrightarrow F_2(\omega)$ and $g(t) \Leftrightarrow G(\omega)$. Then

$$G(\omega) = \frac{1}{2\pi} \int_{-\infty}^{\infty} F_1(\beta) F_2(\omega - \beta) \, d\beta = \frac{1}{2\pi} F_1(\omega) \otimes F_2(\omega) \qquad (1.10\text{-}1b)$$

EXAMPLE 1

Consider the product $g(t) = f_1(t) f_2(t) = \text{Cos}(\omega_1 t) \text{Cos}(\omega_2 t)$ where

$$F_1(\omega) = \pi \delta(\omega - \omega_1) + \pi \delta(\omega + \omega_1) \qquad (1.10\text{-}2a)$$

and

$$F_2(\omega) = \pi \delta(\omega - \omega_2) + \pi \delta(\omega + \omega_2) \qquad (1.10\text{-}2b)$$

The convolution of these two frequency-domain functions is shown in fig. 1.10-1a. The result corresponds to

$$g(t) = \frac{1}{2} \text{Cos}[(\omega_2 + \omega_1)t] + \frac{1}{2} \text{Cos}[(\omega_2 - \omega_1)t] \qquad (1.10\text{-}2c)$$

Geometrically, we can think of this process as taking $F_1(\omega)$, shifting it to the right to be centered around the impulse function at ω_2, and at the same time multiplying it by π, the magnitude of that impulse, and dividing the result by 2π. The same operation is repeated, shifting to the left around $-\omega_2$. ∎

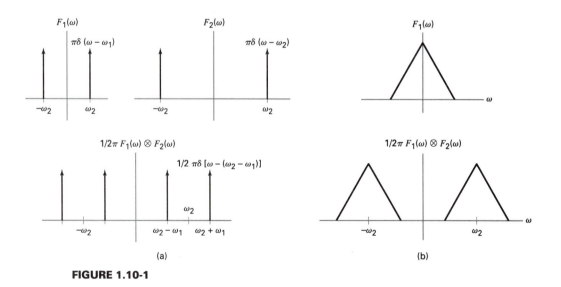

FIGURE 1.10-1

EXAMPLE 2

This time let $f_1(t)$ have the triangular spectrum shown in fig. 1.10-1b and again we are trying to find the spectrum of

$$g(t) = f_1(t) \, \text{Cos}(\omega_2 t) \tag{1.10-3a}$$

The operation that was used in accomplishing this transformation is

$$G(\omega) = \frac{1}{2\pi} \int_{-\infty}^{\infty} F_1(\beta)[\pi\delta(\omega - \beta - \omega_2) + \pi\delta(\omega - \beta + \omega_2)] \, d\beta$$

$$= \frac{1}{2} F_1(\omega - \omega_2) + \frac{1}{2} F_1(\omega + \omega_2) \tag{1.10-3b}$$

Again, we see that convolution of a waveform with an impulse function can be understood geometrically as shifting that waveform to the place that the impulse occurs, taking care to multiply the amplitudes of the waveform and the impulse and dividing by 2π as the frequency convolution integral tells us to do. ■

Proposition 3
In the case of the previous example, where we were interested in the Fourier Transform of

$$g(t) = f_1(t) \, \text{Cos}(\omega_2 t) = \frac{1}{2}(t)[e^{j\omega_2 t} + e^{-j\omega_2 t}] \tag{1.10-4a}$$

we may use the Frequency Shifting Theorem to obtain the same result

$$G(\omega) = \frac{1}{2} F_1(\omega - \omega_2) + \frac{1}{2} F_1(\omega + \omega_2) \tag{1.10-4b}$$

Proposition 4
We finally want to remind ourselves of the Fourier Transform of a periodic signal $f(t)$:

$$\left\{ f(t) = \sum_{n=-\infty}^{\infty} f_n e^{jn\omega_0 t} \right\} \Leftrightarrow \left\{ F(\omega) = 2\pi \sum_{n=-\infty}^{\infty} f_n \delta(\omega - n\omega_0) \right\} \tag{1.7-4d}$$

Now the thing to keep in mind here is that the Fourier Transform of *any periodic signal* consists of uniformly spaced impulse functions occuring at integer multiples of the fundamental frequency. The distinction between different periodic functions having the same period or fundamental frequency is the amplitude of the impulses.

We are now ready to develop the *Sampling Theorem*.

STEP 1 *Natural Sampling*

As shown in fig. 1.10-2a, $f(t) \Leftrightarrow F(\omega)$ is a bandlimited signal having W rad/sec as its highest frequency. It is multiplied by $s(t)$, referred to as a sampling function, which is a periodic function having fundamental frequency $\omega_0 = 2\pi/T$. The Fourier Series of

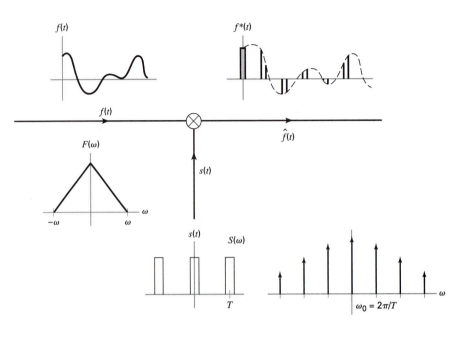

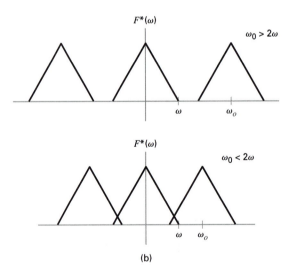

FIGURE 1.10-2

$s(t)$ is:

$$\left\{ s(t) = \sum_{n=-\infty}^{\infty} s_n e^{jn\omega_0 t} \right\}$$
(1.10-5a)

In this example, $s(t)$ consists of narrow, periodic rectangular pulses as shown, but it is important to establish that $s(t)$ could, for purposes of this discussion, be any periodic function having period T. The point is that *every periodic function having period T has essentially the same kind of spectrum*. The spectrum of $s(t) \Leftrightarrow S(\omega)$ consists of impulse functions occurring at integer multiples of the fundamental frequency $n\omega_0$. The areas of these impulse functions are, of course, the Fourier Series coefficents multiplied by 2π.

The sampled signal $f^*(t) = f(t)s(t)$ consists of portions or segments of the signal $f(t)$, as shown in fig. 1.10-2a. Consequently, this form of sampling is referred to as *natural sampling*. Since $f^*(t)$ is the product of two signals, its Fourier Transform is

$$F^*(\omega) = \frac{1}{2\pi} F(\omega) \otimes S(\omega) \tag{1.7-4d}$$

Since $S(\omega)$ consists of impulse functions, $F^*(\omega)$ simply consists of reproductions of $F(\omega)$ weighted by the s_n coefficients of the Fourier Series for $s(t)$. Specifically,

$$F^*(\omega) = \sum_{n=-\infty}^{\infty} s_n F(\omega - n\omega_0) \tag{1.7-4d}$$

The essential issue demonstrated in fig. 1.10-1b is:

If $\omega_0 > 2W$, these shifted versions of $F(\omega)$ do not overlap and the original $F(\omega)$ can be recovered from the sampled signal using a lowpass filter. On the other hand, if $\omega_0 < 2W$, then the shifted versions of $F(\omega)$ do overlap and the original signal cannot be recovered from the samples without distortion. The distortion, caused by overlapping spectra, is called **aliasing**.

Note that the recovery of the signal from its samples is illustrated in the frequency domain by an ideal lowpass filter having cutoff frequencies at $\pm\omega_0/2$.

Another way of saying this is the following: Let $f(t)$ be a signal bandlimited to $B = \pm W/2\pi$ Hz. If we sample this signal with a periodic sampling functions $s(t)$ having a fundamental frequency f_0, *multiplying* $f(t)$ and $s(t)$ together, then we can recover the signal from its samples provided that $f_0 > 2B$. The Nyquist Frequency is defined as B Hz, so another way of saying this is that we must sample at least at twice the Nyquist Frequency.

The problem with natural sampling from an applications point of view is that samples are actually analog slices of an analog signal and must consequently be transmitted through linear analog channels. Taking full advantage of the possibilities inherent in sampling requires their conversion into digital data. In order to accomplish this it is necessary to take two steps: the first step is from natural sampling to ideal sampling, followed by the next step, to flat-top sampling. ■

STEP 2   *Ideal Sampling*

In the limit, as the periodic sampling functions $s(t)$ become periodic impulse functions, natural sampling becomes *ideal sampling*. The Fourier Transform of periodic impulse functions is:

$$\left\{ s(t) = \sum_{n=-\infty}^{\infty} \delta(t - nT) = \frac{1}{T} \sum_{n=-\infty}^{\infty} e^{jn\omega_0 t} \right\}$$

$$\Leftrightarrow \left\{ S(\omega) = \frac{2\pi}{T} \sum_{n=-\infty}^{\infty} \delta(\omega - n\omega_0) = \omega_0 \sum_{n=-\infty}^{\infty} \delta(\omega - n\omega_0) \right\} \tag{1.7-5a}$$

Following eq. 1.10-5c, the spectrum of the sampled signal is

$$F^*(\omega) = \frac{1}{T} \sum_{n=-\infty}^{\infty} F(\omega - n\omega_0) \tag{1.10-6}$$

and the repetitions of the basic spectrum are all equal in amplitude, making it a periodic signal. It should be clear that there is no difference in the issue of how to recover the signal from its samples between natural and ideal sampling. The difference between the two lies in the nature of the samples: the samples in *natural sampling* consist of pieces of the actual waveform of $f(t)$ multiplied by $s(t)$ and therefore must still be considered an analog signal; the samples in *ideal sampling* consist of a sequence of numerical values of the function $f(nT)$ at the sampling points. Although these numerical values are still analog, we may easily turn them into digital quantities. This will be discussed subsequently.

It is of considerable interest to examine what the recovery of an ideally sampled signal from its samples looks like in the time domain. (Recall that in the frequency domain this recovery is acheived by passing the signal through an ideal lowpass filter having cutoff frequencies at $\pm\omega_0/2$.

Since the ideally sampled signal has a periodic spectrum, it can be represented as a Fourier Series in the frequency domain. Making a change of variable from n to $-n$:

$$F^*(\omega) = \frac{1}{T} \sum_{n=-\infty}^{\infty} F(\omega - n\omega_0) = \sum_{n=-\infty}^{\infty} A_n e^{-jnT\omega} \tag{1.10-7a}$$

where

$$A_n = \frac{1}{\omega_0} \int_{-\omega_0/2}^{\omega_0/2} F^*(\omega)e^{jnT\omega}\,d\omega \tag{1.10-7b}$$

but within the limits of integration $F^*(\omega) = F(\omega)$ (note that in this instance the superscript * refers to the spectrum of the sampled signal, not the complex conjugate), so that:

$$A_n = \frac{1}{\omega_0} \int_{-\infty}^{\infty} F(\omega)e^{jnT\omega}\,d\omega. \tag{1.10-7c}$$

Noting that according to the Fourier Transform

$$f(t) = \frac{1}{2\pi} \int_{-\infty}^{\infty} F(\omega)e^{j\omega t}\,d\omega \tag{1.10-7d}$$

It follows by comparison that $A_n = Tf(nT)$ so that

$$F^*(\omega) = \sum_{n=-\infty}^{\infty} Tf(nT)e^{-jnT\omega} \tag{1.10-7e}$$

Now pass $F^*(\omega)$ through an ideal lowpass filter defined by

$$H(\omega) = \begin{cases} 1; & |\omega| \leq \omega_0/2 \\ 0; & \text{otherwise} \end{cases} \tag{1.10-8a}$$

yielding $F(\omega)$. The Inverse Fourier Transform (eq. 1.5-2b) of the result is:

$$f(t) = \frac{1}{2\pi} \int_{-\omega_0/2}^{\omega_0/2} \sum_{n=-\infty}^{\infty} Tf(nT)e^{-jnT\omega}e^{j\omega t}\,d\omega$$

$$= \frac{1}{\omega_0} \sum_{n=-\infty}^{\infty} f(nT) \int_{-\omega_0/2}^{\omega_0/2} e^{j\omega(t-nT)}\,d\omega = \sum_{n=-\infty}^{\infty} f(nT)\,\text{Sinc}\left[\frac{\omega_0}{2}(t-nT)\right] \tag{1.10-8b}$$

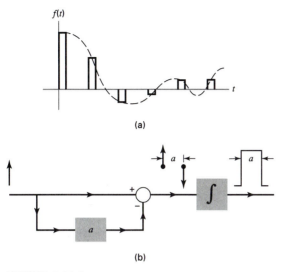

(a)

(b)

FIGURE 1.10-3

This says that the time-domain equivalent of passing $F^*(\omega)$ through an ideal lowpass filter is to reconstruct the signal by multiplying the samples by a sequence of sampling functions. This, by the way, is the justification for calling $\mathrm{Sinc}(x)$ the sampling function.

There are a number of practical and theoretical problems that should be mentioned. On the practical side, the theory that has been presented involves ideal sampling with impulses and an ideal lowpass filter for the for reconstruction of the signal. In practice, since neither ideal impulses nor ideal filters are available, there will be an inevitable distortion. The idea of a bandlimited signal also produces a theoretical anomaly. Although intuitively satisfying, bandlimited signals are, according to theory, infinite in time duration. Similarly, strictly time limited signals have infinite bandwidth, being non-zero out to infinity. Clearly, this is a point where classical Fourier theory departs from reality. Readers interested in pursuing this issue are referred to Slepian [Sle76]. ■

STEP 3   *Flat-Top Sampling*

Flat-top sampling, also known as 'sample and hold,' illustrated in fig. 1.10-3a, is a more realistic approach than ideal impulse sampling. Clearly, the resulting sampled waveform is a distortion of what we would get through ideal sampling, and we would like to get some mathematical appreciation of the nature and amount of that distortion. Notice that the magnitudes of the flat-top samples are equal to those we would get through ideal sampling. A mathematical model of how to generate flat-top samples from ideal samples is shown in fig. 1.10-3b. In this circuit an input impulse function and its delayed and inverted version are integrated to form a rectangular pulse having width a, the delay time. This is often called a first-order-hold circuit or a sample-and-hold circuit. The idea is to view the flat-top sampling process as ideal sampling followed by passage of the resulting impulses through the delay and integration circuit. The spectrum of the flat-top-sampled signal is $\hat{F}^*(\omega) = F^*(\omega)H(\omega)$, where

$$H(s) = \frac{1 - e^{-sa}}{s} \quad \text{or} \quad H(\omega) = \frac{1 - e^{-j\omega a}}{j\omega} = a\,\mathrm{Sinc}\,(\omega a/2)e^{-j\omega a/2} \qquad (1.10\text{-}9a)$$

This issue can best be illustrated with an example. ■

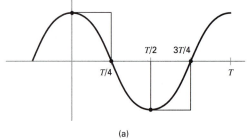

(a)

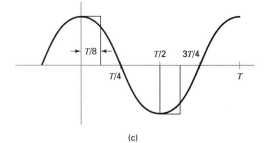

(b)

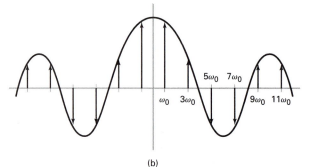

(c)

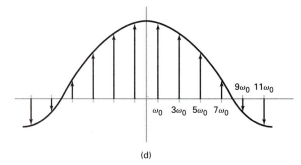

(d)

FIGURE 1.10-4

EXAMPLE 3

Cos($\omega_0 t$) is flat-top-sampled four times in each period at $t = 0$, $T/4$, $T/2$, $3T/4$ and the samples are held for $a = T/4$, as shown in fig. 1.10-4a. It follows that

$$H(\omega) = \frac{T}{4}\text{Sinc}(\omega T/8)e^{-j\omega T/8} = \frac{T}{4}\text{Sinc}\left(\pi\frac{\omega}{4\omega_0}\right)e^{-j\pi\omega/4\omega_0} \qquad (1.10\text{-}9b)$$

This transfer function clearly has zeros at $\omega = \pm 4n\omega_0$ so that, if we neglect the phase term, which simply represents the time delay of $a/2$, the sampled-and-held or flat-top-sampled spectrum is shown in fig. 1.10-4b.

Now let us hold the samples for $a = T/8$ as shown in fig. 1.10-4c, resulting in the spectrum of fig 1.10-4d. ∎

It is clearly possible to recover the sinusoid from its flat-top samples by passing the sampled signal through a lowpass filter. The distortion introduced by flat-top sampling becomes more of an issue when the sampled signal has a continuous spectrum. As a becomes smaller, the larger the crossover frequency $\omega = 2\pi/a$ the less distortion we have and the closer to ideal sampling we get. As a gets larger, the distortion may become a serious issue and require a compensating filter or equalizer. This kind of distortion is often called "aperture distortion" corresponding to the interference pattern generated by passing light through a small opening or the near field of a dipole antenna.

STEP 4   *Quantizing a Flat-Top-Sampled Signal*

The amplitude of a flat-top-sampled signal is still an analog voltage. In order to have digital representation it is necessary to quantize the sampled signal to discrete levels and then associate each level with a digital sequence. This process clearly leads to some distortion, termed *quantization noise*. Computing this noise power will be delayed until the next chapter.

As an example of how all this works, consider the operation of digital telephones and voiceband digital signal processing. The original analog telephone line has been typically characterized as having a bandwidth from 300 Hz to 3300 Hz. For the purposes of digital transmission this voiceband has been characterized as having a 4 kHz bandwidth leading to an 8 kHz sampling rate. The resulting samples are quantized to eight bits, leading to a 64 kHz bit rate for a single telephone conversation. This is an example of PCM (pulse-code modulation).

The 64 kHz bit rate forms the basis for the entire digital transmission hierarchy, which will be explored in subsequent chapters. ∎

CHAPTER 1 PROBLEMS

1.1   Find the average power in the periodic signals shown below:

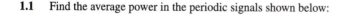

1.2   Prove the following trigonometric identities. With n, m integers and with $\omega_0 = 2\pi/T$,

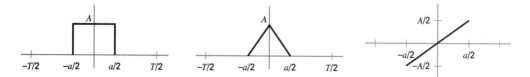

1. $$\int_{t_1}^{t_1+T} \text{Cos}(n\omega_0 t)\, dt = \int_{t_1}^{t_1+T} \text{Sin}(n\omega_0 t)\, dt = 0 \quad \text{for all } t_1$$

2.
$$\int_{t_1}^{t_1+T} \text{Cos}(n\omega_0 t)\,\text{Sin}(m\omega_0 t)\,dt = 0 \quad \text{for all } t_1, \text{ for all } m, n$$

3.
$$\int_{t_1}^{t_1+T} \text{Cos}(n\omega_0 t)\,\text{Cos}(m\omega_0 t)\,dt = \int_{t_1}^{t_1+T} \text{Sin}(n\omega_0 t)\,dt = 0 \quad \text{for all } t_1, \text{ only if } m \neq n$$

1.3   Periodic rectangular pulses having a duty cycle of $T/4$ are applied to a lowpass filter having the transfer function shown below. Find the average power in the output.

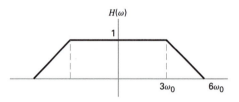

1.4   Find the amplitude spectrum for symmetric rectangular pulses having a duty cycle of 3/8.

1.5   Find the amplitude spectrum for symmetric cosine pulses having a duty cycle of a) 3/8; b) 5/8. Sketch the spectrum and the envelope. Find the first five terms of the trigonometric series.

1.6   Repeat Problem 1-5 for triangular pulses.

1.7   Repeat Problem 1-5 for ramp pulses.

1.8   Consider rectangular, cosine, and triangular pulses each having a duty cycle 1/4. Keeping in mind that the main lobes of the three spectra are different, compute the fraction of the average power in each of these signals that is contained in the main lobe. Note that this fraction is highest for the smoothest signal and smallest for the signal having the sharpest transition.

1.9   Using linearity, superposition, and time-shifting find an expression for the Fourier Series f_n coefficents for the periodic waveforms shown below. For each of them, find the ratio of the power in the dc component and the first five harmonics (including the fundamental) to the average power in the waveform.

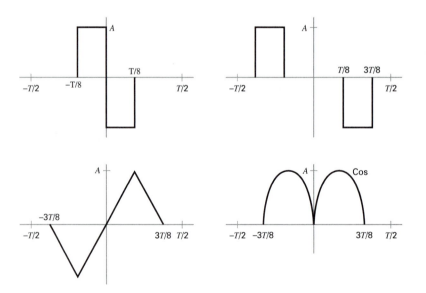

1.10   The periodic waveform shown below is an approximation to a sinusoid.

The object is to find the ratio of A_2/A_1 that will make a particular harmonic go to zero. Find the appropriate ratios for the four cases defined by $n = 3, 5$ and $\lambda = 1/4, 1/3$.

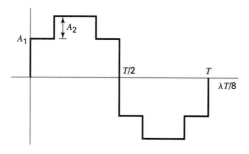

1.11   Using the Theorems of Time Differentiation and Time Integration, find the Fourier Series f_n coefficients for the periodic functions shown below:

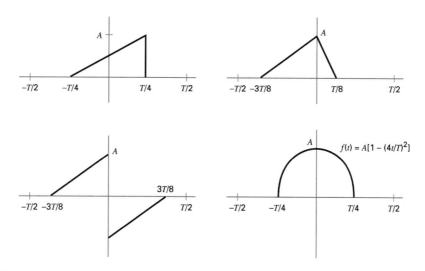

1.12   Let $\omega_1 = 2\pi 2400$, $\omega_2 = 2\pi 600$.

(a) Find the spectrum of $f(t) = \text{Cos } \omega_1 t \text{ Cos } \omega_2 t$. Represent the horizontal axis as frequency rather than radians/sec.

(b) Find the spectrum of $f(t) = \text{Cos } \omega_1 t * \text{“Cos } \omega_2 t$,” where “Cos” refers to the square-wave version. What is it that is of particular interest in this spectrum?

1.13   Repeat the analysis of FSK for $f_0 = 9 f_b$ and $f_1 = 15 f_b$. Make an accurate picture of the amplitude spectrum.

1.14   By direct integration, find the Fourier Transform of $f(t) = e^{at} u(-t)$.

1.15   By direct integration, find the inverse Fourier Transform of

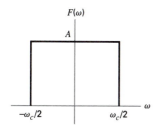

1.16 Provide proofs for the following theorems: a. Linearity and Superposition, b. Time Shifting, c. Frequency Differentiation.

1.17 Show that if $x(t)$ is a solution to the differential equation

$$\frac{d^2x(t)}{dt^2} - t^2x(t) = \lambda x(t)$$

then its Fourier Transform $X(\omega)$ is a solution to the same equation.

1.18 Show that $e^{-\alpha|t|}(a + \alpha|t|) \Leftrightarrow \frac{4\alpha^3}{(\alpha^2+\omega^2)^2}$.

1.19 A function $f(t)$ has a Fourier Transform $F(\omega) = \frac{1+2j\omega}{10 - 3\omega^2 + j4\omega}$. Find the Fourier Transform of

a) $f(2t)$ e) $df(t)/dt$
b) $f(3t - 2)$ f) $tf(t)$
c) $f(t/3)$ g) $f(t - 1) + f(t + 1)$
d) $f(t)\cos(2t)$ h) $f(-2t)$

1.20 (a) Assuming that the functions shown below are time functions, find their Fourier Transforms.

(b) Assuming that the functions shown below are frequency functions, find their inverse Fourier Transforms.

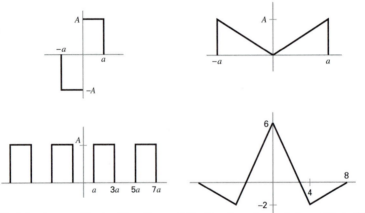

1.21 Find the inverse Fourier Transform of the function shown below. (Hint: think about frequency shifting.)

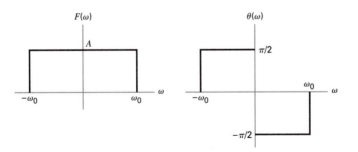

1.22 (a) A signal $f(t)$ which is bandlimited to $(-\omega_c, \omega_c)$ is applied to a lowpass filter having a transfer function that is equal to zero outside the band of the signal. The filter output is

$$g(t) = f(t - T) + \alpha f(t - T + t_0) + \alpha f(t - T - t_0)$$

where $\alpha \ll 1$. Find the amplitude and phase response of the filter. Make a sketch.

(b) Repeat part (a) if $g(t) = f(t - T) + \alpha f(t - T + t_0) - \alpha f(t - T - t_0)$.

1.23 (a) Let $f_1(t) = u(t) - u(t - 2)$ and $f_2(t) = u(t) - u(t - 4)$. Find $f(t) = f_1(t) \otimes f_2(t)$, the convolution of these two time functions. Also find $F(\omega)$, the FT of $f(t)$.

(b) Repeat part (a) if $f_2(t) = e^{-2(t-1)}u(t - 1)$.

1.24 Let

$$f(\omega) = \begin{cases} 2je^{-j\omega T} \operatorname{Sin}(\omega T) & |\omega| \leq \pi/T \\ 0 & \text{elsewhere} \end{cases}$$

Using the defining integral of the Fourier Transform, show that

$$f(t) = \frac{1}{T}\left[\operatorname{Sa}\left(\pi \frac{t}{T}\right) - \operatorname{Sa}\left(\pi \frac{t - 2T}{T}\right)\right]$$

1.25 Using Parseval's Theorem, evaluate

1.
$$\int_{-\infty}^{\infty} \operatorname{Sa}^4\left(\frac{\omega_c t}{4}\right) dt$$

2.
$$\int_{-\infty}^{\infty} \operatorname{Sa}\left(\frac{\omega_c t}{2}\right) \operatorname{Sa}\left(\frac{\omega_c}{2}\left(t - \frac{2\pi}{\omega_c}\right)\right) dt$$

1.26 Let $f_1(t) = u(t) - u(t - 2)$ and $f_2 = (t)e^{-2t}u(t)$. Find the following correlation functions: (a) the autocorrelation of $f_1(t)$, $R_1(\tau)$, (b) the autocorrelation of $f_2(t)$, $R_2(\tau)$, (c) the crosscorrelation $R_{12}(\tau)$, (d) the cross correlation $R_{21}(\tau)$.

1.27 Repeat Problem 1-15 with $f_1(t) = u(t) - 2u(t - 1) + u(t - 2)$ and $f_2(t) = r(t) - r(t - 4) - 4u(t)$.

1.28 (a) Prove that the Fourier Transform of an even function is real and even.

(b) Prove that the Fourier Transform of an odd function is imaginary and odd.

1.29 A complex function $x(t)$ is given as $x(t) = x_1(t) + jx_2(t)$, where $x_1(t)$ and $x_2(t)$ are both real.

(a) Show that the autocorrelation function of $x(t)$ is

$$R_x(\tau) = [R_{x_1}(\tau) + R_{x_2}(\tau)] - j[R_{x_1 x_2}(\tau) - R_{x_2 x_1}(\tau)]$$

where $\operatorname{Re}[R_x(\tau)]$ is even and $\operatorname{Im}[R_x(\tau)]$ is odd.

(b) Now show that $E_x(\omega) \Leftrightarrow R_x(\tau)$ is an entirely real function.

(c) Find the autocorrelation function of $x(t) = e^{-2t}u(t) + je^{-3t}u(t)$ by direct integration.

(d) Now find the energy spectral density $E_x(\omega) \Leftrightarrow R_x(\tau)$.

1.30 Prove the identities of eqs. 1.8-2c,d,e.

1.31 (a) Find the time-average autocorrelation function and the power spectral density for the periodic signal $f(t) = A\operatorname{Cos}(3t) + B\operatorname{Cos}(5t)$. Express the result in both the trigonometric and exponential forms.

(b) Repeat part (a) for the periodic signal rectangular pulses having a duty cycle $\frac{a}{T} = \frac{3}{4}$.

1.32 Consider two periodic functions $f(t)$ and $g(t)$, each having period T, and having Fourier Series coefficients f_n and g_n, respectively. Define the time-average cross-correlation function $\bar{R}_{fg}(\tau)$ and the associated cross power spectral density $P_{fg}(\omega)$.

1.33 The signal $f(t) = 1000\operatorname{Sinc}(1000\pi t)$ is transmitted through a system having a transfer function

$$H_1(\omega) = \begin{cases} (1 - 2000\pi |\omega|); & |\omega| \leq 2000\pi \\ 0; & \text{otherwise} \end{cases}$$

The output $g(t)$ is ideally sampled by uniform impulse functions occurring at a 1 kHz rate. (a) Sketch the spectrum of the sampled signal; (b) The sampled signal $g(t)$ is then transmitted through a second system $H_2(\omega)$. Determine and sketch $H_2(\omega)$ so that its output is the original function $f(t)$.

1.34   The signal $f(t) = \text{Cos}(10\omega_0 t) + \text{Cos}(11\omega_0 t) + \text{Cos}(12\omega_0 t)$ is ideally sampled by uniform impulse functions occurring every $T = 1/8f_0$ sec. The sampled signal is then applied to an ideal lowpass filter having a cutoff frequency of $2.5\omega_0$. Find an expression for the filter output.

1.35   A signal $f(t) = \text{Cos}(2\pi f_m t)$ is ideally sampled by impulse functions eight times each period. These samples are then the input to a first-order-hold circuit having a delay equal to the time between samples.

   (a) Sketch the spectrum of the sampled signal before it is applied to the first-order hold.

   (b) Sketch the transfer function of the first-order hold and the signal spectrum at the output.

   (c) Sketch the time domain function at the output.

1.36   A 'cos' square wave having zero average value and a frequency f_0 is multiplied by $\text{Cos}(12\pi f_0 t) + \text{Cos}(16\pi f_0 t)$ and the resulting signal is the input to an ideal lowpass filter having a cutoff frequency of $2.5f_0$. Find an exact expression for the filter output.

REFERENCES

[Sle76]   Slepian, D., "On Bandwidth," *Proceedings of the IEEE*, **64**, No. 3, March 1976, pp. 292–300.

[Str90]   F.G. Stremler, *Introduction to Communication Systems*, 3rd ed., Addison-Wesley, 1990.

CHAPTER 2

SPECTRAL ANALYSIS OF DATA SIGNALS AND NOISE

The previous chapter introduced some basic ideas about spectral analysis of deterministic signals. As we shall see, both data signals and noise are random waveforms that are also power signals. It is consequently necessary to extend the ideas of spectral analysis, particularly power spectral density and correlation, to these signals. This analysis will include various forms of coding of the signal that come under the generic name of "line coding."

In order to accomplish this, it is useful to begin with a brief review of some of the essential issues in probability theory and random processes that underlie these ideas.

2.1 THE ELEMENTS OF PROBABILITY THEORY

2.1.1 The Probability Space

A classical starting point for the discussion of probability is the idea of an experiment having a discrete number of basic or fundamental outcomes known as *sample points* which together constitute the *sample space*, for example $S = \{s_1, s_2, s_3, s_4, s_5\}$. Each of these sample points is assigned a probability with the restrictions that each probabilty must be positive and the sum of the probabilities must be equal to unity. In mathematical terms:

$$P(s_i) > 0 \quad \text{and} \quad \sum_i P(s_i) = 1 \tag{2.1-1}$$

Continuing the example, we make the probability assignment $P\{s_1\} = P_1 = 1/2$, $P_2 = 1/4$, $P_3 = 1/8$, $P_4 = 1/16$, $P_5 = 1/16$.

An *event* is a defined outcome of the experiment so that any proper subset of S is an event. Each one of the sample points is an event. Since a defined outcome of the experiment may be that either s_1 or s_2 or s_5 occurs, it follows that the subset $A = \{s_1, s_2, s_5\}$ is an event. The probability of an event is simply the sum of the probabilities of the sample points contained in that event, so that $P(A) = P_1 + P_2 + P_5 = 13/16$. The *null event* is the empty set $\phi = \{\}$ containing no sample points so that $P(\phi) = 0$, and the entire space S is the *certain event* so that $P(S) = 1$.

For the purpose of continuing the discussion, define the following events:

$$
\begin{aligned}
A &= \{s_1, s_2, s_5\} & P(A) &= 13/16 \\
B &= \{s_1, s_3, s_4\} & P(B) &= 11/16 \\
C &= \{s_4, s_5\} & P(C) &= 1/8
\end{aligned}
\tag{2.1-2a}
$$

63

It is easy to see that the *union* of two events is an event

$$D = B \cup C = \{s_1, s_3, s_4, s_5\} \qquad P(D) = 3/4 \qquad (2.1\text{-}2b)$$

and that the *intersection* of two events is also an event

$$E = A \cap C = \{s_5\} \qquad P(E) = 1/16. \qquad (2.1\text{-}2c)$$

A straightforward extension is that

$$P(A \cup B) = P(A) + P(B) - P(A \cap B) \qquad (2.1\text{-}3a)$$

so that

$$P(A \cup B) \leq P(A) + P(B) \qquad (2.1\text{-}3b)$$

which is the statement of the *union bound*. If $P(A \cap B) = \phi$ then the events A and B are *mutually exclusive* or *disjoint*.

2.1.2 Joint and Conditional Probabilities

Consider a sample space S divided into two different sets of events $\{A_i\}$ and $\{B_j\}$, where $\bigcup_i A_i = \bigcup_j B_j = S$ and where each set of events is disjoint, i.e., $A_i \cap A_m = \phi, \forall i \neq m$ and $B_j \cap B_n = \phi, \forall j \neq n$. In general, $A_i \cap B_j \neq \phi$. The definition of the *joint probability* of A_i and B_j is

$$P\{A_i, B_j\} = P\{A_i \cap B_j\} \qquad (2.1\text{-}4a)$$

The conditions on the sets of events $\{A_i\}$ and $\{B_j\}$ make it clear that

$$\sum_i P\{A_i, B_j\} = P\{B_j\} \qquad (2.1\text{-}4b)$$

$$\sum_j P\{A_i, B_j\} = P\{A_i\} \qquad (2.1\text{-}4c)$$

The conditional probability is defined as

$$P\{A_i/B_j\} = \frac{P\{A_i, B_j\}}{P\{B_j\}} \qquad (2.1\text{-}5a)$$

and using eq. 2.1-4b,

$$P\{A_i/B_j\} = \frac{P\{A_i, B_j\}}{\sum_i P\{A_i, B_j\}} = \frac{P\{A_i, B_j\}}{\sum_i P\{A_i\}P\{B_j/A_i\}} \qquad (2.1\text{-}5b)$$

which is known as *Bayes' Rule*.

An interesting and useful example of the use of Bayes' Rule is in the computation of probabilities in a communications channel such as the one shown in fig. 2.1-1.

The channel inputs have the probability distribution

$$P(x_1) = 0.2; \quad P(x_2) = 0.3; \quad P(x_3) = 0.5; \qquad (2.1\text{-}6a)$$

and the transmission process is characterized by the conditional probabilities $P(y_j/x_i)$, the probability of receiving y_j given that x_i was transmitted.

$$
\begin{array}{lll}
P(y_1/x_1) = 0.7 & P(y_1/x_2) = 0.2 & P(y_1/x_3) = 0.1 \\
P(y_2/x_1) = 0.2 & P(y_2/x_2) = 0.5 & P(y_2/x_3) = 0.1 \\
P(y_3/x_1) = 0.1 & P(y_3/x_2) = 0.3 & P(y_3/x_3) = 0.8
\end{array} \qquad (2.1\text{-}6b)
$$

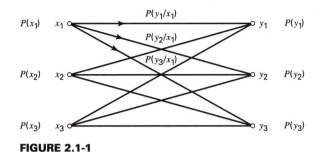

FIGURE 2.1-1

Note that

$$\sum_{j=1}^{3} P(y_j/x_1) = \sum_{j=1}^{3} P(y_j/x_2) = \sum_{j=1}^{3} P(y_j/x_3) = 1 \qquad (2.1\text{-}6c)$$

which can be understood as saying that, for a given value of x, the conditional probabilities of the outputs all add to unity.

We first want to find the probability of each output. This is a simple application of Bayes' rule:

$$P(y_1) = \sum_{i=1}^{3} P(y_1, x_i) = \sum_{i=1}^{3} P(y_1/x_i)P(x_i)$$
$$= (.7)(.2) + (.2)(.3) + (.1)(.5) = .25$$

$$P(y_2) = \sum_{i=1}^{3} P(y_2, x_i) = \sum_{i=1}^{3} P(y_2/x_i)P(x_i) \qquad (2.1\text{-}6d)$$
$$= (.2)(.2) + (.5)(.3) + (.1)(.5) = .24$$

$$P(y_3) = \sum_{i=1}^{3} P(y_3, x_i) = \sum_{i=1}^{3} P(y_3/x_i)P(x_i)$$
$$= (.1)(.2) + (.3)(.3) + (.8)(.5) = .51$$

Having found the output probabilities, we now can easily find the reverse conditional probability $P(x_i/y_j)$, which measures the probability of having transmitted x_i given that y_j was observed at the output. For example

$$P(x_1/y_1) = \frac{P(y_1/x_1)P(x_1)}{P(y_1)} = \frac{(.7)(.2)}{(.25)} = .56$$

$$P(x_2/y_1) = \frac{P(y_1/x_2)P(x_2)}{P(y_1)} = \frac{(.2)(.3)}{(.25)} = .24 \qquad (2.1\text{-}6e)$$

$$P(x_3/y_1) = \frac{P(y_1/x_3)P(x_3)}{P(y_1)} = \frac{(.1)(.5)}{(.25)} = .20$$

where $\sum_{i=1}^{3} P(x_i/y_1) = 1$. The computation of the six remaining conditional probabilities is left to the Problems.

Finally we want to consider the phenomenon of *independence* of events. Two events A_1 and A_2 are independent if $P\{A_1 \cap A_2\} = P\{A_1\}P\{A_2\}$. Three events $A_i, i = 1, 2, 3$ are independent if the following conditions are met:

$$P\{A_i \cap A_j\} = P\{A_i\}P\{A_j\}; \qquad i \neq j \qquad (2.1\text{-}7a)$$

$$P\{A_1 \cap A_2 \cap A_3\} = P\{A_1\}P\{A_2\}P\{A_3\} \qquad (2.1\text{-}7b)$$

The conditions for independence of larger numbers of events is a staightforward extension of these equations.

2.1.3   Random Variables

In order to define a random variable, let us first return to the sample space $S = \{s_1, s_2, s_3, s_4, s_5\}$ and recognize that the sample points are abstract outcomes of an abstract experiment that carry with them certain probabilities. A random variable is an assignment of real (or complex) numbers to these sample points. For example, a definition of a random variable X on the sample space S could be:

S	P(S)	X
s_1	1/2	$x_1 = -1$
s_2	1/4	$x_2 = 2$
s_3	1/8	$x_3 = 1.5$
s_4	1/16	$x_4 = -0.5$
s_5	1/16	$x_5 = 0.8$

$$(2.1\text{-}8)$$

Clearly, many different random variables can be defined on the same probability space S, all of which will carry with them the same probabilities assigned to the sample points. We shall, in fact, return to this issue when we discuss multiple random variables and joint probabilities. Much confusion is generated by the assignment of the term 'random variable' to X because X is really a deterministic function having the sample space as its domain and the real line as its range. The randomness appears only in the outcomes of the sample space. In this, as well as many other things, we must learn to live with the consequences of the history of how things are named.

Random variables are most often described by their statistics, the most important of which are the mean, the mean-square, and the variance. These are defined below along with examples based on the random variable of eq. 2.1-8. In making these definitions, we introduce the expectation operator $E[..]$, which is defined in the specific examples that are given and will be subsequently more generally defined.

Mean

$$m_x = E[X] = \sum_n x_n P[x_n]$$

$$= (-1)\left(\frac{1}{2}\right) + (2)\left(\frac{1}{4}\right) + (1.5)\left(\frac{1}{8}\right) + (-0.5)\left(\frac{1}{16}\right) + (0.8)\left(\frac{1}{16}\right)$$

$$= 0.206$$

$$(2.1\text{-}9a)$$

Mean-Square

$$m_x^2 = E[X^2] = \sum_n x_n^2 P[x_n]$$

$$= (-1)^2\left(\frac{1}{2}\right) + (2)^2\left(\frac{1}{4}\right) + (1.5)^2\left(\frac{1}{8}\right) + (-0.5)^2\left(\frac{1}{16}\right) + (0.8)^2\left(\frac{1}{16}\right)$$

$$= 1.837$$

$$(2.1\text{-}9b)$$

Variance

$$\sigma^2 = E[(X - m_x)^2] = E[X^2 - 2m_x X + m_x^2] = E[X^2] - 2m_x E[X] + m_x^2$$

$$= E[X^2] - m_x^2 = \sum_n [x_n - m_x]^2 P[x_n]$$

$$= 1.837 - (.206)^2 = 1.795 \tag{2.1-9c}$$

More generally, a statistic of the random variable X is defined as

$$E[f(X)] = \sum_n f(x_n) P[x_n] \tag{2.1-10}$$

We may also define the *probability distribution function* $F_x(x)$ as

$$F_x(x) = P[X \leq x]; \quad -\infty < x < \infty \tag{2.1-11a}$$

This function is shown for the random variable of eq. 2.1-8 in fig. 2.1-2a. Note that $F(\infty) = 1$ and $F(-\infty) = 0$ and that the descending probabilities are in relation to the value of the random variable and not the ordering of the sample space. The *probability density function* $p_x(x)$ is defined as

$$p_x(x) = \frac{dF_x(x)}{dx} \tag{2.1-11b}$$

A graph of $p_x(x)$ corresponding to the $F_x(x)$ of fig. 2.1-2a is shown in fig. 2.1-2b. Since $F_x(x)$ is a staircase function corresponding to a discrete random variable, it follows that

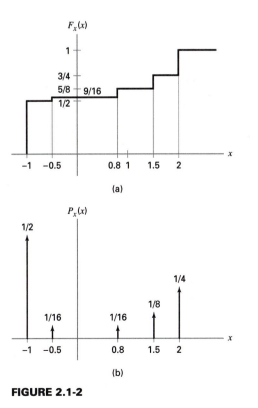

(a)

(b)

FIGURE 2.1-2

$p_x(x)$ will be a collection of impulse functions:

$$p_x(x) = \frac{1}{2}\delta(x+1) + \frac{1}{16}\delta(x+0.5) + \frac{1}{16}\delta(x-0.8) + \frac{1}{8}\delta(x-1.5) + \frac{1}{2}\delta(x-2.0) \quad (2.1\text{-}11c)$$

so it can easily be seen that

$$F_x(x) = \int_{-\infty}^{x} p_x(\alpha)\, d\alpha \qquad (2.1\text{-}11d)$$

Some examples of important discrete random variables are given below.

The Binomial Distribution

The random variable X takes on two values $X = 0, 1$ where $P[X = 0] = p$ and $P[X = 1] = (1 - p)$. We define $Y = \sum_{i=1}^{n} X_i$ where the different values of X_i are independent. It then follows that

$$P[Y = k] = \binom{n}{k} p^k (1 - p)^{n-k} \qquad (2.1\text{-}12a)$$

or, using the probability density function notation,

$$p_Y(y) = \sum_{k=0}^{n} \binom{n}{k} p^k (1 - p)^{n-k}\, \delta(y - k) \qquad (2.1\text{-}12b)$$

and

$$F_Y(y) = \sum_{k=0}^{n} \binom{n}{k} p^k (1 - p)^{n-k}\, u(y - k) \qquad (2.1\text{-}12c)$$

where

$$\binom{n}{k} = \frac{n!}{k!(n - k)!} \qquad (2.1\text{-}12d)$$

The crucial statistics of this distribution are

$$E[Y] = m_Y = np \qquad (2.1\text{-}12e)$$

$$E[Y^2] = np(1 - p) + (np)^2 \qquad (2.1\text{-}12f)$$

$$\sigma^2 = np(1 - p) \qquad (2.1\text{-}12g)$$

The Poisson Distribution

The density and distribution of the Poisson random variable is

$$p_x(x) = e^{-b} \sum_{k=0}^{\infty} \frac{b^k}{k!}\delta(x - k) \qquad (2.1\text{-}13a)$$

and

$$F_x(x) = e^{-b} \sum_{k=0}^{\infty} \frac{b^k}{k!}u(x - k) \qquad (2.1\text{-}13b)$$

where

$$E[X] = m_x = b \qquad (2.1\text{-}13c)$$

$$E[X^2] = b + b^2 \qquad (2.1\text{-}13d)$$

$$\sigma^2 = b \qquad (2.1\text{-}13e)$$

Random variables have been introduced in relation to discrete sample spaces in a way as to make the sample space explicit. We now wish to extend these ideas to continuous random variables and, in so doing, place the now continuous sample space in the background and focus on the probabilty of the value of the random variable itself. Under these circumstances, finding the mean, the mean-squared, and the variance require replacing the summations of eq. 2.1-9a,b,c by integrals:

$$E[X] = m_x = \int_{-\infty}^{\infty} x p_x(x)\, dx \qquad (2.1\text{-}14a)$$

$$E[\overline{X^2}] = \int_{-\infty}^{\infty} x^2 p_x(x)\, dx \qquad (2.1\text{-}14b)$$

$$\sigma^2 = E[\overline{(X - m_x)^2}] = \int_{-\infty}^{\infty} (x - m_x)^2 p_x(x)\, dx \qquad (2.1\text{-}14c)$$

There are many different probability density functions and no effort will be made to list them here. We will very briefly describe three: the uniform, the Rayleigh, and the Gaussian. Some other density functions will be described in the remainder of the text as they are needed.

The Uniform Density Function

The uniform random variable, fig. 2.1-3, has a probability density function

$$p_x(x) = \frac{1}{b - a}; \qquad a \le x \le b \qquad (2.1\text{-}15a)$$

where

$$E[X] = \frac{a + b}{2} \qquad (2.1\text{-}15b)$$

$$\sigma_x^2 = \frac{(b - a)^2}{12} \qquad (2.1\text{-}15c)$$

The Rayleigh Density Function

The Rayleigh random variable, fig. 2.1-4, has a probability density function

$$p_x(x) = \frac{x}{a^2} e^{-x^2/2a^2}; \qquad x \ge 0,\ a > 0 \qquad (2.1\text{-}16a)$$

and a cumulative probability distribution of

$$F_x(x) = 1 - e^{-x^2/2a^2}; \qquad x \ge 0,\ a > 0 \qquad (2.1\text{-}16b)$$

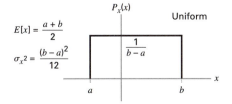

FIGURE 2.1-3

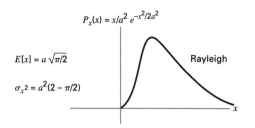

FIGURE 2.1-4

with

$$E[X] = a\sqrt{\pi/2} \tag{2.1-16c}$$

$$\sigma_x^2 = a^2(2 - \pi/2) \tag{2.1-16d}$$

The Gaussian Density Function

Of all the probability functions that will be used in this text, the Gaussian, fig. 2.1-5a, is by far the most important. It will be briefly introduced at this point and more fully developed in the next section. The Gaussian probability density function is given by:

$$p_x(x) = \frac{1}{\sigma\sqrt{2\pi}} e^{-(x-m)^2/2\sigma^2} \tag{2.1-17a}$$

where

$$E[X] = m \quad \text{and} \quad \sigma_x^2 = \sigma^2 \tag{2.1-17b}$$

The area under the tail of the Gaussian density function, normalized to zero mean and unit variance, fig. 2.1-5b, is defined as

$$Q(x) = \int_x^{\infty} \frac{1}{\sqrt{2\pi}} e^{-\alpha^2/2} \, d\alpha \tag{2.1-17c}$$

It is most important in computing the probability of error in communication systems. Consequently $Q(x)$ has been both tabulated (see Appendix 2A) and approximated. We note without proof that

$$Q(x) \approx \frac{1}{x\sqrt{2\pi}} \left[1 - \frac{1}{x^2} \right] e^{-x^2/2} \tag{2.1-17d}$$

Among the more important properties of the Gaussian random variable is that it is the limiting density funtion of the sum of a large number of independent random variables.

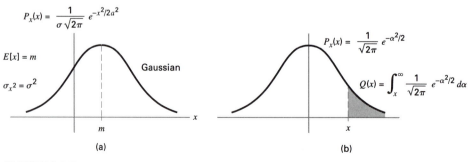

FIGURE 2.1-5

In this regard, the sum of the means of the random variables and the sum of their variances equal the mean and variance of the resulting Gaussian density function.

2.2 PROBABILITIES OF MULTIPLE RANDOM VARIABLES

In analyzing communication systems it will be necessary to find the relationship between different random variables. The idea of correlation of functions has already been introduced in relation to deterministic functions. We want to extend that idea to random variables. In order to accomplish this we will first extend the example of eq. 2.1-8 to multiple random variables.

S	P(S)	X_1	X_2	X_3	
s_1	1/2	$x_{11} = -1$	$x_{21} = 1.2$	$x_{31} = 0.5$	
s_2	1/4	$x_{12} = 2$	$x_{22} = 0.3$	$x_{32} = -0.5$	(2.2-1)
s_3	1/8	$x_{13} = 1.5$	$x_{23} = -1.5$	$x_{33} = 1$	
s_4	1/16	$x_{14} = -0.5$	$x_{24} = 0.5$	$x_{34} = 1.3$	
s_5	1/16	$x_{15} = 0.8$	$x_{25} = 1.3$	$x_{35} = 2$	

It is crucial to note that all three random variables X_1, X_2, X_3, are defined on the same sample space. The mean and variance of each of these random variables can be computed separately, following the examples of eq. 2.1-9a, b, c. The thing that is new is the definition of some additional statistics.

The *correlation* of the two random variables X_1 and X_2 is defined as

$$\rho_{12}^2 = E[X_1 X_2] = \sum_{j=1}^{5} x_{1j} x_{2j} P[s_j] \tag{2.2-2a}$$

and the *covariance* is

$$\sigma_{12}^2 = E[(X_1 - m_1)(X_2 - m_2)] = \sum_{j=1}^{5} (x_{1j} - m_1)(x_{2j} - m_2) P[s_j] \tag{2.2-2b}$$

Note that if the random variables are zero mean then the correlation and the covariance are the same. These definitions can, of course, be extended to any of the pairs of random variables in eq. 2.2-1.

It is also the case that any statistic can be defined in this way. In general,

$$E[f(X_1 X_2 X_3)] = \sum_{j=1}^{5} f(x_{1j} x_{2j} x_{3j}) P[s_j] \tag{2.2-2c}$$

This description of the probabilities on multiple discrete random variables can be taken into the continuous domain with the definition of the joint cumulative distribution function $F[x_1, x_2]$ and the joint pdf (probability density function) $p(x_1, x_2)$ where:

$$F[x_1, x_2] = P[X_1 \leq x_1, X_2 \leq x_2]$$

$$= \int_{-\infty}^{x_1} \int_{-\infty}^{x_2} p(\alpha_1, \alpha_2) \, d\alpha_1 \, d\alpha_2 \tag{2.2-3a}$$

and, alternatively,

$$p(x_1, x_2) = \frac{\partial^2}{\partial x_1 \partial x_2} F[x_1, x_2] \tag{2.2-3b}$$

Integrating the joint pdf with respect to one of the variables yields the marginal density function of the other:

$$\int_{-\infty}^{\infty} p(x_1, x_2)\, dx_2 = p(x_1), \qquad \int_{-\infty}^{\infty} p(x_1, x_2)\, dx_1 = p(x_2) \qquad (2.2\text{-}3c)$$

This may be easily extended to any number of random variables. For example,

$$\int_{-\infty}^{\infty} p(x_1, x_2, x_3)\, dx_3 = p(x_1, x_2) \qquad (2.2\text{-}3d)$$

The idea of a joint probability density function leads directly to that of a conditional density function

$$p(x_1/x_2) = \frac{p(x_1, x_2)}{p(x_2)} \qquad (2.2\text{-}4a)$$

where, for example,

$$P(X_1 \leq x_1/X_2 = x_2) = \frac{\int_{-\infty}^{x_1} p(\alpha_1, x_2)\, d\alpha_1}{p(x_2)} \qquad (2.2\text{-}4b)$$

Finally, if the random variables X_1, X_2, X_3 are independent then

$$p(x_1, x_2, x_3) = p(x_1)p(x_2)p(x_3) \qquad (2.2\text{-}5)$$

2.3 GAUSSIAN PROBABILITY DENSITY FUNCTIONS

Since the idea of the transmission of data in the presence of Gaussian noise is a central one in this text, we shall spend a bit of effort in reviewing some basic ideas of Gaussian random variables and Gaussian random processes.

To recapitulate from earlier in this chapter, a Gaussian random variable X has probability density function

$$p_x(\alpha) = \frac{1}{\sigma\sqrt{2\pi}} e^{-(\alpha-m)^2/2\sigma^2} = \frac{1}{\sigma\sqrt{2\pi}} \exp\left\{-\frac{(\alpha-m)^2}{2\sigma^2}\right\} \qquad (2.3\text{-}1a)$$

where m is the mean value of X and σ^2 is the variance. Note that

$$m = \int_{-\infty}^{\infty} \alpha p_x(\alpha)\, d\alpha \quad \text{and} \quad \sigma^2 = \int_{-\infty}^{\infty} (\alpha-m)^2 p_x(\alpha)\, d\alpha \qquad (2.3\text{-}1b)$$

We are interested in defining the joint probability density function of jointly Gaussian random variables. In order to do this, let us define the vectors

$$\mathbf{X}_N = \begin{bmatrix} x_1 \\ x_2 \\ \cdot \\ \cdot \\ \cdot \\ x_N \end{bmatrix} \qquad \alpha_N = \begin{bmatrix} \alpha_1 \\ \alpha_2 \\ \cdot \\ \cdot \\ \cdot \\ \alpha_N \end{bmatrix} \qquad \mathbf{m}_N = \begin{bmatrix} m_1 \\ m_2 \\ \cdot \\ \cdot \\ \cdot \\ m_N \end{bmatrix} \qquad (2.3\text{-}2a)$$

where $\mathbf{m}_N$ is the vector of the mean values, and the covariance matrix

$$
C_X = \begin{bmatrix} \sigma_1^2 & \rho_{12}\sigma_1\sigma_2 & \rho_{13}\sigma_1\sigma_3 & \cdot & \rho_{1N}\sigma_1\sigma_N \\ \rho_{12}\sigma_1\sigma_2 & \sigma_2^2 & \rho_{23}\sigma_2\sigma_3 & \cdot & \cdot \\ \rho_{13}\sigma_1\sigma_3 & \rho_{23}\sigma_2\sigma_3 & \sigma_3^2 & \cdot & \cdot \\ \cdot & \cdot & \cdot & \cdot & \cdot \\ \rho_{1N}\sigma_1\sigma_N & \cdot & \cdot & \cdot & \sigma_N^2 \end{bmatrix}
\tag{2.3-2b}
$$

The jointly Gaussian density function is

$$
p_{\mathbf{x}_N}(\boldsymbol{\alpha}_N) = \frac{|C_x^{-1}|^{1/2}}{(2\pi)^{N/2}} \exp\left\{ -\frac{[\boldsymbol{\alpha}_N - \boldsymbol{m}_N]^T C_x^{-1}[\boldsymbol{\alpha}_N - \mathbf{m}_N]}{2} \right\}
\tag{2.3-2c}
$$

where

$$
\sigma_n^2 = \int_{-\infty}^{\infty} (\alpha_n - m_n) p_{\mathbf{x}}(\boldsymbol{\alpha})\, d\alpha_1 \ldots d\alpha_N
\tag{2.3-2d}
$$

and

$$
\rho_{nm}\sigma_n\sigma_m = \int_{-\infty}^{\infty} (\alpha_n - m_n)(\alpha_m - m_m) p_{\mathbf{x}}(\boldsymbol{\alpha})\, d\alpha_1 \cdots d\alpha_N
\tag{2.3-2e}
$$

and the *covariance matrix is symmetric*.

When $N = 2$, we get a particularly important joint density function:

$$
p_{x_1,x_2}(\alpha_1, \alpha_2) = \frac{1}{2\pi\sigma_1\sigma_2\sqrt{1 - \rho_{12}^2}} \exp\left\{ -\frac{1}{2(1 - \rho_{12}^2)} \left[\frac{(\alpha_1 - m_1)^2}{\sigma_1^2} \right.\right.
$$
$$
\left.\left. -\frac{2\rho_{12}(\alpha_1 - m_1)(\alpha_2 - m_2)}{\sigma_1\sigma_2} + \frac{(\alpha_2 - m_2)^2}{\sigma_2^2} \right] \right\}
\tag{2.3-4}
$$

In the special case where $\rho_{12} = 0$, this joint density reduces to

$$
p_{x_1,x_2}(\alpha_1, \alpha_2) = \frac{1}{2\pi\sigma_1\sigma_2} \exp\left\{ -\frac{1}{2} \left[\frac{(\alpha_1 - \bar{x}_1)^2}{\sigma_1^2} + \frac{(\alpha_2 - \bar{x}_2)^2}{\sigma_2^2} \right] \right\}
$$
$$
= \frac{1}{\sqrt{2\pi}\sigma_1} \exp\left\{ -\frac{(\alpha_1 - \bar{x}_1)^2}{2\sigma_1^2} \right\} \frac{1}{\sqrt{2\pi}\sigma_2} \exp\left\{ -\frac{(\alpha_2 - \bar{x}_2)^2}{2\sigma_2^2} \right\}
$$
$$
= p_{x_1}(\alpha_1) p_{x_2}(\alpha_2)
\tag{2.3-5}
$$

which is the product of two probability density functions, so that the two random variables are independent. This important result, that uncorrelated Gaussian random variables are independent, can be generalized to multiple Gaussian random variables.

It can be shown that for the jointly Gaussian density function of eq. 2.3-4

$$
\int_{-\infty}^{\infty} p_{x_1,x_2}(\alpha_1, \alpha_2)\, d\alpha_2 = p_{x_1}(\alpha_1)
\tag{2.3-6}
$$

by completing the square in the numerator. The details are left to the Problems. It can also be shown that this operation can be extended to any number of variables by a simple operation on the covariance matrix. For example, with $N = 3$, the covariance matrix is

$$C_x = \begin{bmatrix} \sigma_1^2 & \rho_{12}\sigma_1\sigma_2 & \rho_{13}\sigma_1\sigma_3 \\ \rho_{12}\sigma_1\sigma_2 & \sigma_2^2 & \rho_{23}\sigma_2\sigma_3 \\ \rho_{13}\sigma_1\sigma_3 & \rho_{23}\sigma_2\sigma_3 & \sigma_3^2 \end{bmatrix} \qquad (2.3\text{-}7a)$$

When we perform the operation

$$\int\limits_{-\infty}^{\infty} \int\limits_{-\infty}^{\infty} p_{x_1,x_2,x_3}(\alpha_1, \alpha_2, \alpha_3)\, d\alpha_3 = p_{x_1,x_2}(\alpha_1, \alpha_2) \qquad (2.3\text{-}7b)$$

the covariance matrix of the resulting two-dimensional density function is obtained by simply striking the third row and third column from C_x in eq. 2.3-7a. We are particularly interested in linear transformations of Gaussian random variables. Let A be an $N \times N$ matrix of real numbers and let

$$\mathbf{Y}_N = \mathbf{A}\mathbf{X}_N \qquad (2.3\text{-}8a)$$

where

$$\mathbf{Y}_N = \begin{bmatrix} y_1 \\ y_2 \\ \cdot \\ \cdot \\ y_N \end{bmatrix} \qquad \beta_N = \begin{bmatrix} \beta_1 \\ \beta_2 \\ \cdot \\ \cdot \\ \beta_N \end{bmatrix} \qquad \mathbf{b}_N = \begin{bmatrix} b_1 \\ b_2 \\ \cdot \\ \cdot \\ b_N \end{bmatrix} \qquad (2.3\text{-}8b)$$

and

$$A = \begin{bmatrix} a_{11} & a_{12} & \cdot & a_{1N} \\ a_{21} & a_{22} & \cdot & a_{2N} \\ \cdot & \cdot & \cdot & \cdot \\ a_{N1} & a_{N2} & \cdot & a_{NN} \end{bmatrix} \qquad (2.3\text{-}8c)$$

It is known [Pee93] that this linear transformation also results in a jointly Gaussian density function where

$$p_{\mathbf{Y}_N}(\beta_N) = \frac{|C_Y^{-1}|^{1/2}}{(2\pi)^{N/2}} \exp\left\{ -\frac{[\beta_N - \bar{\mathbf{y}}_N]^T C_Y^{-1}[\beta_N - \bar{\mathbf{y}}_N]}{2} \right\} \qquad (2.3\text{-}8d)$$

and

$$C_Y = AC_x A^T \qquad (2.3\text{-}8e)$$

Since C_x is a symmetric matrix, it is possible to find a matrix A such that $A^T = A^{-1}$ resulting in C_Y being a diagonal matrix. This matrix A, a unitary matrix, is made up of the eigenvectors of C_x. We will deal with the issue of eigenvalues and eigenvectors of matrices in Chapter 5. But it is straightforward to see that a diagonal correlation matrix C_Y corresponds to a set of uncorrelated and therefore independent Gaussian random variables.

It is shown in the Problems that the two-dimensional transformation A that diagonalizes a two-dimensional covariance matrix is a rotation of the axes by an angle θ. The general process of diagonalization of the covariance matrix can be regarded as a multidimensional rotation.

Using linear transformations, we can show that a linear combination of Gaussian random variables is also Gaussian. Let the linear transfomation be of the form

$$
\begin{bmatrix} y \\ x_2 \\ x_3 \\ x_4 \end{bmatrix} = \begin{bmatrix} a_1 & a_2 & a_3 & a_4 \\ 0 & 1 & 0 & 0 \\ 0 & 0 & 1 & 0 \\ 0 & 0 & 0 & 1 \end{bmatrix} \begin{bmatrix} x_1 \\ x_2 \\ x_3 \\ x_4 \end{bmatrix} \tag{2.3-9a}
$$

yielding, according to eq. 2.3-8d, a jointly Gaussian density function

$$
p_{y,x_2,x_3,x_4}(\beta, \alpha_2, \alpha_3, \alpha_4) \tag{2.3-9b}
$$

which can be integrated to give

$$
\int_{-\infty}^{\infty} \int_{-\infty}^{\infty} \int_{-\infty}^{\infty} p_{y,x_2,x_3,x_4}(\beta, \alpha_2, \alpha_3, \alpha_4)\, d\alpha_4\, d\alpha_3\, d\alpha_2 = p_y(\beta) \tag{2.3-9c}
$$

establishing that $y = a_1x_1 + a_2x_2 + a_3x_3 + a_4x_4$ is Gaussian.

The mean value of the random variable y is

$$
\bar{y} = \int_{-\infty}^{\infty} \int_{-\infty}^{\infty} \int_{-\infty}^{\infty} \int_{-\infty}^{\infty} [a_1x_1 + a_2x_2 + a_3x_3 + a_4x_4]
$$

$$
\times p_{x_1,x_2,x_3,x_4}(\alpha_1, \alpha_2, \alpha_3, \alpha_4)\, d\alpha_1\, d\alpha_2\, d\alpha_3\, d\alpha_4
$$

$$
= \sum_{n=1}^{4} \int_{-\infty}^{\infty} a_n x_n p_{x_n}(\alpha_n)\, d\alpha_n = \sum_{n=1}^{4} a_n \bar{x}_n \tag{2.3-10a}
$$

and the variance of y is

$$
\sigma_y^2 = \int_{-\infty}^{\infty} \int_{-\infty}^{\infty} \int_{-\infty}^{\infty} \int_{-\infty}^{\infty} [a_1(x_1 - \bar{x}_1) + a_2(x_2 - \bar{x}_2) + a_3(x_3 - \bar{x}_3) + a_4(x_4 - \bar{x}_4)]^2
$$

$$
\times p_{x_1,x_2,x_3,x_4}(\alpha_1, \alpha_2, \alpha_3, \alpha_4)\, d\alpha_1\, d\alpha_2\, d\alpha_3\, d\alpha_4
$$

$$
= \sum_{n=1}^{4} a_n^2 \sigma_n^2 + 2 \sum_{n=1}^{4} \sum_{m>n}^{4} a_n a_m \rho_{nm} \tag{2.3-10b}
$$

If the random variables are uncorrelated, then the variance of the linear combination simply depends on their individual variances.

Note that these two results on the mean and variance of linear combinations of random variables apply to all joint probability density functions, not simply to Gaussian ones.

2.4 THE BRIEFEST POSSIBLE REVIEW OF RANDOM PROCESSES

In this text we shall be dealing only with random processes where we are able to establish an equivalence between time domain and ensemble statistics; i.e., we shall be dealing with

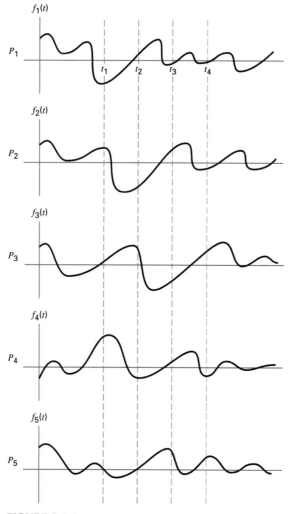

FIGURE 2.4-1

random processes that are, in some sense, stationary and, as well, ergodic. This brief section is for the purpose of guaranteeing a common vocabulary; readers who are not familiar with this material should review this theory in one of the many available textbooks on probability and random processes.

As shown in fig. 2.4-1, a finite random process $f(t)$ consists of a collection of functions $\{f_1(t),\ f_2(t),\ f_3(t),\ f_4(t),\ f_5(t)\}$, each of which we shall assume to be a power signal. Each of these functions is associated with a point in a sample space. We have chosen the sample space to be the same one associated with the random variable defined in eq. 2.1-8 and extended to multiple random variables in eq. 2.2-1. Each function has an associated probability of occurrence; an infinite collection of such functions would be characterized by a probability density function. In this sense, a random process is an extension of a random variable. A random variable associates a number to each point in a sample space and a random process associates a function to each of these points.

At some time t_1 the collection of values $\{f_1(t_1),\ f_2(t_1),\ f_3(t_1),\ f_4(t_1),\ f_5(t_1)\}$ constitute a random variable X_1 having associated probabilities P_1, P_2, P_3, P_4, P_5. The *ensemble expected value* of the random process is

$$E[f(t_1)] = \sum_n P_n f_n(t_1) \tag{2.4-1a}$$

or, if the random process is continuous,

$$E[f(t_1)] = \int_{-\infty}^{\infty} f(t_1) p_f [f(t_1)] \, df \tag{2.4-1b}$$

and is, in general, a function of t_1. The first requirement for the random process to be stationary is that the expected value be independent of the value of t_1, i.e.

$$E[f(t_1)] = E[f(t_2)], \qquad \forall t_1, t_2 \tag{2.4-1c}$$

The *ensemble autocorrelation function* of the random process is defined as

$$E[f(t_1)f(t_2)] = \sum_n P_n f_n(t_1) f_n(t_2) \tag{2.4-2a}$$

or, if the random process is continuous,

$$R_{12}(t_1, t_2) = E[f(t_1)f(t_2)] = \int_{-\infty}^{\infty} \int_{-\infty}^{\infty} f(t_1) f(t_2) p_{f_1 f_2}[f(t_1)f(t_2)] \, df_1 \, df_2 \tag{2.4-2b}$$

The second requirement for the random process to be stationary is that

$$R_{12}(t_1, t_2) = R_{12}(t_2 - t_1) = R(\tau) \tag{2.4-2c}$$

so that this ensemble autocorrelation function depends on time differences rather than on absolute values of time. As a special case, when $\tau = 0$, the mean-squared value of the random process is

$$E[f^2(t_1)] = R(0) = \int_{-\infty}^{\infty} f^2(t_1) p_{f_1}[f(t_1)] \, df_1 \tag{2.4-2d}$$

which is also required to be independent of time.

The variance of the random process is defined to be

$$\sigma_f^2 = E[f^2(t_1)] - E^2[f(t_1)] \tag{2.4-2e}$$

and, again, should be independent of t_1.

A random process that meets these conditions is termed *wide-sense stationary*. For a random process to be *strictly stationary* requires that every conceivable statistic that can be computed on this process may not depend on absolute values of time, but rather must depend only on time differences. For example,

$$E[f(t_1)f(t_2)f^2(t_3)] = R[(t_2 - t_1), (t_3 - t_2)] \tag{2.4-2f}$$

A *Gaussian random process* has the property that the joint probability density function of any set of random variables chosen in the manner described above is Gaussian according to eq. 2.3-2c. Since this joint density function is completely described by the vector of the mean values and the covariance matrix, it follows that a wise-sense stationary Gaussian random process is also strictly stationary.

For each of the ensemble statistics that have been examined above, there is a corresponding time domain quantity that we can compute for a *single one of the sample functions* of the random process. For example:

Time-Average Mean

$$\overline{f_n(t)} = \lim_{T \to \infty} \frac{1}{T} \int_{-T/2}^{T/2} f_n(t)\, dt \tag{2.4-3a}$$

Time-Average Autocorrelation Function

$$\overline{R_n(\tau)} = \lim_{T \to \infty} \frac{1}{T} \int_{-T/2}^{T/2} f_n(t) f_n(t - \tau)\, dt \tag{2.4-3b}$$

The first condition that we are aiming for is to have the time-average mean and the time-average autocorrelation be the same for all sample functions in the random process:

$$\overline{f(t)} = \lim_{T \to \infty} \frac{1}{T} \int_{-T/2}^{T/2} f(t)\, dt = \overline{f_n(t)} \tag{2.4-4a}$$

$$\overline{R_f(\tau)} = \lim_{T \to \infty} \frac{1}{T} \int_{-T/2}^{T/2} f(t) f(t - \tau)\, dt = \overline{R_n(\tau)} \tag{2.4-4b}$$

This is an indicator that all of the sample functions in the random process are alike in the time domain.

The second condition that we are aiming for is to have the ensemble average and the time average statistics be equal:

$$E[f(t)] = \overline{f(t)} \qquad R_f(\tau) = \overline{R_f(\tau)} \tag{2.4-4c}$$

This is the condition of *ergodicity* and, except where specifically noted, all random processes with which we deal will be considered ergodic. This means that we will be able to find the mean value, the variance, and the autocorrelation function either through time averaging or through ensemble averaging, whichever is more convenient.

The time-average autocorrelation function of a periodic function was found in Chapter 1 and it was demonstrated that its Fourier Transform was, in fact, the power spectral density of that signal. We want to extend that relationship to ergodic random processes so that we may find the power spectral density by taking the Fourier Transform of the autocorrelation function.

In order to provide some justification for this operation, consider the arbitrary power signal $f(t)$ shown in fig. 2.4-2a and its truncated version $f_T(t)$ shown in fig. 2.4-2b. Although we cannot find a Fourier Transform for $f(t)$, we can say that

$$f_T(t) \Leftrightarrow F_T(\omega) \tag{2.4-5a}$$

and using Parseval's Theorem

$$\int_{-T/2}^{T/2} |f(t)|^2\, dt = \frac{1}{2\pi} \int_{-\infty}^{\infty} |F_T(\omega)|^2\, d\omega \tag{2.4-5b}$$

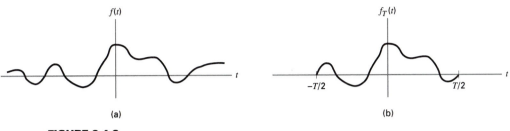

FIGURE 2.4-2

so that the average power is, in the limit,

$$P = \lim_{T \to \infty} \frac{1}{T} \int_{-T/2}^{T/2} |f(t)|^2 \, dt = \lim_{T \to \infty} \frac{1}{T} \frac{1}{2\pi} \int_{-\infty}^{\infty} |F_T(\omega)|^2 \, d\omega \qquad (2.4\text{-}5c)$$

The definition of power spectral density then requires that

$$P = \frac{1}{2\pi} \int_{-\infty}^{\infty} P(\omega) \, d\omega = \frac{1}{2\pi} \int_{-\infty}^{\infty} \left[\lim_{t \to \infty} \frac{1}{T} |F_T(\omega)|^2 \right] d\omega \qquad (2.4\text{-}5d)$$

so that, if the reordering of limit and integration is valid, then

$$P(\omega) = \lim_{T \to \infty} \frac{1}{T} |F_T(\omega)|^2 \qquad (2.4\text{-}5e)$$

We now want to show that the inverse Fourier Transform of $P(\omega)$ is the time-average autocorrelation function of $f(t)$. From eq. 2.4-5e,

$$F^{-1}[P(\omega)] = \frac{1}{2\pi} \int_{-\infty}^{\infty} \left[\lim_{T \to \infty} \frac{1}{T} |F_T(\omega)|^2 \right] e^{j\omega\tau} \, d\omega$$

$$= \lim_{T \to \infty} \frac{1}{2\pi T} \int_{-\infty}^{\infty} F_T^*(\omega) F_T(\omega) e^{j\omega\tau} \, d\omega$$

$$= \lim_{T \to \infty} \frac{1}{2\pi T} \int_{-\infty}^{\infty} \left[\int_{-T/2}^{T/2} f^*(t) e^{j\omega t} \, dt \right] \left[\int_{-T/2}^{T/2} f(t_1) e^{-j\omega t_1} \, dt_1 \right] e^{j\omega\tau} \, d\omega$$

$$= \lim_{T \to \infty} \frac{1}{T} \int_{-T/2}^{T/2} f^*(t) \int_{-T/2}^{T/2} f(t_1) \left[\frac{1}{2\pi} \int_{-\infty}^{\infty} e^{j\omega(t - t_1 + \tau)} \, d\tau \right] dt_1 \, dt$$

$$= \lim_{T \to \infty} \frac{1}{T} \int_{-T/2}^{T/2} f*(t_1) \int_{-T/2}^{T/2} f(t_1) \delta(t - t_1 + \tau) \, dt_1 \, dt$$

$$= \lim_{T \to \infty} \frac{1}{T} \int_{-T/2}^{T/2} f^*(t) f(t + \tau) \, dt = \overline{R_f(\tau)} \qquad (2.4\text{-}6)$$

which is exactly the time-average autocorrelation function.

We have now accomplished the task of establishing the tools that are necessary to analyze data signals and noise.

2.5   DATA SIGNALS: POWER SPECTRAL DENSITY AND AUTOCORRELATION FUNCTION

A data signal is generated by applying a sequence of impulse functions

$$d(t) = \sum_{n=-\infty}^{\infty} d_n\delta(t - nT_s) \tag{2.5-1a}$$

every T_s seconds to a linear system having an impulse response $h(t)$, as shown in fig. 2.5-1. The impulse response is an energy signal that has an autocorrelation function $R_h(\tau) \Leftrightarrow E_h(\omega)$. The magnitudes of the impulse functions at the input are random variables representing serial data. In the simplest case, the data is binary

$$P[d_n = 1] = P[d_n = -1] = 1/2 \tag{2.5-1b}$$

and the successive data symbols are uncorrelated. For the case where the data are not equally probable, see Franks [Fra69].

The resulting random process is:

$$x(t) = \sum_{n=-\infty}^{\infty} d_n h(t - nT_s) \tag{2.5-1c}$$

The ensemble autocorrelation function of this random process is

$$E[x(t)x^*(t - \tau)] = E\left[\sum_{n=-\infty}^{\infty} d_n h(t - nT_s) \sum_{m=-\infty}^{\infty} d_m^* h^*(t - \tau - mT_s)\right]$$

$$= E\left[\sum_{n=-\infty}^{\infty} |d_n|^2 h(t - nT_s)h^*(t - \tau - nT_s)\right]$$

$$+ E\left[\sum_{n\neq m}\sum_{m=-\infty}^{\infty} d_n d_m^* h(t - nT_s)h^*(t - \tau - mT_s)\right]$$

$$= \sum_{n=-\infty}^{\infty} E[|d_n|^2]h(t - nT)h^*(t - \tau - nT)$$

$$+ \sum_{n\neq m}\sum_{m=-\infty}^{\infty} E[d_n d_m^*]h(t - nT_s)h^*(t - \tau - mT_s) \tag{2.5-2a}$$

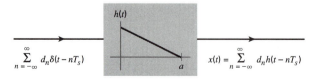

FIGURE 2.5-1

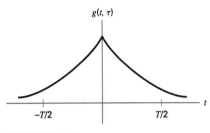

FIGURE 2.5-2

but $E[|d_n|^2] = 1$ and $E[d_n d_m^*] = 0$, so that

$$E[x(t)x^*(t - \tau)] = \sum_{n=-\infty}^{\infty} h(t - nT_s)h^*(t - \tau - nT_s). \qquad (2.5\text{-}2b)$$

Clearly, this ensemble autocorrelation function is not stationary. It is, in fact, a periodic function of t having period T_s. This is made particularly clear if we define

$$g(t, \tau) = h(t)h^*(t - \tau) \qquad (2.5\text{-}2c)$$

so that

$$E[x(t)x^*(t - \tau)] = \sum_{n=-\infty}^{\infty} g(t - nT_s, \tau) \qquad (2.5\text{-}2d)$$

Note that even though $g(t, \tau)$ is periodic with period T_s, it may not be confined to that period, as shown in fig. 2.5-2.

This kind of random process is called *periodically stationary* or *cyclostationary*. It is treated by taking the time average of the ensemble autocorrelation function. Since we propose to average a periodic function, that average may take place over a single period. We therefore have

$$\overline{R_x}(\tau) = \frac{1}{T} \int_{-T/2}^{T/2} \sum_{n=-\infty}^{\infty} g(t - nT_s, \tau)\,dt = \frac{1}{T_s} \int_{-\infty}^{\infty} g(t, \tau)\,dt. \qquad (2.5\text{-}3a)$$

Note that this integral involves taking the tails of time-shifted versions of $g(t, \tau)$ appearing within a single period and assembling them in a single integral. It follows that

$$\overline{R_x}(\tau) = \frac{1}{T} \int_{-\infty}^{\infty} h(t)h^*(t - \tau)\,dt = \frac{1}{T}R_h(\tau) \qquad (2.5\text{-}3b)$$

where $R_h(\tau)$ is the energy autocorrelation function of $h(t)$.

We may now take the Fourier Transform of $R_h(\tau)$ to find the average power spectral density of the random process $x(t)$:

$$\overline{R_x}(\tau) \Leftrightarrow P_x(\omega) = \frac{E_h(\omega)}{T_s} = \frac{|H(\omega)|^2}{T_s} \qquad (2.5\text{-}3c)$$

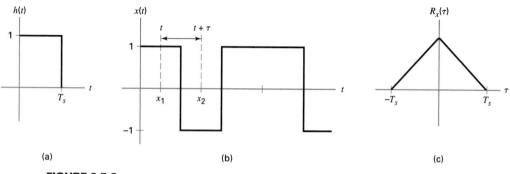

(a)   (b)   (c)

FIGURE 2.5-3

EXAMPLE 1   *The Random Telegraph Signal*

Referring to fig. 2.5-1, the random telegraph signal corresponds to the impulse response $h(t)$ being equal to a rectangular pulse having duration T_s, as shown in fig. 2.5-3a. This means that

$$H(\omega) = T_s \operatorname{Sinc}(\omega T_s/2)e^{-j\omega T_s/2} \tag{2.5-4a}$$

and therefore according to eq. 2.42c,

$$\bar{R}_x(\tau) \Leftrightarrow P_x(\omega) = \frac{E_h(\omega)}{T_s} = \frac{|H(\omega)|^2}{T_s}. \tag{2.5-4b}$$

This result can also be derived by taking the ensemble average of the random process. Under the assumption of ergodicity, this will demonstrate the equivalence of differing ways of finding the autocorrelation function.

A sample function of the random telegraph signal is shown in fig. 2.5-3b. The amplitude of the random process can assume one of two values, $+1$ or -1. Transitions between the two values may occur every T_s sec at designated points called nodes. At each node $P[\text{transition}] = P[\text{no transition}] = 1/2$. Each node operates independently of all of the others

The ensemble autocorrelation function is given by

$$R_x(\tau) = \sum_{x_1=\pm 1} \sum_{x_2=\pm 1} x_1 x_2 P_{x_1 x_2}(x_1, x_2)$$

$$= P_{x_1 x_2}(1, 1) + P_{x_1 x_2}(-1, -1) - P_{x_1 x_2}(-1, 1) - P_{x_1 x_2}(1, -1)$$

$$= P_{x_1}(1)P_{x_2}(1/x_1 = 1) + P_{x_1}(-1)P_{x_2}(-1/x_1 = -1)$$

$$+ P_{x_1}(1)P_{x_2}(-1/x_1 = 1) + P_{x_1}(-1)P_{x_2}(1/x_1 = -1) \tag{2.5-5a}$$

Since $P_{x_1}(1) = P_{x_1}(-1) = 1/2$ and the waveform is symmetric with respect to transitions, it follows that we can simplify eq. 2.5-5a to be

$$R_x(\tau) = P_{x_2}(1/x_1 = 1) - P_{x_2}(-1/x_1 = 1)$$

$$= \left[1 - P_{x_2}(-1/x_1 = 1)\right] - P_{x_2}(-1/x_1 = 1)$$

$$= 1 - 2P_{x_2}(-1/x_1 = 1) \tag{2.5-5b}$$

The problem then reduces to finding one conditional probability. This conditional probability can be written as

$$P_{x_2}(-1/x_1 = 1) = P[\text{node lies in } (t, t + \tau)]P[\text{amplitude change}] \qquad (2.5\text{-}5c)$$

Noting that x_1 is at an arbitrary time and that with respect to x_1 a node can be anywhere, it follows that

$$P[\text{node lies in } (t, t + \tau)] = \begin{cases} |\tau|/Ts; & \text{if } |\tau| < T_s \\ 1; & \text{if } |\tau| \geq T_s \end{cases} \qquad (2.5\text{-}5d)$$

Since $P[\text{amplitude change}] = 1/2$, we conclude that

$$R_x(\tau) = 1 - 2P_{x_2}(-1/x_1 = 1) = \begin{cases} 1 - |\tau|/Ts; & \text{if } |\tau| < T_s \\ 0; & \text{if } |\tau| \geq T_s \end{cases} \qquad (2.5\text{-}5e)$$

which is shown in fig. 2.5-3c.

The Fourier Transform of this autocorrelation function yields exactly the same power spectral density as eq. 2.5-4b. This indicates the equivalence of the two methods.

This technique will also be used in finding the autocorrelation function of a variety of random noise processes. ∎

2.6 HOW CODING CHANGES THE POWER SPECTRAL DENSITY OF DATA SIGNALS

Referring to fig. 2.6-1, it is often the case that the data is encoded before transmission. The data from the data source is considered to be uncorrelated but data at the output of the encoder is correlated. The process of encoding the data, often referred to as line coding, is equivalent to introducing correlation between successive data symbols.

The autocorrelation function of the data symbols at the encoder output is defined as

$$R_b(j) = E[b_n b_{n+j}^*] \qquad j = 0, 1, 2, \ldots \qquad (2.6\text{-}1a)$$

Consequently, the autocorrelation function of the transmitted signal $x(t)$ is:

$$E[x(t)x^*(t - \tau)] = E\left[\sum_{n=-\infty}^{\infty} b_n h(t - nT) \sum_{m=-\infty}^{\infty} b_m^* h^*(t - \tau - mT) \right]$$

$$= \sum_{n=-\infty}^{\infty} E[|b_n|^2]h(t - nT)h^*(t - \tau - nT)$$

$$+ \sum_{n=-\infty}^{\infty} \sum_{j \neq 0} E[b_n b_{n+j}^*]h(t - nT)h^*(t - \tau - nT - jT)$$

$$= \sum_{n=-\infty}^{\infty} \left[R_b(0)h(t - nT)h^*(t - \tau - nT) \right.$$

$$\left. + \sum_{j \neq 0} R_b(j)h(t - nT)h^*(t - \tau - mT) \right]$$

$$= \sum_{n=-\infty}^{\infty} \left[R_b(0)g(t - nT, \tau) + \sum_{j \neq 0} R_b(j)g(t - nT, \tau - jT) \right] \qquad (2.6\text{-}1b)$$

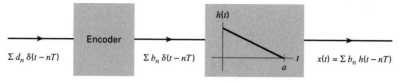

FIGURE 2.6-1

Taking the time average as we did earlier:

$$\overline{R_x}(\tau) = \frac{1}{T_s}\left[R_b(0)R_h(\tau) + \sum_{j\neq 0} R_b(j)R_h(\tau - jT_s) \right] \qquad (2.6\text{-}1c)$$

we get the time-average autocorrelation function of this cyclostationary random process.

In order to find the average power spectral density, we take the Fourier Transform of eq. 2.6-1c:

$$P_x(\omega) = \frac{E_h(\omega)}{T_s}\left[R_b(0) + 2\sum_{m=1}^{\infty} R_b(m)\text{Cos}(m\omega T_s) \right] \qquad (2.6\text{-}1d)$$

Note that the term that modifies the power spectral density of eq. 2.5-4b has the form of a Fourier Series with the components of the autocorrelation function of the data serving as the Fourier coefficients.

This expression is quite general and will be used subsequently in some interesting applications.

EXAMPLE 2 *Manchester (Bi-Phase) Encoding*

In this example the pulse waveform $h(t)$ is the signal shown in fig. 2.6-2. Although there is, according to our definition, no encoding, the signal is called a Manchester or bi-phase coded signal. Because names are historically and culturally, rather than logically, derived we should understand the term 'coding' to refer to this signal as being an alternative to a rectangular pulse. It follows from eq. 2.6-1d that

$$P_x(\omega) = \frac{E_h(\omega)}{T_s} = \frac{4a^2}{T_s}\,\text{Sinc}^2\,(\omega a/2)\text{Sin}^2(\omega a/2) \qquad (2.6\text{-}2)$$

∎

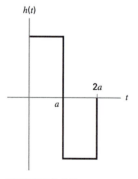

FIGURE 2.6-2

EXAMPLE 3   *Alternate Mark Inversion (AMI) or Bipolar Encoding*

AMI is an authentically encoded signal and the encoding rule is, in fact, nonlinear. That rule is:

$$\text{if } a_k = -1 \quad \text{then} \quad b_k = 0$$
$$\text{if } a_k = +1 \quad \text{then} \quad b_k = \pm 1, \text{ alternately}$$

$$(2.6\text{-}3\text{a})$$

A listing of the possible input and output sequences is given below.

a_k	a_{k+1}	a_{k+2}	b_k	b_{k+1}	b_{k+2}
−1	−1	−1	0	0	0
−1	−1	+1	0	0	+1
−1	+1	−1	0	+1	0
−1	+1	+1	0	+1	−1
+1	−1	−1	+1	0	0
+1	−1	+1	+1	0	−1
+1	+1	−1	+1	−1	0
+1	+1	+1	+1	−1	+1

$$(2.6\text{-}3\text{b})$$

We can now compute the required statistics directly from the table, remembering that each sequence is equally likely.

$$R_b(0) = E[b_k b_k] = \frac{1}{2}(1)^2 = \frac{1}{2}$$

$$R_b(1) = E[b_k b_{k+1}] = -\frac{1}{4}$$

$$(2.6\text{-}3\text{c})$$

$$R_b(2) = E[b_k b_{k+2}] = 0$$

It follows that

$$P_x(\omega) = \frac{E_h(\omega)}{T_s} \left[R_b(0) + 2 \sum_{m=1}^{\infty} R_b(m) \text{Cos}(m\omega T_s) \right]$$

$$= \frac{E_h(\omega)}{T_s} \left[\frac{1}{2} - 2\frac{1}{4} \text{Cos}(\omega T_s) \right] = \frac{E_h(\omega)}{T_s} [1 - \text{Cos}(\omega T_s)] \qquad (2.6\text{-}3\text{d})$$

where the expression for $E_h(\omega)$ comes from the basic waveform $h(t)$ that is transmitted. ∎

EXAMPLE 4   *The Power Spectral Density of a Discrete Random Process*

We may also find that the autocorrelation function of the encoding process is given as a discrete function $R(n)$ having a z-transform $P(z)$. For example, let

$$R_x(n) = \left(\frac{1}{4}\right)^{|n|} \Leftrightarrow P_x(z) = \sum_{n=-\infty}^{\infty} \left(\frac{1}{4}\right)^{|n|} z^{-n}$$

$$= \sum_{n=0}^{\infty} \left(\frac{z}{4}\right)^n + \sum_{n=0}^{\infty} \left(\frac{z^{-1}}{4}\right)^n - 1 = P_x^-(z) + P_x^+(z) - R_x(0)$$

$$= \frac{1}{1 - \frac{z^{-1}}{4}} + \frac{1}{1 - \frac{z}{4}} - 1 = \frac{15}{17 - 8(z + z^{-1})} \qquad (2.6\text{-}4\text{a})$$

The power spectral density in the frequency domain is then

$$P_x(\omega) = P_x(z)|_{z=e^{j\omega T_s}} = \frac{15}{17 - 16\cos(\omega T_s)} \qquad (2.6\text{-}4a)$$

which is, of course, periodic. ■

2.7 THE POWER SPECTRAL DENSITY OF A MARKOV PROCESS

In Example 3, we found the autocorrelation function and power spectral density of an AMI encoded signal using eq. 2.6-1d. This can be accomplished is another way by considering the encoding as a Markov process. We will proceed by example.

EXAMPLE 5   *AMI Coding as a Markov Process*

The AMI encoding scheme may also be represented as the finite state machine shown in fig. 2.7-1. In this machine is a single output associated with each state, making it a Moore machine. (By contrast, in a Mealy machine the output depends on the state and the current input.) The inputs $d_k = \pm 1$ are independent, each having probability $1/2$. Each of the four states has two inputs and can therefore be seen to be equally probable with probability $1/4$. If the states had different numbers of inputs they would not be equally probable. For example, if q_1 had one input and q_2 had three inputs then $p(q_1) = 1/8$ and $p(q_2) = 3/8$ with $p(q_3) = p(q_4) = 1/2$.

Clearly, if the input is random data, the output of this machine will be a random process of AMI encoded data. Many encoding rules can be represented by finite state machines in this way. As we shall see in Chapter 3, intersymbol interference arising from a distorted channel can also be so represented. This kind of representation is one special example of a type of random process known as a Markov process. ■

We shall not deal with the general issue of Markov processes in this book. We shall only deal with a few special cases that bear directly on our subject matter. We are interested, at this point, in treating the finite state machine as an encoding rule and figuring out a way of finding out how that encoding rule shapes the power spectral density of the

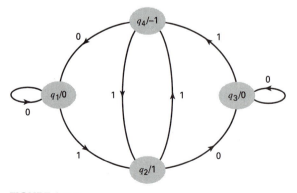

FIGURE 2.7-1

data stream. We shall always assume that the random process is wide-sense stationary and ergodic. In accomplishing this, we shall follow the treatment by Lee and Messerschmitt [Lee94].

The starting point is to find the autocorrelation function of the output. In our example having four states:

a. $f(q_i)$ is the output associated with state q_i (in this example $f(q_i) = 0, \pm 1$)

b. $p_{0,n}(q_i, q_j)$ is the joint probability of the machine being in state q_i at time 0 and state q_j at time n. Since the output is assumed to be stationary, we can say that

$$p_{0,n}(q_i, q_j) = p_{m,m+n}(q_i, q_j) \qquad (2.7\text{-}1a)$$

Consequently the autocorrelation function can be expressed as

$$R_x(n) = \sum_{i=1}^{4} \sum_{j=1}^{4} f(q_i) f(q_j) p_{0,n}(q_i, q_j). \qquad (2.7\text{-}1b)$$

or can be written as the quadratic form

$$R_x(n) = \mathbf{f} P_{0,n} \mathbf{f}^T = [f(q_1) \quad f(q_2) \quad f(q_3) \quad f(q_4)] \begin{bmatrix} \cdot & \cdot & & \cdot \\ \cdot & \cdot & p_{0,n}(q_i, q_j) & \cdot \\ \cdot & \cdot & & \cdot \\ \cdot & \cdot & & \cdot \end{bmatrix} \begin{bmatrix} f(q_1) \\ f(q_2) \\ f(q_3) \\ f(q_4) \end{bmatrix}$$

$$(2.7\text{-}1c)$$

where

$$p_{0,n}(q_i, q_j) = p_{n/0}(q_j/q_i) p_0(q_i) \qquad (2.7\text{-}1d)$$

and where, because the random process is stationary,

$$p_0(q_i) = p(q_i) \qquad (2.7\text{-}1e)$$

the known steady-state probability of q_i. In effect, we are saying that the machine has started long before $n = 0$ so that the transient has died off and the machine is in the steady state. In fact, an already decayed transient is required for the output to be stationary.

Implicitly, we have computed the autocorrelation function in eq. 2.7-1b for $n > 0$. Since the autocorrelation function is symmetric with respect to n, that is sufficient information to construct the entire function. The power spectral density is the z-transform of the autocorrelation function

$$P_x(z) = \sum_{n=-\infty}^{\infty} R_x(n) z^{-n}$$

$$= \sum_{n=0}^{\infty} R_x(n) z^{-n} + \sum_{n=0}^{\infty} R_x(-n) z^{n} - R_x(0)$$

$$= P_x^+(z) + P_x^+(z^{-1}) - P_x^+(\infty) \qquad (2.7\text{-}2)$$

Returning to Example 4, we can start by finding

$$P_x^+(z) = \sum_{n=0}^{\infty} R_x(n)z^{-n} = \sum_{n=0}^{\infty} \left\{ \sum_{i=1}^{4} \sum_{j=1}^{4} f(q_i)f(q_j)p_{0,n}(q_i, q_j) \right\} z^{-n}$$

$$= \sum_{i=1}^{4} \sum_{j=1}^{4} f(q_i)f(q_j) \sum_{n=0}^{\infty} p_{0,n}(q_i, q_j)z^{-n} = \sum_{i=1}^{4} \sum_{j=1}^{4} f(q_i)f(q_j)P_{i,j}(z) \quad \text{(2.7-3)}$$

where $P_{i,j}$ is a 4×4 matrix.

We now want to find the matrix $P_{i,j}(z)$. The basic recursion equation for the finite-state machine of fig. 2.7-1 is given by

$$P_{n+1}(j) = \sum_{i=1}^{4} p(j/i)P_n(i) \quad \text{(2.7-4a)}$$

where $P_n(i) = $ probability of being in state q_i at the nth iteration, $p(j/i) = $ probability of going from q_i to q_j (equal to 1/2 for binary, equally probable input data).

Taking the z-transform of eq. 2.7-4a, we get for our four-state machine

$$P_j(z) = p_0(q_j) + \sum_{i=1}^{4} p(q_j/q_i)z^{-1}P_i(z) \quad \text{(2.7-4b)}$$

In our AMI example of fig. 2.7-1, this results in four equations. Note that since we have two equally probable inputs, the conditional probabilities $p(q_j/q_i) = \frac{1}{2}$ for all i, j.

$$P_1(z) = p(q_1) + \frac{1}{2}z^{-1}P_1(z) + \frac{1}{2}z^{-1}P_4(z)$$

$$P_2(z) = p(q_2) + \frac{1}{2}z^{-1}P_1(z) + \frac{1}{2}z^{-1}P_4(z)$$

$$P_3(z) = p(q_3) + \frac{1}{2}z^{-1}P_2(z) + \frac{1}{2}z^{-1}P_3(z) \quad \text{(2.7-5a)}$$

$$P_4(z) = p(q_4) + \frac{1}{2}z^{-1}P_2(z) + \frac{1}{2}z^{-1}P_3(z)$$

and in matrix form:

$$\begin{bmatrix} 1 - \frac{1}{2}z^{-1} & 0 & 0 & -\frac{1}{2}z^{-1} \\ 1 - \frac{1}{2}z^{-1} & 1 & 0 & -\frac{1}{2}z^{-1} \\ 0 & -\frac{1}{2}z^{-1} & 1 - \frac{1}{2}z^{-1} & 0 \\ 0 & -\frac{1}{2}z^{-1} & -\frac{1}{2}z^{-1} & 1 \end{bmatrix} \begin{bmatrix} P_1(z) \\ P_2(z) \\ P_3(z) \\ P_4(z) \end{bmatrix} = \begin{bmatrix} p(q_1) \\ p(q_2) \\ p(q_3) \\ p(q_4) \end{bmatrix} \quad \text{(2.7-5b)}$$

■

In order to solve for the vector $P(z)$, we must invert the matrix. This will be accomplished using the MATLAB Symbolic Toolbox. In order to simplify the matrix, we have multiplied each term by z. We will subsequently compensate for this operation.

```
A = sym('[(z-1/2), 0, 0, -1/2; -1/2, z, 0, -1/2; 0, -1/2, z-1/2, 0; 0, -
1/2, -1/2, z]')
```

A =
[(z-1/2),   0,   0,   -1/2]
[-1/2,   z,   0,   -1/2]
[0,   -1/2,   z-1/2,   0]
[0,   -1/2,   -1/2,   z]

B = inverse (A)

B =
[1/4/z^2*(4*z^2-2*z-1)/(z-1),   1/4/z^2/(z-1),
1/4/z^2/(z-1),   1/4/z^2*(2*z-1)/(z-1)]
[1/4/z^2*(2*z-1)/(z-1),   1/4*(2*z-1)^2/z^2/(z-1),
1/4/z^2/(z-1),   1/4/z^2*(2*z-1)/(z-1)]
[1/4/z^2/(z-1),   1/4/z^2*(2*z-1)/(z-1),   1/4/z^2*(4*z^2-2*z-
1)/(z-1),   1/4/z^2/(z-1)]
[1/4/z^2/(z-1),   1/4/z^2*(2*z-1)/(z-1),   1/4/z^2*(2*z-
1)/(z-1),   1/4*(2*z-1)^2/z^2/(z-1)]

We now have a matrix equation

$$
\begin{bmatrix} P_1(z) \\ P_2(z) \\ P_3(z) \\ P_4(z) \end{bmatrix} = [B] \begin{bmatrix} p(q_1) \\ p(q_2) \\ p(q_3) \\ p(q_4) \end{bmatrix} \tag{2.7-6a}
$$

where we can interpret the matrix $[B]$ as a matrix of z-transforms of conditional probabilities $P(j/i)$

$$
[B] = \begin{bmatrix} p_{1/1}(z) & p_{1/2}(z) & p_{1/3}(z) & p_{1/4}(z) \\ p_{2/1}(z) & p_{2/2}(z) & p_{2/3}(z) & p_{2/4}(z) \\ p_{3/1}(z) & p_{3/2}(z) & p_{3/3}(z) & p_{3/4}(z) \\ p_{4/1}(z) & p_{4/2}(z) & p_{4/3}(z) & p_{4/4}(z) \end{bmatrix} \tag{2.7-6b}
$$

since, for example, the computation of $P_1(z)$ results from multiplying the first row of the matrix by the column vector of probabilities.

$$
\begin{aligned} P_1(z) &= p_{1/1}(z)p(q_1) + p_{1/2}(z)p(q_2) + p_{1/3}(z)p(q_3) + p_{1/4}(z)p(q_4) \\ &= P_{1,1}(z) + P_{1,2}(z) + P_{1,3}(z) + P_{1,4}(z) \end{aligned} \tag{2.7-6c}
$$

It follows that in order to get the matrix of joint probabilities that we are looking for, we must multiply the matrix $[B]$ by the diagonal matrix

$$
D = \text{diag}[p(q_1), p(q_2), p(q_3), p(q_4)] \tag{2.7-6d}
$$

In our problem of AMI where all the states are equally probable, this simply involves a diagonal matrix $D = \text{diag}[1/4, 1/4, 1/4, 1/4]$. Where the states are not equally probable, the diagonal matrix should be constructed appropriately.

We now want to evaluate $P_x^+(z)$, the z-transform of eq. 2.7-1c.

$$P_x^+(z) = [f(q_1) \quad f(q_2) \quad f(q_3) \quad f(q_4)] \begin{bmatrix} \cdot & \cdot & \cdot & \cdot \\ \cdot & \cdot & p_{i,j}(z) & \cdot \\ \cdot & \cdot & \cdot & \cdot \\ \cdot & \cdot & \cdot & \cdot \end{bmatrix} \begin{bmatrix} f(q_1) \\ f(q_2) \\ f(q_3) \\ f(q_4) \end{bmatrix} \quad (2.7\text{-}7a)$$

In the AMI example, the output vector is

$$[f(q_1) \quad f(q_2) \quad f(q_3) \quad f(q_4)] = [0, \quad 1, \quad 0, \quad -1] \quad (2.7\text{-}7b)$$

Continuing with the MATLAB Symbolic Toolbox

```
f  =  [0, 1, 0, -1]

f =

    0     1     0     -1

res  =  symmul( f, symmul( B, transpose(f)))

res =
(2*z-1)/z^2
```

We now must multiply the quantity res by z to compensate for the earlier factorization by z and this yields

$$P_x^+(z) = z^*\text{res} = z\frac{2z-1}{z^2} \quad (2.7\text{-}8a)$$

and, recognizing that $P_x^+(\infty) = \bar{R}_x(0)$,

$$P_x(z) = P_x^+(z) + P_x^+(z^{-1}) - P_x^+(\infty)$$
$$= \frac{2z-1}{z} + \frac{(2/z)-1}{(1/z)} - 2 = -\frac{z^2-2z+1}{z}$$
$$= -(z - 2 - z^{-1}) \quad (2.7\text{-}8b)$$

In order to go from here to the frequency domain, we allow $z = e^{j\omega T_s}$, so that

$$P_x(e^{j\omega T_s}) = -(e^{j\omega T_s} - 2 + e^{-j\omega T_s}) = 2[1 - \text{Cos}(\omega T_s)] \quad (2.7\text{-}8c)$$

which is exactly the same result that we got in Example 3. A graph of this power spectral density is shown in fig. 2.7-2:

```
psd  =  sym('[2*(1-cos(x))]')

psd =
2*(1-cos(x))

ezplot(psd, [0, 2*pi])
```

The fact that we obtained the same result in two different ways is very encouraging because it provides two different tools, one of which may be much easier to use than the other in a given circumstance.

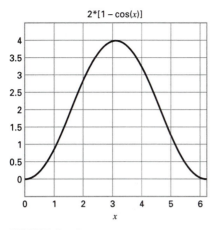

$2*[1 - \cos(x)]$

FIGURE 2.7-2

EXAMPLE 6 *Miller Encoding as a Markov Process*

The Miller Code, also known as delay modulation, has its origins in magnetic recording. Just as AMI, it can be represented as a sequential machine, as shown in fig. 2.7-3a. After the power spectral density is derived we shall be able to discuss its particular characteristics. At this point, we should note that this sequential machine differs in two respects from the machine in the previous example of AMI:

1. The Miller Code machine is, as shown, a Mealy machine rather than a Moore machine. This means that there is not a unique output associated with each state; rather, the output depends on both the current state and the current input. We shall address this problem by first converting the Mealy machine to a Moore machine in which the output depends only on the current state. A general approach to converting a Mealy machine to a Moore machine is found in most introductory textbooks in digital design. In this instance, the conversion is quite direct and the resulting Moore machine, fig. 2.7-3b, also has four states.

2. The outputs in this code are functions of time rather than numbers as they were in the AMI example and as they could be in a different Mealy machine example. These functions and their Fourier Transforms are shown in fig. 2.7-3a. In finding the power spectral density we will have to make sure that we are multiplying the Fourier Transform of one function with the complex conjugate of another in order to obtain the required result.

Finding the z-transform matrix of the joint probabilities follows in exactly the same way as in the previous example. Referring to fig. 2.7-3b, and assuming that $p(0) = p(1) = 1/2$, the z-transform conditional probabilities are given by

$$P_1(z) = \frac{1}{4} + \frac{1}{2}z^{-1}P_3(z) + \frac{1}{2}z^{-1}P_4(z)$$

$$P_2(z) = \frac{1}{4} + \frac{1}{2}z^{-1}P_1(z) + \frac{1}{2}z^{-1}P_3(z)$$

$$P_3(z) = \frac{1}{4} + \frac{1}{2}z^{-1}P_2(z) + \frac{1}{2}z^{-1}P_4(z) \qquad (2.7\text{-}9a)$$

$$P_4(z) = \frac{1}{4} + \frac{1}{2}z^{-1}P_1(z) + \frac{1}{2}z^{-1}P_2(z)$$

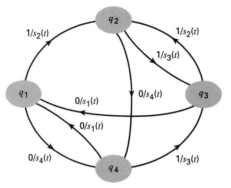

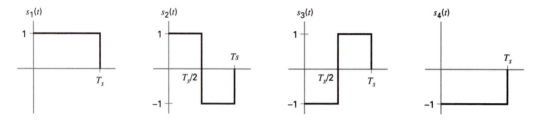

$$s_1(\omega) = \text{sinc}(\omega T_s/4)\, e^{-j\omega T_s/4}\, [1 + e^{-j\omega T_s/2}] = 2s(\omega) \cos\,(\omega T_s/4)\, e^{-j\omega T_s/2}$$

$$s_2(\omega) = \text{sinc}(\omega T_s/4)\, e^{-j\omega T_s/4}\, [1 - e^{-j\omega T_s/2}] = 2js(\omega) \sin\,(\omega T_s/4)\, e^{-j\omega T_s/2}$$

$$s_3(\omega) = -\text{sinc}(\omega T_s/4)\, e^{-j\omega T_s/4}\, [1 - e^{-j\omega T_s/2}] = -2js(\omega) \sin\,(\omega T_s/4)\, e^{-j\omega T_s/2}$$

$$s_4(\omega) = -\text{sinc}(\omega T_s/4)\, e^{-j\omega T_s/4}\, [1 + e^{-j\omega T_s/2}] = -2s(\omega) \cos\,(\omega T_s/4)\, e^{-j\omega T_s/2}$$

$$s(\omega) = S_a(\omega T_s/4)$$

(a)

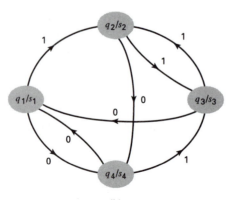

(b)

FIGURE 2.7-3

or in matrix form:

$$\frac{1}{z}\begin{bmatrix} z & 0 & -1/2 & -1/2 \\ -1/2 & z & -1/2 & 0 \\ 0 & -1/2 & z & -1/2 \\ -1/2 & -1/2 & 0 & z \end{bmatrix}\begin{bmatrix} P_1(z) \\ P_2(z) \\ P_3(z) \\ P_4(z) \end{bmatrix} = \frac{1}{4}\begin{bmatrix} 1 \\ 1 \\ 1 \\ 1 \end{bmatrix} \qquad (2.7\text{-}9b)$$

or

$$\frac{1}{z}A\begin{bmatrix} P_1(z) \\ P_2(z) \\ P_3(z) \\ P_4(z) \end{bmatrix} = \frac{1}{4}\begin{bmatrix} 1 \\ 1 \\ 1 \\ 1 \end{bmatrix} \Rightarrow \begin{bmatrix} P_1(z) \\ P_2(z) \\ P_3(z) \\ P_4(z) \end{bmatrix} = \frac{z}{4}[AI]\begin{bmatrix} 1 \\ 1 \\ 1 \\ 1 \end{bmatrix} \qquad (2.7\text{-}9c)$$

Again, using the MATLAB Symbolic Toolbox, we can find the matrix inverse. Now we have to multiply the matrix AI, the inverse of A, by the diagonal matrix of the steady-state probabilities and construct the quadratic form for power spectral density. Thus:

$$P_x^+(z) = z[S_1 \quad S_2 \quad S_3 \quad S_4]\{[AI][\mathrm{diag}(p_1, p_2, p_3, p_4)]\}\begin{bmatrix} S_1 \\ S_2 \\ S_3 \\ S_4 \end{bmatrix}^* \qquad (2.7\text{-}9d)$$

Since all of the states are equally likely, the diagonal matrix simply becomes an identity matrix multiplied by the constant $1/4$.

Handling the output vector requires a bit more thought. Referring to fig. 2.7-3a, we first see that $S_4 = -S_1$ and $S_3 = -S_2$. Next we note that in evaluating the quadratic form of eq. 2.7-9d, each term will be multiplied by the complex conjugate of another term so that the complex exponential $e^{-j\omega Ts}/2$ is effectively cancelled. We finally note that without the complex exponential term $S_1^* = S_1$ and $S_2^* = -S_2$ and eq. 2.7-9d can be simplified to

$$P_x^+(z) = \frac{z}{4}[S_1 \quad S_2 \quad -S_2 \quad -S_1]\{[AI]\}\begin{bmatrix} S_1 \\ -S_2 \\ S_2 \\ -S_1 \end{bmatrix} \qquad (2.7\text{-}9e)$$

Again using the MATLAB Symbolic Toolbox,

```
A = sym('[z, 0, -1/2, -1/2; -1/2, z, -1/2, 0; 0, -1/2, z, -1/2; -1/2,
1/2, 0, z]')
```

$$A = \begin{bmatrix} z, & 0, & -1/2, & -1/2 \\ -1/2, & z, & -1/2, & 0 \\ 0, & -1/2, & z, & -1/2 \\ -1/2, & -1/2, & 0, & z \end{bmatrix}$$

and allowing $a = S_1$ and $b = S_2$ we can set

```
C1 = sym('[a, b, -b, -a]')
```

```
C1  =
[a,  b,  -b,  -a]
```

and

```
C2  =  sym('[a,  -b,  b,  -a]')

C2  =
[a,  -b,  b,  -a]
```

so that

```
Px  =  symmul(C1,  symmul(inverse(A),  transpose(C2)))

Px  =
2*(2*a^2*z+a^2+2*a*b-2*b^2*z-b^2)/(2*z^2+2*z+1)
```

so that, multiplying by z to complete eq. 2.7-9d,

$$P_x^+(z) = 2z \frac{2(a^2 - b^2)z + (a^2 - b^2) + 2ab}{2z^2 + 2z + 1} \tag{2.7-10a}$$

where

$$a = S_1 = 2\,\mathrm{Sinc}\left(\frac{\omega T_s}{4}\right)\mathrm{Cos}\left(\frac{\omega T_s}{4}\right), \qquad b = S_2 = 2j\,\mathrm{Sinc}\left(\frac{\omega T_s}{4}\right)\mathrm{Sin}\left(\frac{\omega T_s}{4}\right) \tag{2.7-10b}$$

so that

$$a^2 - b^2 = 4\,\mathrm{Sinc}^2\left(\frac{\omega T_s}{4}\right), \qquad 2ab = 4j\,\mathrm{Sinc}^2\left(\frac{\omega T_s}{4}\right)\mathrm{Sin}\left(\frac{\omega T_s}{2}\right) \tag{2.7-10c}$$

Since

$$P_x(z) = P_x^+(z) + P_x^+(z^{-1}) - P_x^+(\infty) \tag{2.7-2}$$

it follows that

$$P_x(e^{j\omega T_s}) = \left[\frac{\mathrm{Sin}(\omega T_s/4)}{(\omega T_s/4)}\right]^2 \frac{3 + \mathrm{Cos}(\omega T_s/2) + 2\mathrm{Cos}(\omega T_s) - \mathrm{Cos}(3\omega T_s/2)}{9 + 12\,\mathrm{Cos}(\omega T_s) + 4\,\mathrm{Cos}(2\omega T_s)} \tag{2.7-11}$$

We can also use the MATLAB Symbolic Toolbox to plot this power spectral density, shown in fig. 2.7-4:

```
Px  =  sym('[((sin(x/4)/(x/4))^2)*(3+cos(x/2)+2*cos(x)-
cos(3*x/2))/(9+12*cos(x)+4*cos(2*x))]')

Px  =
((sin(x/4)/(x/4))^2)*(3+cos(x/2)+2*cos(x)-cos(3*x/2))/(9+12*cos(x)+4*cos(2*x))

ezplot(Px,  [0,  2*pi])
```

$[(\sin(x/4)/(x/4))^2]*(3 + \cos(x/2) + \sim\sim\sim$
$(3*x/2))/(9 + 12* \cos(x) + 4* \cos(2*x))]$

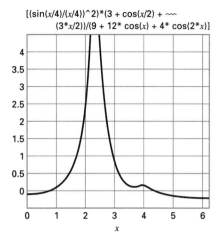

FIGURE 2.7-4

Note that the peak value occurs at $x = \omega T_s = 2.5$, corresponding to a frequency of $f = \frac{1.25}{\pi}$ fs.

A comparison of the power spectral densities of the Miller Code and the AMI code shows that while both of them have small values at low frequencies, corresponding to the magnetic channel, the Miller Code power is more concentrated and at a lower frequency than the AMI. This has made it preferable for the magnetic channel. This will be discussed further in Chapter 3. ∎

2.8 THE GAUSSIAN RANDOM PROCESS

We now want to begin a multistep development of the idea and characterization of a Gaussian random process. The key result is that we will be able to model this process in a manner that is consistent with the model of a data signal shown in fig. 2.6-1. In that model, the information signal is derived from data impulses occurring every T_s sec at the input of a linear system having impulse response $h(t)$. We shall demonstrate that Gaussian noise can be modeled as randomly occurring impulses at that input. These random impulses occur at an average rate of α impulses per second according to a Poisson distribution and have areas that are distributed according to a Gaussian probability density function. We will be finding the ensemble autocorrelation function but since the random processes are assumed

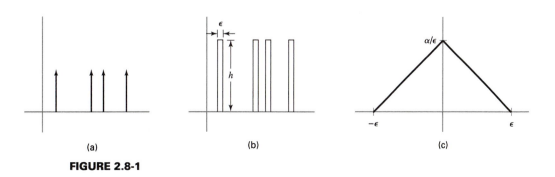

(a)　　　　　　　　　　(b)　　　　　　　　　　(c)

FIGURE 2.8-1

to be ergodic, this will equal the time average autocorrelation function, which, in turn, will lead to the power spectral density.

Following Lathi [Lat68], we will develop this model in three steps.

2.8.1   Step 1. The Autocorrelation Function of Poisson-Distributed Impulses

In fig. 2.8-1a we have Poisson-distributed unit impulse functions occurring at a rate of α impulses per sec. An approximation to theses impulses, shown in fig. 2.8-1b, is narrow rectangular pulses having amplitude h and width ε where $h\varepsilon = 1$. We want to find the autocorrelation function

$$R_x(\tau) = \sum_{x_1} \sum_{x_2} x_1 x_2 p_{x_1 x_2}(x_1, x_2) \tag{2.8-1a}$$

where x_1 and x_2 may have the values $0, h$. The two instants of time t_1 and t_2 are separated by τ.

$\underline{\tau > \varepsilon}$
When $\tau > \varepsilon$, we are looking at two different pulses which may have values $0, h$ independently. Since we multiply these values in finding the autocorrelation function, the result is

$$R_x(\tau) = h^2 p_{x_1 x_2}(h, h) = h^2 p_{x_1}(h) p_{x_2/x_1}(h/x_1 = h) = h^2 p_{x_1}(h) p_{x_2}(h) \tag{2.8-1b}$$

since these impulses occur independently. Since the pulses occur at the rate of α impulses per sec and have width ε, it follows that the probability of a pulse occuring at a given time is $\alpha\varepsilon$. Consequently,

$$R_x(\tau) = h^2 p_{x_1}(h) p_{x_2}(h) = h^2 (\alpha\varepsilon)^2 = \alpha^2 \tag{2.8-1c}$$

$\underline{\tau < \varepsilon}$
When $\tau < \varepsilon$, we are either looking at the same pulse at two different points on the pulse or, since ε is vanishingly small, we are looking at one point on the pulse and one off or both points off. Since points off the pulse have value zero, we conclude that

$$R_x(\tau) = h^2 p_{x_1 x_2}(h, h) = h^2 p_{x_1}(h) p_{x_2/x_1}(h/x_1 = h) = h^2(\alpha\varepsilon)\left(1 - \frac{\tau}{\varepsilon}\right) = \frac{\alpha}{\varepsilon}\left(1 - \frac{\tau}{\varepsilon}\right) \tag{2.8-1d}$$

since $h\varepsilon = 1$. As shown in fig. 2.8-1c, this triangle becomes an impulse function in the limit as $\varepsilon \to 0$.

Consequently, the resulting autocorrelation function for all values of τ is

$$R_x(\tau) = \alpha\delta(\tau) + \alpha^2 = \overline{R_x}(\tau) \tag{2.8-1e}$$

and the power spectral density is

$$P_x(\omega) = \alpha + 2\pi\alpha^2\delta(\omega) \tag{2.8-1f}$$

The impulse function in $R_x(\tau)$ and corresponding constant α in the power spectral density indicate a 'white' component to the signal. The impulse function in $P_x(\omega)$ comes from the dc component corresponding to the strictly positive amplitude of all the impulses.

2.8.2 Step 2. White Noise

We can model white noise by allowing the impulses in the previous example to have amplitudes $\pm h$ with equal probability. Under these circumstances

$$P_{x_1}(h) = P_{x_1}(-h) = P_{x_2}(h) = P_{x_2}(-h) = \frac{\alpha\varepsilon}{2} \quad \text{and} \quad P_{x_1}(0) = P_{x_2}(0) = 1 - \frac{\alpha\varepsilon}{2}$$

$$(2.8\text{-}2a)$$

so that substituting these values into eq. 2.8-1a for $\tau > \varepsilon$ yields $R_x(\tau) = 0$.

For $\tau < \varepsilon$, we have a repetition of the previous example so that the final result is

$$R_x(\tau) = \alpha\delta(\tau) = \overline{R_x}(\tau) \quad \text{and} \quad P_x(\omega) = \alpha \qquad (2.8\text{-}2b)$$

the autocorrelation function and power spectral density for white noise.

2.8.3 Step 3. White Gaussian Noise (AWGN)

We now take the final step of allowing the impulses of the previous example to have amplitudes that are distributed according to the zero-mean, unit-variance Gaussian random variable

$$p_x(x) = \frac{1}{\sqrt{2\pi}}e^{-x^2/2}. \qquad (2.8\text{-}3a)$$

Since this amplitude random variable is continuous, it follows that we must find the autocorrelation function as an integral

$$R_x(\tau) = \int\limits_{-\infty}^{\infty} \int\limits_{-\infty}^{\infty} x_1 x_2 p_{x_1 x_2}(x_1, x_2)\, dx_1\, dx_2 \qquad (2.8\text{-}3b)$$

$\tau > \varepsilon$

Again, for $\tau > \varepsilon$ the two sample points occur independently. If one or both do not lie on pulses, the integral is zero. If both lie on pulses, then the joint density function of the two pulses is the product of the marginal density functions, so the autocorrelation integral becomes

$$R_x(\tau) = \int\limits_{-\infty}^{\infty} x_1 p_{x_1}(x_1)\, dx_1 \int\limits_{-\infty}^{\infty} x_2 p_{x_2}(x_2)\, dx_2 = 0 \qquad (2.8\text{-}3c)$$

since the random variable is zero mean.

$\tau < \varepsilon$

When $\tau < \varepsilon$, if the two sample points lie on a pulse it must be the same pulse, so that the autocorrelation function becomes a single integral with $x_1 = x_2$. Consequently,

$$R_x(\tau) = \int\limits_{-\infty}^{\infty} x^2 p_{x_1 x_2}(x, x)\, dx = \int\limits_{-\infty}^{\infty} x^2 p_{x_1}(x) p_{x_2/x_1}(x/x_1 = x)\, dx$$

$$= \left(1 - \frac{\tau}{\varepsilon}\right) \int\limits_{-\infty}^{\infty} x^2 p_{x_1}(x)\, dx = \left(1 - \frac{\tau}{\varepsilon}\right)(\alpha\varepsilon) \int\limits_{-\infty}^{\infty} x^2 \frac{1}{\sqrt{2\pi}}e^{-x^2/2}\, dx$$

$$= \left(1 - \frac{\tau}{\varepsilon}\right)(\alpha\varepsilon) \rightarrow \alpha\delta(\tau) \qquad (2.8\text{-}3d)$$

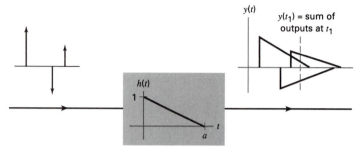

FIGURE 2.8-2

Clearly this noise is white, having the same power spectral density as the previous white-noise example, but also with its constituent pulses having a Gaussian distributed amplitude.

Since our model of AWGN consists of randomly separated impulse functions, the response of a linear system to these impulse functions, shown in fig. 2.8-2, is overlapping, randomly scaled according to a Gaussian density function, and randomly displaced repetitions of the impulse response. At any instant of time, the total output of the linear system is the sum of these impulse responses. In fact, at any instant of time the output is the sum of Gaussian distributed random variables and therefore itself Gaussian.

Finally, recall that a time-average autocorrelation function is written as

$$\overline{R_f}(\tau) = \lim_{T \to \infty} \frac{1}{T} \int_{-T/2}^{T/2} f(t) f^*(t - \tau)\, dt = \overline{f(t) f^*(t - \tau)} \qquad (2.8\text{-}4a)$$

and the average power is

$$P_{av} = \overline{R_f}(0) = \lim_{T \to \infty} \frac{1}{T} \int_{-T/2}^{T/2} f(t) f^*(t)\, dt = \overline{|f(t)|^2} = E[f^2(t)] \qquad (2.8\text{-}4b)$$

Since the process is assumed to be ergodic, the time-average mean-squared value equals the ensemble-average mean-squared value. If $f(t)$ is zero mean, i.e., no dc component, it follows that

$$P_{av} = \overline{|f(t)|^2} = E[f^2] = \int_{-\infty}^{\infty} x^2 p_x(x)\, dx = \int_{-\infty}^{\infty} x^2 \frac{1}{\sigma \sqrt{2\pi}} e^{-x^2/2\sigma^2}\, dx = \sigma^2 \qquad (2.8\text{-}4c)$$

the variance of the Gaussian density function. This is a key result that will allow the computation of probability of detection error in the presence of Gaussian noise.

2.9   TRANSMISSION OF RANDOM SIGNALS THROUGH LINEAR SYSTEMS

We shall now find an input-output relationship for autocorrelation functions and power spectral densities for linear systems. Consider the linear system of fig. 2.9-1 having input

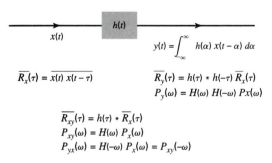

$$\overline{R_x}(\tau) = \overline{x(t)\, x(t-\tau)}$$

$$\overline{R_y}(\tau) = h(\tau) * h(-\tau)\,\overline{R_x}(\tau)$$
$$P_y(\omega) = H(\omega)\, H(-\omega)\, Px(\omega)$$

$$\overline{R_{xy}}(\tau) = h(\tau) * \overline{R_x}(\tau)$$
$$P_{xy}(\omega) = H(\omega)\, P_x(\omega)$$
$$P_{yx}(\omega) = H(-\omega)\, P_x(\omega) = P_{xy}(-\omega)$$

FIGURE 2.9-1

$x(t)$, output $y(t)$, and impulse response $h(t)$. The output $y(t)$ is related to the input $x(t)$ by the convolution relationship

$$y(t) = \int\limits_{-\infty}^{\infty} h(\alpha)x(t-\alpha)\,d\alpha = h(t) * x(t). \tag{2.9-1a}$$

It follows that:

$$\overline{R_y}(\tau) = \overline{y(t)y(t-\tau)} = \overline{\int\limits_{-\infty}^{\infty} h(\alpha)x(t-\alpha)\,d\alpha \int\limits_{-\infty}^{\infty} h(\beta)x(t-\beta)\,d\beta}$$

$$= \overline{\int\limits_{-\infty}^{\infty}\int\limits_{-\infty}^{\infty} h(\alpha)h(\beta)x(t-\alpha)x(t-\tau-\beta)\,d\alpha\,d\beta}$$

$$= \int\limits_{-\infty}^{\infty}\int\limits_{-\infty}^{\infty} h(\alpha)h(\beta)\overline{x(t-\alpha)x(t-\tau-\beta)}\,d\alpha\,d\beta$$

$$= \int\limits_{-\infty}^{\infty}\int\limits_{-\infty}^{\infty} h(\alpha)h(\beta)\overline{R_x}(\tau-\alpha+\beta)\,d\alpha\,d\beta = h(\tau) * h(-\tau) * \overline{R_x}(\tau) \tag{2.9-1b}$$

In the special case that $\overline{R_x}(\tau) = \dfrac{\eta}{2}\delta(\tau)$, then

$$\overline{R_y}(\tau) = \frac{\eta}{2}h(\tau) * h(-\tau) \tag{2.9-1c}$$

We may now take the Fourier Transform of eq. 2.9-1b in order to get an average power spectral density:

$$P_y(\omega) = H(\omega)H(-\omega)P_x(\omega) \tag{2.9-1d}$$

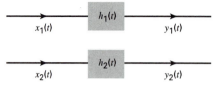

FIGURE 2.9-2

Again referring to fig. 2.9-1, we may find the cross-correlation between input and output:

$$\overline{R_{xy}}(\tau) = \overline{x(t)y(t-\tau)} = \overline{x(t) \int_{-\infty}^{\infty} h(\alpha)x(t-\tau-\alpha)\,d\alpha}$$

$$= \int_{-\infty}^{\infty} h(\alpha)\overline{x(t)x(t-\tau-\alpha)}\,d\alpha = \int_{-\infty}^{\infty} h(\alpha)\overline{R_x}(\tau+\alpha)\,d\alpha = h(-\tau) * \overline{R_x}(\tau)$$

(2.9-2a)

and taking the Fourier Transform:

$$P_{xy}(\omega) = H(-\omega)P_x(\omega). \tag{2.9-2b}$$

Note that the definition of the time-average autocorrelation function leads to the following properties:

$$\overline{R_x}(\tau) = \overline{R_x^*}(-\tau) \Leftrightarrow P_x(\omega) = P_x(-\omega) \tag{2.9-3a}$$

$$\overline{R_{xy}}(\tau) = \overline{R_{yx}^*}(-\tau) \Leftrightarrow P_{xy}(\omega) = P_{yx}(-\omega) \tag{2.9-3b}$$

Continuing with fig. 2.9-2, we can see that correlated inputs of two different circuits lead to correlated outputs of those circuits:

$$\overline{R_{y_1 y_2}}(\tau) = \overline{y_1(t)y_2(t-\tau)} = \overline{\int_{-\infty}^{\infty} h_1(\alpha)x_1(t-\alpha)\,d\alpha \int_{-\infty}^{\infty} h_2(\beta)x_2(t-\tau-\beta)\,d\beta}$$

$$= \overline{\int_{-\infty}^{\infty}\int_{-\infty}^{\infty} h_1(\alpha)h_2(\beta)x_1(t-\alpha)x_2(t-\tau-\beta)\,d\alpha\,d\beta}$$

$$= \int_{-\infty}^{\infty}\int_{-\infty}^{\infty} h_1(\alpha)h_2(\beta)\overline{x_1(t-\alpha)x_2(t-\tau-\beta)}\,d\alpha\,d\beta$$

$$= \int_{-\infty}^{\infty}\int_{-\infty}^{\infty} h_1(\alpha)h_2(\beta)\overline{R_{x_1 x_2}}(\tau-\beta+\alpha)\,d\alpha\,d\beta$$

$$= h_1(-\tau) * h_2(\tau) * \overline{R_{x_1 x_2}}(\tau) \tag{2.9-4a}$$

and

$$\overline{R_{y_2 y_1}}(\tau) = h_2(-\tau) * h_1(\tau) * \overline{R_{x_2 x_1}}(\tau) = \overline{R_{y_1 y_2}}(-\tau) \tag{2.9-4b}$$

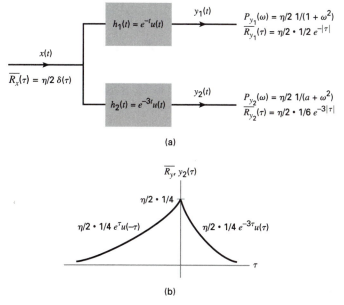

(a)

(b)

FIGURE 2.9-3

where:

$$P_{y_1 y_2}(\omega) = H_1(-\omega)H_2(\omega)P_{x_1 x_2}(\omega) \qquad (2.9\text{-}4c)$$

$$P_{y_2 y_1}(\omega) = H_2(-\omega)H_1(\omega)P_{x_2 x_1}(\omega) = P_{y_1 y_2}(-\omega) \qquad (2.9\text{-}4d)$$

EXAMPLE 7

In fig. 2.9-3a a single AWGN source serves as input to two linear systems having impulse responses and transfer functions

$$h_1(t) = e^{-t}u(t) \Leftrightarrow H_1(\omega) = \frac{1}{1 + j\omega} \qquad (2.9\text{-}5a)$$

$$h_2(t) = e^{-3t}u(t) \Leftrightarrow H_2(\omega) = \frac{1}{3 + j\omega} \qquad (2.9\text{-}5b)$$

The noise source is characterized by:

$$\overline{R_x}(\tau) = \frac{\eta}{2}\delta(t) \Leftrightarrow P_x(\omega) \qquad (2.9\text{-}5c)$$

It follows that:

$$P_{y_1}(\omega) = |H_1(\omega)|^2 P_x(\omega) = \frac{\eta}{2}\frac{1}{1 + \omega^2} \Leftrightarrow \overline{R_{y_1}}(\tau) = \frac{\eta}{2}\frac{1}{2}e^{-|\tau|} \qquad (2.9\text{-}6a)$$

$$P_{y_1}(\omega) = |H_2(\omega)|^2 P_x(\omega) = \frac{\eta}{2}\frac{1}{9 + \omega^2} \Leftrightarrow \overline{R_{y_1}}(\tau) = \frac{\eta}{2}\frac{1}{6}e^{-3|\tau|} \qquad (2.9\text{-}6b)$$

and

$$\overline{R_{y_1 y_2}}(\tau) = h_1(-\tau) * h_2(\tau) * \overline{R_x}(\tau) = \frac{\eta}{2}h_1(-\tau) * h_2(\tau) \qquad (2.9\text{-}6c)$$

According to fig. 2.9-3b,

$$h_1(-\tau) * h_2(\tau) = \int_{-\infty}^{\infty} e^{\lambda} u(-\lambda) e^{-3(\tau-\lambda)} u(\tau - \lambda) \, d\lambda = e^{-3\tau} \int_{-\infty}^{\infty} e^{4\lambda} u(-\lambda) u(\tau - \lambda) \, d\lambda$$

(2.9-4d)

For $\tau \geq 0$

$$h_1(-\tau) * h_2(\tau) = e^{-3\tau} \int_{-\infty}^{0} e^{4\lambda} \, d\lambda u(\tau) = \frac{1}{4} e^{-3\tau} u(\tau)$$

(2.9-4e)

For $\tau \leq 0$

$$h_1(-\tau) * h_2(\tau) = e^{-3\tau} \int_{-\infty}^{\tau} e^{4\lambda} \, d\lambda u(-\tau) = \frac{1}{4} e^{\tau} u(-\tau)$$

(2.9-6f)

∎

Based on these results, we want to find the covariance matrix of the Gaussian random variables

$$y_1(t), \ y_1(t + 2), \ y_2(t + 3), \ y_2(t + 4)$$

(2.9-7a)

$$C = \begin{bmatrix} \sigma_1^2 = \overline{R_{y_1}}(0) & \overline{R_{y_1}}(2) & \overline{R_{y_1 y_2}}(3) & \overline{R_{y_1 y_2}}(4) \\ \overline{R_{y_1}}(-2) & \sigma_1^2 = \overline{R_{y_1}}(0) & \overline{R_{y_1 y_2}}(1) & \overline{R_{y_1 y_2}}(2) \\ \overline{R_{y_2 y_1}}(-3) & \overline{R_{y_2 y_1}}(-1) & \sigma_2^2 = \overline{R_{y_2}}(0) & \overline{R_{y_2}}(1) \\ \overline{R_{y_2 y_1}}(-4) & \overline{R_{y_2 y_1}}(-2) & \overline{R_{y_2}}(-1) & \sigma_2^2 = \overline{R_{y_2}}(0) \end{bmatrix}$$

(2.9-7b)

Since $\overline{R_{y_1 y_2}}(\tau) = \overline{R_{y_2 y_1}}(-\tau)$ and $\overline{R_x}(\tau) = \overline{R_x}(-\tau)$, it follows that this covariance matrix is symmetric.

Finally, we want to construct the covariance matrix of N random variables obtained by uniformly sampling a Gaussian random process every T_s sec, as shown in fig. 2.9-4 and denoted as

$$y(t_1), \ y(t_1 + T_s), \ y(t_1 + 2T_s), \ \ldots, \ y(t_1 + (N - 1)T_s)$$

(2.9-8a)

For simplicity, we shall have $N = 4$. The matrix is then:

$$C = \begin{bmatrix} \sigma_y^2 = \overline{R_y}(0) & \overline{R_y}(T_s) & \overline{R_y}(2T_s) & \overline{R_y}(3T_s) \\ \overline{R_y}(T_s) & \overline{R_y}(0) & \overline{R_y}(T_s) & \overline{R_y}(2T_s) \\ \overline{R_y}(2T_s) & \overline{R_y}(T_s) & \overline{R_y}(0) & \overline{R_y}(T_s) \\ \overline{R_y}(3T_s) & \overline{R_y}(2T_s) & \overline{R_y}(T_s) & \overline{R_y}(0) \end{bmatrix}$$

(2.9-8b)

In addition to being symmetric, this matrix has the property that all of the elements on all of

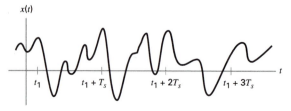

FIGURE 2.9-4

the diagonals are equal, i.e. $C_{ij} = \overline{R_y}(i - j)$. Such a matrix is *Toeplitz*. We shall see more of these matrices in subsequent discussion of adaptive equalization.

2.10 BAND-LIMITED GAUSSIAN NOISE

Let us now consider the special and important case of additive white Gaussian noise having a power spectral density $P_n(\omega) = \eta/2$ passed through an ideal lowpass filter having bandwidth B Hz. The resulting power spectral density and autocorrelation function are shown in figs. 2.10-1a and 2.10-1b. Mathematically, these two quantities are written as:

$$P_n(\omega) = \begin{cases} \eta/2; & |\omega| \le 2\pi B \\ 0; & \text{otherwise} \end{cases} \qquad \overline{R_n}(\tau) = \eta B \, \text{Sinc}(2\pi B\tau) \qquad (2.10\text{-}1)$$

Notice that samples of the noise taken at the Nyquist rate $T_s = 1/2B$ are uncorrelated and, because the noise is Gaussian, are therefore independent. The Sampling Theorem says that any signal $x(t)$ that is bandlimited to B Hz as shown in fig. 2.10-1a may be written as:

$$x(t) = \sum_{n=-\infty}^{\infty} x_n \, \text{Sinc}[2\pi B(t - nT_s)] \qquad (2.10\text{-}2)$$

where the x_n are samples of $x(t)$ taken every T_s seconds. In the case of bandlimited Gaussian noise, these samples are independent, zero-mean, Gaussian random variables each having variance $\sigma^2 = \eta B$. Under these circumstances, the covariance matrix of eq. 2.9-8b becomes:

$$C = \eta B \begin{bmatrix} 1 & 0 & 0 & 0 \\ 0 & 1 & 0 & 0 \\ 0 & 0 & 1 & 0 \\ 0 & 0 & 0 & 1 \end{bmatrix} \qquad (2.10\text{-}3)$$

and the joint density function is:

$$p_x(\alpha) = \frac{1}{(2\pi\eta B)^{N/2}} \exp\left\{ -\frac{\sum_{n=1}^{N} \alpha_n^2}{2\eta B} \right\} \qquad (2.10\text{-}4)$$

The fact that under these circumstances the successive noise samples are uncorrelated will be central to the discussion of the detection process in Chapter 3. We shall also return to this issue in Chapter 8, Channel Capacity, when we find the capacity of a waveform channel.

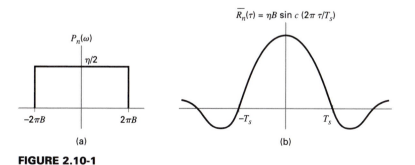

FIGURE 2.10-1

2.11 AN APPLICATION: SPECTRAL LINE TIMING RECOVERY

We have seen that an information signal may be represented as

$$x(t) = \sum_{n=-\infty}^{\infty} d_n h(t - nT_s), \qquad (2.11\text{-}1a)$$

as shown in fig. 2.5-1, where the data signals may or may not be correlated.

For the first part of this discussion we shall assume that they are correlated and that

$$E\lfloor d_n d_{n+k}\rfloor = R_d(k). \qquad (2.11\text{-}1)$$

The object is to perform an operation on $x(t)$ at the receiver that will enable us to extract a timing signal from this waveform that can be used to sample the received signal $x(t)$. As shown in fig. 2.11-1a, a timing signal can be thought of as a square wave having frequency $f_s = 1/T_s$ whose leading edge determines the instant of sampling. In most data-transmission systems a timing signal is not sent along with the data; rather, it is derived from the received signal [Ben58].

Following Aaron [Aar62], in this example the timing signal will be derived by first passing the received data signal through a square law device, generating an output:

$$y(t) = x^2(t) = \left\{ \sum_{n=-\infty}^{\infty} d_n h(t - nT_s) \right\}^2$$

$$= \sum_{n=-\infty}^{\infty} d_n^2 h^2(t - nT_s) + \sum_{n} \sum_{n \neq m} d_n d_m h(t - nT_s) h(t - mT_s) \qquad (2.11\text{-}2)$$

As shown in fig. 2.11-1b, $y(t)$ is applied to a narrow-band filter tuned to f_s that performs both the operations of averaging and frequency selectivity. These two operations will be treated separately. Averaging produces the expected value:

$$E[y(t)] = \sum_{n=-\infty}^{\infty} R_d(0) h^2(t - nT_s) + \sum_{n=-\infty}^{\infty} \sum_{k \neq 0} R_d(k) h(t - nT_s) h(t - (n+k)T_s)$$

$$(2.11\text{-}3)$$

which is clearly periodic with period T_s. Note that $h(t)$ may have duration greater than T_s.

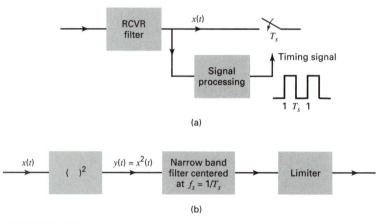

(a)

(b)

FIGURE 2.11-1

Since $E[y(t)]$ is periodic, it may be expanded in a Fourier Series

$$E[y(t)] = \sum_{n=-\infty}^{\infty} f_n e^{jn\omega_s t} \quad \text{where} \quad \omega_s = 2\pi/T_s \qquad (2.11\text{-}4a)$$

where

$$f_n = \frac{1}{T_s} \int_{-T_s/2}^{T_s/2} E[y(t)] e^{-jn\omega_s t} \, dt \qquad (2.11\text{-}4b)$$

The frequency selectivity of the narrowband filter, tuned to f_s, means that we will only be concerned with the value of f_1. Nevertheless, we shall first find a general expression for f_n.

Since we wish to integrate over a single period, the integral of eq. 2.11-4b can, using eq. 2.11-3, be reduced to

$$f_n = \frac{1}{T_s} \left\{ R_d(0) \int_{-\infty}^{\infty} h^2(t) e^{-jn\omega_s t} \, dt + \sum_{k \neq 0} R_d(k) \int_{-\infty}^{\infty} h(t) h(t - kT_s) e^{-jn\omega_s t} \, dt \right\}$$

$$(2.11\text{-}5)$$

If $f_1(t)$ and $f_2(t)$ are real, then Parseval's Theorem says that

$$\int_{-\infty}^{\infty} f_1(t) f_2(t) \, dt = \frac{1}{2\pi} \int_{-\infty}^{\infty} F_1(\omega) F_2(-\omega) \, d\omega \qquad (2.11\text{-}6a)$$

Accordingly, we shall set

$$f_1(t) = h(t) \Rightarrow F_2(-\omega) = H(-\omega)$$
$$f_2(t) = h(t) e^{-jn\omega_s t} \Rightarrow F_1(\omega) = H(\omega + n\omega_s) \qquad (2.11\text{-}6b)$$

and in the second integral

$$f_1(t) = h(t - kT_s) e^{-jn\omega_s t} \Rightarrow F_1(\omega) = H(\omega + n\omega_s) e^{jk\omega_s t} \qquad (2.11\text{-}6c)$$

It follows that

$$f_n = \frac{1}{2\pi T_s} \int_{-\infty}^{\infty} H(-\omega) H(\omega + n\omega_s) \left[R_d(0) + 2 \sum_{k=1}^{\infty} R_d(k) \cos(k\omega T_s) \right] d\omega \qquad (2.11\text{-}6d)$$

where we define

$$C(\omega) = \left[R_d(0) + 2 \sum_{k=1}^{\infty} R_d(k) \cos(k\omega T_s) \right] \qquad (2.11\text{-}7)$$

which is the shaping that the encoding of the data imparts to the power spectral density of the data signal $x(t)$. It follows that:

$$f_n = \frac{1}{2\pi T_s} \int_{-\infty}^{\infty} H(-\omega) H(\omega + n\omega_s) C(\omega) \, d\omega \qquad (2.11\text{-}8a)$$

and since $C(\omega)$ is an even function

$$f_n = \frac{1}{2\pi T_s} \int_{-\infty}^{\infty} H(\omega) H(n\omega_s - \omega) C(\omega) \, d\omega \qquad (2.11\text{-}8b)$$

Finally, since we are concerned only with the fundamental, we have

$$f_1 = \frac{1}{2\pi T_s} \int_{-\infty}^{\infty} H(\omega)H(\omega_s - \omega)C(\omega)\,d\omega \tag{2.11-9}$$

Since the narrow band filter will have negligible phase shift at its center frequency, the filter output will not only be of the appropriate frequency, but also of the proper phase for sampling the received signal.

Consequently, the recovered timing signal is

$$E[y(t)]_{\text{filt}} = 2c_1 \operatorname{Cos} \omega_s t \tag{2.11-10}$$

EXAMPLE 8

a) Random Binary With Rectangular Pulses. Let $h(t)$ be a rectangular pulse having duration T_s so that

$$H(\omega) = T_s \operatorname{Sinc}(\omega T_s/2)$$

Logic tells us that if the input to a square-law device is identical positive and negative rectangular pulses, then the output will be constant and there will be no timing information. Let's show that this is the case. Since the data is not encoded, we can say that $C(\omega) = 1$ and

$$f_1 = \frac{1}{2\pi T_s} \int_{-\infty}^{\infty} T_s \operatorname{Sinc}(\omega T_s/2) T_s \operatorname{Sinc}((\omega - \omega_s)T_s/2)\,d\omega.$$

According to Parseval's Theorem,

$$f_1 = \frac{1}{T_s} \int_{-T_s/2}^{T_s/2} e^{-j\omega_s t}\,dt = 0$$

b) AMI with Rectangular Pulses. Although the pulses in AMI are the same rectangular pulses as those in random binary, we have already seen that there is correlation between sucessive symbols so that

$$C(\omega) = \frac{1}{2}(1 - \operatorname{Cos} \omega T_s)$$

and

$$c_1 = \frac{1}{2\pi T_s} \int_{-\infty}^{\infty} T_s \operatorname{Sinc}(\omega T_s/2) T_s \operatorname{Sinc}((\omega - \omega_s)T_s/2)\frac{1}{2}(1 - \operatorname{Cos} \omega T_s)\,d\omega.$$

This integral, while not easy to evaluate, is not equal to zero. Consequently, a clock may be extracted through square-law timing recovery. ∎

Evaluation of the recovered timing signal for additional pulse shapes is referred to the Problems.

Having found the expected value of the timing signal, which is our desired output, we must also recognize that that this output is a statistical average. The same output also has a variance, called the timing jitter, that has the effect of causing the phase of the timing signal to change rapidly in what appears to be a random manner. Although the analysis of this phenomenon is somewhat arduous, it is worth doing it both for moral reasons and for the insight that it will give us. This discussion follows Franks and Bubrouski [Fra74].

Recalling eq. 2.11-2, we want to find

$$\text{var}[y(t)] = E[y^2(t)] - E^2[y(t)] \tag{2.11-11}$$

In order to make this task reasonable, we will assume that the data is uncorrelated. Subsequently, we shall make the leap of faith that the effect of coding the data can be inserted by making the substitution

$$H(\omega)H(n\omega_s - \omega) \Rightarrow H(\omega)H(n\omega_s - \omega)C(\omega) \tag{2.11-12}$$

Some preliminary notation will be helpful. Since the data are zero-mean, uncorrelated, and stationary

$$E[d_n] = 0; \quad E[d_n^2] = \sigma^2; \quad E[d_n^4] = \beta^4 \tag{2.11-13a}$$

and, in the simplest case where $d_n = \pm 1$. $E[d_n^2] = E[d_n^4] = 1$. In forming $y^2(t)$, we will also make use of the fact that:

$$E[d_n d_m] = \begin{cases} \sigma^2, & \text{if } m = n \\ 0, & \text{if } m \neq n \end{cases} \tag{2.11-13b}$$

and the more complicated

$$E[d_k d_{k+m} d_{k+j} d_{k+j+i}] = \begin{cases} E[d_n^4] = \beta^4; & \text{if } m = j = i = 0 \\ E[d_n^2 d_m^2] = \sigma^4 \end{cases} \tag{2.11-13c}$$

where the second possibility can take place when

1. $\qquad\qquad\qquad m = i = 0; \quad j \neq 0$

2. $\qquad\qquad\qquad m = j \neq 0; \quad i = -j$ $\qquad$ (2.11-13d)

3. $\qquad\qquad\qquad m = i \neq 0; \quad j = 0$

Letting

$$p_n(t) = h(t)h(t - nT_s) \tag{2.11-14a}$$

we have

$$E[y^2(t)] = \sum_k \sum_m \sum_j \sum_i E[d_k d_{k+m} d_{k+j} d_{k+j+i}] p_m(t - kT_s) p_i(t - (k+j)T_s). \tag{2.11-14b}$$

The four circumstances that give non-zero results are, in order:

1. $m = j = i = 0$

$$\Rightarrow \sum_k \beta^4 p_0(t - kT_s) p_0(t - kT_s) = \sum_k \beta^4 p_0^2(t - kT_s) \tag{2.11-14c}$$

2. $m = i = 0; \quad j \neq 0$

$$\Rightarrow \sum_k \sum_{j \neq 0} E[d_k d_k d_{k+j} d_{k+j}] p_m(t - kT_s) p_i(t - (k+j)T_s)$$

$$= \sigma^4 \left\{ \sum_k \sum_j p_0(t - kT_s) p_0(t - (k+j)T_s) - \sum_k p_0(t - kT_s) p_0(t - (k+j)T_s) \right\}$$

$$= \sigma^4 \left\{ \sum_k p_0(t - kT_s) \right\}^2 - \sigma^4 \sum_k p_0^2(t - kT_s) \tag{2.11-14d}$$

where we have added and subtracted the term corresponding to $j = 0$.

3. $m = j \neq 0;$ $\quad i = -j$

$$\Rightarrow \sum_k \sum_{m \neq 0} E[d_k d_{k+m} d_{k+m} d_k] p_m(t - kT_s) p_{-m}(t - (k+m)T_s)$$

$$= \sigma^4 \left\{ \sum_k \sum_m p_m(t - kT_s) p_{-m}(t - (k+m)T_s) - \sum_k p_0^2(t - kT_s) \right\} \quad (2.11\text{-}14e)$$

4. $m = i \neq 0;$ $\quad j = 0$

$$\Rightarrow \sum_k \sum_{m \neq 0} E[d_k d_{k+m} d_k d_{k+m}] p_m(t - kT_s) p_m(t - (k+m)T_s)$$

$$= \sigma^4 \left\{ \sum_k \sum_m p_m(t - kT_s) p_m(t - (k+m)T_s) - \sum_k p_0^2(t - kT_s) \right\} \quad (2.11\text{-}14f)$$

Since $p_n(t) = h(t)h(t - nT_s) = p_{-n}(t)$, it follows that the third and fourth terms are equal. Adding all of these terms together, we get:

$$E[y^2(t)] = \sigma^4 \left\{ \sum_k p_0(t - kT_s) \right\}^2$$

$$+ (\beta^4 - 3\sigma^4) \sum_k p_0^2(t - kT_s) + 2\sigma^4 \left\{ \sum_k \sum_m p_m^2(t - kT_s) \right\} \quad (2.11\text{-}15)$$

According to eq. 2.11-3, and with uncorrelated data

$$E[y(t)] = \sum_{n=-\infty}^{\infty} R_a(0)h^2(t - nT_s) = \sigma^2 \sum_k p_0(t - kT_s) \quad (2.11\text{-}16a)$$

so that

$$\text{var}[y(t)] = E[y^2(t)] - E^2[y(t)]$$

$$= (\beta^4 - 3\sigma^4) \sum_k p_0^2(t - kT_s) + 2\sigma^4 \sum_k \sum_m p_m^2(t - kT_s)$$

$$= \sum_k \left\{ (\beta^4 - 3\sigma^4) p_0^2(t - kT_s) + 2\sigma^4 \sum_m p_m^2(t - kT_s) \right\} \quad (2.11\text{-}16b)$$

Recalling that

$$p_n(t) = h(t)h(t - nT_s) \quad (2.11\text{-}14a)$$

it is worth noting that for non-overlapping $h(t)$ pulses that $p_n(t) = 0$ for $n \neq 0$.

Although eq. 2.11-16b gives us the actual shape of the timing jitter waveform, the information is not in a form that is directly useful to us because it does not easily tell us what the sampling deviation caused by jitter actually is.

To continue the analysis, note that $\text{var}[y(t)]$ is periodic with period T_s so that it may be expressed as a Fourier Series

$$\text{var}[y(t)] = \sum_{n=-\infty}^{\infty} f_n e^{jn\omega_s t} \qquad \text{where } \omega_s = 2\pi / T_s \quad (2.11\text{-}17a)$$

where

$$f_n = \frac{1}{T_s} \int_{-T_s/2}^{T_s/2} \text{var}[y(t)] e^{-jn\omega_s t} \, dt$$

$$= \frac{1}{T_s} \int_{-\infty}^{\infty} \left\{ (\beta^4 - 3\sigma^4) p_0^2(t) + 2\sigma^4 \sum_m p_m^2(t) \right\} e^{-jn\omega_s t} \, dt$$

$$= \frac{1}{T_s} \left\{ (\beta^4 - 3\sigma^4) \int_{-\infty}^{\infty} p_0^2(t) e^{-jn\omega_s t} \, dt + 2\sigma^4 \sum_m \int_{-\infty}^{\infty} p_m^2(t) e^{-jn\omega_s t} \, dt \right\} \qquad (2.11\text{-}17b)$$

Although this formulation of the problem also lacks immediate appeal, the advantage of computing the Fourier series coefficents is that we can determine from them the effective amplitude of the jitter.

At least in principle, we can take the following steps in computing the Fourier coefficients.

1. Using frequency domain convolution:

$$P_m(\omega) = \frac{1}{2\pi} \int_{-\infty}^{\infty} H(u) H(\omega - u) e^{-jm(\omega - u)T_s} \, du$$

$$= \left\{ \frac{1}{2\pi} \int_{-\infty}^{\infty} H(u) H(\omega - u) e^{jmuT_s} \, du \right\} e^{-jm\omega T_s} \qquad (2.11\text{-}18a)$$

where, if the signal is encoded, we may make the substitution

$$H(\omega) \to C^{1/2}(\omega) H(\omega) \qquad (2.11\text{-}18b)$$

2. Then, using Parseval's Theorem in (2.11-15b),

$$f_n = \frac{1}{2\pi T_s} (\beta^4 - 3\sigma^4) \int_{-\infty}^{\infty} P_0(-\omega) P_0(\omega + n\omega_s) e^{jn\omega T_s} \, d\omega$$

$$+ \frac{1}{2\pi T_s} 2\sigma^4 \sum_m \int_{-\infty}^{\infty} P_m(-\omega) P_m(\omega + n\omega_s) e^{jn\omega T_s} \, d\omega \qquad (2.11\text{-}18c)$$

Clearly, computing the jitter is a major undertaking, one which is rarely done. But eq. 2.11-16b can also give us a qualitative understanding of the sources of jitter. The major source of jitter is the overlapping in time of the the the basic pulses $h(t)$. A second source of jitter arises if the signal is multilevel instead of binary, i.e. if $a_n = \pm 1, \pm 3$. If the signal is binary, then $\beta^4 = \sigma^4$ and eqs. 2.11-15b and 2.11-18c simplify considerably. Unfortunately, most of the systems that we will subsequently discuss have are multilevel and have overlapping pulses in order to combat intersymbol interference.

In Chapter 6 we will deal with a number of additional methods to accomplish this task and in the course of doing this we will continue the discussion of timing jitter.

The accumulation of timing jitter when a data signal is passed through a series of repeaters (digital amplifiers) will considerably deteriorate the transmission. This will be discussed in Chapter 3.

2.12   FORWARD-ACTING CARRIER RECOVERY

Double-Sideband, Suppressed Carrier Amplitude Modulation (DSB-SC) was briefly examined in Chapter 1. It is a major form of modulation for data transmission on a wide variety of bandpass channels. We saw that a simple double-sideband, suppressed carrier signal may be expressed as

$$x(t) = a(t)\operatorname{Cos}(\omega_c t + \theta) \qquad (2.12\text{-}1)$$

where $a(t)$ is a data signal. The basic receiver structure is shown in fig. 2.12-1a and it is clear that a major problem is generating a demodulating carrier at the receiver having correct frequency and phase. This problem will be discussed in some generality in Chapter 6 but the approaches developed in this chapter are easily extended to the simple technique of Forward-Acting Carrier Recovery [Fra80].

As shown in fig. 2.12-1b, we square the received signal of eq. 2.12-1, resulting in $y(t)$

$$y(t) = x^2(t) = a^2(t)\operatorname{Cos}^2(\omega_c t + \theta)$$

$$= \frac{a^2(t)}{2}[1 + \operatorname{Cos}(2\omega_c t + 2\theta)] \qquad (2.12\text{-}2)$$

which is then applied to a bandpass filter, eliminating its dc component, and then hard-limited. The resulting square wave is the input to a divide-by-two counter, from which two square-wave phases, 'cos' and 'sin', of the original frequency ω_c may be extracted. The chosen phase may then be filtered to give the demodulating carrier.

The fundamental difficulty with this approach is the effect of $a(t)$, the modulating data signal, on the instantaneous frequency and phase of the derived carrier. In simple binary transmission where each data symbol $d_k = \pm 1$ is multiplied by a pulse $h(t)$,

$$a(t)\sum_n d_n h(t - nT_s) \qquad (2.12\text{-}3a)$$

and

$$a^2(t) = \sum_n \sum_m d_n d_m h(t - nT_s)h(t - mT_s) \qquad (2.12\text{-}3b)$$

then $a^2(t)$ is clearly data dependent. If the pulses $h(t)$ are nonoverlapping, limited to the time interval $[0, T_s]$, then $a^2(t)$ will reduce to the single condition of $n = m$, so that

$$a^2(t) = \sum_n h^2(t - nT_s) \quad \text{nonoverlapping pulses} \qquad (2.12\text{-}3c)$$

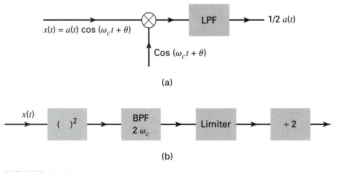

(a)

(b)

FIGURE 2.12-1

which exhibits no data dependency. Under these circumstances it is possible to extract a clean carrier from the bandpass filter.

As we shall see in the next chapter, signalling on bandlimited channels requires data pulses $h(t)$ that do overlap. Under these circumstances the instantaneous value of $a^2(t)$ is very much data-dependent. Gardner [Gar80] has shown that as a consequence of this data dependency, passing

$$y(t) = x^2(t) = a^2(t) \cos^2(\omega_c t + \theta)$$

$$= \frac{1}{2} \sum_n \sum_m d_n d_m h(t - nT_s) h(t - mT_s)[1 + \cos(2\omega_c t + 2\theta)] \quad (2.12\text{-}4)$$

through a bandpass filter generates randomness in the carrier phase of the output signal at $2\omega_c$. Note that this randomness comes from the randomness of the data rather than from any noise accompanying the received signal.

Even with non-overlapping pulses, $a^2(t)$ will have a data dependency if the data is multilevel (e.g., $-d_n = \pm 1, \pm 3$) because the value of $d_n d_m$ will be changing with time.

Forward-acting carrier recovery can be used to reasonable effect in bandpass signals where the information is carried entirely in the phase. A QPSK (quadrature-phase-shift keying) signal can be written as

$$y(t) = \sum_n h(t - nT_s) \cos\left(\omega_c t + d_n \frac{\pi}{2} + \theta\right) \quad (2.12\text{-}5a)$$

where

$$P[d_n = 0] = P[d_n = 1] = P[d_n = 2] = P[d_n = 3] = \frac{1}{4} \quad (2.12\text{-}5b)$$

Raising this signal to the fourth power and filtering the result to recover the component around $4\omega_c$ similarly yields a signal having a jittery phase as a consequence of the randomness of the data. Again, this signal may be hard-limited, divided by 4, and filtered to obtain a demodulating carrier, but it must be emphasized that the result has data-derived phase noise that tends to degrade the performance of the receiver.

This basic limitation of forward-acting carrier recovery will be addressed in Chapter 6.

APPENDIX 2A

TABULATION OF
$$Q(x) = \int_x^\infty \frac{1}{\sqrt{2\pi}} e^{-\alpha^2/2}\, d\alpha$$

From *Probability & Stochastic Processes* by Roy D. Yates and David J. Goodman, John Wiley & Sons, Inc., 1999.

x	Q(x)	x	Q(x)	x	Q(x)	x	Q(x)	x	Q(x)
3.00	$1.35 \cdot 10^{-3}$	3.40	$3.37 \cdot 10^{-4}$	3.80	$7.23 \cdot 10^{-5}$	4.20	$1.33 \cdot 10^{-5}$	4.60	$2.11 \cdot 10^{-6}$
3.01	$1.31 \cdot 10^{-3}$	3.41	$3.25 \cdot 10^{-4}$	3.81	$6.95 \cdot 10^{-5}$	4.21	$1.28 \cdot 10^{-5}$	4.61	$2.01 \cdot 10^{-6}$
3.02	$1.26 \cdot 10^{-3}$	3.42	$3.13 \cdot 10^{-4}$	3.82	$6.67 \cdot 10^{-5}$	4.22	$1.22 \cdot 10^{-5}$	4.62	$1.92 \cdot 10^{-6}$
3.03	$1.22 \cdot 10^{-3}$	3.43	$3.02 \cdot 10^{-4}$	3.83	$6.41 \cdot 10^{-5}$	4.23	$1.17 \cdot 10^{-5}$	4.63	$1.83 \cdot 10^{-6}$
3.04	$1.18 \cdot 10^{-3}$	3.44	$2.91 \cdot 10^{-4}$	3.84	$6.15 \cdot 10^{-5}$	4.24	$1.12 \cdot 10^{-5}$	4.64	$1.74 \cdot 10^{-6}$
3.05	$1.14 \cdot 10^{-3}$	3.45	$2.80 \cdot 10^{-4}$	3.85	$5.91 \cdot 10^{-5}$	4.25	$1.07 \cdot 10^{-5}$	4.65	$1.66 \cdot 10^{-6}$
3.06	$1.11 \cdot 10^{-3}$	3.46	$2.70 \cdot 10^{-4}$	3.86	$5.67 \cdot 10^{-5}$	4.26	$1.02 \cdot 10^{-5}$	4.66	$1.58 \cdot 10^{-6}$
3.07	$1.07 \cdot 10^{-3}$	3.47	$2.60 \cdot 10^{-4}$	3.87	$5.44 \cdot 10^{-5}$	4.27	$9.77 \cdot 10^{-6}$	4.67	$1.51 \cdot 10^{-6}$
3.08	$1.04 \cdot 10^{-3}$	3.48	$2.51 \cdot 10^{-4}$	3.88	$5.22 \cdot 10^{-5}$	4.28	$9.34 \cdot 10^{-6}$	4.68	$1.43 \cdot 10^{-6}$
3.09	$1.00 \cdot 10^{-3}$	3.49	$2.42 \cdot 10^{-4}$	3.89	$5.01 \cdot 10^{-5}$	4.29	$8.93 \cdot 10^{-6}$	4.69	$1.37 \cdot 10^{-6}$
3.10	$9.68 \cdot 10^{-4}$	3.50	$2.33 \cdot 10^{-4}$	3.90	$4.81 \cdot 10^{-5}$	4.30	$8.54 \cdot 10^{-6}$	4.70	$1.30 \cdot 10^{-6}$
3.11	$9.35 \cdot 10^{-4}$	3.51	$2.24 \cdot 10^{-4}$	3.91	$4.61 \cdot 10^{-5}$	4.31	$8.16 \cdot 10^{-6}$	4.71	$1.24 \cdot 10^{-6}$
3.12	$9.04 \cdot 10^{-4}$	3.52	$2.16 \cdot 10^{-4}$	3.92	$4.43 \cdot 10^{-5}$	4.32	$7.80 \cdot 10^{-6}$	4.72	$1.18 \cdot 10^{-6}$
3.13	$8.74 \cdot 10^{-4}$	3.53	$2.08 \cdot 10^{-4}$	3.93	$4.25 \cdot 10^{-5}$	4.33	$7.46 \cdot 10^{-6}$	4.73	$1.12 \cdot 10^{-6}$
3.14	$8.45 \cdot 10^{-4}$	3.54	$2.00 \cdot 10^{-4}$	3.94	$4.07 \cdot 10^{-5}$	4.34	$7.12 \cdot 10^{-6}$	4.74	$1.07 \cdot 10^{-6}$
3.15	$8.16 \cdot 10^{-4}$	3.55	$1.93 \cdot 10^{-4}$	3.95	$3.91 \cdot 10^{-5}$	4.35	$6.81 \cdot 10^{-6}$	4.75	$1.02 \cdot 10^{-6}$
3.16	$7.89 \cdot 10^{-4}$	3.56	$1.85 \cdot 10^{-4}$	3.96	$3.75 \cdot 10^{-5}$	4.36	$6.50 \cdot 10^{-6}$	4.76	$9.68 \cdot 10^{-7}$
3.17	$7.62 \cdot 10^{-4}$	3.57	$1.78 \cdot 10^{-4}$	3.97	$3.59 \cdot 10^{-5}$	4.37	$6.21 \cdot 10^{-6}$	4.77	$9.21 \cdot 10^{-7}$
3.18	$7.36 \cdot 10^{-4}$	3.58	$1.72 \cdot 10^{-4}$	3.98	$3.45 \cdot 10^{-5}$	4.38	$5.93 \cdot 10^{-6}$	4.78	$8.76 \cdot 10^{-7}$
3.19	$7.11 \cdot 10^{-4}$	3.59	$1.65 \cdot 10^{-4}$	3.99	$3.30 \cdot 10^{-5}$	4.39	$5.67 \cdot 10^{-6}$	4.79	$8.34 \cdot 10^{-7}$
3.20	$6.87 \cdot 10^{-4}$	3.60	$1.59 \cdot 10^{-4}$	4.00	$3.17 \cdot 10^{-5}$	4.40	$5.41 \cdot 10^{-6}$	4.80	$7.93 \cdot 10^{-7}$
3.21	$6.64 \cdot 10^{-4}$	3.61	$1.53 \cdot 10^{-4}$	4.01	$3.04 \cdot 10^{-5}$	4.41	$5.17 \cdot 10^{-6}$	4.81	$7.55 \cdot 10^{-7}$
3.22	$6.41 \cdot 10^{-4}$	3.62	$1.47 \cdot 10^{-4}$	4.02	$2.91 \cdot 10^{-5}$	4.42	$4.94 \cdot 10^{-6}$	4.82	$7.18 \cdot 10^{-7}$
3.23	$6.19 \cdot 10^{-4}$	3.63	$1.42 \cdot 10^{-4}$	4.03	$2.79 \cdot 10^{-5}$	4.43	$4.71 \cdot 10^{-6}$	4.83	$6.83 \cdot 10^{-7}$
3.24	$5.98 \cdot 10^{-4}$	3.64	$1.36 \cdot 10^{-4}$	4.04	$2.67 \cdot 10^{-5}$	4.44	$4.50 \cdot 10^{-6}$	4.84	$6.49 \cdot 10^{-7}$
3.25	$5.77 \cdot 10^{-4}$	3.65	$1.31 \cdot 10^{-4}$	4.05	$2.56 \cdot 10^{-5}$	4.45	$4.29 \cdot 10^{-6}$	4.85	$6.17 \cdot 10^{-7}$
3.26	$5.57 \cdot 10^{-4}$	3.66	$1.26 \cdot 10^{-4}$	4.06	$2.45 \cdot 10^{-5}$	4.46	$4.10 \cdot 10^{-6}$	4.86	$5.87 \cdot 10^{-7}$
3.27	$5.38 \cdot 10^{-4}$	3.67	$1.21 \cdot 10^{-4}$	4.07	$2.35 \cdot 10^{-5}$	4.47	$3.91 \cdot 10^{-6}$	4.87	$5.58 \cdot 10^{-7}$
3.28	$5.19 \cdot 10^{-4}$	3.68	$1.17 \cdot 10^{-4}$	4.08	$2.25 \cdot 10^{-5}$	4.48	$3.73 \cdot 10^{-6}$	4.88	$5.30 \cdot 10^{-7}$
3.29	$5.01 \cdot 10^{-4}$	3.69	$1.12 \cdot 10^{-4}$	4.09	$2.16 \cdot 10^{-5}$	4.49	$3.56 \cdot 10^{-6}$	4.89	$5.04 \cdot 10^{-7}$
3.30	$4.83 \cdot 10^{-4}$	3.70	$1.08 \cdot 10^{-4}$	4.10	$2.07 \cdot 10^{-5}$	4.50	$3.40 \cdot 10^{-6}$	4.90	$4.79 \cdot 10^{-7}$
3.31	$4.66 \cdot 10^{-4}$	3.71	$1.04 \cdot 10^{-4}$	4.11	$1.98 \cdot 10^{-5}$	4.51	$3.24 \cdot 10^{-6}$	4.91	$4.55 \cdot 10^{-7}$
3.32	$4.50 \cdot 10^{-4}$	3.72	$9.96 \cdot 10^{-5}$	4.12	$1.89 \cdot 10^{-5}$	4.52	$3.09 \cdot 10^{-6}$	4.92	$4.33 \cdot 10^{-7}$
3.33	$4.34 \cdot 10^{-4}$	3.73	$9.57 \cdot 10^{-5}$	4.13	$1.81 \cdot 10^{-5}$	4.53	$2.95 \cdot 10^{-6}$	4.93	$4.11 \cdot 10^{-7}$
3.34	$4.19 \cdot 10^{-4}$	3.74	$9.20 \cdot 10^{-5}$	4.14	$1.74 \cdot 10^{-5}$	4.54	$2.81 \cdot 10^{-6}$	4.94	$3.91 \cdot 10^{-7}$
3.35	$4.04 \cdot 10^{-4}$	3.75	$8.84 \cdot 10^{-5}$	4.15	$1.66 \cdot 10^{-5}$	4.55	$2.68 \cdot 10^{-6}$	4.95	$3.71 \cdot 10^{-7}$
3.36	$3.90 \cdot 10^{-4}$	3.76	$8.50 \cdot 10^{-5}$	4.16	$1.59 \cdot 10^{-5}$	4.56	$2.56 \cdot 10^{-6}$	4.96	$3.52 \cdot 10^{-7}$
3.37	$3.76 \cdot 10^{-4}$	3.77	$8.16 \cdot 10^{-5}$	4.17	$1.52 \cdot 10^{-5}$	4.57	$2.44 \cdot 10^{-6}$	4.97	$3.35 \cdot 10^{-7}$
3.38	$3.62 \cdot 10^{-4}$	3.78	$7.84 \cdot 10^{-5}$	4.18	$1.46 \cdot 10^{-5}$	4.58	$2.32 \cdot 10^{-6}$	4.98	$3.18 \cdot 10^{-7}$
3.39	$3.49 \cdot 10^{-4}$	3.79	$7.53 \cdot 10^{-5}$	4.19	$1.39 \cdot 10^{-5}$	4.59	$2.22 \cdot 10^{-6}$	4.99	$3.02 \cdot 10^{-7}$

CHAPTER 2 PROBLEMS

2.1 Compute the six remaining conditional probabilities for the example in Sec. 2.1B.

2.2 For the random variable of eq. 2.1-8, find $E[x^2 + 3x + 7]$.

2.3 (a) Find the correlation and the covariance of the two random variables X_1 and X_2 in eq. 2.2-1.
(b) Find the correlation matrix of the three random variables of eq. 2.2-1.

2.4 Using the data of Appendix 2A, find the values of x that make $Q(x) = 10^{-4}, 10^{-5}, 10^{-6}$.

2.5 The covariance matrix of the zero-mean jointly Gaussian random variables x_1 and x_2 is

$$C_x = \begin{bmatrix} 2 & 1 \\ 1 & 4 \end{bmatrix}$$

The random variables y_1, y_2 are defined by the matrix equation

$$\begin{bmatrix} y_1 \\ y_2 \end{bmatrix} = \begin{bmatrix} 3 & -2 \\ 2 & 5 \end{bmatrix} \begin{bmatrix} x_1 \\ x_2 \end{bmatrix}$$

Find the covariance matrix C_Y.

2.6 The covariance matrix of two zero-mean gaussian random variables X_1 and X_2 is

$$C_x = \begin{bmatrix} \sigma_1^2 & \rho_{12}\sigma_1\sigma_2 \\ \rho_{12}\sigma_1\sigma_2 & \sigma_2^2 \end{bmatrix}$$

(a) Find the variance of $X_3 = \alpha X_1 + \beta X_2$.
(b) If the random variables X_1 and X_2 are linearly transformed by the matrix

$$A = \begin{bmatrix} \text{Cos}\,\theta & \text{Sin}\,\theta \\ -\text{Sin}\,\theta & \text{Cos}\,\theta \end{bmatrix}$$

then show that if

$$\theta = \frac{1}{2} \tan^{-1} \left[\frac{2\rho_{12}\sigma_1\sigma_2}{\sigma_1^2 - \sigma_2^2} \right]$$

then the resulting covariance matrix is diagonal.

2.7 The autocorrelation function of the encoded data in fig. 2.10 is
(a) $R_b(n) = (1/2)^{|n|} \quad |n| \le 3$
(b) $R_b(n) = (1/2)^{|n|} \quad$ for all n
How does this encoding affect the power spectral density of the signal?

2.8 An encoding device has the form of a four-state machine

State/output	Input $= 0$	Input $= 1$
$q_0/+1$	q_1	q_2
$q_1/-1$	q_2	q_3
$q_2/+1$	q_1	q_0
$q_3/-1$	q_2	q_1

How does this encoding change the power spectral density of the data?

2.9 The autocorrelation function of a zero-mean Gaussian random process is given below:

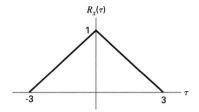

(a) Find the covariance matrix for the jointly Gaussian random variables $x(1)$, $x(3)$, $x(6)$.

(b) Repeat **(a)** for $x(1)$, $x(-2)$, $x(0)$.

2.10 A wide-sense stationary Gaussian random process $x(t)$ having autocorrelation function $\overline{R_x(\tau)}$ is passed through a square-law device. If $y(t) = x^2(t)$, find $\overline{R_y(\tau)}$.

2.11 Referring to the jointly Gaussian density function of eq. 2.3-4.

(a) Show (with $m_1 = m_2 = 0$) by completing the square in the exponent that

$$\int_{-\infty}^{\infty} p_{x_1 x_2}(\alpha_1, \alpha_2)\, d\alpha_2 = p_{x_1}(\alpha_1)$$

is Gaussian.

(b) Show that the conditional probability density function

$$p_{x_1/x_2}(\alpha_1/x_2 = \alpha_2) = \frac{p_{x_1 x_2}(\alpha_1, \alpha_2)}{p_{x_2}(\alpha_2)}$$

is Gaussian. Find the mean and variance.

2.12 A random process $x(t)$ having autocorrelation function $\overline{R_x(\tau)}$ is applied to a transversal filter: Find $\overline{R_y(\tau)}$.

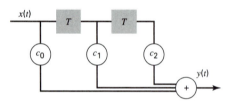

2.13 Show that if two random processes have no spectral overlap, they are uncorrelated.

2.14 **(a)** For the system of fig. 2.9-3b, find numerical values for the covariance matrix for the random variables

$$y_1(t - 1),\ y_1(t + 2),\ y_2(t + 1),\ y_2(t + 4)$$

(b) Repeat part **(a)** if $\overline{R_x(\tau)} = 2e^{-2|\tau|}$.

2.15 [Lat68]

Let $x_1(t)$ and $x_2(t)$ be two independent random processes having autocorrelation functions $\overline{R_{x_1}(\tau)}$ and $\overline{R_{x_2}(\tau)}$. Two new random processes $y_1(t)$ and $y_2(t)$ are defined by

$$\begin{bmatrix} y_1(t) \\ y_2(t) \end{bmatrix} = \begin{bmatrix} 1 & 1 \\ 2 & 3 \end{bmatrix} \begin{bmatrix} x_1(t) \\ x_2(t) \end{bmatrix}$$

Find $\overline{R_{y_1}(\tau)}$, $\overline{R_{y_2}(\tau)}$, $\overline{R_{y_1 y_2}(\tau)}$, $\overline{R_{y_2 y_1}(\tau)}$ in terms of $\overline{R_{x_1}(\tau)}$ and $\overline{R_{x_2}(\tau)}$.

2.16 Consider the Random Telegraph Signal of fig. 2.5-3. Find the autocorrelation funtion and power spectral density if the probability of a state transition is 0.4 instead of 0.5.

2.17 [Lat68]

The random square wave shown below has equally probable values $\pm A$. Transitions between these values are independent and random (therefore Poisson distributed). Find the autocorrelation function and power spectral density of this process.

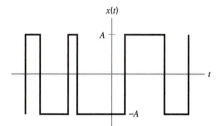

2.18 [Lat68]

The rectangular pulses shown below alternate sign every T sec and have height equal to the random variable x, having a probability density function

$$p_x(x) = \frac{1}{2}|x|e^{-|x|}$$

The heights of the pulses are independent of all past history. Find the autocorrelation function and the power spectral density.

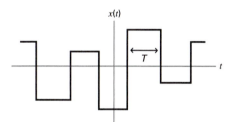

REFERENCES

[Aar62] M.R. Aaron, "PCM in the exchange plant," *BSTJ*, **41**, Nov. 1962, pp. 99–141.

[Ben58] W.R. Bennett, "Statistical properties of regenerative digital transmission," *Bell System Technical J.*, **37**, Nov. 1958, pp. 1501–1542.

[Fra69] L. Franks, *Signal Theory*, Prentice-Hall, 1969, Chapter 8.

[Fra74] L. Franks and J.P. Bubrouski, "Statistical properties of timing jitter in a PAM timing recovery scheme," *IEEE Trans. Communications*, **Com-22**, No. 7, July 1974, pp. 913–920.

[Fra80] L.E. Franks, "Carrier and bit synchronization in data communication—a tutorial review," *IEEE Trans. Communications*, **Comm-28**, No. 8, August 1980, pp. 1107–1121.

[Gar80] F.M. Gardner, "Self-noise in synchronizers," *IEEE Trans. Communications*, **Comm-28**, August 1980, pp. 1159–1163.

[Lat68] B.P. Lathi, *An Introduction to random Signals and Communication Theory*, International, 1968.

[Lee94] E. Lee and D. Messerschmitt, *Digital Communication*, 2nd ed., KAP, 1994.

[Pee93] P.Z. Peebles Jr., *Probability, Random Variables, and Random Signal Principles*, 3rd ed., McGraw-Hill, 1993.

BASEBAND DATA TRANSMISSION

This chapter develops the basic model of a baseband digital communications system. This model will provide the framework for our discussions throughout the text. We develop the ideas of probabilty of error, matched filter transmission, the correlation receiver, and intersymbol interference. In doing this we will use the basic results of Chapter 1 and Chapter 2.

3.1 TRANSMITTING AND RECEIVING A DATA SIGNAL

As shown in fig. 3.1-1, the basic digital transmission system consists of a data source, a transmitter filter (XMIT FILT), a channel, a noise source, a receiver filter (RCVR FILT), a sampler, and a decision device.

A bit of data is represented mathematically by an impulse function having area $\pm\sqrt{E_s}$ corresponding to a digital 1 or 0. Successive pieces of data are, in this model, independent and therefore uncorrelated. Data impulses occur regularly at intervals equal to T_s, the *symbol rate* or *baud rate*. In this introductory discussion each symbol or impulse contains one bit of data, but as the model is expanded each symbol may contain multiple bits of data. For example, if each impulse function may assume the values $\pm\sqrt{E_s}, \pm3\sqrt{E_s}$, then there are four possibilities or two bits associated with each impulse or symbol. For this reason it is important to think about the impulse function representing the data as being a symbol rather than a bit.

A binary data stream is represented by

$$d(t) = \sum_{n=-\infty}^{\infty} d_n \delta(t - nT_s) \qquad (3.1\text{-}1a)$$

where each data symbol has a given probabilty. We shall define

$$\begin{aligned} P_1 &= P(d_n = +\sqrt{E_s}) \\ P_0 &= P(d_n = -\sqrt{E_s}) = 1 - P_1 \end{aligned} \qquad (3.1\text{-}1b)$$

where $P_1 + P_0 = 1$. It will generally be the case that $P_1 = P_0 = 1/2$.

Even though the data is digital, each piece of data is transmitted through a channel as an analog waveform. The purpose of the XMIT FILT is to convert the piece of digital data, represented by the impulse function, into this waveform. Consequently, the XMIT FILT is characterized by its unit impulse response $s(t) \Leftrightarrow S(\omega)$. In fig. 3.1-1, $s(t)$ is shown as being a rectangular pulse. This is only for purposes of illustration and, as we develop this discussion, we shall be using other waveforms for $s(t)$ as well.

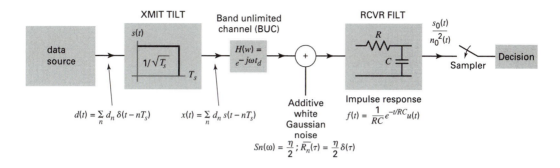

FIGURE 3.1-1

There are two parts to the representation of $s(t)$ that are important to note. The first is that the energy contained in $s(t)$ is always equal to unity so that, according to Parseval's Theorem,

$$\int\limits_{-\infty}^{\infty} s^2(t)\, dt = \frac{1}{2\pi} \int\limits_{-\infty}^{\infty} |S(\omega)|^2\, d\omega = 1 \tag{3.1-2}$$

Thus the response of XMIT FILT to a data impulse having area $\pm\sqrt{E_s}$ is the transmitted pulse $\pm\sqrt{E_s}s(t)$, which has energy E_s. The second point is that the pulse has duration T_s, which is the time between symbols. We will subsequently see that some of the most interesting pulse shapes are not time-limited in this fashion but, for the moment, it is important to stay with this restriction.

The transmitted waveform or analog signal is therefore

$$x(t) = \sum_{n=-\infty}^{\infty} d_n s(t - nT_s). \tag{3.1-3}$$

Although our primary concern in this text will be the transmission of data on band-limited channels, the channel of fig. 3.1-1 is shown as being all-pass or having an infinite bandwidth and a time delay t_d or linear phase shift $-\omega t_d$; i.e., the transfer function is $H(\omega) = e^{-j\omega t_d}$. The consequence is that the data pulses are transmitted through the channel without distortion. This is very important for setting the stage for subsequent developments, including incorporating into the model an ideal bandlimited channel, and then a bandlimited channel that also introduces amplitude and phase distortion into the signal.

Although we are transmitting an analog signal on the channel, the output of the receiver filter will be sampled in order to reconstruct the original digital signal. The purpose of the receiver filter is twofold. One required function of the RCVR FILT is to maximize the value of its output at the sampling instant (often called the *sampling epoch*). This requires a wideband filter. Another required function is to minimize the noise power at the filter output, which requires a narrowband filter. The problem is how to satisfy these two contradictory requirements in an optimum manner. As we shall see, the solution to the problem is the matched filter.

For the purposes of the present analysis, we will consider the signal output from the RCVR FILT for only a single data pulse. The reason for this is that for the present we wish to avoid the problems presented by the possible overlapping of output pulses. For the lowpass RC circuit and the rectangular data pulse of fig. 3.1-1, the output waveform $s_0(t)$

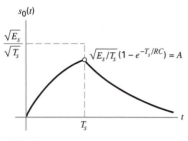

FIGURE 3.1-2

is shown in fig. 3.1-2. Clearly the peak value of $s_0(t)$ occurs at $t = T_s$ and

$$s_0(T_s) = \pm A = \pm\sqrt{\frac{E_s}{T_s}}\Big[1 - e^{-T_s/RC}\Big]. \tag{3.1-4}$$

In addition to this output signal, the RCVR FILT also has a noise output. Following the classic modeling of this system, additive white Gaussian noise (AWGN) having a two-sided power spectral density (psd) $S_n(\omega) = \eta/2$ watts/radian per sec and time-average autocorrelation function $\overline{R_n}(\tau) = \frac{\eta}{2}\delta(t)$ is added to the transmitted data at the input to the RCVR FILT.

The impulse response of the RCVR FILT is $f(t)$ and its transfer function is $F(\omega)$, a Fourier Transform pair $f(t) \Leftrightarrow F(\omega)$. According to Parseval's Theorem, the output noise power of the filter is

$$\overline{n_0^2(t)} = \frac{1}{2\pi}\frac{\eta}{2}\int_{-\infty}^{\infty} |F(\omega)|^2 \, d\omega. \tag{3.1-5a}$$

In the case of the lowpass RC circuit shown in fig. 3.1-1 where $F(\omega) = 1/(1+j\omega RC)$, it follows that the output noise power is

$$\overline{n_0^2(t)} = \frac{1}{2\pi}\frac{\eta}{2}\int_{-\infty}^{\infty} \frac{1}{1 + (\omega RC)^2} \, d\omega = \frac{1}{2\pi}\frac{\eta}{2}\frac{1}{RC}\int_{-\infty}^{\infty} \frac{dx}{1 + x^2}$$

$$= \frac{1}{2\pi}\frac{\eta}{2}\frac{1}{RC} \tan^{-1}(x)\Big|_{-\infty}^{\infty} = \frac{\eta}{2}\frac{1}{2RC} = \sigma_n^2. \tag{3.1-5b}$$

The output of the RCVR FILT is sampled in order to determine whether the transmitted data pulse was positive or negative. It should be clear that the best place to sample this waveform is at its peak, corresponding to $t = T_s$. If there were no noise, there would be no decision errors. But with the existence of noise added to the sampled signal, it is possible that even though a positive pulse was sent, the sampled value could be negative and vice versa. In this way, a decision error can be made. We are interested in the probability of such an error.

The key to approaching this problem is to recognize that the sample of the output noise of the RCVR FILT is a zero-mean Gaussian random variable having a variance σ_n^2, the output noise power of the filter, as shown in eq. 3.5b. This extremely important result was derived in Chapter 2.

3.2 COMPUTING THE PROBABILITY OF ERROR

We are now dealing with a signal sample that is $s_0(T_s) = \pm A$ plus a noise sample that is a zero-mean Gaussian random variable having variance σ_n^2. This noisy sample is denoted as

$$y = \pm A + n \tag{3.2-1}$$

where n is the noise sample.

As shown in fig. 3.2-1, this results in two different conditional probability density functions:

$$p_{Y/1}(y) = \frac{1}{\sigma_n\sqrt{2\pi}}e^{-(y-A)^2/2\sigma_n^2}; \quad \text{if a 1 was transmitted}$$

$$\tag{3.2-2}$$

$$p_{Y/0}(y) = \frac{1}{\sigma_n\sqrt{2\pi}}e^{-(y+A)^2/2\sigma_n^2}; \quad \text{if a 0 was transmitted}$$

We want to look at the value of y and decide whether the original piece of data was a 1 or a 0. This elementary decision-theory problem is shown in fig. 3.2-1. We choose a decision boundary a and if $y > a$, we decide that a 1 was sent; if $y < a$, we decide that a 0 was sent. Now we want to compute the probability of making an error. Using some elementary rules of probability theory,

$$
\begin{aligned}
P_e &= P[\text{error, '1' sent}] + P[\text{error, '0' sent}] \\
&= P_1 P[\text{error}/1] + P_0 P[\text{error}/0] \\
&= P_1 \int_{-\infty}^{a} p_{Y/1}(y)\,dy + P_0 \int_{a}^{\infty} p_{Y/0}(y)\,dy \\
&= \frac{1}{\sigma_n\sqrt{2\pi}} \left\{ P_1 \int_{-\infty}^{a} e^{-(y-A)^2/2\sigma_n^2}\,dy + P_0 \int_{a}^{\infty} e^{-(y+A)^2/2\sigma_n^2}\,dy \right\}
\end{aligned}
\tag{3.2-3}
$$

We now want to choose the decision boundary a in order to minimize the probability of error P_e. In order to do this, we will set the derivative of P_e with respect to a equal to zero:

$$\frac{dP_e}{da} = \frac{1}{\sigma_n\sqrt{2\pi}} \left\{ P_1 \exp\left(-(a-A)^2/2\sigma_n^2\right) - P_0 \exp\left(-(a+A)^2/2\sigma_n^2\right) \right\} = 0$$

$$\tag{3.2-4a}$$

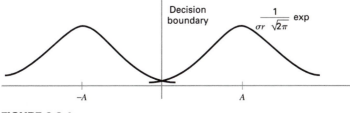

FIGURE 3.2-1

yielding:

$$\frac{P_1}{P_0} = \frac{\exp\left(-(a+A)^2\big/2\sigma_n^2\right)}{\exp\left(-(a-A)^2\big/2\sigma_n^2\right)} = \exp(-[(a+A)^2 - (a-A)^2])/2\sigma_n^2 = \exp\left(-2aA/\sigma_n^2\right)$$

(3.2-4b)

and taking the natural logarithm:

$$a = \frac{\sigma_n^2}{2A}\ln\left[\frac{P_0}{P_1}\right]$$

(3.2-4c)

If $P_1 = P_0 = 1/2$, then the decision boundary clearly becomes $a = 0$. Under these circumstances, for equally likely transmissions,

$$P_e = \frac{1}{2}\frac{1}{\sigma_n\sqrt{2\pi}}\int_{-\infty}^{0} e^{-(y-A)^2/2\sigma_n^2}\,dx + \frac{1}{2}\frac{1}{\sigma_n\sqrt{2\pi}}\int_{0}^{\infty} e^{-(y+A)^2/2\sigma_n^2}\,dx$$

$$= \frac{1}{\sigma_n\sqrt{2\pi}}\int_{A}^{\infty} e^{-y^2/2\sigma_n^2}\,dy = \frac{1}{\sqrt{2\pi}}\int_{A/\sigma_n}^{\infty} e^{-x^2/2}\,dx$$

$$= Q(A/\sigma_n) = Q(\sqrt{\rho})$$

(3.2-3)

where $\rho = \frac{A^2}{\sigma_n^2} = \frac{s_0^2(T_s)}{\sigma_n^2}$.

The Q function, which gives the area under the tail of a normalized Gaussian probability density function, was discussed in Chapter 2.

Clearly, the probability of error decreases as ρ increases. As an example, if $\rho = 16$ then we have $P_e = Q(4) = 3.17 \times 10^{-5}$. Further, the translation of the numerical value of $\rho = 16$ into db yields $\rho_{\mathrm{db}} = 10\log_{10}\rho = 12.04$ db.

3.3   MINIMIZING THE PROBABILITY OF ERROR WITH AN RC RECEIVER FILTER

We now want to examine the lowpass RC receiver filter of fig. 3.1-1 in relation to the probability of error. Decreasing the filter bandwidth will decrease the noise power out of the filter but will also decrease the peak value of the output signal $s_0(T_s)$. Similarly, increasing the bandwidth will increase both of these quantities. It stand to reason that there should be an optimum value of the bandwidth that results in a minimum probability of error and, equivalently, a maximum value of ρ.

From eqs. 3.1-4 and 3.1-5,

$$\rho = \frac{s_0^2(T_s)}{\sigma_n^2} = \frac{\frac{E_s}{T_s}\left[1 - e^{-T_s/RC}\right]^2}{\frac{\eta}{2}\frac{1}{2RC}} = 2\frac{\left[1 - e^{-T_s/RC}\right]^2}{T_s/RC}\frac{2E_s}{\eta}$$

$$= \left[2\frac{[1-e^{-x}]^2}{x}\right]\frac{2E_s}{\eta} = [g(x)]\rho_M$$

(3.3-1)

where $\rho_M = 2E_s/\eta$. We shall see that this quantity is of essential importance.

It is left to the Problems to show that the maximum value of $g(x)$ occurs at $x_0 = 3.2564$ and that $g(x_0) = 0.8145$. It follws that the minimum value of the probability of error for this system with a lowpass RC circuit as the receiver filter is $P_e = Q(\sqrt{\rho}) = Q(\sqrt{0.8145\rho_M})$.

3.4 THE OPTIMUM RECEIVER FILTER— THE MATCHED FILTER

The fundamental problem that we now wish to address is the design of the RCVR FILT, in particular how to choose the receiver filter impulse response $f(t)$ in order to maximize the value of ρ. The filter that accomplishes this is called the *matched filter*.

Referring to fig. 3.1-1, the signal output of the RCVR FILT can be wriiten as the inverse Fourier Transform of the output spectrum:

$$s_0(t) = \frac{1}{2\pi} \int\limits_{-\infty}^{\infty} S(\omega) F(\omega) e^{j\omega t} \, d\omega \qquad (3.4\text{-}1a)$$

so that at $t = t_m$:

$$s_0(t_m) = \frac{1}{2\pi} \int\limits_{-\infty}^{\infty} S(\omega) F(\omega) e^{j\omega t_m} \, d\omega \qquad (3.4\text{-}1b)$$

where $f(t) \Leftrightarrow F(\omega)$ and $s(t) \Leftrightarrow S(\omega)$ are Fourier Transform pairs.

It follows that:

$$\rho = \frac{s_0^2(t_m)}{n_0^2(t)} = \frac{\left| \frac{1}{2\pi} \int_{-\infty}^{\infty} S(\omega) F(\omega) e^{j\omega t_m} \, d\omega \right|^2}{\frac{\eta}{2} \frac{1}{2\pi} \int_{-\infty}^{\infty} |F(\omega)|^2 \, d\omega} \qquad (3.4\text{-}2)$$

In order to maximize this quantity, we will have to make use of the Schwarz Inequality. The statement and proof of the inequality follows.

SCHWARZ INEQUALITY

Let t be a real variable and

$$\int\limits_a^b |f_1(t)|^2 \, dt < \infty \quad \text{and} \quad \int\limits_a^b |f_2(t)|^2 \, dt < \infty \qquad (3.4\text{-}3a)$$

Then

$$\left| \int\limits_a^b f_1(t) f_2(t) \, dt \right|^2 \leq \int\limits_a^b |f_1(t)|^2 \, dt \int\limits_a^b |f_2(t)|^2 \, dt \qquad (3.4\text{-}3b)$$

with equality when $f_1(t) = -k f_2(t)$. ∎

Proof:
Form the integral

$$\int\limits_a^b \{[k f_1(t) + f_2(t)][k f_1^*(t) + f_2^*(t)]\} \, dt = k^2 A + k(B + B^*) + C \geq 0$$

Since this integral is non-negative, it may not have real roots except for a possible double root corresponding to a value of the integral equal to zero. According to the quadratic formula, the condition for this is

$$(B + B^*)^2 \leq 4AC$$

or

$$\left\{ \int_a^b [f_1^*(t)f_2(t) + f_1(t)f_2^*(t)] \, dt \right\}^2 \le 4 \int_a^b |f_1(t)|^2 \, dt \int_a^b |f_2(t)|^2 \, dt$$

which is the inequality for complex functions or, if the functions are real,

$$\left| \int_a^b f_1(t)f_2(t) \, dt \right|^2 \le \int_a^b |f_1(t)|^2 \, dt \int_a^b |f_2(t)|^2 \, dt.$$

The condition for equality corresponds to the double root, causing the integral to become zero, a situation that will exist when the derivative of the function is zero or

$$kA + B = \int_a^b \left[kf_1^2(t) + f_1(t)f_2(t) \right] dt = 0$$

or $f_1(t) = -kf_2(t)$. When the functions are complex $f_1(t) = -kf_2^*(t)$.   *end of proof*

According to the Schwarz Inequality, the value of ρ in eq. 3.4-2 is maximized when

$$F(\omega) = S^*(\omega)e^{-j\omega t_m} \tag{3.4-4a}$$

or, using the theorems of Fourier Transforms,

$$f(t) = ks^*(t_m - t). \tag{3.4-4b}$$

If $s(t)$, the transmitted signal, is a real function then these relationships reduce to

$$F(\omega) = S(-\omega)e^{-j\omega t_m} \tag{3.4-4c}$$
$$f(t) = ks(t_m - t) \tag{3.4-4d}$$

This then becomes the definition of the matched filter.

As illustrated in fig. 3.4-1a, $f(t)$, the impulse response of the matched filter, is the time-reversed and time-shifted version of the transmitted signal $s(t)$. If t_m is made sufficiently large, the matched filter will be causal and it is reasonable to set $t_m = T_s$, the duration of the transmitted symbol. This is the sampling instant or, as it will be known, the sampling or timing epoch. It will be assumed that the timing epoch is known at the receiver. The issue of how to learn the timing epoch, i.e. the timing recovery problem, was initially addressed in Chapter 2 and will be addressed in more detail in Chapter 6.

The output of the receiver filter is the convolution of the input and the impulse response:

$$s_0(t) = \int_{-\infty}^{\infty} s(\lambda)f(t - \lambda) \, d\lambda = \int_{-\infty}^{\infty} s(\lambda)s(T_s - t + \lambda) \, d\lambda$$

$$= \int_{-\infty}^{\infty} s(\lambda)s(\lambda - (t - T_s)) \, d\lambda = R_s(t - T_s) \tag{3.4-5a}$$

where

$$R_s(\tau) = \int_{-\infty}^{\infty} s(t)s(t - \tau) \, dt \tag{3.4-5b}$$

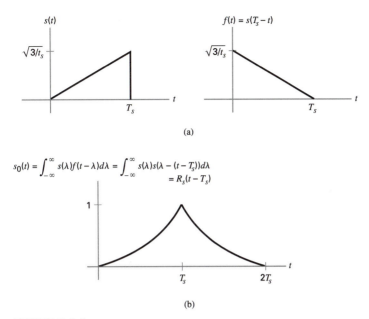

FIGURE 3.4-1

is the autocorrelation function of the transmitter waveform $s(t)$. As shown in fig. 3.4-1b, the output is even around its center. It is also clear that at $t = T_s$, the sampling instant, that

$$s_0(T_s) = \int_{-\infty}^{\infty} s(\lambda)s(\lambda) \, d\lambda = 1 \tag{3.4-5c}$$

Note that if the transmitted signal is $\pm\sqrt{E_s}s(t)$, then

$$s_0(T_s) = \pm \int_{-\infty}^{\infty} \sqrt{E_s}s(\lambda)s(\lambda) \, d\lambda = \pm\sqrt{E_s}. \tag{3.4-5d}$$

In the frequency domain

$$S_0(\omega) = S(\omega)F(\omega) = S(\omega)S^*(\omega)e^{-j\omega T_s} = |S(\omega)|^2 e^{-j\omega T_s} \tag{3.4-6}$$

where $E_s(\omega) = |S(\omega)|^2$ is the energy spectral density of $s(t)$.

We may now return to eq. 3.4-2 to find the value of ρ corresponding to the matched filter receiver. Recognizing that for this system the transmitted signal is $s(t)$ where $\pm\sqrt{E_s}s(t)$ $\Leftrightarrow \pm\sqrt{E_s}S(\omega)$ and that for the receiver filter $F(\omega) = S^*(\omega)e^{-j\omega T_s}$, then

$$\rho = \frac{s_0^2(T_s)}{n_0^2(t)} = \frac{\left|\frac{1}{2\pi}\int_{-\infty}^{\infty} S(\omega)F(\omega)e^{j\omega T_s} \, d\omega\right|^2}{\frac{\eta}{2}\frac{1}{2\pi}\int_{-\infty}^{\infty} |S(\omega)|^2 \, d\omega}$$

$$= \frac{\left|\frac{1}{2\pi}\int_{-\infty}^{\infty}[\pm\sqrt{E_s}S(\omega)][S^*(\omega)e^{-j\omega T_s}]e^{j\omega T_s} \, d\omega\right|^2}{\frac{\eta}{2}\frac{1}{2\pi}\int_{-\infty}^{\infty} |S(\omega)|^2 \, d\omega}$$

$$= \frac{2E_s}{\eta}\frac{\left[\frac{1}{2\pi}\int_{-\infty}^{\infty} |S(\omega)|^2 \, d\omega\right]^2}{\frac{1}{2\pi}\int_{-\infty}^{\infty} |S(\omega)|^2 \, d\omega} = \frac{2E_s}{\eta}\frac{1}{2\pi}\int_{-\infty}^{\infty} |S(\omega)|^2 \, d\omega = \frac{2E_s}{\eta} = \rho_M \tag{3.4-7}$$

where we have referred to eq. 3.1-2.

Note that the noise output of the matched filter is

$$\overline{n_0^2(t)} = \frac{\eta}{2} \frac{1}{2\pi} \int\limits_{-\infty}^{\infty} |S(\omega)|^2 \, d\omega = \frac{\eta}{2} \qquad (3.4\text{-}8)$$

A crucial result of this development is that the value of ρ for the binary, matched filter communication system described above is, *independent of the particular transmitted signal $s(t)$*,

$$\rho = \rho_M = \frac{2E_s}{\eta} \qquad (3.4\text{-}9a)$$

and the probability of error for this system is

$$P_e = Q(\sqrt{\rho_M}) \qquad (3.4\text{-}9b)$$

Note the comparison between the error probability of the matched filter system and that of the optimized RC filter receiver: $P_e = Q(\sqrt{0.8145\rho_M})$. This is an illustration of the result that the matched filter receiver is the best possible linear receiver for minimizing the probability of error.

The quantity $\rho_M = 2E_s/\eta$, which is understood as the energy in each transmitted symbol divided by the power spectral density of the noise, is known as the SNR, or signal-to-noise ratio. Subsequently in this chapter, in dealing with bandlimited systems, we shall see that this definition is equivalent to the better-known ratio of signal power to noise power.

In order to make the point that the performance of the matched filter system is independent of the particular shape of $s(t)$, provided that $s(t)$ has unit energy, we show in fig. 3.4-2 four different possible waveforms for $s(t)$ thar satisfy eq. 3.1-2. It is left to the Problems to confirm that for each, the system probability of error is given by eq. 3.4-9b.

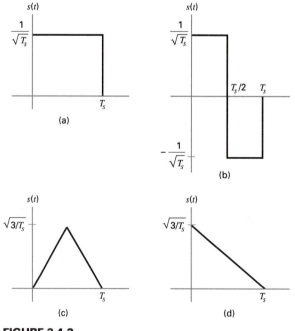

FIGURE 3.4-2

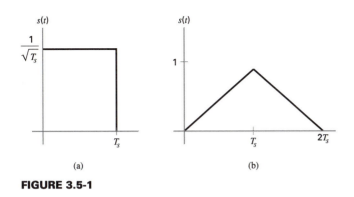

FIGURE 3.5-1

3.5 CONTINUOUS DATA STREAMS
AND INTERSYMBOL INTERFERENCE (ISI)

At this point it is possible to relax the artificial restriction that only a single isolated data pulse is transmitted and, instead, consider that the data stream of eq. 3.1-3 is transmitted:

$$x(t) = \sum_{n=-\infty}^{\infty} d_n s(t - nT_s).$$ (3.1-3)

where $s(t)$ is the rectangular pulse of fig. 3.5-1a. The corresponding matched receiver filter output $s_0(t)$ is the triangular pulse of fig. 3.5-1b.

   According to fig. 3.5-2a, the overlapping outputs of the matched filter at the receiver do not interfere with the sampling process because the values of adjacent pulses at the instant of sampling are zero. Contrast this with the output of the lowpass RC circuit operating as a receiver filter in fig. 3.5-2b, where the outputs from successive data symbols do overlap. Such overlapping is called *intersymbol interference (ISI)*. ISI has the effect of, on the average, lowering the threshold against noise and degrading the performance of the system. Among

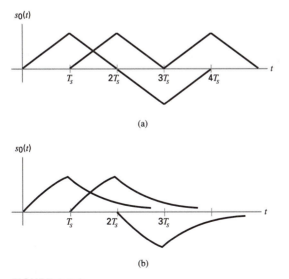

FIGURE 3.5-2

the properties of the matched filter receiver for the model that we have established is that there is zero ISI. Recall that this model includes a transmitted waveform that is time-limited to the symbol interval T_s and a wideband (band-unlimited) channel that guarantees undistorted transmission of the data signal. This distortionless transmission is independent of the value of T_s. Therefore there is no restriction on the speed of data transmission in this system.

Subsequently we shall see that a bandlimited channel restricts the possible speed of data transmission and a distorted channel generates ISI that may be compensated by channel equalization. We may also characterize ISI as giving the channel some memory because the value of the detected signal clearly depends not only on the current data but also on past data symbols. A zero-ISI system is therefore characterized as memoryless and the successive signal samples are uncorrelated.

With the received signal coming from a data stream, it is also important to establish that successive samples of the noise are uncorrelated so that successive samples of signals plus noise can be processed independently. In order to demonstrate this, note that the autocorrelation function of the noise output of the RCVR FILT is the convolution of the impulse response of the RCVR FILT with itself in turn convolved with the autocorrelation function of the input noise:

$$\overline{R_0}(\tau) = f(\tau) \otimes f(\tau) \otimes \overline{R_n}(\tau) \tag{3.5-1a}$$

and since $\overline{R_n}(\tau) = \eta/2\delta(\tau)$ and $f(t) = s(t_m - t)$, it follows that

$$\overline{R_0}(\tau) = \frac{\eta}{2} R_s(\tau) \tag{3.5-1b}$$

where $R_s(\tau)$ is the autocorrelation function of $s(t)$ and $R_s(\tau) = s_0(\tau)$, the matched filter output. In the model that we have developed (see fig. 3.5-2a) this means that

$$\overline{R_0}(\tau) = 0, \qquad \text{for } \tau \geq T_s. \tag{3.5-1c}$$

so that successive noise samples are indeed uncorrelated. With zero ISI and successive noise samples uncorrelated, it follows that we can deal with a steady stream of data as if it were a collection of isolated data pulses. There is no interaction between the symbols.

3.6   THE CORRELATION RECEIVER

A particularly important version of the matched filter receiver is the correlation receiver. Recall that the output of the matched filter receiver is:

$$s_0(t) = \int\limits_{-\infty}^{\infty} s(\lambda)s(\lambda - (t - T_s))\, d\lambda \tag{3.4-5a}$$

If $s(t)$ is a finite-time waveform extending from $t = 0$ to $t = T_s$, the limits of this integral can be changed so that

$$s_0(t) = \int\limits_{0}^{T_s} s(\lambda)s(\lambda - (t - T_s))\, d\lambda \tag{3.6-1a}$$

The key issue here is that we are interested only in the output at the instant of sampling. We really don't care what the output is at any other time. If the sampling takes place at $t = T_s$

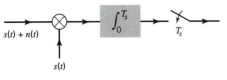

FIGURE 3.6-1

then

$$s_0(T_s) = \int_0^{T_s} s(\lambda)s(\lambda)\,d\lambda \qquad (3.6\text{-}1b)$$

which is shown implemented in fig. 3.6-1. If the transmitted is a rectangular pulse it becomes unneccesary to multiply by $s(t)$ at the receiver. The resulting structure is called "integrate and dump." The difference between the signal output for the matched filter and the correlation receiver is shown in fig. 3.6-2 for two different transmitted signals.

We shall now prove that the value of the noise output of the two receiver structures is the same. Recall that the input noise $n(t)$ has an autocorrelation function $\overline{R_n}(\tau) = \frac{\eta}{2}\delta(t)$. The noise output of the correlation receiver at the sampling instant $t = T_s$ is:

$$n_0(T_s) = \int_0^{T_s} n(\alpha)s(\alpha)\,d\alpha \qquad (3.6\text{-}2a)$$

so that

$$\overline{n_0^2(T_s)} = \overline{\int_0^{T_s} n(\alpha)s(\alpha)\,d\alpha \int_0^{T_s} n(\beta)s(\beta)\,d\beta}$$

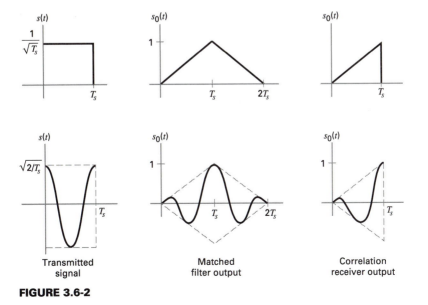

Transmitted signal   Matched filter output   Correlation receiver output

FIGURE 3.6-2

$$= \int\limits_{0}^{T_s} \int\limits_{0}^{T_s} \overline{n(\alpha)n(\beta)}s(\alpha)s(\beta)\,d\alpha\,d\beta = \int\limits_{0}^{T_s} \int\limits_{0}^{T_s} \overline{R_n}(\alpha - \beta)s(\alpha)s(\beta)\,d\alpha\,d\beta$$

$$= \int\limits_{0}^{T_s} \int\limits_{0}^{T_s} \frac{\eta}{2}\delta(\alpha - \beta)s(\alpha)s(\beta)\,d\alpha\,d\beta = \frac{\eta}{2}\int\limits_{0}^{T_s} s^2(\alpha)\,d\alpha = \frac{\eta}{2}. \qquad (3.6\text{-}2b)$$

This is clearly the same result as for the matched filter receiver. We can now use these two structures interchangably.

3.7 A DSP-ORIENTED RECEIVER STRUCTURE

We are also in a position to extend the receiver model of fig. 3.1-1 to show some important details of a DSP-oriented receiver structure that has both conceptual and practical importance in subsequent developments. As shown in fig. 3.7-1, the sampler is implemented by an analog switch, activated by a squarewave timing waveform, followed by a storage capacitor an an isolating op-amp. One can imagine the timing waveform being generated by passing the matched filter output through the square-wave timing recovery system discussed in Chapter 2. When the square wave is negative, the switch is closed and the matched filter output is tracked by the capacitor voltage. On the positive leading edge of the timing waveform, corresponding to the sampling instant, the analog switch opens and the matched filter output, peak signal plus noise, is held by the capacitor. This arrangement is called a "track and hold" circuit.

In order to bring this sampled signal into a microprocessor for further processing, we use the same positive leading edge to start an ADC (analog-to-digital converter). When the ADC completes its cycle, we use the EOC (end of conversion) signal to interrupt the microprocessor. In this simple model, the interrupt service routine serves as the decision

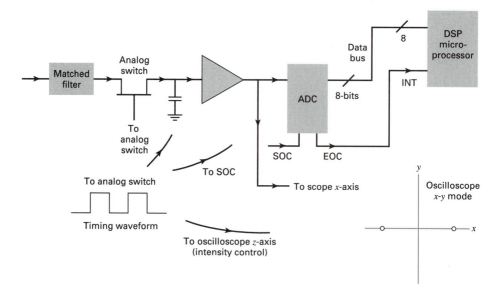

FIGURE 3.7-1

device, as indicated in the following pseudocode:

```
Interrupt: input data from bus
          if data > 0 then d = +1
                     else d = -1
       return
```

Although at first glance this may seem a puzzling way to use a microprocessor, we shall see that it forms the framework for a DSP implementation of the whole modem.

We also want to display the track-and-hold output on an oscilloscope. The scope is set up in the X-Y mode with sampled output on the X axis. The timing waveform is applied to the Z axis in order to control the intensity of the beam; when the waveform is positive and the output voltage is held, the beam is enabled and when the waveform is negative the beam is cut off. Consequently, in the noiseless situation, two dots, corresponding to the two possible signal points having amplitudes $\pm\sqrt{E_s}$, appear on the scope. When noise is added to the system the dots expand into lines.

Not only is this model quite important for visualizing the system, it also represents the actual implementation strategy of digital communications receivers. We shall be using this model consistently. When we introduce two-dimensional signaling in Chapter 4, one dimension will be displayed on the X axis and the other on the Y axis.

3.8 MULTILEVEL SIGNALING WITH A MATCHED FILTER RECEIVER

We start by assuming a matched filter receiver. Each piece of data d_n is described as a symbol, cautioning the reader that each such symbol might contain more than one bit of information. Until this point each symbol has had only two possible values $d_n = \pm\sqrt{E_s}$ involving one bit per transmitted symbol. If each data symbol were to contain two bits of information, it would require four distinct values. In order to minimize the probability of error at the receiver, these values are made symmetric:

$$d_n = \pm a\sqrt{E_s}, \quad \pm 3a\sqrt{E_s}. \tag{3.8-1a}$$

These points are shown in fig. 3.8-1 as the oscilloscope output of fig. 3.7-1. Clearly these four signals have differing energies and in order to establish consistency with the binary case we will choose the value of a in order to have the *average energy per transmitted symbol* be equal to E_s. Since the symbols are equally likely, it follows that

$$E_s = \frac{1}{4}2(a^2 + 9a^2)E_s \quad \text{or} \quad a = \frac{1}{\sqrt{5}} \tag{3.8-1b}$$

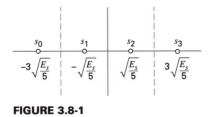

FIGURE 3.8-1

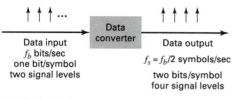

FIGURE 3.8-2

so that the transmitted data symbols are

$$d_n = \pm\frac{1}{\sqrt{5}}\sqrt{E_s}, \quad \pm\frac{3}{\sqrt{5}}\sqrt{E_s}. \tag{3.8-1c}$$

In order to generate this multilevel data from a serial bit stream, we require a converter, as shown in fig. 3.8-2, which accepts binary inputs at the bit rate f_b, gathers them up in groups of two bits (dibits) and outputs them at the symbol rate $f_s = \frac{1}{2}f_b$.

In order to compute the probability of error, let's look again at fig. 3.8-1. Clearly the probability of error for the individual points are different. The inner points each have two ways of making an error while the outer points have only one. It therefore makes sense to think of an average probability of error per transmitted symbol. The basic error consists of the Gaussian noise with variance $\eta/2$ exceeding a threshold of $\sqrt{E_s/5}$. This is equal to $Q(\sqrt{\rho_m/5})$, where $\rho_m = 2E_s/\eta$. Since there are four transmitted points and six ways of making an error, it follows that for this four-level system

$$\overline{P_e} = \frac{3}{2}Q(\sqrt{\rho_m/5}). \tag{3.8-2}$$

Another way of getting the same result, one that is more general, is to note that:

$$P(\text{error}) = 1 - P(\text{correct}) = 1 - \sum_{i=0}^{3} P_i P(\text{correct}/P_i)$$

$$= 1 - \frac{1}{4}[(1 - Q(a)) + (1 - 2Q(a)) + (1 - 2Q(a)) + (1 - Q(a))]$$

$$= 3/2Q(a) \tag{3.8-3}$$

where $a = \sqrt{\rho_M/5}$

It is not hard to extend this to a system where we have L levels, where L is an even number. In that general case:

$$\overline{P_e} = 2\frac{L-1}{L}Q(\sqrt{3\rho_M/(L^2-1)}). \tag{3.8-4}$$

This result is shown for various values of L in fig. 3.8-3. Clearly for a given value of ρ_M the value of $\overline{P_e}$ increases with L. This is because the decision boundaries get closer.

The basic tradeoff in increasing the value of L is that of bit rate vs. performance if the symbol rate and the average energy per transmitted symbol are kept constant. For example, if the symbol rate is 1000 baud (symbols/sec) then a binary ($L = 2$) system will transmit 1000 bits/sec (1 bit/symbol) and a four-level ($L = 4$) system will transmit 2000 bits/sec (2 bits/symbol). The penalty is that faster transmission under these constraints will result in a larger probability of error.

Although it should be intuitively obvious why average energy per transmitted symbol should be a constraint, the basis for the constraint on symbol rate will become clear only as we discuss bandlimited channels.

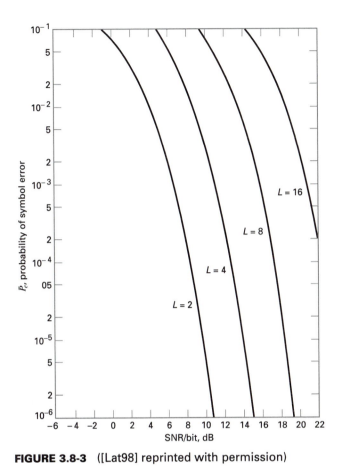

FIGURE 3.8-3 ([Lat98] reprinted with permission)

3.9 BANDLIMITED CHANNELS AND NYQUIST SIGNALING

The discussion up to this point has involved the transmission of data through band-unlimited channels. There are, of course, no such things, although there are circumstances in which the bandwidth of the signal is sufficiently small compared to the bandwidth of the channel that the approximation is reasonable. In general, however, we must accept the reality of a bandlimited channel and we shall now discuss optimum ways of transmitting data on such a channel.

Our initial discussion will deal with an ideal lowpass channel as shown in fig. 3.9-1. An ideal bandpass channel will be discussed in the next chapter. We will subsequently deal with the issue of distorted channels. Our fundamental concern in transmitting on a bandlimited channel is that of intersymbol interference (ISI). Our central task in this section is to demonstrate that it is possible to find strictly bandlimited signals that satisfy the condition of zero ISI. Consider first the impulse response of an ideal lowpass filter having a cutoff of $\omega_N = \omega_s/2 = \pi/T_s$ as shown in fig. 3.9-2. This filter, known as a brick-wall filter, has an impulse response

$$h(t) = \text{Sinc}(\omega_N t) = \text{Sinc}\left(\pi \frac{t}{T_s}\right) \tag{3.9-1}$$

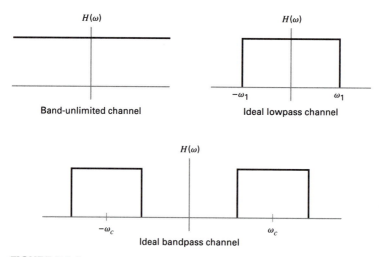

FIGURE 3.9-1

that generates zero intersymbol interference for data impulses transmitted every T_s seconds. The frequency f_N is called the Nyquist frequency and the idea is that we can transmit data with zero ISI at frequency $f_s = 1/T_s$, twice the Nyquist frequency.

In some ideal sense, we can imagine a communication system built around this impulse response. The XMIT FILT and RCVR FILT are each brick-wall filters with a cutoff equal to the Nyquist frequency and an amplitude equal to $\sqrt{\pi/\omega_N}$. This satisfies the requirement that these two filters be matched. The channel, also an ideal lowpass filter having the same bandwidth, has unit gain. The noise power out of the receiver filter is

$$P_n = \frac{\eta}{2}\frac{1}{2\pi}\int_{-\infty}^{\infty}|H(\omega)|^2\,d\omega = \frac{\eta}{2}\frac{1}{2\pi}\frac{\pi}{\omega_N}2\omega_N = \frac{\eta}{2} \tag{3.9-2}$$

equivalent to our previous matched filter result. As well, an impulse function at the input having amplitude $\sqrt{E_s}$ will result in a transmitted energy of E_s and a peak value at the sampler of $\sqrt{E_s}$, so that the performance is identical to that of a matched filter receiver.

This system is, however, unrealistic because the filters are not possible to build. Even if they could be built, the discussion of square-law timing recovery in the previous chapter indicates that, at least with that method, it would be very hard to derive an appropriate timing signal.

While continuing to assume that the channel is an ideal lowpass filter having a cutoff frequency ω_1, we can back off somewhat and make the Nyquist frequency somewhat less

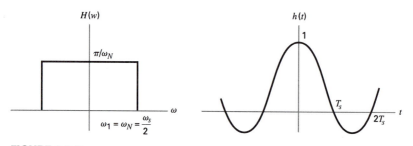

FIGURE 3.9-2

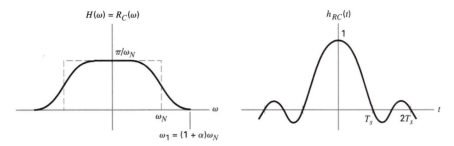

FIGURE 3.9-3

than the cutoff frequency of the channel. This affords us the possibility of using the "raised-cosine" spectrum and its corresponding impulse response, shown in fig. 3.9-3,

$$h(t) = \text{Sinc}(\omega_N t) \frac{\text{Cos}(\alpha \omega_N t)}{1 - \left(\frac{2}{\pi} \alpha \omega_N t\right)^2} \qquad (3.9\text{-}3)$$

for data transmission. Note that the zero crossings at the symbol times are identical to those of the brick-wall case. The intermediate zeros cause no problems.

Although the raised cosine spectrum is also theoretically not realizable because it goes to zero over a band of frequencies, it can be very closely approximated. The difference between the cutoff frequency and the Nyquist frequency is called the "excess bandwidth" and is equal to $\omega_1 - \omega_N = \alpha \omega_N$, where α is called the "rolloff" and ordinarily $0 \le \alpha \le 1$. The analytic representation of this spectrum is:

$$\begin{aligned}
\text{RC}(\omega) &= \frac{\pi}{\omega_N}; \qquad |\omega| \le (1 - \alpha)\omega_N \\
&= 0; \qquad |\omega| \ge (1 + \alpha)\omega_N \\
&= \frac{\pi}{\omega_N} \frac{1}{2}\left(1 + \text{Sin}\left(\frac{\pi}{2} \frac{\omega + \omega_N}{\alpha \omega_N}\right)\right); \qquad \begin{aligned} -(1 + \alpha)\omega_N &\le \omega \\ &\le -(1 - \alpha)\omega_N \end{aligned} \\
&= \frac{\pi}{\omega_N} \frac{1}{2}\left(1 - \text{Sin}\left(\frac{\pi}{2} \frac{\omega - \omega_N}{\alpha \omega_N}\right)\right); \qquad (1 - \alpha)\omega_N \le \omega \le (1 + \alpha)\omega_N \quad (3.9\text{-}4)
\end{aligned}$$

We now want to combine this result with matched filter theory. The idea is that we want both the XMIT FILT and the RCVR FILT to have the same amplitude characteristic and have phase characteristics that combine to give a linear phase or constant delay. Putting aside the issue of the phase characteristic, we can accomplish the amplitude part by letting each have the characteristic of $\sqrt{\text{RC}}(\omega)$, the square root of raised cosine. This structure is shown in fig. 3.9-4. In the actual design of these filters, we first design for the amplitude characteristic and then add all-pass sections to generate the desired linear phase. In order to avoid confusion, it is important to state explicitly that the impulse response of neither one of the two filters is that of eq. 3.9-3. Rather, each filter has an impulse response that is the inverse Fourier Transform of $\sqrt{\text{RC}}(\omega)$ which we may denote as $h_{\sqrt{\text{RC}}}(t)$. It follows that $h_{\sqrt{\text{RC}}}(t) * h_{\sqrt{\text{RC}}}(t) = h_{\text{RC}}(t)$, where $*$ denotes convolution.

Note that the noise output of the RCVR FILT is given by

$$P_n = \frac{\eta}{2} \frac{1}{2\pi} \int_{-\infty}^{\infty} |\sqrt{\text{RC}}(\omega)|^2 \, d\omega = \frac{\eta}{2} \qquad (3.9\text{-}5)$$

and that the peak value of the normalized system impulse response is unity. It follows that

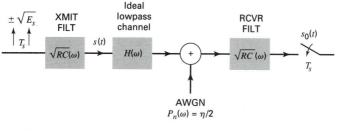

FIGURE 3.9-4

if the input impulse function has area equal to $\sqrt{E_s}$, then the energy per transmitted symbol equals E_s.

This means that the bandlimited system that we have described has exactly the same performance as the original matched filter system. The only difference is that in the wideband case there is no constraint on the speed of transmission, whereas in the narrowband case the speed is constrained by the system bandwidth.

We say that the end-to-end characteristic of the system is raised cosine, combining the characteristics of XMIT FILT and RCVR FILT.

The rolloff of the raised cosine characteristics can vary considerably in different modems, from a low of $\alpha = 0.1$ to a high of $\alpha = 1.0$. The larger the rolloff, the easier the filter is to build, but a large rolloff uses less of the available bandwidth for transmission. It is shown in the Problems that as α increases so does the strength of the square-law timing recovery signal. Clearly, then, the choice of an appropriate value of α is an engineering decision rather than an analytical result. We will examine some commercial modems in the next chapter to see how these choices are made.

Finally, we want to note the equivalence in bandlimited transmission of ρ_M, the defined SNR (signal-to-noise ratio) in matched filter theory, and the standard definition of SNR. Typically SNR is defined as the average power of the transmitted signal divided by the noise power output of the $\sqrt{RC}(\omega)$ receiver filter. Therefore:

$$\rho_M = \frac{E_s}{\eta/2} = \frac{E_s/T_s}{(\eta/2)(1/T_s)} = \frac{E_s/T_s}{(\eta/2)(2B_N)} = \frac{P_{av}}{P_{noise}} = \text{SNR} \qquad (3.9\text{-}6)$$

where the noise power out of the receiver filter is measured by twice the Nyquist bandwidth of the raised cosine, the equivalent noise bandwidth of that filter.

We now wish to explore and generalize the conditions that made the raised-cosine an appropriate end-to-end characteristic for this system. These conditions are known as the Nyquist Criteria.

The basic issue in choosing a waveform for data transmission on a bandlimited channel is that the sampled output has zero intersymbol interference. This means that in fig. 3.9-4,

$$s_0(kT_s) = \begin{cases} 1; & \text{for } k = 0 \\ 0; & \text{for } k \neq 0 \end{cases} \qquad (3.9\text{-}7a)$$

The spectrum of the sampled output is

$$S_0^*(\omega) = \sum_{n=-\infty}^{\infty} S(\omega + n\omega_s). \qquad (3.9\text{-}7b)$$

and we want to find the conditions on this spectrum that will produce the condition of zero,

ISI as indicated in eq. 3.9-7a. Since

$$s_0(t) = \frac{1}{2\pi} \int_{-\infty}^{\infty} S_0(\omega)e^{j\omega t} \, d\omega \tag{3.9-7c}$$

it follows that

$$s_0(kT_s) = \frac{1}{2\pi} \int_{-\infty}^{\infty} S_0(\omega)e^{j\omega kT_s} \, d\omega$$

$$= \frac{1}{2\pi} \sum_{n=-\infty}^{\infty} \int_{-(2n-1)\frac{\pi}{T_s}}^{(2n-1)\frac{\pi}{T_s}} S_0(\omega)e^{j\omega kT_s} \, d\omega$$

$$= \frac{1}{2\pi} \sum_{n=-\infty}^{\infty} \int_{-\pi/T_s}^{\pi/T_s} S_0\left(u + n\frac{2\pi}{T_s}\right)e^{jukT_s} \, du \tag{3.9-7d}$$

where $\omega = u + (2\pi/T_s)$. Exchanging summation and integration and doing a bit of scaling, we get

$$s_0(kT_s) = \frac{1}{2\pi} \int_{-\pi/T_s}^{\pi/T_s} \left[\sum_{n=-\infty}^{\infty} S_0\left(u + n\frac{2\pi}{T_s}\right) \right] e^{jukT_s} \, du$$

$$= \frac{1}{2\omega_N} \int_{-\omega_N}^{\omega_N} \left[\sum_{n=-\infty}^{\infty} S_0(u + 2n\omega_N) \right] e^{jukT_s} \, du \tag{3.9-7e}$$

Since the term inside the brackets is periodic, the integral produces a Fourier Series coefficient. Examining this from the point of view of eq. 3.9-7a, it follows that this Fourier Series is to have only a single term, that is, a constant. The only way that this can happen is if the function inside the brackets is a constant, i.e.:

$$\left[\sum_{n=-\infty}^{\infty} S_0(u + n2\omega_N) \right] = 1 \tag{3.9-7f}$$

This result, that the "folded spectrum" or "sampled spectrum" must be a constant for data transmission with zero ISI, is *Nyquist's First Criterion*. This is illustrated in fig. 3.9-5 for the brick-wall filter, the raised cosine, and an arbitrary spectrum satisfying the criterion.

Another way of thinking about this issue is as follows: if the transmitted data signal were an impulse and the channel were ideal, then the output would be the same impulse delayed in time. The spectrum of the received impulse is flat. In a Nyquist system the data impulse is transmitted as an analog waveform having the characteristic that, *after sampling*, has a flat amplitude spectrum and, as a sampled waveform, looks the same as an impulse.

Representing $S_0(\omega) = R(\omega) + jX(\omega)$, eq. 3.22 is the same as

$$R(\omega) + R(-\omega + 2\omega_N) = 1 \tag{3.9-7g}$$
$$X(\omega) + X(-\omega + 2\omega_N) = 0 \tag{3.9-7h}$$

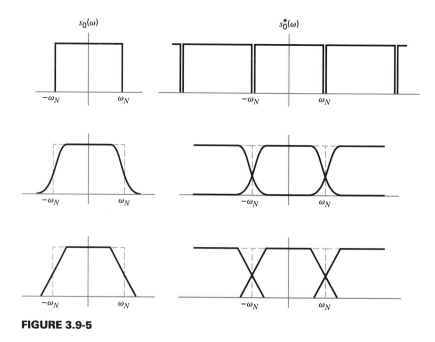

FIGURE 3.9-5

There are other aspects of Nyquist's theory that are not of direct concern to us at this point. Interested readers may consult Nyquist's original paper [Nyq28] or Bennett [Ben70].

3.10 EXAMPLES OF BASEBAND DATA TRANSMISSION

3.10.1 The ISDN (Integrated Services Digital Network)

The telephone system consists of a large number of local telephone exchanges (central offices) that are connected on one side (fig. 3.10-1) to a geographically compact collection of telephone subscribers and, on the other side, to means for the exchanges to communicate with each other. Although communication between the exchanges was originally analog, organized in frequency division multiplexed 4 khz channels, this communication is now typically digital, using a variety of high-speed data links including coaxial cable, microwave, satellite, and fiber optics.

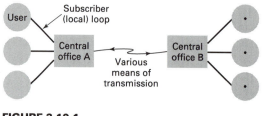

FIGURE 3.10-1

The connection from the telephone exchange to the individual subscriber is a twisted pair of 24- or 26-gauge copper wire called the subscriber loop or local loop. These individual connections, which are of considerably different lengths, are typically bundled into cables.

Until recently this connection has only been used for the single 4 khz channel corresponding to traditional analog telephone voice transmission. Standard data modems, from the original Bell 103 300 bps modem to the V.34 operating at 28.8 kb/sec and the current 56 kb/sec modems, all operate within this 4 khz bandwidth.

The ISDN originated as a system (Lin and Tzeng [L&T88]) for using the subscriber loop to transmit voice digitally. As we noted in Chapter 1, the 4 khz voice channel must be sampled at least at 8 khz to avoid aliasing. An industry standard fixes this 8 khz as the voiceband sampling rate and also specifies that each analog sample be quantized to 8 bits. A digitized voice signal therefore requires 64 kbps. The ISDN (Integrated Services Digital Network) provides for two such B channels and also provides for a D channel of 16 kbs for a variety of differents kinds of ancillary signalling (hence 2B + D) for a total of 144 kbps. An additional 16 kbps channel carries framing, maintenance, and control signalling, making a total of 160 kbps. There has been considerable discussion of how to do the line coding for the ISDN. This discussion is well summarized in Lechleider [Lec89]. In the end, the decision was made to use the simple four-level signal, two bits/symbol, described in this chapter. This code has been designated 2B1Q and the symbol rate is 80 khz. There is active discussion of increasing the symbol rate to 400 khz, supporting a data rate of 800 kbits/sec.

The system is required to be full duplex, transmitting at 160 kbps in both directions simultaneously. In order to see how this is accomplished it is necessary to take a closer look at the local loop, fig. 3.10-2a. The user modem's transmitter and receiver are connected to the two-wire twisted pair through a device called a hybrid. Ideally, a hybrid would prevent any of the transmitted signal from appearing at the receiver and would direct the received signal to the receiver. A hybrid is similarly used at the central office. Real hybrids are, regrettably,

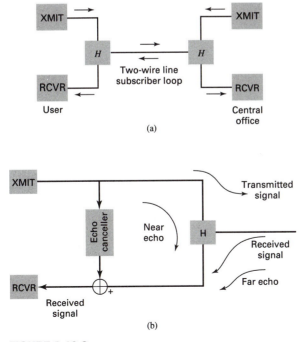

FIGURE 3.10-2

quite leaky and also generate reflections. The transmitted signal will leak through the hybrid and enter the receiver, typically at a higher power level than the received signal. It will also be reflected back from the hybrid at the central office along with the users received signal.

Both these echos are therefore distorted versions of the transmitted signal and, together, they come into the receiver along with the received signal. As shown in fig. 3.10-2b, a modem has a device called an echo canceller that adaptively creates a replica of this total received echo and simply subtracts the replica from the received signal. This leaves only the portion of the received signal that actually was transmitted at the far end.

The subscriber loop has a variety of impairments. The attenuation and frequency response depend on the length of the wire, which varies from a very short distance up to two miles. The bundling of these twisted pairs into cables leads to crosstalk between cables, which can be characterized as a form of noise. This crosstalk has been denoted NEXT (near-end crosstalk) and FEXT (far-end crosstalk). The existence of spurs on the line, called bridge taps, leads to mutiple reflections that can produce nulls in the transfer function. These impairments will be discussed in some detail in Chapter 7. They have been summarized in Ahmed et al. [ABG81]. Adaptive equalization is required to eliminate the substantial line distortion.

Adaptive echo cancellation, adaptive equalization of channel distortion, and timing recovery will all be discussed in subsequent chapters with applications to the ISDN as well as to other systems. Although they will be cited in these subsequent discussions, the interested reader may now consult van Gerwen et al. [GVC84] and Messerschmitt [Mes86].

Although originally intended for digital telephone to the home, the ISDN is currently also being used as a higher-speed Internet connection.

We should take note that there are also ways of using the subscriber loop for much higher data rates, up to 6 Mbits/sec; see the *IEEE Journal on Selected Areas in Communications*, August 1991: Special Issue on High-Speed Digital Subscriber Lines. These approaches involve bandpass data-transmission techniques and will be discussed in subsequent chapters.

3.10.2   The Digital Hierarchy

Even when transmission on the local loop is analog, corresponding to the alloted voiceband bandwidth of 4 khz, modern transmission between central offices is all digital. At the central office an incoming analog telephone signal is bandlimited, sampled at 8 khz and quantized to eight bits, one byte, leading to a 64 kbps digital transmission, as described above and in Chapter 1. Each of these bytes, referred to as DS0, requires 125 μsec for transmission. This is the basis for modern digital telephony and in order to support this transmission a digital hierarchy has been developed.

The digital hierarchy is constructed through the process of time multiplexing successively larger numbers of these 64 khz signals (Lathi and Wright [L&W97] and Couch [Cou97]). A frame is constructed by time multiplexing 24 of these digitized analog signals and adding a framing bit yielding 193 bits. These frames must be transmitted at an 8 khz rate, yielding the well known transmission rate of 1.544 Mbps, constituting DS1 transmission. It is a baseband system that uses AMI line coding (discussed in Chapter 2) to be able easily to extract a clock using square-law timing recovery. The medium of transmission is twisted pair with regenerative repeaters placed every mile. In order to have full duplex transmission, two of the twisted pairs are used for each connection, constituting a T1 line.

DS1 signals are also time multiplexed together to obtain higher-order transmissions as shown below. These transmission standards also incorporate overhead bits whose purpose is not relevant to this discussion. This constitutes the North American (and Japanese) Digital Hierarchy.

Level	Bit rate	#DS0 circuits
DS0	64 kbps	1
DS1/T1	1.544 Mbps	24
DS1C/T1C	3.152 Mbps	48 (2 DS1's)
DS2/T2	6.312 Mbps	96 (4 DS1's)
DS3/T3	44.736 Mbps	672 (28 DS1's)
DS4/T4	274.176 Mbps	4032 (422 DS1's)

The CCITT standard, used in most of the rest of the world, is somewhat different. It first multiplexes 30 DS0s, resulting in a bit rate of 2.048 Mbps, and then does recursive multiplexing of four resulting groups, leading to a hierarchy of bit rates of 8.448 Mbps, 34.386 Mbps, and 139.264 Mbps, all including overhead.

Any of these data systems can be transmitted using baseband techniques and using various line codes by coaxial cable or by fiber optics. A few of them have been designated as well for line-of-sight digital microwave transmission. This approach will be discussed in the Chapter 4, Bandpass Data Transmission.

3.10.3 Accumulation of Timing Jitter on Baseband Repeatered Systems

The use of twisted pair or coaxial cable for digital transmission is limited by their severe attenuation characteristics, as discussed in Chapter 7. Briefly, however, the attenuation in both cases increases rapidly with frequency and with the line length. This requires that in order to prevent the digital signal from deteriorating, it must be detected and reconstructed using repeaters periodically.

The difficulty associated with this operation is that the timing jitter associated with the signal increases as it goes through each successive repeater. Timing jitter causes the sampling phase on successive symbols to vary randomly. This means that the variation in symbol amplitude at the different sampling phases looks like an additional component of noise. The basic references on this issue are Rowe [Row58] and Byrne et al. [BKR63]. An accessible summary is contained in [Bell70].

In a chain of repeaters, fig. 3.10-3a, the effect of each repeater on the spectrum of the timing jitter at its input is

$$\theta_0(f) = \frac{1}{1 + j \, f/B} \theta_i(f) \tag{3.10-1}$$

where B is the half bandwidth of the filter used in a square-law timing recovery circuit as described in Chapter 2.

The repeaters add equal jitters at each stage, so that at the end of N stages

$$\theta_N(f) = \sum_{n=1}^{N} \left[\frac{1}{1 + j \, f/B} \right]^n \theta_i(f) \tag{3.10-2a}$$

This is a geometric series so that

$$\theta_N(f) = B \left[1 - \left[\frac{1}{1 + j \, f/B} \right]^N \right] \theta_i(f) \tag{3.10-2b}$$

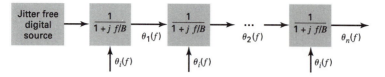

(a) Jitter accumulation in a chain of repeaters

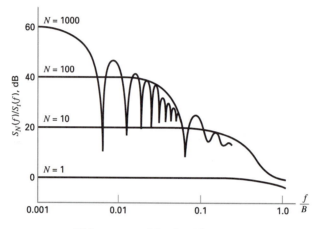

(b) Power spectral density of jitter

FIGURE 3.10-3   ([Smi93] reprinted with permission)

Assuming that $\theta_i(f)$ has a flat spectrum, the proper measure of the jitter is

$$|\theta_N(f)|^2 = B^2 \left| \left[1 - \left[\frac{1}{1 + j\, f/B} \right]^N \right] \right|^2 \tag{3.10-2c}$$

This expression is plotted in fig. 3.10-3b. An examination of the plot indicates that the mean square jitter increases linearly with N for low frequencies and is considerably attenuated at higher frequencies.

It has also been shown that the average power in the output jitter is

$$\overline{\theta_N} = N B \overline{\theta_i} \tag{3.10-2d}$$

indicating that jitter can be reduced by having a very narrow filter in the timing recovery.

3.11   EQUALIZATION, NOISE AMPLIFICATION, AND INTERSYMBOL INTERFERENCE (ISI)

If the channel is not ideal, as shown in fig. 3.11-1, the system becomes considerably more complicated. The first thing that is apparent is that the square-root-of-raised-cosine receiver filter is no longer matched to the received signal. This may be addressed by placing an analog filter or equalizer after the receiver to compensate for the effects of the channel. If the compensation is exact, then the receiver filter plus equalizer are then matched to the received signal and the system is once again linearly optimal. Any intersymbol interference

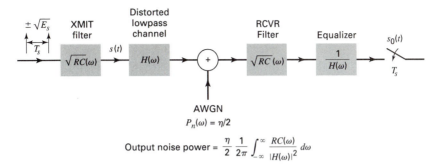

FIGURE 3.11-1

that has been generated as a consequence of the smearing of the pulse in the channel has been corrected by the equalizer.

Unfortunately, the noise input at the receiver must also go through the equalizer and is consequently amplified and degrades the performance of the system. This fundamental issue in the design of high-performance communication systems will be explored and developed in great detail in Chapter 10. At this point we may note that if the channel transfer function is $H(\omega)$ and the receiver filter is $\sqrt{RC(\omega)}$, the square-root of a raised-cosine filter, and we have AWGN, then the noise amplification in a perfectly equalized system is given as

$$\text{amp} = 10 \log_{10} \left[\int_{-(1+\alpha)\omega_N}^{(1+\alpha)\omega_N} \frac{RC(\omega)}{|H(\omega)|^2} \, d\omega \right] \text{db} \qquad (3.11\text{-}1)$$

The effect of the noise amplification can be summarized in the following way: if the value of the SNR (signal to noise ratio) ρ_M is 10 db and the noise amplification is 1 db, then the slicing or decision is made based on a new SNR of 9 db. This, of course, increases the probability of error or, equivalently, degrades the performance.

In systems where the specific characteristics of the medium of transmission are known as part of the design process (such as magnetic recording) or are fixed after installation (such as a dedicated line modem), it is possible to implement equalization with an analog filter. In most systems, particularly the telephone line, where the particular channel is not known in advance or where it is impractical to tailor a modem to each one of many known lines (such as the subscriber loop), we must turn to techniques of adaptive equalization. This, in turn, requires digital rather than analog filters to acheive the equalization because the mechanism for adaptation in analog filters is simply impractical. On the other hand, digital methods of implementing a modem are, as we shall see, quite accessible with current VLSI technology.

This means that we should begin to think about the effects of channel distortion resulting in ISI (intersymbol interference) in discrete rather than continuous terms. In order to accomplish this we shall introduce the idea of the sampled impulse response of a channel. The continuous inpulse response of the system shown in fig. 3.11-1 (with the analog equalizer $1/H(\omega)$ removed) is shown in fig. 3.11-2a. It is applied to a sampler at the symbol rate and the output of this sampler, called the sampled impulse response, is shown in fig. 3.11-2b. As shown, this sampled impulse response can be divided into three components: the main sample, those ISI terms that precede the main sample (referred to as leading echoes or precursors), and those ISI terms that follow the main sample (referred to as trailing echoes or postcursors).

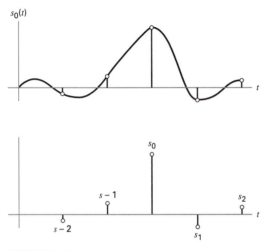

FIGURE 3.11-2

The length of the sampled impulse response, measured in the number of symbol intervals T_s, is denoted as M and is characterized as the memory of the channel.

When the transmitted data is uncorrelated the main sample is modified by a random sum of the ISI terms corresponding to data symbols that occur both before and after the piece of data corresponding to the main sample. In adding to the main sample, this ISI makes the main sample either closer to or further away from the decision threshold. When the ISI moves the main sample closer to the decision boundary, the noise may more easily cause an error.

In an unequalized system where the ISI terms close to the main sample may be large, it is difficult to estimate with precision the effect of ISI on the average probability of error. A number of upper bounds are described in Gitlin, Hayes, and Weinstein [GH&W92] (art. 4.9.4 and 4.9.5). The interested reader is referred to this reference. When the ISI terms are not too large and when there are a relatively large number of them, their random sum becomes a zero-mean Gaussian random variable having a variance equal to the sum of the squares of the ISI terms. This will be further explored in the Problems.

3.12 THE EYE PATTERN

Since the analog output signal of the receiver filter is the superposition of end-to-end channel impulse responses corresponding to random data at the transmitter input, it follows that this output depends on the particular data pattern. If it is applied to the vertical axis of an oscilloscope, with the negative edge of the timing signal used to trigger the horizontal sweep (the positive edge is used for sampling), and the time base adjusted appropriately, we will get the *eye pattern* of fig. 3.12-1 on the screen.

ISI will make the vertical opening of the eye smaller, indicating the degree to which the system is sensitive to noise. The width of the eye pattern indicates the sensitivity to timing error. In a distortionless raised-cosine system, the crossover of the eye pattern is dependent on the data pattern and is therefore smeared except when the rolloff $\alpha = 1$. In this case the crossover always takes place at the same point independent of the data pattern, corresponding to a maximum and stable width of the eye. This corresponds to Nyquist's

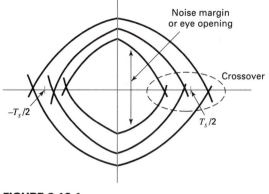

FIGURE 3.12-1

Second Criterion [Ben70]. Under these circumstances a stable timing can be derived from the zero crossings of the data pattern. Although ISI degrades this property, $\alpha = 1$ still corresponds to maximum stability of the crossovers. The eye pattern for multilevel data is simply a vertical collection of eyes.

Historically, examination of the eye pattern has been a very useful way getting a qualitative understanding of the performance of the system.

3.13 INTERSYMBOL INTERFERENCE AS A MARKOV PROCESS

It is of considerable interest to model the sampled impulse response of a distorted channel as a finite-state machine or Markov process. As we shall see in Chapter 5, this provides the basis for overcoming the effects of ISI through the use of the Viterbi Algorithm.

We shall proceed through example. Consider the sampled impulse response of fig. 3.13-1, which consists of one leading echo $h_{-1} = -0.3$, a main sample $h_0 = 1$, and one trailing echo $h_1 = 0.2$. The model is a transversal filter having two storage or delay elements for previous inputs. The state space of the system is the set of ordered pairs of previous inputs (u_{k-1}, u_{k-2}). In this example we will assume a binary input $u_k = \pm 1$ so that the set of states can be defined as

	u_{k-1}	u_{k-2}
q_0	$+1$	$+1$
q_1	$+1$	-1
q_2	-1	$+1$
q_3	-1	-1

$$(3.13\text{-}1a)$$

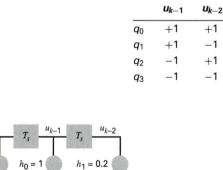

FIGURE 3.13-1

and we can easily generate a state transition table

u_k	$+1$	-1
q_0	q_0/a_0	q_2/a_1
q_1	q_0/a_2	q_2/a_3
q_2	q_1/a_4	q_3/a_5
q_3	q_1/a_6	q_3/a_7

(3.13-1b)

where the outputs are

$$a_0 = +h_{-1} + h_0 + h_1 = 0.9$$
$$a_1 = -h_{-1} + h_0 + h_1 = 1.5$$
$$a_2 = +h_{-1} + h_0 - h_1 = 0.5$$
$$a_3 = -h_{-1} + h_0 - h_1 = 1.1$$
$$a_4 = +h_{-1} - h_0 + h_1 = -1.1$$
$$a_5 = -h_{-1} - h_0 + h_1 = -0.5$$
$$a_6 = +h_{-1} - h_0 - h_1 = -1.5$$
$$a_7 = -h_{-1} - h_0 - h_1 = -0.9$$

(3.13-1c)

Note that this structure is a Mealy machine in which the output depends both on the current state and the current input. Note, as well, that if the input to this structure was $L = 4$ levels ($\pm 1, \pm 3$) we would have $4^2 = 16$ states and if the channel memory were $M = 3$ we would have $4^3 = 64$ states. Generally, the number of states in this machine is L^M. For multilevel systems having large memory this rapidly becomes a very large number of states.

This characterization of a distorted channel also has the advantage of helping to create a conceptual unity. We already have characterized the coding of data (also called line coding) as a Markov process and we are about to characterize Partial Response Signaling as a Markov process as well.

3.14 PARTIAL RESPONSE SIGNALING

We have just seen that if we want to transmit data on a bandlimited channel at a rate of f_s symbols/sec, then we must have a channel having a (one-sided) bandwidth of $(1 + \alpha)f_s/2$ or $(1 + \alpha)f_N$ in order to guarantee having some rolloff. Under these circumstances we can acheive zero ISI but we have excess bandwidth. Such systems are called *full response*.

In a *partial response* system the required channel bandwidth is reduced to exactly the Nyquist bandwidth but we accomplish this by introducing a *known* ISI that we can subsequently remove at the receiver.

Partial Response Signaling has been used [Pas77] in PCM, the digital transmission of analog telephone signals, and in type T1 telephone lines. The major current application of partial response signaling, which we will discuss, is in magnetic recording.

3.14.1 Class-1 Partial Response (Duobinary)

A block diagram that illustrates a mathematically correct but stucturally abstract Duobinary system is shown in fig. 3.14-1. Data impulses are applied to the Duobinary encoder, which has an impulse response

$$g(t) = \delta(t) + \delta(t - T_s)$$

(3.14-1a)

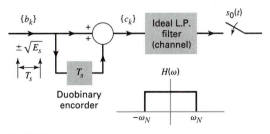

FIGURE 3.14-1

and a transfer function

$$G(z) = 1 + z^{-1} \qquad (3.14\text{-}1b)$$

and with $z = e^{j\omega T_s}$

$$G\left(e^{j\omega T_s}\right) = 1 + e^{-j\omega T_s} = 2\mathrm{Cos}\left(\frac{\omega T_s}{2}\right)e^{-j\omega T_s/2} \qquad (3.14\text{-}1c)$$

which is clearly periodic. Note that the effect on the power spectral density of the data resulting from this encoding is given by

$$\left|G\left(e^{j\omega T_s}\right)\right|^2 = 4\mathrm{Cos}^2\left(\frac{\omega T_s}{2}\right) = 2(1 + \mathrm{Cos}(\omega T_s)) \qquad (3.14\text{-}1d)$$

This result will be verified using the approach outlined in eq. 2.6-1d.

$$P_x(\omega) = \frac{E_h(\omega)}{T_s}\left[R_h(0) + 2\sum_{n=1}^{\infty} R_h(n)\mathrm{Cos}(n\omega T_s)\right] \qquad (2.6\text{-}1d)$$

We will undertake to find the autocorrelation function of the Duobinary encoder output by first exhautively listing all of the data patterns of length 4 because the encoder clearly has a memory of 1. We shall see that this is sufficient.

b_{k-1}	b_k	b_{k+1}	b_{k+2}	c_k	c_{k+1}	c_{k+2}	
+1	−1	−1	−1	0	−2	−2	
	−1	−1	+1	0	−2	0	
	−1	+1	−1	0	0	0	
	−1	+1	+1	0	0	+2	
	+1	−1	−1	+2	0	−2	
	+1	−1	+1	+2	0	0	
	+1	+1	−1	+2	+2	0	
	+1	+1	+1	+2	+2	+2	(3.14-2a)
−1	−1	−1	−1	−2	−2	−2	
	−1	−1	+1	−2	−2	0	
	−1	+1	−1	−2	0	0	
	−1	+1	+1	−2	0	+2	
	+1	−1	−1	0	0	−2	
	+1	−1	+1	0	0	0	
	+1	+1	−1	0	+2	0	
	+1	+1	+1	0	+2	+2	

We can now compute the values of the autocorrelation function of the data:

$$R_b(0) = E[c_k c_k] = (0)^2 \frac{1}{2} + (2)^2 \frac{1}{4} + (-2)^2 \frac{1}{4} = 2$$

$$R_b(1) = E[c_k c_{k+1}] = (-2)(-2)\frac{1}{8} + (2)(2)\frac{1}{8} = 1 \tag{3.14-2b}$$

$$R_b(2) = E[c_k c_{k+2}] = (-2)(-2)\frac{1}{16} + (2)(-2)\frac{1}{16} + (-2)(2)\frac{1}{16} + (2)(2)\frac{1}{16} = 0$$

Substituting these values, we get

$$P_x(\omega) = \frac{E_h(\omega)}{T_s}\left[R_h(0) + 2\sum_{n=1}^{\infty} R_h(n)\operatorname{Cos}(n\omega T_s)\right] = \frac{2E_h(\omega)}{T_s}[1 + \operatorname{Cos}(\omega T_s)] \tag{3.14-2c}$$

which is the same result as eq. 3.14-1d.

This result can also be confirmed by interpreting the encoding rule as a state machine and using the techniques of Markov processes to find the power spectral density. This is referred to the Problems.

The encoded data is passed through the ideal baseband channel, which cuts off at the Nyquist frequency, and the output is sampled at the receiver. We can obtain the overall end-to-end system impulse response by convolving the impulse response with the impulse response of the channel:

$$s_0(t) = [\delta(t) + \delta(t - T_s)] \otimes \left[\frac{1}{T_s}\operatorname{Sinc}\left(\pi\frac{t}{T_s}\right)\right]$$

$$= \frac{1}{T_s}\left[\frac{\operatorname{Sin}\left(\pi\frac{t}{T_s}\right)}{\pi\frac{t}{T_s}} + \frac{\operatorname{Sin}\left(\pi\frac{(t-T_s)}{T_s}\right)}{\pi\frac{(t-T_s)}{T_s}}\right]$$

$$= \frac{1}{T_s}\operatorname{Sin}\left(\pi\frac{t}{T_s}\right)\left[\frac{1}{\pi\frac{t}{T_s}} - \frac{1}{\pi\frac{(t-T_s)}{T_s}}\right] = T_s\frac{\operatorname{Sin}\left(\pi\frac{t}{T_s}\right)}{\pi t(T_s - t)} \tag{3.14-3}$$

This impulse response is shown in fig. 3.14-2a. Note that the samples at $t = 0$ and $t = T_s$ are each equal to $1/T_s$ but that the samples at all other multiples of T_s are equal to zero. This is the controlled ISI. The end-to-end frequency response of the system is shown in fig. 3.14-2b. Note that the system bandwidth is the Nyquist frequency and that the transfer function is quite sharp and not strictly realizable. It can, however, be approximated. Assuming that the timing of the sampler is correct, the impulse response seen after the sampler, the sampled impulse response, is shown in fig. 3.14-2c.

Although this spectrum has been created through a delay and add operation, it could just as well have been created by applying the data impulses to a filter that produced the same spectrum. In that case we would be able to split the filters equally between the transmitter and the receiver according to rules of matched filter theory. This version of the receiver is shown in fig. 3.14-3a, where we have introduced scaling to make the energy of the transmitted signal equal to E_s when the data impulse function has amplitude $\pm\sqrt{E_s}$. Note that this scaling causes the sample points on the impulse response to be equal to $\pm\frac{\pi}{4}\sqrt{E_s}$.

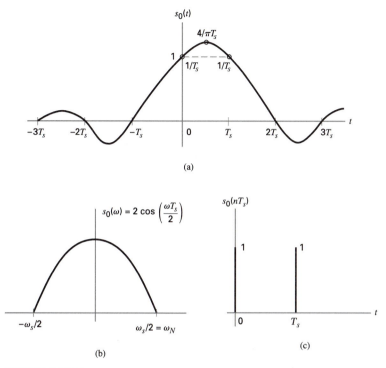

(b)

FIGURE 3.14-2

Because the successive signals overlap in this system, the sample of the signal will be $0, \pm 2(\frac{\pi}{4})\sqrt{E_s}$, as shown in fig. 3.14-3b. The noise power out of the properly scaled receiver filter is:

$$\sigma^2 = \frac{\eta}{2}\frac{1}{2\pi}\int_{-\omega_s/2}^{\omega_s/2} |G(\omega)|\, d\omega = \frac{\eta}{2}\frac{1}{2\pi}\int_{-\omega_s/2}^{\omega_s/2} \frac{\pi}{2}T_s\cos\left(\frac{\omega T_s}{2}\right) d\omega = \frac{\eta}{2} \qquad (3.14\text{-}4)$$

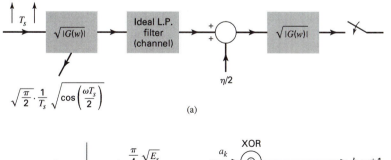

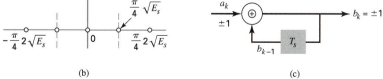

FIGURE 3.14-3

We shall set the decision boundaries at $\pm\frac{\pi}{4}\sqrt{E_s}$ so that the probability of the noise exceeding a boundary is

$$Q\left(\frac{\pi}{4}\sqrt{2E_s/\eta}\right) = Q\left(\frac{\pi}{4}\sqrt{\rho_m}\right). \tag{3.14-5}$$

Note that the choice of decision boundaries is not a simple one because the three possible received points do not have equal probability. This issue will be left to the Problems.

To compute the probability of a decision error, we must recognize that the data has a memory of one symbol. Characterizing the original data as $+1, -1$:

$$
\begin{aligned}
P_e &= P(\text{err}/+1, +1)P(+1, +1) + P(\text{err}/+1, -1)P(+1, -1) \\
&\quad + P(\text{err}/-1, +1)P(-1, +1) + P(\text{err}/-1, -1)P(-1, -1) \\
&= \frac{1}{4}[P(\text{err}/+1, +1) + P(\text{err}/+1, -1) + P(\text{err}/+1, -1) + P(\text{err}/-1, -1)] \\
&= \frac{1}{2}[P(\text{err}/+1, +1) + P(\text{err}/+1, -1)] \\
&= \frac{1}{2}[2 + 1]Q\left(\frac{\pi}{4}\sqrt{\rho_m}\right) = \frac{3}{2}Q\left(\frac{\pi}{4}\sqrt{\rho_m}\right)
\end{aligned} \tag{3.14-6}
$$

Note that the performance is worse than for a comparable binary matched filter system. In order to acheive the same probability of error it is necessary to have a SNR that is 2.1 db greater. In Chapter 10 we shall see that we can recover most of the 2.1 db through the use of a detection process called maximum likelihood sequence estimation (MLSE) which is implemented with a technique called the Viterbi Algorithm.

The performance loss, however, is not the fundamental problem with this system. The real problem is that if a decision error is made, that error either propogates or hangs up the system. An example of this is given below:

transmit	+1	+1	+1	−1	−1	+1	−1	+1	−1	−1
receive		+2	+2	0	−2	0	0	0	0	−2
slicer error		+2	+2	0	−2	+2*	0	0	0	−2
decode		+1	+1	−1	−1	#	hangup			

This fundamental problem of partial response systems is dealt with by precoding the data. The precoding for the Duobinary system is shown in fig. 3.14-3c. It is a mirror image of the encoding rule, except that it uses mod(2) rather than regular addition. The rule is:

$$b_k = a_k \oplus b_{k-1} = \begin{cases} -1 & \text{if } a_k = b_{k-1} \\ +1 & \text{if } a_k \neq b_{k-1} \end{cases} \tag{3.14-7a}$$

where -1 is a substitute for 0 in the exclusive-OR operation. The operation of this rule can be easily demonstrated by example.

a_k		+1	+1	−1	+1	−1	−1	+1	−1	+1
b_k	+1	−1	+1	+1	−1	−1	−1	+1	+1	−1
c_k		0	0	+2	0	−2	−2	0	+2	0
data		+1	+1	−1	+1	−1	−1	+1	−1	+1

The decoding rule here is quite straightforward.

$$
\begin{aligned}
&\text{if } c_k = 0 \quad \text{then data} = +1 \\
&\text{if } c_k = \pm 2 \quad \text{then data} = -1
\end{aligned} \tag{3.14-7b}
$$

A small bit of experimentation will reveal that a decision error, one that in effect makes a wrong choice among 0, +2, −2, does not propagate.

Finally, this precoding does not affect the power spectral density of the signal. This can be seen by again invoking our technique of exhaustively listing the possible combinations of input and output.

a_k	a_{k+1}	a_{k+2}	b_{k-1}	b_k	b_{k+1}	b_{k+2}	
−1	−1	−1	+1	+1	+1	+1	
−1	−1	+1		+1	+1	−1	
−1	+1	−1		+1	−1	−1	
−1	+1	+1		+1	−1	+1	(3.14-7c)
+1	−1	−1		−1	−1	−1	
+1	−1	+1		−1	−1	+1	
+1	+1	−1		−1	+1	+1	
+1	+1	+1		−1	+1	−1	

A close examination of these results reveals that all of the input seqences are reproduced at the output, although not in the same order. This means that the power spectral density will not be changed. This result can easily be verified arithmetically to show that $R(0) = 1$ and for all other values $R(n) = 0$.

We promised that it would be possible to transmit data on a channel faster using partial response signalling than using raised cosine (Nyquist) signaling. If we have an ideal lowpass channel with a cutoff frequency of ω_1, then in duobinary we can make the Nyquist frequency ω_N equal to ω_1. For a raised cosine system, the cutoff frequency of the channel is $\omega_1 = (1 + \alpha)\omega_N$ where α may be as large as $\alpha = 1$. Depending on the value of α, the data rate of PRS may be up to twice that of raised cosine transmission.

3.14.2 Class-4 Partial Response (Modified Duobinary)

In this variant of a partial response system shown in fig. 3.14-4, the encoding rule is

$$g(t) = \delta(t) - \delta(t - 2T_s) \tag{3.14-8a}$$

generating a transfer function

$$G(z) = 1 - z^{-2} = (1 - z^{-1})(1 + z^{-1}) \tag{3.14-8b}$$
$$G(e^{j\omega T_s}) = 1 - e^{-j\omega 2T_s} = 2j\mathrm{Sin}\,(\omega T_s)e^{-j\omega T_s} \tag{3.14-8c}$$

which is clearly periodic. For future reference, the term $(1 - z^{-1})$ causes the transfer function to go to zero at dc. Note that the effect on the power spectral density of the data resulting from this encoding is given by

$$|G(e^{j\omega T_s})|^2 = 4\mathrm{Sin}^2(\omega T_s) = 2(1 - \mathrm{Cos}\,(2\omega T_s)) \tag{3.14-8d}$$

This result will be verified using the approach outlined in eq. 2.6-1d. We will undertake to find the autocorrelation function of the Class-4 encoder output by first exhautively listing all of the data patterns of length 5 because the encoder clearly has a memory of 2. We shall

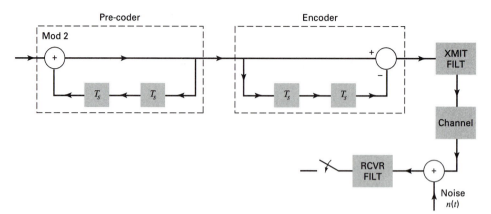

FIGURE 3.14-4

see that this is sufficient.

b_{k-2}	b_{k-1}	b_k	b_{k+1}	b_{k+2}	c_k	c_{k+1}	c_{k+2}
+1	+1	−1	−1	−1	−2	−2	0
		−1	−1	+1	−2	−2	+2
		−1	+1	−1	−2	0	0
		−1	+1	+1	−2	0	+2
		+1	−1	−1	0	−2	−2
		+1	−1	+1	0	−2	0
		+1	+1	−1	0	0	−2
		+1	+1	+1	0	0	0
+1	−1	−1	−1	−1	−2	0	0
		−1	−1	+1	−2	0	−2
		−1	+1	−1	−2	+2	0
		−1	+1	+1	−2	+2	+2
		+1	−1	−1	0	0	−2
		+1	−1	+1	0	0	0
		+1	+1	−1	0	+2	−2
		+1	+1	+1	0	+2	0

(3.14-8e)

The reader may undertake to complete the table.

We can now compute the values of the autocorrelation function of the data:

$$R_c(0) = E[c_k c_k] = (0)^2 \frac{1}{2} + (2)^2 \frac{1}{4} + (-2)^2 \frac{1}{4} = 2$$

$$R_c(1) = E[c_k c_{k+1}] = 0 \qquad (3.14\text{-}8f)$$

$$R_c(2) = E[c_k c_{k+2}] = -1$$

Substituting these values, we get

$$P_x(\omega) = \frac{E_h(\omega)}{T_s}\left[R_h(0) + 2\sum_{n=1}^{\infty} R_h(n)\cos(n\omega T_s)\right] = \frac{2E_h(\omega)}{T_s}[1 - \cos(2\omega T_s)]$$

(3.14-8g)

which is the same result as eq. 3.14-8d.

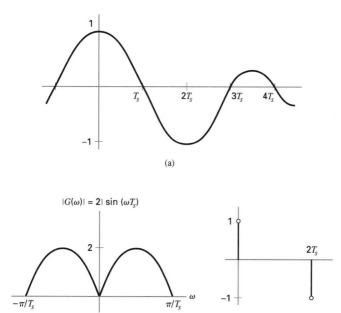

(b)   (c)

FIGURE 3.14-5

The encoded data is passed through the ideal baseband channel, which cuts off at the Nyquist frequency, and the output is sampled at the receiver. We can obtain the overall end-to-end system impulse response by convolving the impulse response with the impulse response of the channel:

$$s_0(t) = [\delta(t) - \delta(t - 2T_s)] \otimes \left[\frac{1}{T_s} \text{Sinc}\left(\pi \frac{t}{T_s}\right)\right]$$

$$= \frac{1}{T_s} \left[\frac{\text{Sin}\left(\pi \frac{t}{T_s}\right)}{\pi \frac{t}{T_s}} - \frac{\text{Sin}\left(\pi \frac{(t - 2T_s)}{T_s}\right)}{\pi \frac{(t - 2T_s)}{T_s}}\right]$$

$$= \frac{1}{T_s} \text{Sin}\left(\pi \frac{t}{T_s}\right) \left[\frac{1}{\pi \frac{t}{T_s}} - \frac{1}{\pi \frac{(t - 2T_s)}{T_s}}\right] = 2T_s \frac{\text{Sin}\left(\pi \frac{t}{T_s}\right)}{\pi t(2T_s - t)} \qquad (3.14\text{-}9)$$

This impulse response is shown in fig. 3.14-5a. Note that the sample at $t = 0$ is equal to $+1$ and the sample at $t = 2T_s$ is -1 but that the samples at all other multiples of T_s are equal to zero. This is the controlled ISI. The end-to-end frequency response of the system is shown in fig. 3.14-5b. Note that, in contrast to the Duobinary spectrum, the Class-4 spectrum does not include dc and places relatively little emphasis on the lower frequencies. The sampled impulse response is shown in fig. 3.14-5c.

Just as in the Duobinary case, this spectrum could just as well have been created by applying the data impulses to an appropriate filter. Again, we are able to split the filters equally between the transmitter and the receiver according to rules of matched filter theory. This version of the receiver, the same as the duobinary but with different filters, is shown in fig. 3.14-3a. We will scale this system in the same way that we scaled duobinary.

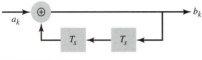

FIGURE 3.14-6

Therefore the sample of the signal will be $0, \pm 2\frac{\pi}{4}\sqrt{E_s}$ as shown in fig. 3.14-3b. The noise power out of the receiver filter is:

$$\sigma^2 = \frac{1}{2\pi}\frac{\eta}{2}\frac{\pi}{4}\int_{-\omega_s/2}^{\omega_s/2}|G(\omega)|\,d\omega = \frac{2}{2\pi}\frac{\eta}{2}\frac{\pi}{4}\int_0^{\omega_s/2}2T_s\mathrm{Sin}\,(\omega T_s)\,d\omega = \frac{\eta}{2} \qquad (3.14\text{-}10)$$

We shall set the decision boundaries at $\pm\frac{\pi}{4}\sqrt{E_s}$ so that the probability of the noise exceeding a boundary is

$$Q\left(\frac{\pi}{4}\sqrt{2E_s/\eta}\right) = Q\left(\frac{\pi}{4}\sqrt{\rho_m}\right) \qquad (3.14\text{-}11)$$

Following the procedure for the Duobinary case, we again find that

$$P_e = \frac{3}{2}Q\left(\frac{\pi}{4}\sqrt{\rho_m}\right) \qquad (3.14\text{-}12)$$

Class-4 partial response, like all partial response systems, has the problem that a decision error tends to propogate or hang up the system because the obvious decoding rule depends on past decisions. The data is precoded to avoid this problem. The precoding rule is always a mirror image, with an XOR operation, of the partial response encoding rule. For Class 4, this precoding is:

$$b_k = a_k \oplus b_{k-2} = \begin{cases} -1 & \text{if } a_k = b_{k-2} \\ +1 & \text{if } a_k \neq b_{k-2} \end{cases} \qquad (3.14\text{-}10)$$

and is shown in fig. 3.14-6.

The operation of this rule is essentially the same as for Duobinary:

a_k			+1	−1	+1	−1	−1	+1	−1	+1	
b_k	+1	−1	−1	−1	+1	−1	+1	+1	+1	−1	
c_k			−2	0	+2	0	0	+2	0	−2	(3.14-11a)
data		+1	+1	−1	+1	−1	−1	+1	−1	+1	

and again the decoding rule is quite straightforward:

$$\begin{aligned} \text{if } c_k = 0 \qquad & \text{then data} = -1 \\ \text{if } c_k = \pm 2 \qquad & \text{then data} = +1 \end{aligned} \qquad (3.14\text{-}11b)$$

Again, a decision error, one that in effect makes a wrong choice among $0, +2, -2$, does not propogate.

Finally, this precoding does not affect the power spectral density of the signal. This can be seen by again invoking our technique of exhaustively listing the possible combinations of input and output. This will be left to the Problems.

3.14.3   Generalized Partial Response Systems

It is quite straightforward to expand the discussion of Class-1 and Class-4 partial response systems to formulate a more general approach to partial response systems. For the purposes

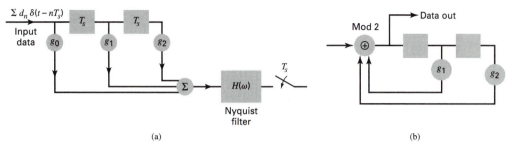

FIGURE 3.14-7

of this discussion the basic polynomial of the PRS system will be confined to the form

$$G(z) = (1 - z^{-1})^m (1 + z^{-1})^k \qquad (3.14\text{-}12a)$$

which can be expanded to

$$G(z) = \sum_{n=0}^{N-1} g_n z^{-n} \qquad (3.14\text{-}12b)$$

where $z = e^{j\omega T_s}$. This polynomial can always be implemented as a transversal filter, as shown in fig. 3.14-7a. When $m = 0$ and $k = 1$ we get the Class-1 (Duobinary) polynomial $G(z) = 1 + z^{-1}$ and when $m = k = 1$ we get the Class-4 polynomial $G(z) = 1 - z^{-2}$.

The output of the system at the sampler is then

$$s_0(t) = \sum_{n=0}^{N-1} g_n h(t - nT_s) \qquad (3.14\text{-}12c)$$

where $h(t)$ is the impulse response of the system filter $H(\omega)$.

The effect of the $(1 - z^{-1})$ term is to cause $G(\omega) = G(e^{j\omega T_s})$ to be equal to zero at $\omega = 0$ and the effect of the $(1 + z^{-1})$ term is to cause $G(\omega)$ to be equal to zero at $\omega = \omega_N = \omega_S/2$, the Nyquist frequency.

We shall continue to assume that, as shown in fig. 3.14-7a, $H(\omega)$ is an ideal lowpass filter having a cutoff at the Nyquist frequency. In all of these systems the precoding will be constructed in a manner similar to that of our two previous examples. This is shown in fig. 3.14-7b. In all cases, as in the examples, the precoding will not affect the power spectral density of the transmitted signal but will act against error propogation.

We will examine some of the properties of generalized partial response in the Problems and subsequently in relation to their use in magnetic recording.

3.15 MAGNETIC RECORDING AND PARTIAL RESPONSE SIGNALING

3.15.1 Basic Ideas of Magnetic Recording

The basic structure of a magnetic recording channel is shown in fig. 3.15-1. The digital data $\{a_k = \pm 1\}$ at the input to the system is differentially encoded to a data sequence $\{b_k = \pm 1\}$ (also referred to as NRZI encoding) according to the rule

$$\begin{aligned} a_k = 0 &\Rightarrow b_k = b_{k-1} \\ a_k = 1 &\Rightarrow b_k = -b_{k-1} \end{aligned} \qquad (3.15\text{-}1)$$

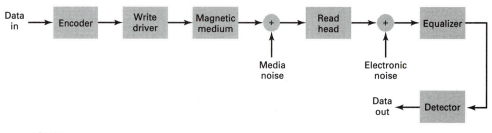

FIGURE 3.15-1

Positive and negative values of b_k cause the write driver current to be positive or negative (fig. 3.15-2) in turn inducing a saturated magnetic flux $m(t)$ in one direction or the other onto the magnetic disc. Note that according to eq. 3.15-1 the data $a_k = 0$ corresponds to the absence of a transition and $a_k = 1$ corresponds to the presence of any transition in the direction of the flux. The response of the read head to a single transition in the "positive" direction of the flux is referred to as the step response $s(t)$ (fig. 3.15-2d) which is typically characterized as the "modified Lorentzian family"

$$s(t) = \frac{dm(t)}{dt} = \frac{1}{1 + (2t/\tau)^{2x}} \tag{3.15-2}$$

where τ is the "pulse-width-50" and the parameter x defines a family of shapes. The value $x = 1$ corresponds to the Lorentz pulse shape and it has been shown that $1 < x < 1.5$ yields more accurate characterizations of the actual step response. Note that the geometry of the read head will yield a step response that is symmetric in time around the actual transition.

For a given disc track velocity, the symbol interval T_s is directly related to the minimum distance between flux transitions. We shall first consider the circumstance where these transitions are sufficiently far from each other that ISI (intersymbol interference) is absent from the head response. A prototypical receiver configuration for this relatively low-density recording is shown in fig. 3.15-3 [Cio90, fig. 7].

This receiver has a data path and a timing path. In the data path the signal is first full-wave rectified, since the data encoding of eq. 3.15-1 establishes that any transition corresponds to $a_k = 1$. It is then applied to a threshold detector in order to shape the pulse.

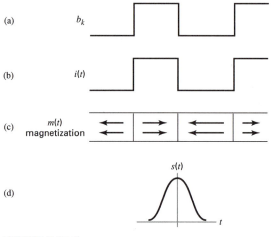

(a) b_k

(b) $i(t)$

(c) $m(t)$ magnetization

(d)

$s(t)$

FIGURE 3.15-2

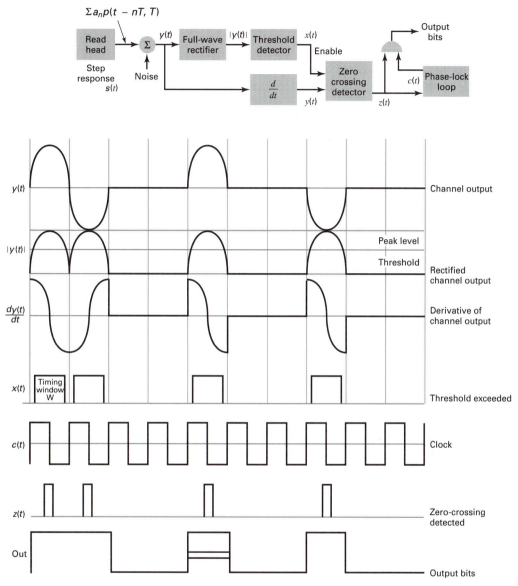

FIGURE 3.15-3   ([Cio90] reprinted with permission)

In the timing path the received signal is differentiated in order to convert the peaks of the received signal into zero crossings. The zero crossing detector, enabled by the output of the threshold detector, generates narrow pulses that occur only with a transition. This means that the pulses do not occur when $a_k = 0$. Accordingly, a clock is established by using these random pulses as input to a digital phase-lock loop. The output data is established by using the zero crossing pulses to gate the clock.

3.15.2   Partial Response Signaling for High-Density Magnetic Recording

When the density of the magnetic flux transitions becomes too great, when T_s becomes too small, the resulting ISI makes it necessary to use more sophisticated data communications

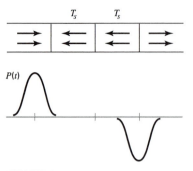

FIGURE 3.15-4

techniques to record and detect the data. It is in this context that Kobayashi and Tang [K&T70] considered Class-4 Partial Response (Modified Duobinary) for magnetic recording.

In order to understand how Class 4 can be used in this way, first consider fig. 3.15-4, which depicts the read head output voltage corresponding to two flux transitions occurring $2T_s$ sec apart. This output is called the pulse response $p(t)$ where

$$p(t) = s(t) - s(t - 2T_s) \qquad (3.15\text{-}3a)$$

is a positive pulse at $t = 0$ followed by a zero at $t = T_s$ and a negative pulse at $t = 2T_s$, corresponding to the basic structure of Class-4 Partial Response as given in eq. 3.14-8a. If the data stream $b_k = \pm 1$ is applied to the system every $2T_s$ seconds, the read head output voltage is given by

$$v(t) = \sum_k b_k p(t - k2T_s) \qquad (3.15\text{-}3b)$$

If two succesive values of b_k, occurring at intervals of $2T_s$, are the same then the output voltage at the sample point will be zero; if they are different, the sampled output voltage will be ± 2 depending on the direction of the difference. This exactly corresponds to the magnetic channel encoding approach of eq. 3.15-1. Note that the development given above does not directly use the fact that $p(T_s) = 0$. This succession of zeros, occurring at all odd multiples of T_s, enables the insertion of a second data stream having exactly the same characteristics of that of eq. 3.15-5b but quite independent of it.

It is fairly straightforward to recognize (confirmation will be left to the Problems) that the combination of the two interleaved encoded data streams is equivalent to the Class-4 Partial Response System of fig. 3.14-4 where the precoding takes the place of the NRZI coding of eq. 3.15-1.

Of course, the response of the magnetic channel is not exactly that of Class-4 Partial Response; rather, it is a distorted version of that response requiring some sort of equalization. We will be discussing a variety of approaches of implementing this equalization in Chapters 10 and 11. The interested reader is also referred to Mallinson [Mal75] and Kobayashi [Kob71].

3.15.3 Codes for Magnetic Recording

There is a considerable literature on the development of line codes for magnetic recording. One of the issues is the shaping of the signal to the specific characteristics of the magnetic channel. Another issue arises from the use of the zero-crossings of the received signal it-self to generate a receiver clock. A long string of zeros in the data stream will result in a loss of clock information. A class of RLL (run-length-limited) codes will prevent too many consecutive zeros from appearing. Discussion of these codes is beyond the scope of the text. The interested reader is referred to Siegel and Wolfe [S&W91] and Bergmans [Ber96].

3.16 HIGHER-ORDER PARTIAL RESPONSE SYSTEMS FOR MAGNETIC RECORDING

In considering PRS polynomials of order higher than those of Class 1 and Class 4, we will be looking for those characteristics that make the system easier to implement. For example, if we think about implementation in relation to fig. 3.14-3a we see that the transmitter and receiver filters for Class 4 will be relatively difficult to implement because they have such sharp rolloff. Thapar and Patel [T&P87] have proposed that for magnetic channels we should look at the class of encoding rules

$$G(z) = (1 - z^{-1})(1 + z^{-1})^k \qquad (3.16\text{-}1)$$

where $k = 1$ corresponds to Class 4. Every member of this class exhibits the essential characteristic of Class 4, namely the positive and negative peaks at $t = 0, 2T_s$ and the zero at $t = T_s$. The higher-order members of this class carry some advantages and a few disadvantages.

As an example let us consider $k = 2$, so that

$$G(z) = (1 - z^{-1})(1 + z^{-1})^2 = (1 + z^{-1} - z^{-2} - z^{-3}) \qquad (3.16\text{-}2a)$$

and

$$G\left(e^{-j\omega T_s}\right) = \left[2j\,\mathrm{Sin}\left(\frac{\omega T_s}{2}\right)e^{-j\omega T_s/2}\right]\left[2\,\mathrm{Cos}^2\left(\frac{\omega T_s}{2}\right)e^{-j\omega T_s}\right] \qquad (3.16\text{-}2b)$$

$$\left|G\left(e^{-j\omega T_s}\right)\right| = \left|4\,\mathrm{Sin}\left(\frac{\omega T_s}{2}\right)\mathrm{Cos}^2\left(\frac{\omega T_s}{2}\right)\right| \quad \text{for } |\omega| \le \omega_N. \qquad (3.16\text{-}2c)$$

We can see from fig. 3.16-1 that this filter is easier to implement than Class 4 but has the same essential characteristics at $\omega = 0$ and $\omega = \omega_N$.

A second advantage of the $k = 2$ filter over Class 4 has to do with the decay of the overall system impulse response $s_0(t)$. It has been shown in the Problems that if $S_0(\omega)$ and and its first $K - 1$ derivatives are continuous and the Kth derivative is not, then the magnitude of $s_0(t)$ decays asymptotically as $1/|t|^{K+1}$. For a nonideal channel this is very desirable because it tends to reduce the resulting unintended intersymbol interference. As an example, consider Class 4, where

$$|S_0(\omega)| = \begin{cases} 2T_s\,\mathrm{Sin}(\omega T_s); & |\omega| \le \omega_N \\ 0; & \text{otherwise} \end{cases} \qquad (3.16\text{-}3a)$$

so that

$$\frac{d|S_0(\omega)|}{d\omega} = \begin{cases} 2T_s^2\,\mathrm{Cos}(\omega T_s); & |\omega| \le \omega_N \\ 0; & \text{otherwise} \end{cases} \qquad (3.16\text{-}3b)$$

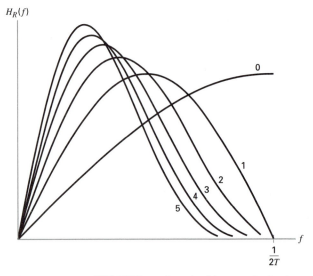

FIGURE 3.16-1 ([T&P87] reprinted with permission)

At $\omega = \omega_N$ this derivative is discontinuous, so that in this case $s_0(t)$ decays as $1/t$. We will see in the Problems that with $k = 2$ the first derivative is continuous but the second derivative is not so it decays as $1/t^2$.

These two criteria, ease of filter constuction and rate of decay, would seem to point in the direction of increasing the value of k. However as k is increased the number of levels in the received signal also increases. In Class 1 and Class 4 we had three levels but in the example given above with $k = 2$ we get five levels. As the number of levels increases there results an increase in the probability of error. This is the basic limiting factor.

3.17 DECISION FEEDBACK DETECTION OF PARTIAL RESPONSE SIGNALS

An alternative to precoding the PRS signal and using symbol-by-symbol detection is the use of the DF (decision feedback) structure of fig. 3.17-1. A partial response signal may be thought of as consisting of a main sample followed by a collection of controlled ISI terms making up the memory of the system. The feedback taps in the DF are the same as

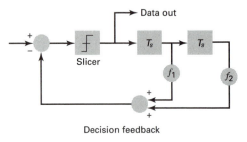

FIGURE 3.17-1

these ISI terms. Consequently a decision made on the main sample, which occurs first, is fed back, multiplied by the appropriate tap, to cancel the following controlled ISI terms exactly.

One can think of precoding and decision feedback as complementary ways of dealing with the memory of the partial response signal. Precoding provides anticipation which effectively cancels the PRS memory to allow symbol-by-symbol detection. Decision feedback incorporates the PRS memory into the detector.

The difficulty with the DF approach is that a decision error may cause error propagation. Unlike the original PRS system without precoding, this error propagation does not result in system hangup. Typically the system recovers after a string of errors but that string has the effect of increasing the average bit probability of error.

Computing this error probability is not an easy task. Kabal and Pasupathy [K&P75] find an upper bound to this quantity and the interested reader is referred to that paper.

A little thought will lead to the observation that DF can more generally be used to cancel trailing echos of any signal with ISI and therefore be used as an equalizer. We will give considerable attention to this subsequently. Of course, some other means will have to be used to equalize leading echos before the signal is applied to the DFE. In particular, it has been noted (Ahmed et al. [ABG81]) that the impulse response of twisted pair in the ISDN has only trailing echos. This makes DF the equalizer of choice for that system.

Finally, the alternative between precoding and DFE will manifest itself on a higher level when we consider a generalization of precoding called Tomlinson-Harashima coding that serves as an alternative to equalization in high-performance modems. This topic will be discussed in Chapter 10.

3.18 SCRAMBLERS

In subsequent discussions of timing recovery, adaptive equalization, and adaptive echo cancellation, it is necessary to guarantee that the transmitted data are essentially uncorrelated. Since voice, fax, video, and other forms of data are strongly correlated, it becomes necessary to decorrelate the data for transmission and then restore the data as the final step in the receiver. This is accomplished with a class of devices known as scramblers together with their associated descramblers (Gitlin and Hayes [G&H75]).

In order to understand the functioning of a scrambler, first consider the shift register based sequential machine of fig. 3.18-1a. If all of the feedback coefficients are either 0 or 1 and input summation is modulo-2 then this circuit can be shown to be linear (Golomb [Gol67], Gill [Gil67]) and can be defined by the polynomial

$$G(D) = 1 + c_1 D + c_2 D^2 + \cdots + c_m D^m \qquad (3.18\text{-}1a)$$

where D is a delay operator. (We note that in relation to the z-transform, $D = z^{-1}$). Note that the modulo-2 operation can be implemented by exclusive-OR gates. If the input data sequence is denoted u_n and the output data sequence is denoted y_n then

$$y_n = u_n \oplus \sum_{i=1}^{m} c_i y_{n-i} \qquad (3.18\text{-}1b)$$

The behavior of the sequential machine is strongly dependent on the polynomial $G(D)$. Using algebraic techniques beyond the scope of this treatment but accessible through the

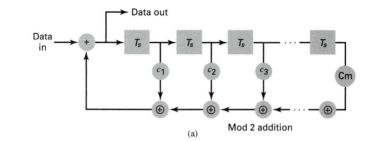

(a)

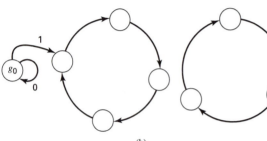

(b)

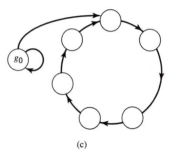

(c)

N	G(D)	N	G(D)
1	$1 + D$	18	$1 + D^7 + D^{18}$
2	$1 + D + D^2$	19	$1 + D + D^2 + D^5 + D^{19}$
3	$1 + D + D^3$	20	$1 + D^3 + D^{20}$
4	$1 + D + D^4$	21	$1 + D^2 + D^{21}$
5	$1 + D^2 + D^5$	22	$1 + D + D^{22}$
6	$1 + D + D^6$	23	$1 + D^5 + D^{23}$
7	$1 + D^3 + D^7$	24	$1 + D + D^2 + D^7 + D^{24}$
8	$1 + D^2 + D^3 + D^4 + D^8$	25	$1 + D + D^{25}$
9	$1 + D^4 + D^9$	26	$1 + D + D^2 + D^{26}$
10	$1 + D^3 + D^{10}$	27	$1 + D + D^2 + D^5 + D^{27}$
11	$1 + D^2 + D^{11}$	28	$1 + D^2 + D^{28}$
12	$1 + D + D^4 + D^6 + D^{12}$	29	$1 + D^3 + D^{29}$
13	$1 + D + D^3 + D^4 + D^{13}$	30	$1 + D + D^2 + D^{30}$
14	$1 + D + D^6 + D^{10} + D^{14}$	31	$1 + D^3 + D^{31}$
15	$1 + D + D^{15}$	32	$1 + D + D^2 + D^{23}$
16	$1 + D + D^3 + D^{12} + D^{16}$	33	$1 + D^{10} + D^{33}$
17	$1 + D^3 + D^{17}$	34	$1 + D + D^2 + D^{27} + D^{34}$

(d)

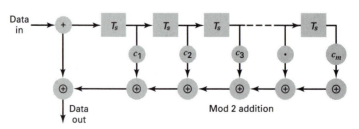

(e)

FIGURE 3.18-1

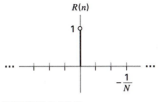

FIGURE 3.18-2

references, it can be shown that if $G(D)$ can be factored into multiple polynomials having coefficents 0 or 1 then the behavior of the circuit will be that of two or more state machines plus the isolated zero state, as shown in fig. 3.18-1b. If, on the other hand, $G(D)$ cannot be factored in such a way and also divides $(D^m + 1)$ then, except for the isolated zero state, the sequential machine will move in a single cycle of $N = 2^m - 1$ states, as shown in fig. 3.18-1c. A machine of this sort is called a maximal-length shift register or a scrambler and $G(D)$ is called irreducible. Extensive tables of such polynomials are given in Ziemer and Peterson [Z&P85] and in Peterson and Weldon [P&W72]. Some examples of these polynomials are given in fig. 3.18-1d.

It has been demonstrated by Savage [Sav67] and by Gitlin and Hayes [G&H75] that connecting the output of the scrambler of fig. 3.18-1a to the input to the descrambler of fig. 3.18-1e results in the reproduction of the original input sequence unless, of course, an error in the transmission system between the two has occurred. Under those circumstances the error will be regenerated by each nonzero shift register tap, thereby multiplying the number of errors by the number of such taps.

A salient feature of these scrambler or maximal length shift register sequences is that their autocorrelation functions closely resemble that of white noise. Recall that for white noise

$$P_n(\omega) = \eta/2 \Rightarrow \overline{R_n}(\tau) = \pi\eta\delta(\tau) \tag{3.18-2a}$$

It can be shown (Ziemer and Peterson [Z&P85]) that the autocorrelation function of the scrambler sequence is periodic with period $N = 2^{M-1}$ and that a single period (fig. 3.18-2) is of the form

$$R(n) = \begin{cases} 1; & n = 0 \\ -\frac{1}{N}; & n \neq 0, \end{cases} \qquad -\frac{N-1}{2} < n < \frac{N-1}{2} \tag{3.18-2b}$$

This demonstrates the uncorrelating effect that the scrambler has on input data.

It can further be demonstrated (Sarwate and Pursely [S&P80]) that two different scrambler sequences are essentially uncorrelated. This means that if the output of scrambler 1 characterized by $G_1(D)$ is applied to the input of descrambler 2 characterized by $G_2(D)$, then the autocorrelation function of the descrambler output is equal to $-1/N$ for all n.

This last fact is of particular importance for full-duplex modems using echo cancelling to separate the transmit and receive data streams, as takes place in the ISDN and in telephone line modems to be discussed in the next chapter. Under these circumstances the two data streams use two different scramblers, one specified for the transmitter of the originating modem and another for the transmitter of the answering modem. Echo cancelling will be discussed in Chapter 11.

CHAPTER 3 PROBLEMS

3.1 In eq. 3.2-1 the signals are set symmetrically around the origin on the grounds that such a placement minimizes the energy in the signal. Justify this assertion.

3.2 The result in eq. 3.2-3 assumes that the decision boundary is halfway between adjacent signal points. Show that this choice of decision boundary minimizes the probability of error.

3.3 (a) In eq. 3.2-4b, find the probability of error if the probability of the two pieces of data are not equal:

$$P[d_n = +\sqrt{E_s}] = p$$
$$P[d_n = -\sqrt{E_s}] = 1 - p$$

(b) Show that the probability of error is maximized for $p = 1/2$.

3.4 Confirm that the maximum of eq. 3.3-1 lies at $x = 0.8145$. A numerical or graphical solution is best. Qualitatively explain why ρ has a maximum value. In doing this, you should adopt two points of view: varying the symbol period T_s while holding RC constant and varying RC while holding T_s constant.

3.5 For each of the signal pulses of fig. 3.4-2:

(a) Find the output waveform of a matched filter receiver;

(b) Show that each of the systems has equivalent performance.

3.6 For each of the systems of Problem 3-4, assume that the sampling of the matched filter output occurs at $0.1T_s$ away from the peak. How does the perfomance change? Are all of the systems equivalent at such non-peak peformance? Which is the best of them?

3.7 This problem introduces the issue of the effect of intersymbol interference on receiver performance.

(a) Consider the situation in which the output of the receiver filter is that of fig. 3.5-2b with $RC = 5T_s$. The data is assumed to be a steady pattern of alternating bits ..010101.. . Find the effect of the ISI so generated on the sampled value of the filter output and, consequently, on the system performance.

(b) Now think about ISI generated from random data. One way of approaching this, although not one that would be approved of by mathematicians, is to add the squares of the values of the ISI terms to arrive at an a 'variance' of an equivalent 'noise' caused by ISI. For the system under question, compute an approximate value for the probability of error caused by ISI.

(c) Finally, consider the situation where a receiver without ISI but with noise operates at a probability of error $P_e = 10^{-4}$. ISI having a variance equal to twice that of the noise is then added to the signal plus noise. Find the new probability of error.

3.8 Two signals having equal probability

$$s_0(t) = -s_1(t) = \begin{cases} e^{-t}; & t \geq 0 \\ 0; & t < 0 \end{cases}$$

are used on an infinite-bandwidth ideal channel having additive white Gaussian noise. A correlation receiver is used.

The receiver bases its decision on an operation on the received signal plus noise over the time interval $0 \leq t \leq 2$. Using an appropriate decision rule, find the probability of error in terms of the Q function. Compare the result with that of an optimum receiver that processes the entire signal.

3.9 Verify eq. 3.8-4 when we have (a) 3 bits/symbol; (b) 4 bits/symbol.

3.10 We want to transmit 16,000 bits/sec on a baseband channel having a one-sided bandwidth of 3,000 hz. Using raised-cosine shaping, how many signal levels should be used to get a design that has a rolloff of at least 1/3?

Using the same shaping, now increase the bit rate to 32,000 bps. How many signal levels are required?

3.11 (a) For AMI in a white gaussian noise channel, show that the optimum decision boundaries are given by "choose zero if $\cosh(2y_n/\eta) < \exp(\rho_M)$," where y_n is the output of the sampler.

(b) Show that this decision problem is equivalent to that of Class-1 Partial Response.

(c) Further, show that in the absence of any knowledge of the noise level, placing the decision boundary halfway between adjacent points makes sense.

3.12 A four-level Nyquist data signal is transmitted at 10 kb with an average transmitted power of 1 mw in AWGN. If the average symbol probability of error is $P_e = 10^{-5}$, what is the average power of the noise out of the receiver filter? What is the value of the power spectral density of the signal?

3.13 The input to the channel of fig. 3.13-1 consists of uncorrelated data impulses which have a flat power spectral density.

(a) Thinking of the channel as a transversal filter, find the power spectral density of the output.

(b) Thinking about the channel as a Markov process, find the power spectral density of the output.

3.14 (a) Find a state machine representation of the Class-1 Partial Response *encoding* rule. Using Markov process techniques, verify how this encoding changes the power spectral density of the transmitted signal.

(b) Repeat for Class-4 Partial Response.

3.15 (a) Find a state machine representation of the Class-1 Partial Response *precoding* rule. Using Markov process techniques, verify how this encoding changes the power spectral density of the transmitted signal.

(b) Repeat for Class-4 Partial Response.

3.16 (a) Referring to the section on square-law timing recovery in Chapter 2, show that if $H(\omega)$ is a brick-wall filter having a cutoff at the Nyquist frequency, then this approach to timing recovery in partial response systems does not work.

(b) Show that Class-4 encoding together with $H(\omega)$ being a raised-cosine having rolloff α still results in zeros at $\omega = 0$, ω_N and that square-law timing recovery is now possible. Find the strength of the timing signal as a function of α.

3.17 For the PRS polynomial

$$G(z) = (1 - z^{-1})(1 + z^{-1})^2 = (1 + z^{-1} - z^{-2} - z^{-3})$$

with $H(\omega)$ a brick-wall filter having a cutoff at the Nyquist frequency, find (a) the number of levels, (b) a state machine representation of the encoding rule, (c) the precoding rule, (d) the probability of error, (e) the decay rate of the peaks of $s_0(t)$.

3.18 Show that Class-4 partial response is equivalent to interleaved AMI pulses.

3.19 A function $f(t)$ has a Fourier Transform $F(\omega)$. Show that if the first $K - 1$ derivatives of $F(\omega)$ are continuous and the Kth derivative is discontinuous, then the magnitude of $f(t)$ decays asymptotically as $1/|t|^{K-1}$.

REFERENCES

[ABG81] S.V. Ahmed, P.P. Bohn, and N.L. Gottfried, "A tutorial on two-wire digital transmission in the loop plant," *IEEE Trans. Communications*, **Com-29**, No. 11, Nov. 1981, pp. 1554–1564.

[BKR63] C.J. Byrne, B.J. Karafin, D.B. Robinson, "Systematic jitter in a chain of digital regenerators," *Bell System Technical J.*, **42**, Nov. 1963, pp. 2679–2714.

[Bell70] Bell Telephone Laboratories, *Transmission Systems for Communications*, 1970.

[Ben70] W. Bennett, *Introduction to Signal Transmission*, McGraw-Hill, 1970.

[Ber96]   J. Bergmans, *Digital Baseband Transmission and Recording*, Kluwer, 1996.

[Cio90]   J. Cioffi et al., "Adaptive equalization in magnetic-disc storage channels," *IEEE Comm. Magazine*, February 1990, pp. 14–29.

[Cou97]   Leon Couch, *Digital and analog communication systems*, 5th ed., Prentice Hall, 1997.

[G&H75]   R.D. Gitlin and J.F. Hayes, "Timing recovery and scramblers in data transmission," *Bell System Technical J.*, **54**, No. 2, March 1975.

[GH&W92]   R.D. Gitlin, J.F. Hayes, and S.B. Weinstein, "Data communications principles," Plenum, 1992.

[GVC84]   P.J. van Gerwen, N.A.M. Verhoeckx, and T.A.C.M. Claasen, "Design considerations for a 144 kbit/s digital transmission unit for the local telephone network," *IEEE J. Selected Areas of Communication*, **SAC-2**, No. 2, March 1984, pp. 314–323.

[Gil67]   A. Gill, *Linear sequential circuits*, McGraw-Hill, 1967.

[Gol67]   S.W. Golomb, *Shift Register Sequences*, Holden-Day, 1967.

[K&P75]   P. Kabal and S. Pasupathy, "Partial-response signaling," *IEEE Trans. Commun.*, **COM-23**, No. 9, Sept. 1975, pp. 921–934.

[K&T70]   H. Kobayashi and D.T. Tang, "Application of partial response channel coding to magnetic recording systems," *IBM J. Research and Development*, July 1970, pp. 368–375.

[Kob71]   H. Kobayashi, "A survey of coding schemes for transmission or recording of digital data," *IEEE Trans. Commun. Technol.*, **COM-19**, No. 6, Dec. 1971, pp. 1087–1100.

[L&T88]   N.S. Lin and C.P.J. Tzeng, "Full-duplex data over local loops," *IEEE Comm. Magazine*, February 1988, pp. 31–42.

[L&W97]   B.P. Lathi and M.A. Wright, "The digital hierarchy," in Gibson (ed.), *The Communications Handbook*, CRC Press, 1997.

[Lat98]   B.P. Lathi, "Modern Digital and Analog Communication Systems," Oxford, 1998.

[Lec89]   Joseph W. Lechleider, "Line codes for digital subscriber lines," *IEEE Comm. Magazine*, Sept. 1989, pp. 25–32.

[Len63]   A. Lender, "The duobinary technique for high-speed data tranmission," *IEEE Trans. Commun. Electron.*, **82**, May 1963, pp. 214–218.

[Mal75]   J.C. Mallinson, "A unified view of high density digital recording theory," *IEEE Trans. Magnetics*, **MAG-11**, No. 5, Sept. 1975.

[Max96]   Kim Maxwell, "Asymmetric digital subscriber line: interim technology for the next forty years," *IEEE Comm. Magazine*, Oct. 1996, pp. 100–106.

[Mes86]   D.G. Messerschmidt, "Design issues in the ISDN U-interface transceiver," *IEEE J. Selected Areas of Communication*, **SAC-4**, No. 8, Nov. 1986, pp. 314–323.

[Nyq28]   H. Nyquist, "Certain topics in telegraph transmission theory," *Trans. AIEE*, **47**, April 1928, pp. 617–644.

[Pas77]   S. Pasupathy, "Correlative coding: a bandwidth efficient signaling scheme," *IEEE Comm. Magazine*, **15**, No. 4, July 1977, pp. 4–11.

[P&W72]   W.W. Peterson and E.J. Weldon, Jr., *Error-Correcting Codes*, (2d ed.), MIT Press, 1972.

[Row58]   H.E. Rowe, "Timing in a long chain of binary regenerative repeaters," *Bell System Technical J.*, **37**, Nov. 1958, pp. 1543–1598.

[S&P80]   D.V. Sarwate and M.B. Pursely, "Crosscorrelation properties of pseudorandom and related sequences," *Proc. IEEE*, May 1980.

[Sav67]   J.E. Savage, "Some simple self-synchronizing digital data scramblers," *Bell System Technical J.*, **46**, No. 2, February 1967, pp. 449–487.

[S&W91] P.H. Siegel and J.K. Wolf, "Modulation and coding for information storage," *IEEE Comm. Magazine*, Dec. 1991, pp. 68–86.

[Smi93] D. Smith, "Digital Transmission Systems," Van Nostrand, 1993.

[T&P87] H.K. Thapar and A.M. Patel, "A class of partial response systems for increasing storage density in magnetic recording," *IEEE Trans. Magnetics.*, **MAG-23**, No. 5, Sept. 1987, pp. 3666–3668.

[Z&P85] R.E. Ziemer and R.L. Peterson, *Digital Communications and Spread Spectrum Systems*, Macmillan, 1985.

CHAPTER 4

BANDPASS DATA TRANSMISSION

In Chapter 3 we dealt with baseband data transmission, which is essentially transmission without a carrier frequency. In this chapter we will begin to address data transmission techniques that involve a carrier frequency and some kind of modulation. Such approaches find their application in standard modems for the telephone line, in digital microwave transmission, satellite communications, and digital cellular telephony.

The most important of these approaches at the current time is Quadrature Amplitude Modulation or QAM, which is the main subject of this chapter. QAM is the general technique that is used in all standard modems for the telephone line and microwave links. It can be understood as an extension of baseband Nyquist signaling to the passband.

As examples of the use of QAM in these systems we shall review the basic characteristics of telephone line modems as well as some representative modems for digital radio.

We will also briefly discuss some other techniques for bandpass data transmission that can be understood as deriving from QAM. These include OQPSK (Offset Quadrature Phase-Shift Keying, also known as SQAM, Staggered QAM) and its variant, MSK (Minimum-Shift Keying). Finally, we will present some fundamental issues of multicarrier modulation, which is currently being used for very high-speed data transmission of up to 6 Mbps on the subscriber loop in a mode called ADSL (Asymmetric Digital Subscriber Loop).

4.1 THE BASIC STRUCTURE OF QUADRATURE AMPLITUDE MODULATION (QAM)

We want to define the idea of Quadrature Amplitude Modulation as an extension of baseband data transmission. Consider the system of fig. 4.1-1. We have, in effect, taken two binary baseband systems, multiplied one with a Sin carrier and the other with a Cos carrier, and added them together. One data symbol consists of two impulse functions simultaneously appearing on the a_n and b_n data lines. Since each of these impulses has amplitude $\pm\sqrt{E_s/2}$, we then have a dibit, two bits per transmitted symbol. The carrier frequency is $\omega_c = n\omega_s = n2\pi/T_s$, making the carrier frequency an integer multiple of the symbol frequency. The transmit filter on each data line is a rectangular pulse $p(t)$ having duration T_s, the symbol interval, and amplitude $\sqrt{1/T_s}$, so that the energy in $p(t)$ normalized to a unit impulse at the input is unity.

With the scaling shown, the transmitted signal in each symbol interval is:

$$s(t) = \pm\sqrt{E_s/2}\, p(t)\, \sqrt{2}\, \text{Cos}(n\omega_s t) \pm \sqrt{E_s/2}\, p(t)\, \sqrt{2}\, \text{Sin}(n\omega_s t) \qquad (4.1\text{-}1)$$

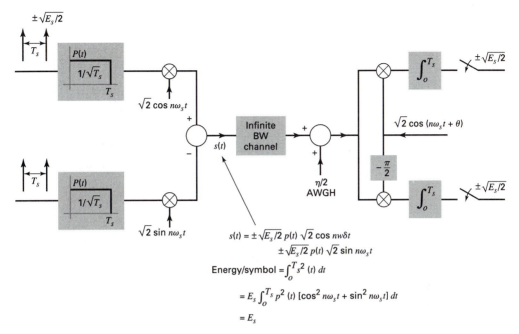

FIGURE 4.1-1

and the energy per transmitted symbol is

$$\text{energy/symbol} = \int_0^{T_s} s^2(t)\, dt = E_s \tag{4.1-2}$$

Our model has the form of a correlation receiver rather than a matched filter receiver.

The scale of the demodulating carrier in the receiver is chosen so that the sampled values of the output signal are $\pm\sqrt{E_s/2}$. The phase of this carrier must be set to be equal to that of the incoming carrier. This is the job of the carrier-phase recovery system, which will be discussed in Chapter 6.

It is, however, useful to discuss the kinds of impairments that the carrier-phase recovery system must address. The received signal will have a carrier that has an angle

$$\varphi(t) = \omega_c t + 2\pi f_{\text{off}}\, t + A_j \frac{2\pi}{360} \text{Cos}(2\pi f_j t) + \varphi_0 \tag{4.1-3}$$

where ω_c is the nominal carrier frequency; f_{off}, the frequency offset, is the difference between the transmitter and receiver oscillators; f_j is the carrier phase-jitter frequency (60 and 120 hz in the United States and 50 and 100 hz in Europe) and A_j is the amplitude, in degrees, of the jitter; φ_0 is the phase offset between transmitter and receiver. The jitter arises from power-line frequencies and their harmonics mixing with the signal during transmission.

Note that if θ, the phase of the demodulating carrier, is properly adjusted, then the upper leg of the receiver will reject the Sin carrier and the lower leg will reject the Cos carrier. The proof of this fact is left to the Problems.

In fig. 4.1-2 we have modified the structure to provide for narrowband signaling using a raised-cosine shaping. First we will make sure that the transmitted signal is properly scaled

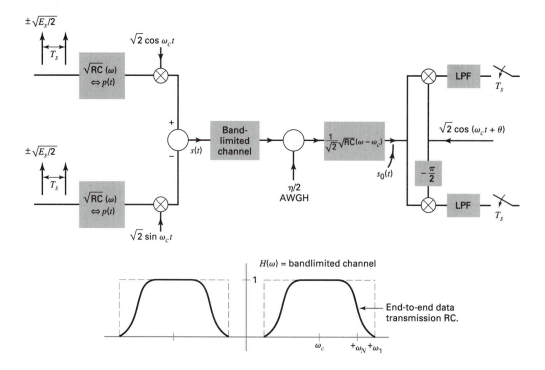

FIGURE 4.1-2

so that the energy per transmitted symbol is E_s. Let $h_{\sqrt{RC}}(t)$ be the impulse response of a filter having the characteristic of $\sqrt{RC}(\omega)$ so that the transmitted signal is

$$s(t) = a_n\sqrt{E_s/2}\, h_{\sqrt{RC}}(t)\sqrt{2}\,\text{Cos}\,\omega_c t - b_n\sqrt{E_s/2}\, h_{\sqrt{RC}}(t)\sqrt{2}\,\text{Sin}\,\omega_c t$$

where

$$a_n, b_n = \pm 1 \tag{4.1-4}$$

It is relatively easily demonstrated that the energy per transmitted symbol in eq. 4.1-4 is equal to E_s.

The bandpass filter at the receiver has a $\sqrt{RC}(\omega)$ characteristic shifted to $\pm\omega_c$ and scaled by $1/\sqrt{2}$. This means that the output of this filter is

$$s_0(t) = a_n\sqrt{E_s/2}\, h_{RC}(t)\,\text{Cos}(\omega_c t) - b_n\sqrt{E_s/2}\, h_{RC}(t)\,\text{Sin}(\omega_c t)$$

where

$$a_n, b_n = \pm 1 \tag{4.1-5}$$

where $h_{RC}(t)$ is the impulse response of a full raised cosine filter having a sample value at the peak equal to unity. Note that the power of the noise $n(t)$ at the output of this filter is $P_n = \eta/2$. By demodulating with $2\,\text{Cos}(\omega_c t)$ and $2\,\text{Sin}(\omega_c t)$ as shown, we establish that the sampled values of the output are equal to $\pm\sqrt{E_s/2}$. The lowpass filters in the receiver eliminate the double frequency terms that result from multiplying the received signal with one or another of the demodulating carriers.

Note that in the wideband correlation form of the receiver in fig. 4.1-1, it was necessary to have the carrier frequency be an integer multiple of the symbol frequency f_s in order to separate the Sin and Cos channels. The carrier frequency can be arbitrarily chosen in the narrowband system of fig. 4.1-2 where the lowpass filtering performs the same function of carrier separation. The proof of this fact is also left to the Problems.

In dealing with the noise, we shall anticipate and assert a result that will be proved in the next section. The noise output of the receiver filter has average power $P_n = \eta/2$ and can be represented as

$$n(t) = n_I(t)\,\mathrm{Cos}(\omega_c t) - n_Q(t)\,\mathrm{Sin}(\omega_c t) \tag{4.1-6}$$

where each of the components $n_p(t)$ and $n_q(t)$ each also have average power equal to $\eta/2$ and the two components are statistically independent of each other. The first part makes sense if we consider that the average power in $n(t)$ is

$$\overline{n^2(t)} = \frac{1}{2}\overline{n_I^2(t)} + \frac{1}{2}\overline{n_Q^2(t)}. \tag{4.1-7}$$

We again assume an appropriate phase for the demodulating carrier so that, multiplying $n(t)$ with $2\,\mathrm{Cos}(\omega_c t)$ and putting the result through a lowpass filter, we get

$$2n(t)\,\mathrm{Cos}(\omega_c t) = n_I(t)\,2\,\mathrm{Cos}^2(\omega_c t) - n_Q(t)\,2\,\mathrm{Cos}(\omega_c t)\,\mathrm{Sin}(\omega_c t) \xrightarrow{\text{LPF}} n_I(t) \tag{4.1-8a}$$

and similarly multiplying by Sin in the other leg of the receiver

$$2n(t)\,\mathrm{Sin}(\omega_c t) = n_I(t)\,2\,\mathrm{Sin}(\omega_c t)\,\mathrm{Cos}(\omega_c t) - n_Q(t)\,2\,\mathrm{Sin}^2(\omega_c t) \xrightarrow{\text{LPF}} n_Q(t) \tag{4.1-8b}$$

This means that on each of the two outputs we have a signal sample equal to $\pm\sqrt{E_s/2}$ and a sample of Gaussian noise having a variance $\sigma^2 = \eta/2$.

We are prepared to assert, with formal proof given in Chapter 5, that we may extend fig. 3.7-1 to two dimensions and properly place the samples on the two receiver output lines onto the X and Y axes of the oscilloscope, once again using the timing signal (which we assume to be available) to blank the oscilloscope forming the two-dimensional constellation shown in fig. 4.1-3a. This constellation is called Quadrature Phase-Shift Keying (QPSK) and is one special case of QAM. Notice that the energy per transmitted symbol E_s is the magnitude-squared of the length of the vector to the signal point. In this system we have two bits per symbol.

In finding the probability of error, the independence of the two components of the noise means that we can treat the two axes independently. Let

$$P = \frac{1}{\sigma\sqrt{2\pi}} \int_{\sqrt{E_s/2}}^{\infty} e^{-\alpha^2/2\sigma^2}\,d\alpha = Q\left(\sqrt{\rho_m/2}\right) \qquad \text{where} \quad \rho_m = 2E_s/\eta \tag{4.1-9}$$

Since all of the signal points are equally likely, the probability of being correct in the I direction is

$$P_I[\text{correct}] = 1 - P \tag{4.1-10a}$$

and similarly in the Q direction

$$P_Q[\text{correct}] = 1 - P \tag{4.1-10b}$$

so that the overall probability of being correct is

$$P[\text{correct}] = P_I[\text{correct}]P_Q[\text{correct}] = (1 - P)^2 \tag{4.1-10c}$$

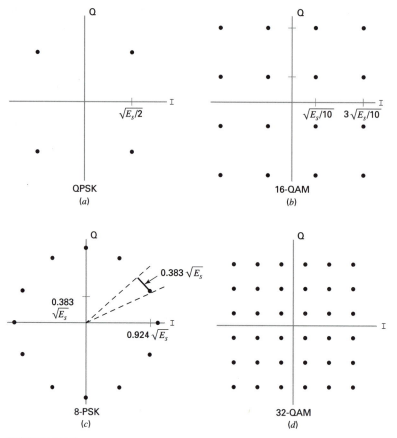

FIGURE 4.1-3

and

$$P_e = P[\text{error}] = 1 - P[\text{correct}] = 1 - (1 - P)^2 = 2P - P^2 \qquad (4.1\text{-}10d)$$

For small values of P we may make the approximation

$$P_e = 2Q\left(\sqrt{\rho_m/2}\right) \quad \text{for QPSK} \qquad (4.1\text{-}10e)$$

We may increase the number of bits per symbol by making each of the two channels multilevel. In fig. 4.1-3b, we show 16-QAM with four bits per symbol (a **quadbit**). The levels in each dimension are $\pm\sqrt{\frac{E_s}{10}}, \pm 3\sqrt{\frac{E_s}{10}}$. We can see that although each individual symbol does not have a transmitted energy of E_s, the average energy per transmitted symbol is E_s. This will be proved in the Problems.

To find the probability of error, note that the decision boundaries are at $0, \pm 2\sqrt{\frac{E_s}{10}}$ in each direction. Following the approach developed for QPSK and recognizing that we must now deal with average probabilities, we see that

$$P_I[\text{correct}] = P_Q[\text{correct}] = \frac{1}{4}[2(1 - P) + 2(1 - 2P)] = 1 - \frac{3}{2}P \qquad (4.1\text{-}11a)$$

where

$$P = \frac{1}{\sigma\sqrt{2\pi}} \int_{\sqrt{E_s/10}}^{\infty} e^{-\alpha^2/2\sigma^2} \, d\alpha = Q\left(\sqrt{\rho_m/10}\right) \qquad \text{where} \qquad \rho_m = 2E_s/\eta \quad \text{(4.1-11b)}$$

It follows that

$$P_e = 1 - \left(1 - \frac{3}{2}P\right)^2 = 3P - \frac{9}{4}P^2 \approx 3Q\left(\sqrt{\rho_m/10}\right) \qquad \text{(4.1-11c)}$$

for 16-QAM.

Similarly, the constellation of fig. 4.1-3c shows eight possible transmitted symbols, or three bits per symbol (a **tribit**), arranged in a circle. This signal set is called 8-PSK and can be generated by the technique of QAM. The points are a subset of $\pm.383\sqrt{E_s}$, $\pm.924\sqrt{E_s}$.

In this case, we cannot find a closed form expression for the probability of error because that would require integrating over a 45° wedge. But we can find an upper bound. Let

$$P = \frac{1}{\sigma\sqrt{2\pi}} \int_{.383\sqrt{E_s}}^{\infty} e^{-\alpha^2/2\sigma^2} \, d\alpha = Q\left(\sqrt{\rho_M/6.8}\right) \qquad \text{where} \qquad \rho_M = 2E_s/\eta$$

$$\text{(4.1-12a)}$$

so that, using the union bound of probability theory noted in Chapter 2,

$$P(AB) \le P(A) + P(B) \qquad \text{(2.1-3b)}$$

we get

$$P_e \le 2P = 2Q\left(\sqrt{\rho_M/6.8}\right) \qquad \text{(4.1-12b)}$$

Finally, in fig. 4.1-3d we see a 32-point signal set. The scaling of this set to give an average energy per transmitted symbol of E_s is left to the Problems, as is the computation of the probability of error.

4.2   DIFFERENTIAL CODING OF QAM SIGNALS

The phase ambiguity of the received signal set requires special consideration in QAM type signals. Examination of the signal constellations in fig. 4.1-3 reveals that with the receiver structure developed so far there is no way to distinguish among the four different quadrants. This fourfold phase ambiguity requires either the transmission of pilot signals along with the data stream, a costly alternative because of the resulting decrease in signal power, or the introduction of differential encoding.

Differential coding and decoding will be introduced in relation to PSK (phase-shift keying), which has a 180° (or twofold) phase ambiguity. As we can see from fig. 4.2-1a, the transmitted signal is

$$s(t) = \pm\sqrt{\frac{2}{T_s}} \, \text{Cos}(n\omega_s t); \qquad 0 \le t \le T_s; \qquad \text{where } \omega_s = \frac{2\pi}{T_s} \quad \text{(4.2-1a)}$$

and the receiver is composed of a demodulator followed by an integrate and dump operation, corresponding to the rectangular baseband pulse. Since PSK has a single carrier, compared to the two carriers in QAM, it may be regarded as lying somewhere between a baseband system and a true passband system.

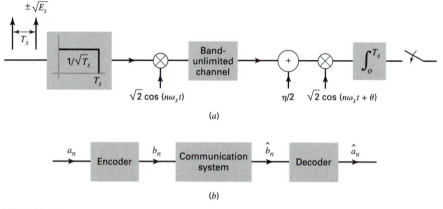

FIGURE 4.2-1

The demodulating signal is denoted as $\sqrt{\frac{2}{T_s}}\,\mathrm{Cos}(n\omega_s t + \theta)$ because the actual phase of the transmitter carrier is unknown at the receiver and some kind of carrier-phase recovery device must be incorporated. Without such a carrier-phase recovery, the output would be:

$$s_o(t) = \sqrt{E_s}\,\mathrm{Cos}\,\theta \qquad (4.2\text{-}2b)$$

while the noise would remain the same, thereby increasing the probability of error.

However, even if the carrier-phase recovery system is working perfectly, there exists an essential 180° ambiguity in the signal set. That is, it simply is not possible to determine whether or not the carrier should be shifted by an additional 180°.

In order to address this issue, we use differential coding and differential decoding on the data (fig. 4.2-1b). The *encoding rule* is very simple:

$$
\begin{aligned}
&\text{if } a_n = +1 \quad \text{then} \quad b_n = b_{n-1}\\
&\text{if } a_n = -1 \quad \text{then} \quad b_n = -b_{n-1}
\end{aligned}
\qquad (4.2\text{-}3a)
$$

The *decoding rule* is equally simple:

$$
\begin{aligned}
&\text{if } \hat{b}_{n-1} = \hat{b}_n \quad \text{then} \quad \hat{a}_n = +1\\
&\text{if } \hat{b}_{n-1} = -\hat{b}_n \quad \text{then} \quad \hat{a}_n = -1
\end{aligned}
\qquad (4.2\text{-}3b)
$$

A simple example will illustrate this procedure:

input data a_n		+1	−1	+1	+1	−1	+1	−1	−1	+1
encoded data b_n	+1	+1	−1	−1	−1	+1	+1	−1	+1	+1

$(4.2\text{-}3c)$

where we must start with an arbitrary initial encoded data bit. At the receiver, the process is reversed:

channel data $\hat{b}_n$	+1	+1	−1	−1	−1	+1	+1	−1	+1	+1
output data $\hat{a}_n$		+1	−1	+1	+1	−1	+1	−1	−1	+1

$(4.2\text{-}3d)$

Since the encoded data represents the *phase changes* in the original data, rather than the phases themselves, it is clear that this method of encoding resolves the phase ambiguity.

There is, however, a small penalty. If there is a decision error in the channel data, that error will manifest itself as two errors in the output data. This can be seen by changing one bit in the channel data sequence given above and continuing the decoding rule:

channel data $\hat{b}_n$	+1	+1	−1	{+1}		−1	+1	+1	−1	+1	+1
output data $\hat{a}_n$		+1	−1	{−1}	{−1}	−1	+1	−1	−1	+1	

$$(4.2\text{-}3e)$$

As a consequence, the probability of error for a differentially encoded PSK system is

$$P_e = 2Q(\sqrt{\rho_m}) \qquad (4.2\text{-}3f)$$

twice that of a binary baseband system.

The final question is how differential encoding affects the power spectral density of the transmitted data. Using the techniques developed in Chapters 2 and 3, we will undertake to find the autocorrelation function of the data by first exhautively listing all of the data patterns of length 3. We shall see that this is suffcient.

a_k	a_{k+1}	a_{k+2}	b_{k-1}	b_k	b_{k+1}	b_{k+2}
−1	−1	−1	+1	−1	+1	+1
−1	−1	+1		−1	+1	+1
−1	+1	−1		−1	−1	+1
−1	+1	+1		−1	−1	−1
+1	−1	−1		+1	−1	+1
+1	−1	+1		+1	−1	−1
+1	+1	−1		+1	+1	−1
+1	+1	+1		+1	+1	+1

$$(4.2\text{-}4a)$$

We can now compute the required statistics directly from the table, remembering that each sequence is equally likely:

$$R_b(0) = E[b_k b_k] = \sum_{k=0}^{7} b_k b_k P_k(b_k b_k) = \sum_{k=0}^{7} \frac{1}{8}(1) = 1$$

$$R_b(1) = E[b_k b_{k+1}] = \sum_{k=0}^{7} b_k b_{k+1} P_k(b_k b_{k+1}) = 0 \qquad (4.2\text{-}4b)$$

$$R_b(2) = E[b_k b_{k+2}] = \sum_{k=0}^{7} b_k b_{k+2} P_k(b_k b_{k+2}) = 0$$

Accordingly, the power spectral density is unchanged. It is worth noting that a way of understanding this is that the encoded sequences are simply a reshuffling (permutation) of the original sequences, so on the average nothing different is happening.

It is quite important to emphasize what has been achieved by differential coding, resolving phase ambiguity without changing the spectrum of the transmitted signal. This will be used many times as this development of digital communications unfolds.

We may now proceed to describe differential encoding for the QAM examples of fig. 4.1-3. In the QPSK system of fig. 4.2-2, the incoming dibits designate *phase changes* in the transmitted signal. A transmitted quadrant is denoted by the pair of data symbols $a_n, b_n = \pm 1, \pm 1$ so, for example, the second quadrant is denoted in the standard way $a_n = -1, b_n = +1$. We may start by transmitting any quadrant, say quadrant 2. We then examine the current dibit and if it is equal to $+1, -1$, corresponding to 180°, we next transmit quadrant 4. If the subsequent dibit is $+1, +1$, corresponding to $270°$, we transmit quadrant 3, and so on. At the receiver, we remember the previous received quadrant, compare it to the current received quadrant, determine the phase change, and assign the appropriate dibit to the output.

The 16-QAM system of fig. 4.2-3 divides the four data bits in its quadbit into two phase bits followed by two amplitude bits. The two phase bits follow exactly the same rules as the QPSK system for how the quadrants are to be transmitted. The amplitude bits then determine exactly which point in the quadrant is to be chosen. Note that the assignment of amplitude bits to points in the quadrant is undertaken so that a 90-degree phase rotation moves a signal point in one quadrant into a corresponding point in an adjacent quadrant having the same amplitude bits. In decoding the sliced data, we find the amplitude bits directly, but we do differential decoding to find the phase bits. The 8-PSK system of fig. 4.2-4 operates on the same principle of phase encoding but it transmits octants rather than quadrants.

The 32-QAM system introduced in the previous section has a differential coding scheme that will not be discussed just yet. A proper discussion of this topic will take place in the context of a discussion of trellis coding, the subject of Chapter 9.

One final note on all of these forms of differential coding is that they do not change the power spectral density of the transmitted signal. The fundamental reason for this, which we can glean from our previous discussions of this topic, is that they maintain each possible output as being equally likely.

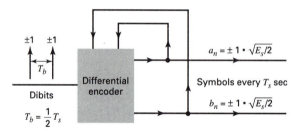

Dibit phase encoding				Current dibit	Previous transmitted quadrant		Current transmitted quadrant	
$-1, -1$	$90°$		$90°$	$-1, -1$	$+1, +1$ ①		$-1, +1$ ②	
$-1, +1$	$0°$				$-1, +1$ ②		$-1, -1$ ③	
$+1, +1$	$270°$				$-1, -1$ ③		$+1, -1$ ④	
$+1, -1$	$180°$				$+1, -1$ ④		$+1, +1$ ①	
			$0°$	$-1, +1$	$+1, +1$ ①		$+1, +1$ ①	
					$-1, +1$ ②		$-1, +1$ ②	
					$-1, -1$ ③		$-1, -1$ ③	
					$+1, -1$ ④		$+1, -1$ ④	
Quadrant encoding			$180°$	$+1, -1$	$+1, +1$ ①		$-1, -1$ ③	
					$-1, +1$ ②		$+1, -1$ ④	
					$-1, -1$ ③		$+1, +1$ ①	
					$+1, -1$ ④		$-1, +1$ ②	
			$270°$	$+1, +1$	$+1, +1$ ①		$+1, -1$ ④	
					$-1, +1$ ②		$+1, +1$ ①	
					$-1, -1$ ③		$-1, +1$ ②	
					$+1, -1$ ④		$+1, -1$ ③	

FIGURE 4.2-2

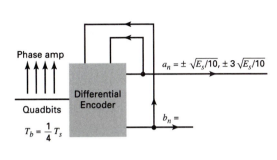

$a_n = \pm\sqrt{E_s/10}, \pm 3\sqrt{E_s/10}$

$b_n =$

$T_b = \dfrac{1}{4}T_s$

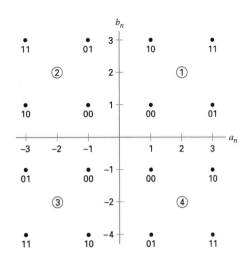

	Current quadbit		Previous quadrant	Current quadrant	XMIT signal	
	Phase	Amp			a_n	b_n
90°	−1, −1	+1, +1	1	2	−1	+1
		−1, +1	1	2	−1	+3
		−1, −1	1	2	−3	+1
		+1, −1	1	2	−3	+3
		− −	2	3	−1	−1
		− +	2	3	−3	−1
		+ −	2	3	−1	−3
		+ +	2	3	−3	−3
		− −	3	4	+1	−1
		− +	3	4	+1	−3
		+ −	3	4	+3	−1
		+ +	3	4	+3	−3
		− −	4	1	+1	+1
		− +	4	1	+3	+1
		+ −	4	1	+1	+3
		+ +	4	1	+3	+3
Repeat with	0°	−1, +1				
	270°	+1, +1				
	180°	+1, −1				

FIGURE 4.2-3

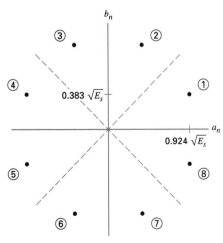

Tribit	Current octant	New octant	XMIT data		Phase encoding	
			a_n	b_n	Tribit	Phase
− − −	1	1	0.924	0.383	− − −	0°
·	2	2	0.383	0.924	− − +	45°
·	3	3	−0.383	0.924	− + −	90°
·	4	4	−0.924	0.383	− + +	135°
·	5	5	−0.924	−0.383	+ − −	180°
·	6	6	−0.383	−0.924	+ − +	225°
	7	7	0.383	−0.924	+ + −	270°
− − −	8	8	0.924	−0.383	+ + +	315°
	Continue					

0.383 $\sqrt{E_s}$

0.924 $\sqrt{E_s}$

FIGURE 4.2-4

4.3 SOME TELEPHONE LINE MODEM EXAMPLES

The structures that have been discussed so far are directly translated into specific modem standards for the telephone line. The ordinary telephone line (POTS—plain old telephone system), introduced in Chapter 2, is outlined in fig. 4.3-1a. It consists of twisted pair, a two-wire line, going from the telephone to the central office and a four-wire path, one pair for transmission and one for reception between central offices. The path between central offices can involve all kinds of communication links including microwave, satellite, fiber optic, etc. Although the twisted pair is relatively wideband and passes dc, the overall channel is shown in fig. 4.3-1b as a bandpass channel nominally going between 300 hz and 3300 hz with a nominal center frequency of 1800 hz.

This system allows for data transmission in one direction at a time, half-duplex, and simultaneous transmission in both directions, full duplex. Half-duplex transmission is quite straightforward. Some standard systems, all using a carrier of 1800 hz, are

V.26	QPSK	2400 bps	1200 baud	$f_N = 600$ hz	$\alpha = 1.0$
V.27	8-PSK	4800 bps	1600 baud	$f_N = 800$ hz	$\alpha = 0.5$
V.29	QAM	9600 bps	2400 baud	$f_N = 1700$ hz	$\alpha = 0.15$

The V.29 modem actually operates with a carrier frequency of 1700 hz and the diamond-shaped 16-point signal constellation shown in fig. 4.3-2. These are choices that reflect the fact that this modem was designed before the practical availability of digital signal processing technology. DSP technology places a premium on having a simple integer relationship between the carrier frequency and the symbol rate. Were the V.29 a more modern design, the carrier frequency would have undoubtedly been chosen to be 1800 hz despite the fact that 1700 hz is closer to the center of the actual telephone band and gives somewhat better performance. In an analog-based design there is no particular constraint on the choice of frequencies. Similarly, the original design of the modem did not allow for a sufficiently effective carrier phase recovery system to address carrier-phase jitter. Consequently the signal constellation was chosen to maximize the immunity to such jitter.

Full-duplex transmission is considerably more complicated because of the two-wire local loop. Two basic methods, frequency division multiplexing (FDM) and echo canceling (EC), have been devised to accomplish the task.

The V.22(bis) modem transmits 2400 bps full-duplex with a 16-QAM constellation using frequency division multiplexing. The carrier frequencies for the two channels, (fig. 4.3-3a) are 1200 hz and 2400 hz. The symbol rate on each channel is 600 baud, so the Nyquist frequency is 300 hz and the excess bandwidth can be up to 300 hz. Typically that bandwidth is restricted to 400 hz in order to make channel separation easier. A hybrid is used at each end of the local loop as a transition from a two-wire line to a four-wire line. Theoretically, the hybrid prevents the transmit signal, which is typically at 0 dbm from entering the receiver. The receive signal may be as low as −45 db. As shown in fig. 4.3-3b, the

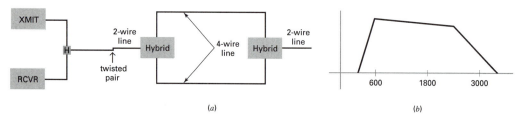

(a)

(b)

FIGURE 4.3-1

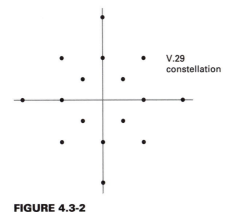

FIGURE 4.3-2

receiver filter must, in addition to providing the required shaping for the passband, provide at least 85 db of attenuation across the transmitter band to compensate for the inability of the hybrid to provide proper attenuation. We therefore have a substantial filter design problem. Frequency division multiplexing has the additional disadvantage of not using the middle of the band, which is the least distorted portion.

The V.32 modem does 9600 bps full-duplex using the entire channel and achieving separation between the transmit and receive signals through echo cancellation. As shown in fig. 4.3-4, the transmitted signal appears at the local receiver in two ways: by direct leakage through the hybrid and by reflections from other points in the transmission system. This sum of distorted and time-shifted versions of the transmitted signal appears at the receiver along with a much smaller receive signal. The echo canceler adaptively produces a replica of this distorted transmitter signal as it appears at the receiver. The replica is subtracted from the total signal, leaving the desired received data signal. In order to accomplish this, the transmitted and the received signals must be uncorrelated, requiring different scramblers for the two. Implementation of echo cancelers will be discussed in a subsequent chapter.

In the V.32 modem the signal has a carrier of 1800 hz and a Nyquist frequency of 1200 hz, allowing a symbol rate of 2400 baud. Consequently, the rolloff is required to be quite low, a maximum of 15%. The signal constellation is 16-QAM or, if trellis coding is used, 32-QAM. The V.32(bis) achieves a data rate of 14,400 bps by transmitting a trellis-coded 128 QAM constellation representing six bits/symbol. Note that the larger signal constellation does not increase the data rate because the extra bit is used for redundancy in order to make communication more reliable. This will be explained in Chapter 9.

The V.34 modem is an even more complex enterprise supporting a large number of bit rates, from 28.8 kbps up to 33.6 kbps. The V.90 achieves up to 56 kbps. It is the first widely accepted modem that adapts itself to the actual connected telephone line to determine

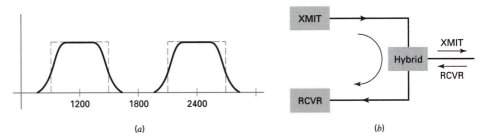

(a)   (b)

FIGURE 4.3-3

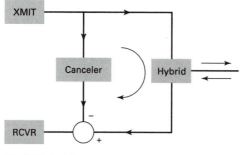

FIGURE 4.3-4

symbol rate and carrier frequency in order to maximize the bit rate for an acceptable performance level. In order to adapt (Forney et al. [For96]), the channel is sounded as part of the handshaking process. Sounding involves transmitting sinusoids having a number of different frequencies and receiving information about their amplitudes and phases. This enables the transmitter automatically to choose the best carrier frequency, symbol rate, and bit rate and transmit that choice to the receiver. It also allows the introduction of partial equalization of the channel into the transmitter through an algorithm called Tomlinson-Harashima precoding, which will be discussed in Chapter 10.

We shall see in Chapter 8 that the capacity of the traditional analog telephone line was set at around 25 kbps. The possibility of transmitting beyond that rate rests on the fact that much of the telephone system has been digitized and the actual available channel is often wider than the traditional one.

The choices of symbol rates and carrier frequencies are given in fig. 4.3-5. The basic and lowest symbol rate is 2400 hz and the available symbol rates are $(a/b) \times 2400$. For example, the highest possible symbol rate is $3429 = (10/7) \times 2400$. Corresponding to each possible symbol rate S are two possible carriers, a low carrier and a high carrier, which are determined by $(c/d) \times S$. For example, the low carrier for $S = 2400$ is $f_c = (2/3) \times 2400 = 1600$ and, for the same value of S, the high carrier is $f_c = (3/4) \times 2400 = 1800$.

Depending on the choice of symbol rate, there are multiple ways of obtaining particular bit rates. For example, 28.8 kbits/s can be obtained from a symbol rate of 2400 hz with 12 bits/QAM symbol or from a symbol rate of 3200 hz with 9 bits/QAM symbol. With a symbol rate of 3000 hz, we require 9.6 bits/symbol, which translates into 10 bits/symbol with stuffing provided by software. Like the V.32 modem, the V.90 uses trellis coding to achieve a certain level of noise immunity, requiring a larger signal set for transmission than even that required to implement 12 bit/symbol. Further, this coding is done on a four-dimensional signal set. All this will be explained in Chapter 10. Finally, the V.90 modem is downward-compatible with the V.32 and the V.34.

Symbol rate			Low Carrier			High Carrier		
S	a	b	Frequency (Hz)	c	d	Frequency (Hz)	c	d
2400	1	1	1600	2	3	1800	3	4
2743	8	7	1646	3	5	1829	2	3
2800	7	6	1680	3	5	1867	2	3
3000	5	4	1800	3	5	2000	2	3
3200	4	3	1829	4	7	1920	3	5
3429	10	7	1959	4	7	1959	4	7

Symbol and Carrier Frequencies in V. 90

FIGURE 4.3-5 ([For 96] reprinted with permission)

It is generally accepted that the V.34 is the fastest modem that will be used on telephone lines that are intended primarily for speech. We have seen that the subscriber loop can be used for much higher data rates. The ISDN, which we discussed in Chapter 3, operates at 160 kbps and, as we shall see later in this chapter, multicarrier can be used to extend signaling on the DSL (digital subscriber loop) to up to 10 Mbps.

4.4 BANDPASS SIGNALS AND NOISE

Having introduced the idea of QAM-type transmission, we now want to place that process on a more theoretical basis.

Let $s_I(t)$ and $s_Q(t)$ each be baseband data signals, bandlimited to $\pm\omega_1$, as described in Section 4.1. A bandpass signal has the form

$$s(t) = s_I(t)\,\mathrm{Cos}(\omega_c t) - s_Q(t)\,\mathrm{Sin}(\omega_c t)$$
$$= \mathrm{Re}[(s_I(t) + js_Q(t))(\mathrm{Cos}(\omega_c t) + j\,\mathrm{Sin}(\omega_c t))]$$
$$= \mathrm{Re}\big[\tilde{s}(t)e^{j\omega_c t}\big] \tag{4.4-1}$$

where $\tilde{s}(t)$ is called the *complex envelope* of $s(t)$.

4.4.1 Hilbert Transforms

As shown in fig. 4.4-1a, the Hilbert Transform of a signal involves a -90-degree phase shift of all frequencies contained in that signal. The Hilbert Transform of a signal $f(t)$ is denoted $\hat{f}(t)$. We note that

$$\mathrm{HT}[\mathrm{Cos}(\omega_c t)] = \mathrm{Cos}(\omega_c t - 90°) = \mathrm{Sin}(\omega_c t)$$
$$\mathrm{HT}[\mathrm{Sin}(\omega_c t)] = \mathrm{Sin}(\omega_c t - 90°) = -\mathrm{Cos}(\omega_c t) \tag{4.4-2}$$

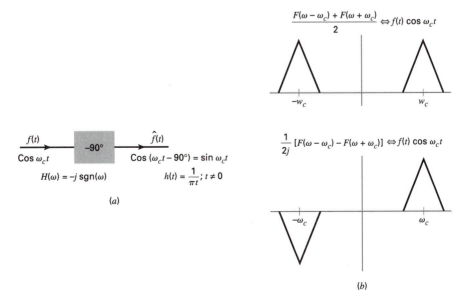

FIGURE 4.4-1

It follows that we can find the transfer function of the Hilbert Transformer by taking the Fourier Transform of both the Cos input and the Sin output:

$$[\pi\delta(\omega - \omega_c) + \pi\delta(\omega + \omega_c)] = H(\omega)\frac{1}{j}[\pi\delta(\omega - \omega_c) - \pi\delta(\omega + \omega_c)] \tag{4.4-3a}$$

so that

$$H(\omega) = -j\,\text{sgn}(\omega) \qquad \text{where} \qquad \text{sgn}(\omega) = \begin{cases} 1 & \text{for } \omega > 0 \\ -1 & \text{for } \omega < 0 \end{cases} \tag{4.4-3b}$$

and using the symmetry theorem of Fourier Transforms, the inverse Fourier Transform of $H(\omega)$, or impulse response $h(t)$, is

$$h(t) = \frac{1}{\pi t}; \qquad t \neq 0 \tag{4.4-3c}$$

FACT

If $f(t)$ is a bandlimited signal as shown in fig. 4.4-1b, then

$$\begin{aligned} HT[f(t)\cos(\omega_c t)] &= f(t)\sin(\omega_c t) \\ HT[f(t)\sin(\omega_c t)] &= -f(t)\cos(\omega_c t) \end{aligned} \tag{4.4-4}$$

∎

Proof:
Using the definition of the Hilbert Transform and the frequency convolution theorem of Fourier Transforms:

$$HT[f(t)\cos(\omega_c t)] \Leftrightarrow -j\,\text{sgn}(\omega)\left\{\frac{1}{2\pi}F(\omega) \otimes [\pi\delta(\omega - \omega_c) + \pi\delta(\omega + \omega_c)]\right\}$$

$$= \frac{1}{2j}[F(\omega - \omega_c) - F(\omega + \omega_c)] \Leftrightarrow f(t)\sin(\omega_c t)$$

The proof of the second proposition follows similarly. Note that these facts do not hold unless $f(t)$ is bandlimited and ω_c is sufficiently high. *end of proof*

Using this fact, we can now return to the bandpass signal of eq. 4.4-1

$$\begin{aligned} s(t) &= s_I(t)\cos(\omega_c t) - s_Q(t)\sin(\omega_c t) \\ &= \text{Re}[(s_I(t) + js_Q(t))(\cos(\omega_c t) + j\sin(\omega_c t))] \end{aligned} \tag{4.4-5a}$$

and take its Hilbert Transform

$$\begin{aligned} \hat{s}(t) &= s_Q(t)\cos(\omega_c t) - s_I(t)\sin(\omega_c t) \\ &= \text{Im}[(s_I(t) + js_Q(t))(\cos(\omega_c t) + j\sin(\omega_c t))] \end{aligned} \tag{4.4-5b}$$

We are now able to define

$$s_+(t) = s(t) + j\hat{s}(t) = \tilde{s}(t)e^{j\omega_c t} \tag{4.4-5c}$$

as the *complex analytic extension* of $s(t)$.

4.4.2 The Hilbert Transform Receiver

We shall now define a variant of the standard receiver for QAM signals called the Hilbert Transform Receiver shown in fig. 4.4-2a. This receiver is defined for two reasons: first, the structure allows us to determine some fundamental properties of noise in a bandpass system, necessary for us to determine the system performance; second, the structure is particularly suited to a DSP implementation under many circumstances, an issue that will subsequently be developed.

The output of the bandpass filter is

$$s(t) = s_I(t)\operatorname{Cos}(\omega_c t + \theta) - s_Q(t)\operatorname{Sin}(\omega_c t + \theta)$$
$$= \operatorname{Re}[(s_I(t) + j s_Q(t))(\operatorname{Cos}(\omega_c t + \theta) + j\operatorname{Sin}(\omega_c t + \theta))]$$
$$= \operatorname{Re}\left[\tilde{s}(t)e^{j(\omega_c t + \theta)}\right] = \operatorname{Re}[s_+(t)] = \operatorname{Re}[s(t) + j\hat{s}(t)] \qquad (4.4\text{-}6)$$

where θ is the unknown phase of the incoming carrier. Subsequently, a Hilbert Transform filter generates $\hat{s}(t)$ and the two signals $s(t)$ and $\hat{s}(t)$ form two "rails," the real and imaginary parts of the complex analytic signal $s_+(t)$.

The demodulation process takes place in two stages. First we do the complex multiplication

$$s_+(t)e^{-j\omega_c t} = [s(t) + j\hat{s}(t)]e^{-j\omega_c t}$$
$$= \left[\tilde{s}(t)e^{j(\omega_c t + \theta)}\right]e^{-j\omega_c t} = (s_I(t) + j s_Q(t))e^{j\theta} \qquad (4.4\text{-}7a)$$

which operates as a "free-running" demodulator. We then use an estimate of θ called $\hat{\theta}$ to

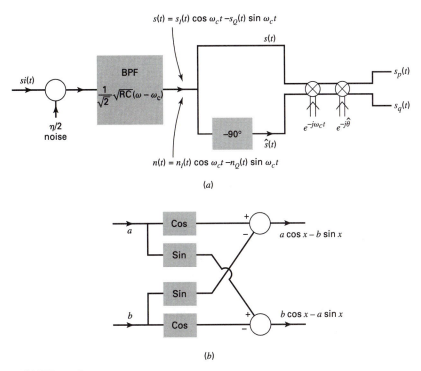

(a)

(b)

FIGURE 4.4-2

do the "rotation" of the signal into its required orientation for slicing, so that

$$[(s_I(t) + js_Q(t))e^{j\hat{\theta}}]e^{-j\hat{\theta}} \approx (s_I(t) + js_Q(t)). \tag{4.4-7b}$$

The implementation of the operation of multiplying a complex number by $e^{j\alpha}$ is based on

$$(a + jb)(\text{Cos}\,\alpha + j\,\text{Sin}\,\alpha) = (a\,\text{Cos}\,\alpha - b\,\text{Sin}\,\alpha) + j(a\,\text{Sin}\,\alpha + b\,\text{Cos}\,\alpha) \tag{4.4-8a}$$

or the same operation based on the rotational matrix

$$[a \quad b]\begin{bmatrix} \text{Cos}\,\alpha & \text{Sin}\,\alpha \\ -\text{Sin}\,\alpha & \text{Cos}\,\alpha \end{bmatrix} = \begin{bmatrix} a\,\text{Cos}\,\alpha - b\,\text{Sin}\,\alpha \\ b\,\text{Cos}\,\alpha + a\,\text{Sin}\,\alpha \end{bmatrix} \tag{4.4-8b}$$

as shown in fig. 4.4-2b. The output of the rotater is the same as output of the lowpass filters in the receiver shown in fig. 4.1-2.

4.4.3 Bandpass Noise

Now let us turn to the noise. The noise entering the bandpass filter at the receiver input is assumed to be AWGN and is modeled as randomly occurring impulse functions, as described in Chapter 2. The output of the bandpass filter, which is called *bandpass noise*, can therefore be thought of as being the superposition of randomly occurring filter impulse responses of random amplitude. Since the filter impulse response will oscillate or "ring" at the center frequency ω_c, we feel justified in arguing that an appropriate model for bandpass noise is:

$$n(t) = n_I(t)\,\text{Cos}(\omega_c t) - n_Q(t)\,\text{Sin}(\omega_c t) \tag{4.4-9a}$$

where the noise has the same structure as a bandpass signal. The random processes $n_I(t)$ and $n_Q(t)$ are baseband Gaussian random processes. Our goal is to determine the statistical properties of these various random processes.

Since the Hilbert Transform Receiver treats the noise in the same way that it does the signal, the Hilbert Transform of the bandpass noise $n(t)$ is:

$$\hat{n}(t) = n_I(t)\,\text{Sin}(\omega_c t) + n_Q(t)\,\text{Cos}(\omega_c t) \tag{4.4-9b}$$

and the complex analytic extension of $n(t)$ is

$$n_+(t) = n(t) + j\hat{n}(t) = (n_I(t) + jn_Q(t))(\text{Cos}(\omega_c t) + j\,\text{Sin}(\omega_c t)) \tag{4.4-9c}$$

THEOREM 1

Keeping in mind that $n(t)$ and $\hat{n}(t)$ are real and $n_+(t)$ is complex, define the autocorrelation functions:

1.
$$\overline{R_n(\tau)} = \overline{n(t)n(t - \tau)} \tag{4.4-10a}$$

2.
$$\overline{R_{\hat{n}}(\tau)} = \overline{\hat{n}(t)\hat{n}(t - \tau)} \tag{4.4-10b}$$

3.
$$\overline{R_+(\tau)} = \overline{n_+(t)n_+^*(t - \tau)} \tag{4.4-10c}$$
Then:

$$\overline{R_+(\tau)} = 2[\overline{R_n(\tau)} + j\overline{\hat{R}_n(\tau)}] \tag{4.4-10d}$$

■

Proof:

The time-average autocorrelation function of $n_+(t)$ is

$$\overline{R_+(\tau)} = \overline{n_+(t)n_+^*(t-\tau)} = \overline{[n(t)+j\hat{n}(t)][n(t-\tau)-j\hat{n}(t-\tau)]}$$
$$= \overline{[n(t)n(t-\tau)]} + \overline{[\hat{n}(t)\hat{n}(t-\tau)]} - j\overline{[n(t)\hat{n}(t-\tau)]} + j\overline{[\hat{n}(t)n(t-\tau)]}$$
$$= \overline{R_n(\tau)} + \overline{R_{\hat{n}}(\tau)} - j\overline{R_{n\hat{n}}(\tau)} + j\overline{R_{n\hat{n}}(\tau)}$$

We may now see that

1.
$$\overline{R_n(\tau)} \Leftrightarrow |N(\omega)|^2$$
$$\overline{R_{\hat{n}}(\tau)} \Leftrightarrow |-j\,\mathrm{sgn}(\omega)N(\omega)|^2 = |N(\omega)|^2$$

so that

$$\overline{R_n(\tau)} = \overline{R_{\hat{n}}(\tau)} \tag{4.4-11a}$$

2. According to Parseval's Theorem,

$$\int_{-\infty}^{\infty} f_1(t)f_2^*(t)\,dt = \frac{1}{2\pi}\int_{-\infty}^{\infty} F_1(\omega)F_2^*(\omega)\,d\omega$$

and using the Time-Shifting Theorem

$$\int_{-\infty}^{\infty} f_1(t)f_2^*(t-\tau)\,dt = \frac{1}{2\pi}\int_{-\infty}^{\infty} F_1(\omega)F_2^*(\omega)e^{j\omega\tau}\,d\omega$$

it follows that

$$\overline{R_{n\hat{n}}(\tau)} = \overline{[n(t)\hat{n}(t-\tau)]} = \lim_{T\to\infty} \frac{1}{T}\int_{-\infty}^{\infty} n(t)\hat{n}(t-\tau)\,dt$$

$$= \frac{1}{2\pi}\int_{-\infty}^{\infty} N(\omega)[-j\,\mathrm{sgn}(\omega)N(\omega)]^* e^{j\omega\tau}\,d\omega$$

$$= \frac{1}{2\pi}\int_{-\infty}^{\infty} j\,\mathrm{sgn}(\omega)|N(\omega)|^2 e^{j\omega\tau}\,d\omega \tag{4.4-11b}$$

and:

$$\overline{R_{\hat{n}n}(\tau)} = \overline{[\hat{n}(t)n(t-\tau)]} = \frac{1}{2\pi}\int_{-\infty}^{\infty} -j\,\mathrm{sgn}(\omega)N(\omega)N^*(\omega)e^{j\omega\tau}\,d\omega \tag{4.4-11c}$$
$$= -\overline{R_{n\hat{n}}(\tau)} = \overline{\hat{R}_n(\tau)}$$

It follows that:

3.
$$\overline{R_+(\tau)} = \overline{R_n(\tau)} + \overline{R_{\hat{n}}(\tau)} - j\overline{R_{n\hat{n}}(\tau)} + j\overline{R_{\hat{n}n}(\tau)}$$
$$= 2[\overline{R_n(\tau)} + j\overline{\hat{R}_n(\tau)}] \tag{4.4-11d}$$

end of proof

THEOREM 2

The autocorrelation functions of the in-phase and quadrature portions of the noise are given by:

$$\overline{R_I}(\tau) = \overline{R_Q}(\tau) = \overline{R_n}(\tau)\,\mathrm{Cos}(\omega_c t) + \overline{\hat{R}_n}(\tau)\,\mathrm{Sin}(\omega_c t) \qquad (4.4\text{-}12a)$$

and the cross-correlation functions are:

$$\overline{R_{QI}}(\tau) = -\overline{R_{IQ}}(\tau) = \overline{R_n}(\tau)\,\mathrm{Sin}(\omega_c t) - \overline{\hat{R}_n}(\tau)\,\mathrm{Cos}(\omega_c t) \qquad (4.4\text{-}12b)$$

■

Proof:
Note that $\mathrm{Re}(x) = \frac{1}{2}(x + x^*)$ and $\mathrm{Im}(x) = \frac{1}{2j}(x - x^*)$. From Theorem 1:

1. $\overline{R_I}(\tau) = \overline{n_I(t)n_I(t - \tau)}$

$$= \frac{1}{4}\overline{\left[n_+(t)e^{-j\omega_c t} + n_+^*(t)e^{j\omega_c t}\right]\left[n_+(t - \tau)e^{-j\omega_c(t-\tau)} + n_+^*(t - \tau)e^{j\omega_c t(t-\tau)}\right]}$$

$$= \frac{1}{4}\overline{\left[\begin{array}{l} n_+(t)n_+(t - \tau)e^{-j\omega_c(2t-\tau)} + n_+^*(t)n_+^*(t - \tau)e^{j\omega_c(2t-\tau)} \\ + n_+^*(t)n_+(t - \tau)e^{-j\omega_c\tau} + n_+(t)n_+^*(t - \tau)e^{j\omega_c\tau} \end{array}\right]}$$

$$= \frac{1}{4}\overline{\left[n_+^*(t)n_+(t - \tau)e^{-j\omega_c\tau} + n_+(t)n_+^*(t - \tau)e^{j\omega_c\tau}\right]} \qquad \text{since } \overline{e^{j2\omega_c t}} = 0$$

$$= \frac{1}{2}\mathrm{Re}\overline{\left[n_+(t)n_+^*(t - \tau)e^{j\omega_c\tau}\right]} = \frac{1}{2}\mathrm{Re}\left[\overline{n_+(t)n_+^*(t - \tau)}e^{j\omega_c\tau}\right]$$

$$= \frac{1}{2}\mathrm{Re}\left[\overline{R_+}(\tau)e^{j\omega_c\tau}\right] = \frac{1}{2}\mathrm{Re}\left[2[\overline{R_n}(\tau) + j\overline{\hat{R}_n}(\tau)]e^{j\omega_c\tau}\right]$$

or:

$$\overline{R_I}(\tau) = \overline{R_Q}(\tau) = \overline{R_n}(\tau)\,\mathrm{Cos}(\omega_c t) + \overline{\hat{R}_n}(\tau)\,\mathrm{Sin}(\omega_c t)$$

proving the first part of the theorem.

2. $\overline{R_{IQ}}(\tau) = \overline{n_I(t)n_Q(t - \tau)}$

$$= \frac{1}{4j}\overline{\left[n_+(t)e^{-j\omega_c t} + n_+^*(t)e^{j\omega_c t}\right]\left[n_+(t - \tau)e^{-j\omega_c(t-\tau)} - n_+^*(t - \tau)e^{j\omega_c t(t-\tau)}\right]}$$

$$= \frac{1}{4j}\overline{\left[\begin{array}{l} n_+(t)n_+(t - \tau)e^{-j\omega_c(2t-\tau)} - n_+^*(t)n_+^*(t - \tau)e^{j\omega_c(2t-\tau)} \\ + n_+^*(t)n_+(t - \tau)e^{-j\omega_c\tau} - n_+(t)n_+^*(t - \tau)e^{j\omega_c\tau} \end{array}\right]}$$

$$= \frac{1}{4j}\overline{\left[n_+^*(t)n_+(t - \tau)e^{-j\omega_c\tau} - n_+(t)n_+^*(t - \tau)e^{j\omega_c\tau}\right]} \qquad \text{since } \overline{e^{j2\omega_c t}} = 0$$

$$= \frac{1}{2}\mathrm{Im}\overline{\left[n_+(t)n_+^*(t - \tau)e^{-j\omega_c\tau}\right]} = \frac{1}{2}\mathrm{Im}\left[\overline{n_+(t)n_+^*(t - \tau)}e^{-j\omega_c\tau}\right]$$

$$= \frac{1}{2}\mathrm{Im}\left[\overline{R_+}(\tau)e^{-j\omega_c\tau}\right] = \frac{1}{2}\mathrm{Im}\left[2[\overline{R_n}(\tau) + j\overline{\hat{R}_n}(\tau)]e^{-j\omega_c\tau}\right]$$

or:

$$\overline{R_{QI}}(\tau) = -\overline{R_{IQ}}(\tau) = \overline{R_n}(\tau)\,\mathrm{Sin}(\omega_c t) - \overline{\hat{R}_n}(\tau)\,\mathrm{Cos}(\omega_c t)$$

which proves the second part. *end of proof*

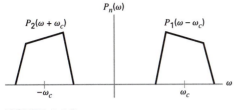

FIGURE 4.4-3

We shall continue to a most important result. The object now is to take the Fourier Transform of eqs. 4.4-12a and 4.4-12b and interpret the results. Let:

$$\overline{R_n}(\tau) \Leftrightarrow P_n(\omega)$$

$$\overline{\hat{R}_n}(\tau) \Leftrightarrow -j \, \text{sgn}(\omega) P_n(\omega)$$

$$\overline{R_I}(\tau) \Leftrightarrow P_I(\omega) = P_Q(\omega) \quad (4.4\text{-}13)$$

$$\overline{R_{IQ}}(\tau) \Leftrightarrow P_{IQ}(\omega)$$

and recall that:

$$\text{Cos}(\omega_c t) \Leftrightarrow \pi[\delta(\omega - \omega_c) + \delta(\omega + \omega_c)]$$

$$\text{Sin}(\omega_c t) \Leftrightarrow \frac{\pi}{j}[\delta(\omega - \omega_c) - \delta(\omega + \omega_c)]$$

The power spectral density of the bandpass noise $n(t)$, $P_n(\omega)$, is shown in fig. 4.4-3, where $P_1(\omega - \omega_c)$ and $P_2(\omega + \omega_c)$ are the upper and lower sidebands of $P_n(\omega)$. Noting that multiplication in the time domain requires convolution in the frequency domain, eq. 4.4-12a implies:

$$\begin{aligned}
P_I(\omega) = P_Q(\omega) &= \frac{1}{2}\{[P_n(\omega - \omega_c) + P_n(\omega + \omega_c)] - [\text{sgn}(\omega - \omega_c)P_n(\omega - \omega_c) \\
&\quad - \text{sgn}(\omega + \omega_c)P_n(\omega + \omega_c)]\} \\
&= \frac{1}{2}\{P_n(\omega - \omega_c)[1 - \text{sgn}(\omega - \omega_c)] + P_n(\omega + \omega_c)[1 + \text{sgn}(\omega + \omega_c)]\} \\
&= P_1(\omega) + P_2(\omega) \quad (4.4\text{-}14a)
\end{aligned}$$

and eq. 4.4-12b implies:

$$\begin{aligned}
j P_{IQ}(\omega) = -j P_{QI}(\omega) &= \frac{1}{2}\{[P_n(\omega - \omega_c) - P_n(\omega + \omega_c)] - [\text{sgn}(\omega - \omega_c)P_n(\omega - \omega_c) \\
&\quad + \text{sgn}(\omega + \omega_c)P_n(\omega + \omega_c)]\} \\
&= \frac{1}{2}\{P_n(\omega - \omega_c)[1 - \text{sgn}(\omega - \omega_c)] \\
&\quad - P_n(\omega + \omega_c)[1 + \text{sgn}(\omega + \omega_c)]\} \\
&= P_1(\omega) - P_2(\omega) \quad (4.4\text{-}14b)
\end{aligned}$$

These operations are illustrated in fig. 4.4-4. There are some important and even fundamental conclusions to be drawn from all of this.

1. The power spectral densities of the in-phase and quadrature components of the noise $n_I(t)$ and $n_Q(t)$, $P_I(\omega)$ and $P_Q(\omega)$, are equal and are even functions of ω. The total

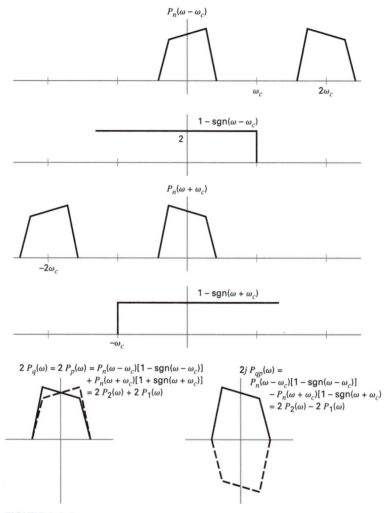

FIGURE 4.4-4

noise power in each of these components is equal to the power P_n contained in $n(t)$. This corresponds to

$$n(t) = n_I(t)\,\text{Cos}(\omega_c t) - n_Q(t)\,\text{Sin}(\omega_c t) \tag{4.1-6}$$

and

$$\overline{n^2(t)} = \frac{1}{2}\overline{n_I^2(t)} + \frac{1}{2}\overline{n_Q^2(t)}. \tag{4.1-7}$$

2. With the transmit signal and the receiver filter scaled as in fig. 4.1-2, the total power in the noise is $P_n = \eta/2$. With a raised cosine shaping, the autocorrelation function of each of the components of the noise is zero at multiples of the sampling time T_s so that successive samples of the noise are uncorrelated.

3. $P_{IQ}(\omega)$ is an odd function of ω and so is $R_{IQ}(\tau)$. This means that samples of the in-phase and quadrature noise taken at the same instant of time are uncorrelated.

Consequently,

$$pn_I n_Q(\alpha_I, \alpha_Q) = \frac{1}{2\pi\sigma^2} \exp\left(-\left(\alpha_I^2 + \alpha_Q^2\right)/2\sigma^2\right) \quad \text{where} \quad \sigma^2 = \eta/2. \quad (4.4\text{-}15)$$

If it happens that the power spectral density of the noise is symmetric around its center frequency, then $P_{IQ}(\omega)$ and $R_{IQ}(\tau)$ are both equal to zero. Consequently, under these circumstances we may consider the in-phase and quadrature noise components to be independent.

As a result of these arduous mathematical manipulations, we have verified all of the assumptions about how to detect the received signal that we made in Section 4.1.

4.5 IMPLEMENTATION OF THE HILBERT TRANSFORM RECEIVER

Although the classic QAM receiver structure of fig. 4.1-2 continues to be used at very high carrier frequencies and data rates corresponding to digital radio, the availability of low-cost DSP hardware for voiceband processing has made the Hilbert Transform Receiver into the structure of choice for those applications.

As shown in fig. 4.5-1, the absence of any elements with memory after the analog Hilbert Transform filtering allows us to sample the signal at that point at the symbol rate $f_s = 1/T_s$ and do all of the subsequent processing digitally. For the voiceband modems described above, this means sampling at a maximum of 2400 hz.

If the Hilbert Transform filtering is to be done digitally we must sample at least at twice the highest frequency contained in the received signal but at a multiple of the symbol frequency in order to allow for decimation. For a modem with a 2400 hz symbol rate, an initial sampling rate of 7200 hz or, preferably, 9600 hz is used.

4.5.1   The Free-Running Demodulator

The key issue in this circuit is the digital demodulator. Recall that we have broken the demodulation process into two parts, a 'free-running' demodulator at the nominal carrier frequency and a rotater having the purpose of compensating for any phase deviations. The rotater is adaptive and is the output of a carrier-phase recovery circuit that will be discussed

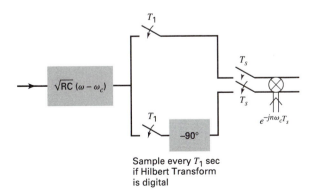

Sample every T_1 sec
if Hilbert Transform
is digital

FIGURE 4.5-1

in a subsequent chapter. The free-running demodulator multiplies the sampled complex signal by

$$e^{-jn\omega_c T_s} = e^{-jn2\pi f_c/f_s}$$

(4.5-1)

If the ratio f_c/f_s is the ratio of two integers, then the value of the exponential will repeat periodically and all of the values may be stored in a lookup table. Through the use of a pointer, the contents of the lookup table are accessed sequentially with each successive symbol rate sample. Upon reaching the end of the table, the pointer returns to the beginning. The construction of these lookup tables is best illustrated by example.

EXAMPLE 1 *V.27 Modem*

In this modem $f_c = 1800$ hz and $f_s = 1600$ hz so that

$$e^{-jn\omega_c T_s} = e^{-jn2\pi f_c/f_s} = e^{-jn2\pi 9/8}$$

Here the repetition pattern is $n = 0, 1, 2, 3, 4, 5, 6, 7$ and the lookup table is

Cos	Sin
1	0
.707	.707
0	1
−.707	.707
−1	0
−.707	−.707
0	−1
.707	−.707

∎

EXAMPLE 2 *V.22(bis) Modem*

In this modem $f_c = 1200$ hz or 2400 hz and $f_s = 600$ hz so that

$$e^{-jn\omega_c T_s} = e^{-jn2\pi f_c/f_s} = e^{-jn2\pi} = 1$$

and in this wonderfully simple case, the process of sampling also accomplishes the demodulation. ∎

EXAMPLE 3 *Fictitious V.29 Type Modem with 1800 hz Carrier*

In this modem $f_c = 1800$ hz and $f_s = 2400$ hz so that

$$e^{-jn\omega_c T_s} = e^{-jn2\pi f_c/f_s} = e^{-jn2\pi 3/4}$$

so that in four successive samples, corresponding to $n = 0, 1, 2, 3$ the digital demodulator will generate

$$e^{-j0} = 1 + j0$$

$$e^{-j1\times 2\pi 3/4} = \text{Cos}\,\frac{3\pi}{2} - j\,\text{Sin}\,\frac{3\pi}{2} = 0 + j1$$

$$e^{-j2\times 2\pi 3/4} = \text{Cos}\,3\pi - j\,\text{Sin}\,3\pi = -1 + j0$$

$$e^{-j3\times 2\pi 3/4} = \text{Cos}\,\frac{9\pi}{2} - j\,\text{Sin}\,\frac{9\pi}{2} = 0 - j1.$$

Consequently, the required lookup table is

Cos	Sin
1	0
0	1
−1	0
0	−1

An example involving the actual V.29 modem is left to the Problems. ■

These examples also demonstrate the Bandpass Sampling Theorem, which states that bandpass signals need be sampled at a rate equal to twice the bandwidth of the signal rather than twice the highest frequency. Each sample, however, must be a complex number repesenting both amplitude and phase information.

4.5.2   An Analog Approach to Designing a Hilbert Transform Filter

If we want to implement the receiver filter and Hilbert Transform in analog, typically with a switched-capacitor filter, we may consider the configuration of fig. 4.5-2a, where $F_1(\omega)$ and $F_2(\omega)$ are all-pass filters having phase characteristics $\theta_1(\omega)$ and $\theta_2(\omega)$. The idea is that in order to approximate a Hilbert Transform filter we should make

$$\theta_1(\omega) - \theta_2(\omega) \approx 90° \qquad (4.5\text{-}2a)$$

over the band of interest.

In order to accomplish this, consider the two all-pass transfer functions

$$F_1(s) = \left(\frac{s - \omega_1}{s + \omega_1}\right)\left(\frac{s - \omega_3}{s + \omega_3}\right) \quad \text{and} \quad F_2(s) = \left(\frac{s - \omega_2}{s + \omega_2}\right) \qquad (4.5\text{-}2b)$$

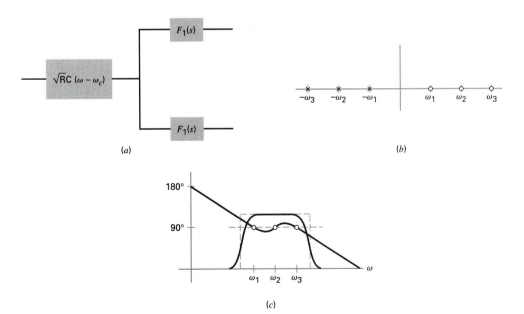

FIGURE 4.5-2

where $\omega_1 < \omega_2 < \omega_3$. The pole-zero plot is shown in fig. 4.5-2b. It follows that

$$\theta_1(\omega) = 180° - 2\tan^{-1}\left(\frac{\omega}{\omega_1}\right) + 180° - 2\tan^{-1}\left(\frac{\omega}{\omega_3}\right) \qquad (4.5\text{-}2c)$$

and

$$\theta_2(\omega) = 180° - 2\tan^{-1}\left(\frac{\omega}{\omega_2}\right) \qquad (4.5\text{-}2d)$$

so that

$$\theta(\omega) = \theta_1(\omega) - \theta_2(\omega) = 180° - 2\left[\tan^{-1}\left(\frac{\omega}{\omega_1}\right) - \tan^{-1}\left(\frac{\omega}{\omega_2}\right) + \tan^{-1}\left(\frac{\omega}{\omega_3}\right)\right] \qquad (4.5\text{-}2e)$$

Since $\tan^{-1}(x)$ goes from 0 to 90 degrees as x goes from 0 to infinity, it follows that $\theta(\omega)$ goes from 180 to 0 degress as ω goes from 0 to ∞. We therefore have the phase curve shown in fig. 4.5-2c. We can choose the values of ω_1, ω_2, ω_3 in order to obtain the best possible approximation over the band of interest. Keeping in mind that by the time that the signal gets to the phase splitter it is shaped by a full raised cosine, it is appropriate to have an rms measure of the phase error weighted by that shaping. The error can therefore be defined as

$$\varepsilon = \int RC^2(\omega)\left[\frac{\pi}{2} - \theta(\omega)\right]^2 d\omega \qquad (4.5\text{-}3)$$

The design of the filter should make the power of the error sufficiently below the power of the noise that is anticipated for the desired performance level. The error can be reduced further by increasing the numbers of poles and zeros in the filters according to the same principle. The next step in complexity would be

$$F_1(s) = \left(\frac{s-\omega_1}{s+\omega_1}\right)\left(\frac{s-\omega_3}{s+\omega_3}\right)\left(\frac{s-\omega_5}{s+\omega_5}\right) \quad \text{and} \quad F_2(s) = \left(\frac{s-\omega_2}{s+\omega_2}\right)\left(\frac{s-\omega_4}{s+\omega_4}\right) \qquad (4.5\text{-}4)$$

with $\omega_1 < \omega_2 < \omega_3 < \omega_4 < \omega_5$.

4.5.3   Digital Approaches to Designing a Hilbert Transform Filter

As shown in fig. 4.5-1, a digital implementation of a Hilbert Transform filter requires sampling at least at twice the highest frequency in the received signal. In our example for the telephone line that frequency is 9600 hz, four times the symbol rate, in order to allow for easy decimation. Under these circumstances, it makes sense also to do the $\sqrt{RC}(\omega - \omega_c)$ receiver shaping digitally as well. If we do this, we will have to precede the sampler with a sharp analog filter that serves the dual purpose of noise limiting and antialiasing.

In this implementation the -90 degree phase shift will be implemented entirely on one rail of the receiver through the use of a transversal filter (fig. 4.5-3a) having negative symmetry on its tap coefficients. The transfer function of this filter is

$$H(\omega) = -c_2 - c_1 e^{-j\omega T_1} + c_0 e^{-j2\omega T_1} + c_1 e^{-j3\omega T_1} + c_2 e^{-j4\omega T_1}$$

$$= \left[c_0 - 2j\sum_{n=1}^{2} c_n \operatorname{Sin}(n\omega T_1)\right]e^{-j2\omega T_1} \qquad (4.5\text{-}5a)$$

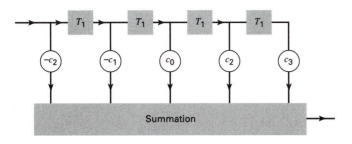

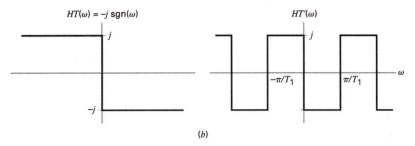

(b)

FIGURE 4-5-3

or, in general,

$$H(\omega) = \left[c_0 - 2j \sum_{n=1}^{N} c_n \, \mathrm{Sin}(n\omega T_1) \right] e^{-jN\omega T_1} \qquad (4.5\text{-}5b)$$

where the transversal has $(2N + 1)$ multiplying taps and the exponential is simply a time delay that can be compensated for by inserting an equal delay in the upper rail of the circuit.

We will use this transfer function, which is periodic with period $\omega_1 = 2\pi/T_1$, to approximate the frequency-domain representation of the Hilbert Transform, fig. 4.5-3b. The approximation requires that

$$c_n = \begin{cases} 0, & n \text{ even;} \\ 4/n\pi, & n \text{ odd;} \end{cases} \qquad (4.5\text{-}5c)$$

representing a Fourier Series in the frequency domain. The mean squared error between the two is

$$\varepsilon_N = 1 - \frac{1}{2} \sum_{n \text{ odd}}^{N} \left(\frac{4}{n\pi} \right)^2 = 1 - \frac{8}{\pi^2} \sum_{n \text{ odd}}^{N} \frac{1}{n^2}. \qquad (4.5\text{-}5d)$$

Hilbert Transform filtering may also be implemented with IIR structures having far fewer delay elements but involving a more demanding design process. The reader may consult Oppenheim and Schafer [O&S89] for details.

4.6 THE BASEBAND EQUIVALENT OF A BANDPASS SYSTEM

Both for analytic purposes and in order to provide the theoretical basis for computer simulation of the system, we wish to use the ideas of the Hilbert Transform receiver in order to develop the baseband equivalent of the bandpass systems discussed in this chapter.

At this point we will assume that the channel is ideal, so that the only impairment that the channel will take into account is noise. In Chapter 6, Carrier Phase Recovery, the model will be extended to include carrier-phase distortion and means for addressing that issue. In Chapter 9, the model will be extended to include trellis coding as a method of increasing immunity to noise. In Chapter 10 the model will be extended yet again to include amplitude and phase distortion in the channel and various structures designed to equalize the resulting intersymbol interference.

In fig. 4.4-2a the received signal that appears after the square-root-raised-cosine bandpass filtering and before the Hilbert transformer is given by

$$s(t) = s_I(t)\,\mathrm{Cos}(\omega_c t + \theta) - s_Q(t)\,\mathrm{Sin}(\omega_c t + \theta)$$
$$= \mathrm{Re}[(s_I(t) + js_Q(t))(\mathrm{Cos}(\omega_c t + \theta) + j\,\mathrm{Sin}(\omega_c t + \theta))]$$
$$= \mathrm{Re}\big[\tilde{s}(t)e^{j(\omega_c t + \theta)}\big] \tag{4.6-1}$$

where the carrier-phase distortion has been included. Note that after the Hilbert transform the receiver has two 'rails,' one carrying the received signal and the other carrying its Hilbert transform.

The first step in developing our model is to eliminate the need for the Hilbert Transform by transmitting (from eq. 4.4-6) a complex signal on two 'rails' so that the complex received signal (fig. 4.6-1a) is

$$s_+(t) = [s(t) + j\hat{s}(t)] = \big[\tilde{s}(t)e^{j\theta}e^{j\omega_c t}\big] = (s_I(t) + js_Q(t))e^{j\theta}e^{j\omega_c t} \tag{4.6-2a}$$

Note that the model has the noise entering the system *after* the filtered signal as opposed to entering the system before the receiver filter. A bit of thought should make it clear that the assumption that we have a matched filter system makes valid that sleight of hand in the model.

The effect of the free-running demodulator is to create the signal

$$\big[\tilde{s}(t)e^{j\theta}e^{j\omega_c t}\big]e^{-j\omega_c t} = (s_I(t) + js_Q(t))e^{j\theta} \tag{4.6-2b}$$

which is in the baseband.

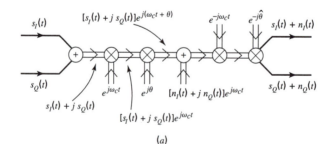

(a)

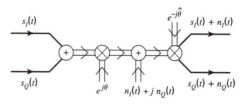

(b)

FIGURE 4-6-1

The idea of the model is to discard both the original modulation and the free-running demodulator so that the resulting system is entirely in the baseband. We then use an estimate of θ called $\hat{\theta}$ to do the "rotation" of the signal into its required orientation for slicing, so that

$$[(s_I(t) + js_Q(t))e^{j\theta}]e^{-j\hat{\theta}} \approx (s_I(t) + js_Q(t)). \tag{4.6-2c}$$

In making the transition from fig. 4.6-1a to fig. 4.6-1b we also note that the bandpass noise $n(t)$ is decomposed into the complex baseband sum of the in-phase and quadrature components.

We now wish to add the channel to the baseband equivalent model. It has already been established in eq. 4.4-1 that a bandpass signal has the form

$$
\begin{aligned}
s(t) &= s_I(t)\,\text{Cos}(\omega_c t) - s_Q(t)\,\text{Sin}(\omega_c t) \\
&= \text{Re}[(s_I(t) + js_Q(t))(\text{Cos}(\omega_c t) + j\,\text{Sin}(\omega_c t))] \\
&= \text{Re}[s_+(t)] = \text{Re}[s(t) + j\hat{s}(t)] = \text{Re}\big[\hat{s}(t)e^{j\omega_c t}\big]
\end{aligned} \tag{4.4-1}
$$

We will assert that the impulse response $h(t)$ of the channel, which is represented by a bandpass filter in fig. 4.6-2 has a similar form:

$$
\begin{aligned}
h(t) &= h_I(t)\,\text{Cos}(\omega_c t) - h_Q(t)\,\text{Sin}(\omega_c t) \\
&= \text{Re}[(h_I(t) + jh_Q(t))(\text{Cos}(\omega_c t) + j\,\text{Sin}(\omega_c t))] \\
&= \text{Re}[h_+(t)] = \text{Re}[h(t) + j\hat{h}(t)] = \text{Re}\big[\tilde{h}(t)e^{j\omega_c t}\big]
\end{aligned} \tag{4.6-3a}
$$

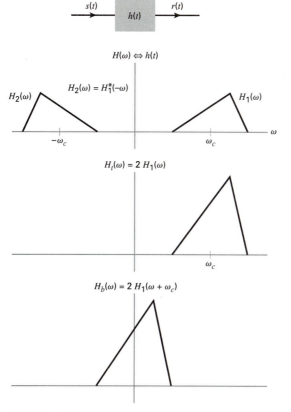

FIGURE 4-6-2

and the response $r(t)$ of the channel to an input $s(t)$ is also a bandpass signal:

$$
\begin{aligned}
r(t) &= r_I(t)\,\mathrm{Cos}(\omega_c t) - r_Q(t)\,\mathrm{Sin}(\omega_c t) \\
&= \mathrm{Re}[(r_I(t) + jr_Q(t))(\mathrm{Cos}(\omega_c t) + j\,\mathrm{Sin}(\omega_c t))] \\
&= \mathrm{Re}[r_+(t)] = \mathrm{Re}[r(t) + j\hat{r}(t)] = \mathrm{Re}\left[\hat{r}(t)e^{j\omega_c t}\right]
\end{aligned}
\tag{4.6-3b}
$$

Referring to (fig. 4.6-2), we shall make an equivalence between the time-domain operations shown above and their corresponding frequency-domain operations. First note that if $h(t)$ is to be a real function of time then it is necessary for

$$
H_1(\omega) = H_2^*(-\omega) \quad \text{or} \quad H_1(\omega + \omega_c) = H_2^*(-\omega + \omega_c).
\tag{4.6-4}
$$

Now the Fourier Transform of

$$
h_+(t) = h(t) + j\hat{h}(t)
\tag{4.6-5a}
$$

yields

$$
H_+(\omega) = H(\omega) + j[-j\,\mathrm{sgn}(\omega)H(\omega)] = [1 + \mathrm{sgn}(\omega)]H(\omega) = 2H_1(\omega)
\tag{4.6-5b}
$$

Using the equivalent notation $\tilde{h}(t) = h_b(t)$,

$$
h_b(t) = h_+(t)e^{-j\omega_c t} \qquad H_b(\omega) = 2H_1(\omega + \omega_c)
\tag{4.6-5c}
$$

or

$$
h_b(t) = h_I(t) + jh_Q(t) \qquad H_b(\omega) = H_I(\omega) + jH_Q(\omega)
\tag{4.6-5d}
$$

In the frequency domain, it clearly follows that

$$
R_+(\omega) = H_+(\omega)S_+(\omega)
\tag{4.6-6a}
$$

and

$$
R_+(\omega + \omega_c) = H_+(\omega + \omega_c)S_+(\omega + \omega_c)
\tag{4.6-6b}
$$

and further,

$$
R_I(\omega) + jR_Q(\omega) = (H_I(\omega) + jH_Q(\omega))(S_I(\omega) + jS_Q(\omega))
\tag{4.6-6c}
$$

so that

$$
R_I(\omega) = H_I(\omega)S_I(\omega) - H_Q(\omega)S_Q(\omega)
\tag{4.6-6d}
$$

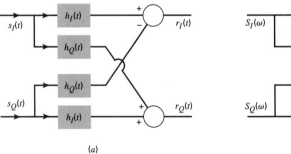

(a)

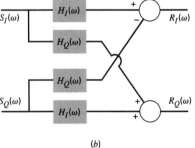

(b)

FIGURE 4-6-3

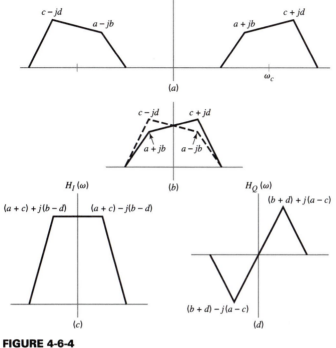

FIGURE 4-6-4

and

$$R_Q(\omega) = H_I(\omega)S_Q(\omega) + H_Q(\omega)S_I(\omega) \qquad (4.6\text{-}6e)$$

In the time domain this corresponds to the complex convolution

$$r_I(t) + jr_Q(t) = (h_I(t) + jh_Q(t)) \otimes (s_I(t) + js_Q(t)) \qquad (4.6\text{-}7a)$$

or

$$r_I(t) = h_I(t) \otimes s_I(t) - h_Q(t) \otimes s_Q(t) \qquad (4.6\text{-}7b)$$

and

$$r_Q(t) = h_I(t) \otimes s_Q(t) + h_Q(t) \otimes s_I(t). \qquad (4.6\text{-}7c)$$

These operations, both in the time and the frequency domains, are illustrated as lattice operations in figs. 4.6-3a and 4.6-3b. The final step in this development is to show how to find the baseband channel model from the bandpass transfer function. From eq. 4.6-5c,d we have

$$H_b(\omega) = 2H_I(\omega + \omega_c) = H_I(\omega) + jH_Q(\omega) \qquad (4.6\text{-}8a)$$

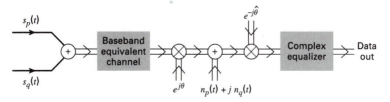

FIGURE 4-6-5

and since

$$H_1(\omega) = H_2^*(-\omega) \quad \text{or} \quad H_1(\omega + \omega_c) = H_2^*(-\omega + \omega_c) \qquad (4.6\text{-}8b)$$

it follows that

$$H_b^*(-\omega) = 2H_2(\omega - \omega_c) = H_I(\omega) - jH_Q(\omega) \qquad (4.6\text{-}8c)$$

so that by adding and subtracting eqs. 4.43a and 4.43b, we get

$$H_I(\omega) = [H_1(\omega + \omega_c) + H_2(\omega + \omega_c)] \qquad (4.6\text{-}9a)$$

$$H_Q(\omega) = -j[H_1(\omega + \omega_c) - H_2(\omega - \omega_c)] \qquad (4.6\text{-}9b)$$

These operations are illustrated in fig. 4.6-4. Note that $H_I(\omega)$ is an even function and $H_Q(\omega)$ is odd. According to Fourier Transform theory, this means that the corresponding time functions are respectively even and odd. As we shall see, in real systems where the channel is not strictly bandlimited, the time-domain baseband impulse responses will not be strictly even or odd but will tend in those directions.

With this, we are able to expand the baseband equivalent model of fig. 4.6-1b to include a baseband equivalent channel as shown in fig. 4.6-5. This baseband equivalent channel is the lattice structure of fig. 4.6-3a, indicating a complex convolution of the transmitted signal with the complex envelope of the channel impulse response. A box called complex equalizer is included in this model in order to show where the equalizer is placed. This box will typically be implemented digitally and will be adaptive. The analysis and design of the DSP-based equalizers will be the subject of Chapters 10 and 11.

This baseband equivalent model of a QAM modem serves as the basis for a computer simulation of modem performance. The basic structure of the program is developed in Appendix 4A. It is designed to simulate performance for various modulation techniques over an ideal channel. In subsequent chapters additional program functions will be added to include carrier recovery and equalization.

4.7 OQPSK (OFFSET QPSK) AND MSK (MINIMUM-SHIFT KEYING)

The spectrum of the transmitted signal in a basic QAM system (fig. 4.1-1) is that of the baseband pulse $p(t)$ translated to the carrier frequency (Amoroso [Amo80]). If $p(t)$ is a rectangular pulse having duration $T_s = T_b/2$, the power spectral density has the form

$$P(\omega) = Sa^2((\omega - \omega_c)T_s/2) \qquad (4.7\text{-}1)$$

The use of raised cosine shaping in QAM absolutely limits the bandwidth of the transmitted signal, but there remain very high-frequency applications where this shaping is difficult to achieve and a time-limited baseband pulse is preferred.

High-frequency applications also place a premium on having the envelope of the transmitted signal being constant because of nonlinearities in the transmitting tube of these systems. In order to derive maximum power from the transmitter it is necessary to operate in the saturation region, and consequently signals having envelope variation are distorted unevenly, yielding a sharp deterioration in performance. This consideration leaves multi-level QAM at a disadvantage. QPSK, a variant of QAM, has a constant envelope so that it addresses the nonlinearity problem, but with rectangular baseband pulses is a relatively wideband signal. That envelope is distorted by the bandpass filtering required in order to control out-of-band emissions before the transmitted signal is applied to the nonlinear output amplifier. This transmitter nonlinearity as well as the hardlimiting of the signal at the input to a repeater causes the regeneration of out-of-band energy, which can also be understood as causing ISI. The problem with QPSK in this regard is that the phase shifts that carry the data, $0°, \pm 90°, 180°$, cause discontinuities that lead to a lot of signal energy relatively far away from the carrier. The largest phase shift, $180°$, causes the biggest problem.

We shall see that OQPSK (Offset QPSK, also characterized as SQAM, Staggered QAM), a variant of QPSK, limits the phase discontinuity to $\pm 90°$ and that MSK (Minimum-Shift Keying), a variant of OQPSK, has a continuous phase with no discontinuities at all (Pasupathy [Pas79]). It turns out that the response of OQPSK to bandpass filtering is less severe than that of standard QPSK. MSK, in turn, performs better than OQPSK in this regard.

The distinction between QPSK and OQPSK (Chang [Cha66]) is in the alignment of the data streams on the I and Q channels. As shown in fig. 4.7-1, the data on the two channels in QPSK is aligned but in OQPSK the data on the Q channel is displaced from that on the I channel by half of a symbol interval, $T_s/2$. Since the data is aligned in QPSK, changes in the data stream may occur simultaneously, allowing all four possible phase changes. However, OQPSK allows only one of the two data streams to change at any given transition point. As shown in fig. 4.7-2, this limits the allowable phase changes to $0°, \pm 90°$. Pasupathy points out that testing has demonstrated that, despite the fact that the power spectral densities of these two signalling schemes are the same (power spectral density takes no account of phase), hardlimiting a bandlimited version of the two yields considerably different results. The bandlimited OQPSK system retains its spectral shaping while the QPSK spectrum is spread out again.

MSK has the same basic structure as OQPSK except that the baseband pulse $p(t)$ is a cosine pulse (fig. 4.7-3):

$$p(t) = \text{Cos}\left(\frac{\pi t}{T_s}\right); \qquad -\frac{T_s}{2} \leq t \leq \frac{T_s}{2} \qquad (4.7\text{-}2a)$$

instead of a rectangular pulse. The transmitted signal is then

$$s(t) = \sum_n a_n p(t - nT_s)\,\text{Cos}(2\pi f_c t) + b_n p(t - T_s/2 - nT_s)\,\text{Sin}(2\pi f_c t) \qquad (4.7\text{-}2b)$$

or

$$s(t) = \sum_n a_n \text{Cos}\left(\frac{\pi(t - nT_s)}{T_s}\right)\text{Cos}(2\pi f_c t) + b_n \text{Sin}\left(\frac{\pi(t - nT_s)}{T_s}\right)\text{Sin}(2\pi f_c t) \qquad (4.7\text{-}2c)$$

where both $p(t)$ and its delayed version are understood to be time-limited pulses. A straightforward way of comparing the spectra of QPSK and OQPSK on the one hand and MSK on the other is to focus on the different pulses that modulate the carrier. The Fourier Transform of the rectangular pulse that multiplies the two carriers in QPSK and OQPSK is

$$\text{rectangular pluse: } F_r(\omega) = T_s \text{Sa}\left(\frac{\omega T_s}{2}\right) \qquad (4.7\text{-}3a)$$

and the Fourier Transform of the cosine pulse that multiplies the carriers in MSK is

$$\text{cosine pluse: } F_c(\omega) = T_s \frac{\pi}{2} \frac{\text{Cos}(\omega T_s/2)}{(\pi/2)^2 - (\omega T_s/2)^2} \qquad (4.7\text{-}3b)$$

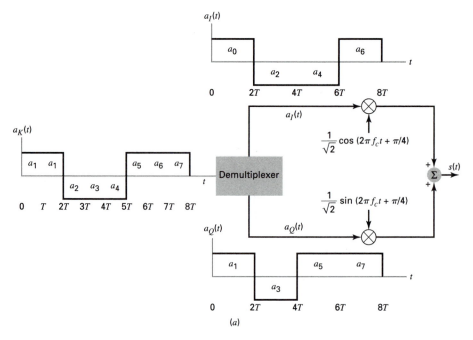

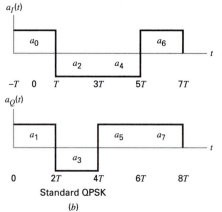

Standard QPSK

(b)

FIGURE 4-7-1 ([Pas79] reprinted with permission)

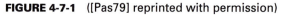

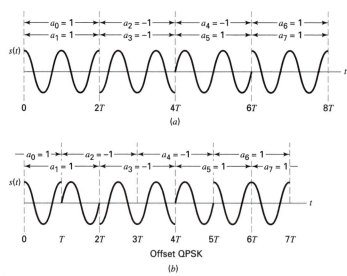

Offset QPSK

(b)

FIGURE 4-7-2 ([Pas79] reprinted with permission)

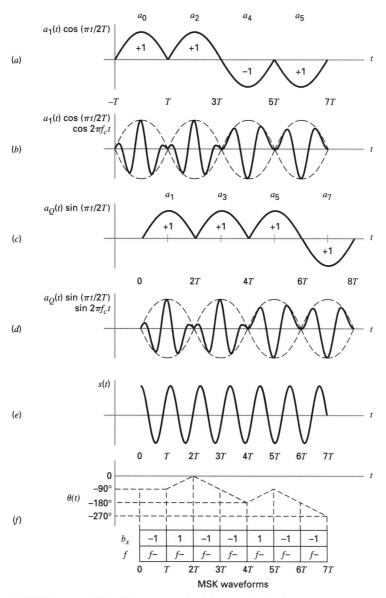

FIGURE 4-7-3   ([Pas79] reprinted with permission)

As shown in fig. 4.7-4 (Amoroso [Amo80, fig. 6]), the main lobe of the spectrum of the cosine pulse is wider than that of the rectangular pulse, having its first zero crossing at $\omega = 3\pi/T_s$ as compared to $\omega = 2\pi/T_s$. Nevertheless, the spectrum of the cosine pulse decreases faster than that of the rectangular pulse, $1/\omega^2$ compared to $1/\omega$. This makes the MSK spectrum more compact than the QPSK spectrum.

Using a small bit of trigonometry, the expression for MSK in eq. 4.7-2c can be rewritten as

$$s(t) = \text{Cos}\left(2\pi f_c t + a_n b_n \frac{\pi t}{T_s} + \varphi_n\right), \qquad nT_s \le t \le (n+1)T_s \qquad (4.7\text{-}4)$$

where φ_n is 0 or π corresponding to $a_n = 1$ or -1. As shown in fig. 4.7-3, the phase trajectory of the MSK signal is continuous.

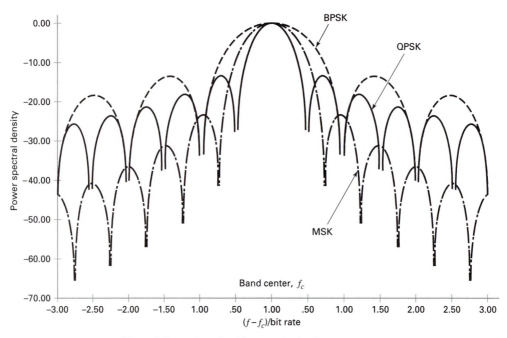

FIGURE 4-7-4 ([Amo80] reprinted with permission)

In Chapter 5 we will see that MSK is also a special case of FSK (Frequency-Shift Keying), another approach to data transmission that is derived from frequency modulation. The interested reader is also referred to [G&H75], [Aus83], [L&P93], and [Bin66].

4.8 DIFFERENTIAL PHASE-SHIFT KEYING (DPSK)

The analysis of performance in the presence of noise for all of the QAM-type systems that we have considered so far has assumed coherent detection. It will be recalled that coherent detection is based on the knowledge of the instantaneous phase of the carrier at the receiver. It was stated that this knowledge could be obtained through a carrier-phase recovery circuit. One example of forward-acting carrier recovery was given in Chapter 2. Considerable discussion will be devoted to decision-directed carrier-phase recovery using phase-locked loops in Chapter 6.

We have also discussed the use of differential coding of QAM signals as a way of resolving the fourfold phase-ambiguity problem in coherent detection of QAM.

We shall now restrict the discussion to those QAM-type signals that take the form of phase-shift keying, such as BPSK, 4-PSK, 8-PSK, or, in general, M-PSK. All of these signals may be differentially encoded in a manner that has been described earlier in this chapter. The combination of M-PSK with differential encoding leads to another receiver strategy, noncoherent detection, that does not require the generation of a demodulating carrier having the correct phase. This approach is called Differential Phase-Shift Keying (DPSK).

A block diagram of a DPSK receiver is shown in fig. 4.8-1a. Note that the front end of the receiver is similar to that of a coherent receiver except that the demodulating carrier

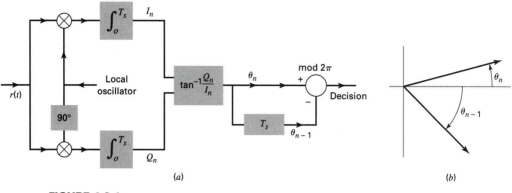

FIGURE 4-8-1

is simply generated by a local oscillator with no effort to determine any appropriate phase. At each received symbol interval, the outputs of the integrate and dump circuits (or lowpass filters) on the I and Q lines are used to determine an angle θ_n by computing

$$\theta_n = \tan^{-1}\left(\frac{Q_n}{I_n}\right) \tag{4.8-1}$$

Since, in a differentially encoded PSK system, the data is in the differences between the successive angles rather than the angles themselves, the idea is to subtract θ_n from θ_{n-1} and make a decision based upon that difference.

Both the absence of the need to generate a coherent reference and the relative simplicity the decision circuitry typically make building the DPSK receiver a good deal easier than building a coherent receiver. There are, however, some downsides. The first is that M-QAM is a more efficient system than M-PSK. For example, the performance of (coherent) 16-QAM having a probability of error $3Q(\sqrt{\rho_M/10})$ is superior to that of coherent 16-PSK, which has an error probability of $2Q(\text{Sin }\frac{\pi}{16}\sqrt{\rho_M})$.

The crucial comparison, however, is between coherent PSK and the noncoherent DPSK. The mathematical analysis is quite involved and the interested reader is referred to Simon, Hinedi, and Lindsey [SHL95], Arthurs and Dym [A&D62], and Prabhu and Salz [P&S81].

We can see from a qualitative point of view in fig. 4.8-1b that the decision is made on the difference between two noisy received angles. In effect, this operation doubles the amount of noise involved in the decision. In heuristic terms we would expect that this would degrade the performance by approximately 3 db, the effective increase of noise power. We have already seen that for

$$\text{Coherent MPSK} \qquad P_e \approx 2Q\left(\text{Sin }\frac{\pi}{M}\sqrt{\rho_M}\right) \tag{4.8-2a}$$

and the references show that a reasonable upper bound on performance for

$$\text{Noncoherent M-DPSK} \qquad P_e \approx 2Q\left(\text{Sin }\frac{\pi}{M\sqrt{2}}\sqrt{\rho_M}\right) \tag{4.8-2b}$$

It is left to the Problems to demonstrate that these two systems in fact have a performance difference of 3 db.

A slight variation of 4-DPSK called $\pi/4$-DPSK is the basis for the IS-54 Standard for digital cellular transmission. This approach uses two different QPSK signal sets that are offset from each other by 45°, switching from one to another in sequential transmissions. The interested reader may consult Mermelstein [Mer97] and Stuber [Stu97].

4.9 DIGITAL MICROWAVE TRANSMISSION

Digital Microwave Transmission is a technology of digital communication that is in relation to the digital hierarchy for telephone transmission discussed in Chapter 3. It will be recalled that DS3 transmission is 44.736 Mbps and the CCITT data rates include 34.386 Mbps and 139.264 Mbps. These data rates, typically rounded to 45 Mbps, 34 Mbps, and 140 Mbps, figure prominently in Digital Microwave. The FCC has assigned (Taylor and Hartmann [T&H86]) carrier bands at 4-, 6-, and 11-Ghz with channel spacings of 20 Mhz, 29.65 Mhz, and 40 Mhz respectively for passband transmission of these data signals using line-of-sight microwave links. In order to minimize interference, adjacent channels operate with orthogonal polarizations.

In this section only the issues relating to the choice of a signal constellation will be addressed. Carrier recovery, timing recovery, and adptive equalization for these systems will be discussed in the appropriate chapters. The microwave channel, which is characterized by fading as well as attenuation, will be discussed in Chapter 7.

The FCC framework requires that each channel must accommodate at least 1152 voiceband circuits. For the transmission to be consistent with the U.S. digital hierarchy, this translates into two DS3 signals, together having a bit rate of 90 Mbps. The problem, [Nog86], is how to transmit this data in 20 Mhz, 30 Mhz, and 40 Mhz channels. In order to discuss this further, it is appropriate to introduce the measure of bits/hz of bandwidth. For the three channels under question this measure becomes 4.5 bits/hz, 3 bits/hz, and 2.25 bits/hz. As a reference, note that 16QAM, 8PSK, and QPSK are 4, 3, and 2 bits/hz respectively.

A major issue in digital radio is the choice of the appropriate modulation for different transmissions. In addition to full response systems such as those indicated above, quadrature partial-response systems (QPRS) have also been used [Kur77]. QPRS consists of two partial-response signals modulated onto sine and cosine carriers. Like baseband partial response, its major advantage is bandwidth efficiency and its major disadvantage with respect to full response systems is the additional 2.1 db of power required for the same error rate. As we have seen, full-response systems require raised cosine filter shaping in order to combat ISI and facilitiate timing recovery. This filter shaping will also make it easier to meet FCC out-of-band radiation restrictions. Typically, an approximate rolloff of $\alpha = 0.5$ is used as a compromise between maximum spectral efficiency and difficulty of construction for a smaller value of α and excessive loss of bandwidth for a larger value.

As an example, consider the problem of transmitting two DS3 signals totaling 90 Mbps in the three different channels described above:

1. A 40 Mhz channel will support 90 Mbps using 8PSK with $\alpha = 0.33$;
2. A 30 Mhz channel will support 90 Mbps using 16 QAM with $\alpha = 0.33$;
3. A 20 Mhz channel will support 90 Mbps using 64 QAM with $\alpha = 0.33$.

The unending quest for more efficient utilization of bandwidth has led to efforts to increase the number of DS3 signals in a single channel, in turn leading to the use of more elaborate

TABLE 4.1.

	Bit rate	BW/20 Mhz	BW/30 Mhz	BW/40 Mhz
Europe:	34 Mbps	4–PSK	4–PSK	4–PSK
	68 Mbps	16–QAM	8–PSK	4–PSK
	140 Mbps	256–QAM	64–QAM	16–QAM
	280 Mbps	—	1024–QAM	256–QAM
USA/Japan:	90 Mbps	64–QAM	16–QAM	8–PSK
	135 Mbps	256–QAM	64–QAM	16–QAM
	180 Mbps	1024–QAM	256–QAM	64–QAM
	270 Mbps	—	1024–QAM	256–QAM

modulation techniques up to 1024 QAM. These are summarized in Table 4.1 below both for Euopean systems and those of the United States and Japan.

The spectral efficiency of these higher-order QAM systems is accompanied with some penalties. The most important of these is that the multiple levels lead to a requirement for linearity in the transmitter. By way of contrast, M-PSK signalling can tolerate high levels of transmitter nonlinearity because all of the transmitted signals have the same amplitude. However, M-PSK signaling requires considerably greater power to achieve the same error rate as a comparable QAM system. The other major issue in QAM is the requirement of a stable carrier recovery system for coherent demodulation. M-PSK signaling, as previously noted, can substitute noncoherent differential detection at the cost of up to 3 db in performance loss. Clearly, as Table 4.1 indicates, the direction is that of higher-order QAM with its spectral efficiency.

Although they will be referred to in subsequent chapters on channel characterization, equalization, and carrier recovery, some additional references on the subject of digital radio are [Din80], [Yam81], [Yam87], and the Special Issue on Digital Radio, *IEEE Transactions on Communications*, December 1979.

4.10 MULTICARRIER DATA TRANSMISSION

An underlying assumption in the previous discussions in this chapter has been that a single data signal, some variant of QAM, occupies the entire allocated bandwidth. We shall now begin the process of examining the use of FDM (frequency division multiplexing) to transmit data on the same channel.

Although this is not a new idea, it is being used in a new application: high-speed data transmission on the Digital Subscriber Loop. The DSL has been described in Chapter 3 in relation to ISDN transmission, which uses the band only up to 200 khz. Above this band, up to around 1 Mhz, the transmission is expected to be at 6 Mbps for most subscribers ([Max96], [Cio97]) and up to 52 Mbps on short lines. These technologies include HDSL, ADSL, and VDSL. We shall more to say about this channel in Chapters 7 and 8. A basic reference for all of these technologies is [Bin00].

The current interest in such high-speed transmission to the home is primarily based on the growing access to the Internet and the demand for constantly higher speeds to support this medium. The attractiveness of the using the DSL for this transmission is the basic fact that the wires are there and there is aleady almost universal connectivity, particularly in Western countries.

The basic idea (Bingham [Bin90]) fig. 4.10-1a, is to establish N_c equally spaced subchannels, each having bandwidth Δf, defined by carrier frequencies

$$f_{c,n} = n\Delta f \quad \text{for } n = n_1 \text{ to } n_2 \tag{4.10-1a}$$

where

$$N_c = n_2 - n_1 + 1 \tag{4.10-1b}$$

Each of these subchannels uses QAM to transmit data at the same symbol rate but possibly at different bit rates. We will see that this makes sense if the channel is distorted, as in fig. 4.10-1b. For example, one channel might use 32 QAM, another 16 QAM, and yet another QPSK. Consequently a serial-to-parallel converter divides the input data stream of f_b bits/sec into N_c parallel data streams that each operate at the symbol rate f_s but different numbers of bits/symbol m_n. Clearly

$$f_b = f_s \sum_{n=n_1}^{n_2} m_n \tag{4.10-1c}$$

The potential advantage of multicarrier over single-carrier shows itself only if the channel is distorted. If we have an ideal bandlimited channel then a simple example shows that single-carrier QAM and multiple-carrier QAM have the same overall bit rate and performance.

Let an ideal bandlimited bandpass channel have a two-sided bandwidth of 1000 hz so that with single-carrier QPSK we have 2 bits per symbol and 1000 symbols per second (or 1 msec per symbol), yielding a bit rate of 2000 bits per second. Now let the same channel be divided into 10 subchannels, each having a two-sided bandwidth of 100 hz, each able to

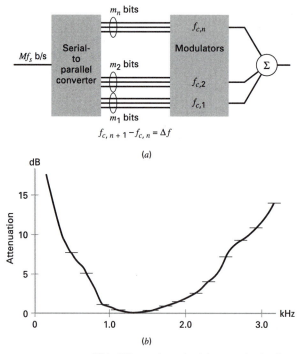

FIGURE 4-10-1 ([Bin90] reprinted with permission)

transmit 100 QPSK symbols per second (or 10 msec per symbol). Clearly the two systems have the same overall bit rate. We shall discuss how to accomplish such transmission below.

With respect to performance, let us have the same average transmitted power P_s in both systems. In the single-carrier system, the average energy per transmitted symbol is $E_s = (P_s)$ (1 msec). When the power is divided equally among all 10 of the subchannels in multiple-carrier, the average energy per transmitted symbol remains the same because the symbol duration increases to 10 msec. The probability of error, or performance, depends on the ratio $\rho_M = 2E_s/\eta$ where $\eta/2$ is the power spectral density of the noise and E_s is the energy per transmitted symbol. Since these quantities are equal for the two systems, then their performance is equal.

When the channel is not ideal, the idea is to use more power in subchannels where the attenuation is smallest and less in those where it is largest. With the goal of making the error probability equal in all the channels, more bandwidth-efficient modulations like 32-QAM are used in the subchannels that have a greater signal-to-noise ratio and less efficient ones like QPSK in worse-performing subchannels. Under these circumstances a multiple-carrier system has the potential of performing significantly better than a single carrier system. In Chapter 8, Channel Capacity, we will look at the capacity of a multicarrier channel. It should be noted, however, that making use of this approach requires knowledge of the channel characteristics so that the power may be allocated and the modulation schemes for each subchannel may be chosen. When the channel is initially unknown, its characteristics may be obtained by a process known as 'sounding.' In it simplest form, successive bursts of sinusoid at the center frequencies of the subchannels are sent from modem A to modem B. Modem B can determine the relative amplitudes and delays of these signals and transmit that information back to modem A. In turn, modem A is able to decide on the power allocation and modulation scheme for each sub-channel and send that information back to modem B so that it can have the proper slicer and level adjustment in each channel. All this takes place before data transmission is begun.

In fact, channel sounding is also performed in 56 kbps modems for the telephone line that use single-carrier QAM. In this instance, knowledge of the channel allows pre-equalization to be performed on the transmitted signal. This issue will be discussed in Chapter 10.

There are two distinct approaches to implementing a multicarrier system. One, by Hirosaki ([Hir80], [Hir81], [Hir86]), who extended a path begun by Chang [Cha66] and Saltzberg [Sal67]. The other is traced from Weinstein and Ebert [W&E71] to Peled and Ruiz [P&R80].

We will begin with Saltzberg, who divides the transmission band, fig. 4.10-2, into parallel overlapping raised-cosine channels separated by the symbol rate $f_s = 2f_N$, where the Nyquist frequency on each subchannel is $f_s/2$. The problem is obvious: how is it possible to transmit overlapping spectra and then separate them at the receiver? The solution that is advanced is to arrange the spectra of the adjacent channels so that they are in some sense orthogonal by using a variant of OQAM as shown in fig. 4.10-3.

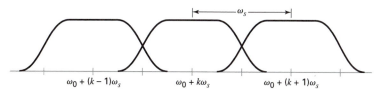

$\omega_0 + (k - 1)\omega_s$ $\omega_0 + k\omega_s$ $\omega_0 + (k + 1)\omega_s$

FIGURE 4-10-2

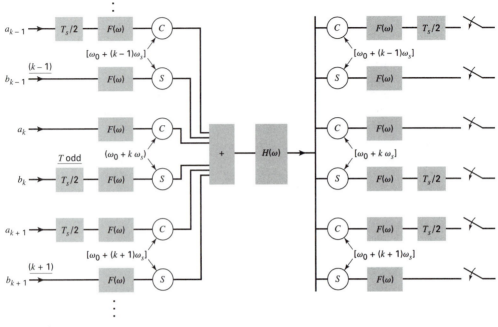

FIGURE 4-10-3

In the kth channel of this system, the data on the Sin channel is delayed in time from the data on the Cos channel by $T_s/2$, corresponding to OQAM. By way of contrast, on the adjacent $(k-1)$th channel and $(k+1)$th channel it is the data on the Cos channel that is delayed in time by $T_s/2$ from the data on the Sin channel. In each channel, both the transmitter filter and the receiver filter are $F(\omega)$, which is the square root of a raised cosine. Consequently each channel, taken by itself, is a standard raised-cosine bandpass channel. We are going to examine the response of the kth channel to simultaneous transmission in channel $k, k+1, k-1$ with the understanding that, since the other channels do not overlap in frequency, they will be of no concern. In this we will assume that the channel is undistorted, i.e. $H(\omega) = 1$.

We will examine the output on the kth Cos channel including contributions to that output from the kth Cos and Sin channels and from the adjacent $(k-1)$ and $(k+1)$ Cos and Sin channels. This will be sufficient to make the point of how this system works. In each case we will assume a single unit impulse being applied to the channel whose contribution to the output we are examining.

4.10.1 Transmit an Impulse on the kth Cos Channel

Note that

$$\mathrm{Cos}(\omega_k t) \Leftrightarrow \pi[\delta(\omega - \omega_k) + \delta(\omega + \omega_k)] \qquad (4.10\text{-}2a)$$

so that the spectrum at the input to the receiver is

$$S_1(\omega) = \frac{1}{2} H(\omega)[F(\omega - \omega_k) + F(\omega + \omega_k)] \qquad (4.10\text{-}2b)$$

After multiplying this received spectrum by $2\cos(\omega_k t)$ and filtering with $F(\omega)$, thereby removing the double frequency terms, we get

$$R_1(\omega) = \frac{1}{2}F^2(\omega)[H(\omega + \omega_k) + H(\omega - \omega_k)] \qquad (4.10\text{-}2c)$$

Since $F^2(\omega)$ is a full raised cosine and $H(\omega) = 1$, it follows that $R_1(\omega)$ is Nyquist and therefore contains no ISI.

4.10.2 Transmit an Impulse on the *k*th Sin Channel

Note that

$$\text{Sin}(\omega_k t) \Leftrightarrow \frac{\pi}{j}[\delta(\omega - \omega_k) - \delta(\omega + \omega_k)] \qquad (4.10\text{-}3a)$$

so that the spectrum at the input to the receiver is

$$S_2(\omega) = \frac{1}{2}H(\omega)\left[F(\omega)e^{-j\omega T_s/2} \otimes \frac{\pi}{j}[\delta(\omega - \omega_k) - \delta(\omega + \omega_k)]\right]$$

$$= \frac{j}{2}H(\omega)\left[F(\omega + \omega_k)e^{-j(\omega+\omega_k)T_s/2} - F(\omega - \omega_k)e^{-j(\omega-\omega_k)T_s/2}\right] \qquad (4.10\text{-}3b)$$

After multiplying this received spectrum by $2\cos(\omega_k t)$ and filtering with $F(\omega)$, thereby removing the double frequency terms, we get

$$R_2(\omega) = \frac{j}{2}F^2(\omega)[H(\omega - \omega_k) - H(\omega + \omega_k)]e^{-j\omega T_s/2} \qquad (4.10\text{-}3c)$$

Since $H(\omega) = 1$, the result is zero, corresponding to what we have known about the relationship of the Cos and Sin channels at a given carrier frequency.

4.10.3 Transmit an Impulse on the (*k* + 1)th Cos Channel

Note that

$$\cos(\omega_k + \omega_s)t \Leftrightarrow \pi[\delta(\omega - \omega_k - \omega_s) + \delta(\omega + \omega_k + \omega_s)] \qquad (4.10\text{-}4a)$$

so that the spectrum at the input to the receiver is

$$S_3(\omega) = \frac{1}{2}H(\omega)[F(\omega)e^{-j\omega T_s/2} \otimes \pi[\delta(\omega - \omega_k - \omega_s) + \delta(\omega + \omega_k + \omega_s)]]$$

$$= \frac{1}{2}H(\omega)\left[F(\omega - \omega_k - \omega_s)e^{-j(\omega-\omega_k-\omega_s)T_s/2} + F(\omega + \omega_k + \omega_s)e^{-j(\omega+\omega_k+\omega_s)T_s/2}\right]$$

$$(4.10\text{-}4b)$$

After multiplying by $2\cos(\omega_s t)$ and filtering with $F(\omega)$, we get

$$R_3(\omega) = \frac{1}{2}F(\omega)\{H(\omega + \omega_k)F(\omega - \omega_s)e^{-j(\omega-\omega_s)T_s/2} + H(\omega - \omega_k)F(\omega + \omega_s)e^{-j(\omega+\omega_s)T_s/2}\}$$

$$(4.10\text{-}4c)$$

but since $e^{\pm j\omega_s T_s/2} = e^{\pm j\pi} = -1$ and $H(\omega) = 1$, then

$$R_3(\omega) = -\frac{1}{2}F(\omega)\{F(\omega - \omega_s) + F(\omega + \omega_s)\}e^{-j\omega T_s/2} \qquad (4.10\text{-}4d)$$

Since $F(\omega)\{F(\omega - \omega_s) + F(\omega + \omega_s)\}$ is an even function of ω, it follows that the inverse Fourier Transform of eq. 4.10-9d is

$$r_3(t) = -\frac{1}{4\pi} \int_{-\infty}^{\infty} F(\omega)\{F(\omega - \omega_s) + F(\omega + \omega_s)\}e^{j\omega(t - T_s/2)}\, d\omega$$

$$= -\frac{1}{2\pi} \int_{0}^{\omega_s} F(\omega)F(\omega - \omega_s)\operatorname{Cos}[\omega(t - T_s/2)]\, dt \qquad (4.10\text{-}4e)$$

and letting $u = \omega - \omega_s/2$,

$$= -\frac{1}{2\pi} \int_{-\omega_s/2}^{\omega_s/2} F(u + \omega_s/2)F(u - \omega_s/2)\operatorname{Cos}[(u + \omega_s/2)(t - T_s/2)]\, dt$$

$$= -\frac{1}{2\pi} \int_{-\omega_s/2}^{\omega_s/2} F(u + \omega_s/2)F(u - \omega_s/2)\operatorname{Sin}[ut + \omega_s/2t - uT_s/2)]\, dt \quad (4.10\text{-}4f)$$

and *at the sampling instants* $t = nT_s$

$$= -\frac{1}{2\pi} \int_{-\omega_s/2}^{\omega_s/2} F(u + \omega_s/2)F(u - \omega_s/2)\operatorname{Sin}\left[u\left(n - \frac{1}{2}\right)T_s + n\pi\right]\, dt \quad (4.10\text{-}4g)$$

and $F(u + \omega_s/2)F(u - \omega_s/2)$ is an even function of u while Sin is an odd function of u. It follows that

$$r_3(nTs) = 0 \qquad (4.10\text{-}4h)$$

Therefore, at the sampling instants, there is no interference from this channel.

4.10.4   Transmit an Impulse on the $(k + 1)$th Sin Channel

Note that

$$\operatorname{Sin}(\omega_k + \omega_s)t \Leftrightarrow \frac{\pi}{j}[\delta(\omega - \omega_k - \omega_s) - \delta(\omega + \omega_k + \omega_s)] \qquad (4.10\text{-}5a)$$

so that the received signal is

$$S_4(\omega) = \frac{1}{2} H(\omega)\left[F(\omega) \otimes \frac{\pi}{j}[\delta(\omega - \omega_k - \omega_s) - \delta(\omega + \omega_k + \omega_s)]\right]$$

$$= \frac{j}{2} H(\omega)[F(\omega + \omega_k + \omega_s) - F(\omega - \omega_k - \omega_s)] \qquad (4.10\text{-}5b)$$

After multiplying by $2\operatorname{Cos}(\omega_k t)$ and filtering with $F(\omega)$, and recognizing that $H(\omega) = 1$, we get

$$R_4(\omega) = \frac{j}{2} F(\omega)[F(\omega + \omega_s) - F(\omega - \omega_s)] \qquad (4.10\text{-}5c)$$

It follows that the inverse Fourier Transform is

$$r_4(t) = \frac{j}{2}\frac{1}{2\pi} \int_{-\infty}^{\infty} \{F(\omega)[F(\omega + \omega_s) - F(\omega - \omega_s)]\}e^{j\omega t}\, d\omega \qquad (4.10\text{-}5d)$$

Since the term inside the braces is odd, it follows that

$$r_4(t) = \frac{j^2}{2}\frac{1}{2\pi} \int_{-\infty}^{\infty} \{F(\omega)[F(\omega + \omega_s) - F(\omega - \omega_s)]\} \operatorname{Sin} \omega t \, d\omega$$

$$= \frac{1}{2\pi} \int_{0}^{\omega_s} \{F(\omega)F(\omega - \omega_s)\} \operatorname{Sin} \omega t \, d\omega \qquad (4.10\text{-}5\text{e})$$

and letting $u = \omega - \omega_s/2$

$$= \frac{1}{2\pi} \int_{-\omega_s/2}^{\omega_s/2} \left\{ F\left(u + \frac{\omega_s}{2}\right) F\left(u - \frac{\omega_s}{2}\right) \right\} \operatorname{Sin}\left(u + \frac{\omega_s}{2}\right) t \, du \qquad (4.10\text{-}5\text{f})$$

and sampling at $t = nT_s$ yields

$$= \frac{1}{2\pi} \int_{-\omega_s/2}^{\omega_s/2} \left\{ F\left(u + \frac{\omega_s}{2}\right) F\left(u - \frac{\omega_s}{2}\right) \right\} \operatorname{Sin}(u_n T_s + \pi) \, du \qquad (4.10\text{-}5\text{g})$$

Since the term inside the brackets is an even function and the Sin is an odd function, it follows that the integral is zero. Therefore

$$r_4(nT_s) = 0 \qquad (4.10\text{-}5\text{h})$$

4.10.5 Transmit an Impulse on the $(k-1)$th Cos Channel

This proof is left to the Problems.

4.10.6 Transmit an Impulse on the $(k-1)$th Sin Channel

This proof is left to the Problems.

Hirosaki's contribution is to implement Saltzberg's approach using DSP techniques. In order to accomplish this, he proposes to generate each of the transmit baseband signals using DSP at one sampling rate and then taking the DFT at another sampling rate. This method fully utilizes the bandwidth of the channel. The difficulty with the Saltzberg-Hirosaki approach lies with the fact that separation of the adjacent data signals is achieved only at the sampling points and then only if the channel is undistorted. This makes the system very sensitive to both timing and equalization.

We shall next consider the approach begun by Weinstein and Ebert. The multicarrier transmitter shown in fig. 4.10-4 shows a serial data stream of complex data symbols $d_n = a_n + jb_n$ at a rate of $f_s = 1/T_s$ symbols/second. A group of N of these symbols is applied to a serial-to-parallel converter whose output consists of N separate channels in parallel, where each channel is transmitting a single complex data symbol $d_n = a_n + jb_n, n = 0, 1, \ldots N - 1$. The duration of each of these symbols is NT_s and the object is to transmit

$$s(t) = 2 \sum_{n=0}^{N-1} [a_n \operatorname{Cos}[(\omega_c + n\omega_N)t] + b_n \operatorname{Sin}[(\omega_c + n\omega_N)t]] \quad \text{where } 0 \le t \le NT_s,$$

$$\omega_N = 2\pi/NT_s \qquad (4.10\text{-}6)$$

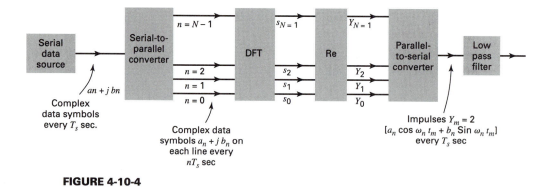

FIGURE 4-10-4

It turns out that this signal may be relatively easily derived by doing a Discrete Fourier Transform [Har82] on a vector $\mathbf{d}_n$ consisting of N complex data symbols d_n, passing the result through an appropriate lowpass filter, and then modulating up to the desired carrier frequency. The DFT of the data vector is another vector $\mathbf{S}_m$ where:

$$S_m = \sum_{n=0}^{N-1} 2d_n e^{-j(2\pi nm/N)} = 2\sum_{n=0}^{N-1} d_n e^{-j(\omega_n t_m)}; \quad m = 0, 1, \ldots, N-1$$

and

$$\omega_n = n\omega_N = n\frac{2\pi}{NT_s} \quad \text{and} \quad t_m = mT_s \tag{4.10-7}$$

The real part of this vector has components

$$Y_m = 2\sum_{n=0}^{N-1} [a_n \cos(\omega_n t_m) + b_n \sin(\omega_n t_m)] \tag{4.10-8}$$

and when we apply the components of the vector sequentially, every T_s sec, to a lowpass filter and modulate to the desired carrier, we get the result of eq. 4.10-6.

At the receiver, shown in fig. 4.10-5, the signal is first demodulated and then sampled every $T_s/2$ seconds, twice the highest frequency contained in the baseband signal. This gives us $2N$ samples, corresponding to $n = 0, 1, 2, \ldots, 2N-1$, which enables us to perform another DFT that will generate the values of $a_n + jb_n$ for $n = 0, 1, \ldots, N-1$.

In the ideal system where we have a distortionless channel, the receiver is that simple. If the channel is real and consequently does have some distortion, some additional processing

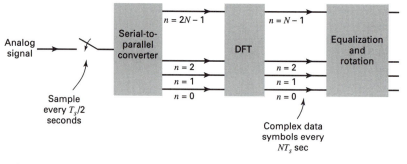

FIGURE 4-10-5

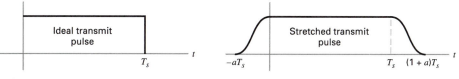

FIGURE 4-10-6

is required. There are two different issues arising from channel distortion that we are required to address:

1. the amplitude and phase distortion that affects the different carriers in the multicarrier system,
2. the problem that a distorted channel prevents the carriers from appearing immediately at the receiver, that a rise time is generated.

The second issue is addressed by lengthening the duration of the transmitted data pulse as shown in fig. 4.10-6 in order to allow for the rise time of the channel. The transmitted pulse is also allowed to rise and fall gradually. Therefore instead of an abrupt pulse having duration NT_s, the pulse has duration $(1 + 4a)NT_s$. The actual transmitted pulse is then a small variation of eq. 4.10-2:

$$\hat{s}(t) = 2g_a(t) \sum_{n=0}^{N-1} [a_n \, \mathrm{Cos}[(\omega_c + n\omega_N)t] + b_n \, \mathrm{Sin}[(\omega_c + n\omega_N)t]] \quad \text{where}$$

$$0 \le t \le (1 + 4a)NT_s, \quad \omega_N = 2\pi/NT_s \quad (4.10\text{-}9a)$$

and

$$g_a(t) = \begin{cases} \dfrac{1}{2}\left[1 + \mathrm{Cos}\dfrac{\pi t}{2aNT_s}\right], & -2aNT_s \le t \le 0 \\[2mm] 1, & 0 \le t \le NT_s \\[2mm] \dfrac{1}{2}\left[1 + \mathrm{Cos}\dfrac{\pi(t - T_s)}{2aNT_s}\right], & NT_s \le t \le (1 + 2a)T_s \end{cases} \quad (4.10\text{-}9b)$$

This, of course, reduces somewhat the actual data rate along the channel.

Although we postpone dealing with the general problem of transmission of data on distorted channels until Chapter 10, we shall deal with this issue now in relation to multicarrier because it is so straightforward. The key point is that when the channel is divided into subchannels, as shown in fig. 4.10-1b, the attenuation and the phase shift of the carrier in each subchannel are approximately constant. The larger the number of subchannels, the better the approximation. The values of the attenuation and phase shift in each subchannel are determined by the 'sounding' of the channel as mentioned above. Within each subchannel, the compensation required simply involves a multiplication and a rotation.

So in the nth subchannel we would take the receiver DFT output

$$z_n = x_n - jy_n \quad (4.10\text{-}10a)$$

and multiply it by

$$w_n = \frac{1}{H_n}[\mathrm{Cos}\,\theta_n - j \, \mathrm{Sin}\,\theta_n] \quad (4.10\text{-}10b)$$

in order to obtain

$$\tilde{a}_n + j\tilde{b}_n = z_n w_n \qquad (4.10\text{-}10\text{c})$$

which is then sliced to get the data estimate $\hat{a}_n + j\hat{b}_n$, where H_n and θ_n are the attenuation and phase of the subchannel carrier. Under these circumstances Weinstein and Ebert show that interference between adjacent channels vanishes. If the subchannels are not sufficiently narrow, it may be that a linear rather than a constant approximation to attenuation and phase is more appropriate. In this case there may be adjacent channel interference.

Extending the transmitted pulse in the time domain in order to minimize adjacent channel interference has been refined by Peled and Ruiz [P&R80] into the idea of a cyclic extension to each of these pulses. The problem that is being addressed can be understood in the time domain in the following way: the transmitted signal $s(t)$ (eq. 4-10.6) is NT_s seconds long and will be sampled every T_s sec at the receiver. If the channel impulse response $h(t)$ has duration KT_s sec then the convolution of the two will have length $N + K - 1$. The tail of this convolution will overlap the beginning of the subsequent convolution of the next transmitted signal. This is the time-domain expression of adjacent channel interference.

The cyclic extension consists of the concatenation of the first K samples of the transmitted signal to the end of the transmission. Consequently the frame of $L = N + K - 1$ samples at the receiver appears to be taken from the convolution of two periodic sequences. This is the condition for the continuous-frequency-domain product of two transfer functions $S(\omega)H(\omega)$ to be taken into the discrete frequency domain $S(\omega_n)H(\omega_n)$. The receiver then discards the last K samples before taking the DFT to restore the original data.

It is then easy to see how increasing N, the number of subchannels, decreases the overhead of compensating for adjacent channel interference for a given impulse response K. As Chow, Tu, and Cioffi [CTC91] point out, however, the impulse response of the DSL may be quite long. This would require N to be very large in order to prevent the overhead associated with the cyclic extension from becoming too large. As a solution to this problem they demonstrate that a special-purpose equalizer can be used to shorten the impulse response without substantial loss of performance.

COMPUTER MODELING OF THE BASEBAND EQUIVALENT OF A QAM MODEM

We want to be able to test the performance of QAM-type modems through computer simulation. Having established the baseband model of the QAM system, writing the program is a fairly straightforward task. The first step will be to develop a program called SNRTEST, which simply allows measurement of probability of error vs. SNR (signal-to-noise ratio) for an ideal channel having no carrier phase impairment. This program is written in C, and is available on the website http://www.com/college/kurzweil.

In Chapter 6 this program will be expanded to include the simulation of carrier-phase recovery systems and then in Chapter 11 it will be further expanded to include various techniques for the adaptive equalization of distorted channels.

The program follows the structure of the Baseband Equivalent System of fig. 4.6-1b. As a baseband equivalent, it can be simulated by symbol rate samples implemented by the block diagram of fig. 4A-1.

The data input to the simulation model is steady mark, a constant stream of 1's, which is then applied to a scrambler, (fig. 3.18-1a), which has been discussed in Chapter 3. We have chosen the (14, 17) scrambler used in the V.22(bis) modem. The scrambler output is a virtually uncorrelated stream of 1s and 0s. This output, applied directly to the input of the descrambler (fig. 3.18-1e) will return a constant stream of 1s. A detection error, caused by noise, will cause a 'wrong' input to the descrambler, resulting in the appearance of some zeros at the output.

The program has a variable called 'bits_per_baud' which may have the values (1, 2, 3, 4) according to the modulation scheme that has been chosen (BPSK, QPSK, 8-PSK, and 16-QAM). The scrambler output may be groups of 1, 2, 3, or 4 bits which are then applied to a modulator. The descrambler is the reciprocal of the scrambler, accepting groups of 1, 2, 3, or 4 bits and, in the absence of error, has at its output a steady stream of ones. Consequently, any zero at the descrambler output is a bit error. In the program SNRTEST the system performance is measured by symbol errors rather than bit errors because the feedback in the descrambler causes errors to propagate and because the differential coding used in modulation makes the relationship between a symbol error and a bit error uncertain.

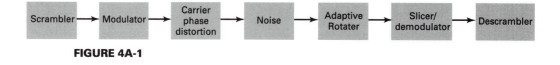

FIGURE 4A-1

4A.1   THE SCRAMBLER-DESCRAMBLER PAIR

The code for the scrambler-descrambler pair is shown below:

```
/*****************************************************************/
void scrambler( int bits_per_baud )
{
    static int sr1[18] = {0,0,0,0,0,0,0,0,0,0,0,0,0,0,0,0,0,0};
    int i, j;
    for( i = 0; i < bits_per_baud; i++ )
        {
        sr1[0] = ONE^(sr1[14]^sr1[17]);
        baud[i] = sr1[0];
        for( j = 17; j > 0; j-- )
                sr1[j] = sr1[j - 1];
        }
}
/*************************************************/
void descrambler( int bits_per_baud )
{
    static int sr2[18] = {0,0,0,0,0,0,0,0,0,0,0,0,0,0,0,0,0,0};
    int i, j;
    for( i = 0; i < bits_per_baud; i++ )
        {
        DOUT[i] = baud[i]^( sr2[14]^sr2[17] );
        sr2[0] = baud[i];
        for( j = 17; j > 0; j-- )
                sr2[j] = sr2[j - 1];
        }
}
/*************************************************/
```

The scrambler code generates an array baud[i] that may have 1, 2, 3, or 4 elements depending on the value of bits_per_baud. In effect, this array is gathering bits, dibits, tribits, or quadbits, depending on the modulation scheme that is intended. The descrambler code accepts an array baud[i] that is the same size as the scrambler output and itself produces a bit stream as its output.

The modulator-demodulator pair is also selected by the variable 'bits_per_baud'. The modulator input and demodulator output are structured to be consistent with the scrambler output and descrambler input respectively. For example, for a 16-QAM system corresponding to bits_per_baud = 4, the output of the scrambler and the corresponding modulator input is the 4-bit array baud[4]. Similarly, the 16-QAM demodulator delivers another four-bit array baud[4] to the descrambler. The modulator output is a pair of real numbers, PQ[0] and

PQ[1], corresponding to the baseband equivalent in-phase and quadrature components of the transmitted signal. These numbers come from the various signal sets shown in fig. 4.1-3. In each modulator-demodulator pair the modulator incorporates the appropriate differential encoding and the demodulator does the differential decoding as well as the slicing.

As shown in fig. 4.6-1b, and fig. 4A-1, the output pair PQ[0] and PQ[1] is carried through each successive block (carrier-phase distortion, channel distortion, additive noise, equalization, and adaptive rotation) until it is finally applied to the demodulator. Note that SNRTEST does not include channel distortion, adaptive rotation, or channel equalization. The demodulator consists of a slicer and a differential decoder. The demodulator output is a group bits that is then applied to the descrambler.

The code for the QPSK and the 16-QAM systems will be explained in some detail.

4A.2   QPSK MODULATOR-DEMODULATOR

The code for the QPSK modulator-demodulator pair is given below:

```
/****************************************************************/
void QPSK_MODULATOR( void )
{
static int quadrant = 0;
int test;
static int XMIT_QUAD[4] = {1, 0, 2, 3};
static float P_XMIT[4] = {.707, -.707, -.707, .707};
static float Q_XMIT[4] = {.707, .707, -.707, -.707};
float *P_ptr, *Q_ptr;
int *QUAD_ptr;
P_ptr = &P_XMIT[0];
Q_ptr = &Q_XMIT[0];
QUAD_ptr = &XMIT_QUAD[0];

if( baud[0] == 1 )
        QUAD_ptr += 2;
if( baud[1] == 1 )
        QUAD_ptr += 1;
quadrant += *QUAD_ptr;
quadrant %= 4;


for( test = quadrant; test > 0; test-- )
        {
        P_ptr += 1;
        Q_ptr += 1;
        }

PQ[0] = *P_ptr;
PQ[1] = *Q_ptr;
}
/****************************************************************/
```

The array XMIT_QUAD[4] = {1, 0, 2, 3} is keyed to the operation of differential coding. According to fig. 4.2-2, the differential coding rule for QPSK is

Baud[0]	Baud[1]	Phase change	Quadrant change
0	0	90°	1
0	1	0°	0
1	0	180°	2
1	1	270°	3

QUAD_ptr, which is always initialized to XMIT_QUAD[0], is advanced according to the values of baud[0] and baud[1] in order to determine the required difference between the previously transmitted quadrant and the new transmission. The quadrant that is to be transmitted is established by adding this result, modulo 4, to the previously transmitted quadrant.

The number of the quadrant to be transmitted is used to increment a pointer into a joint array that holds the required in-phase and quadrature components:

$$\text{static float P\_XMIT[4]} = \{.707, -.707, -.707, .707\};$$
$$\text{static float Q\_XMIT[4]} = \{.707, .707, -.707, -.707\};$$

The resulting numbers are transmitted as PQ[0] and PQ[1].

```
/*****************************************************************/
void QPSK_SLICER( void )
{
static float P_RCV[4] = { .707, .707, -.707, -.707 };
static float Q_RCV[4] = { .707, -.707, .707, -.707 };
static int quad[4] = { 0, 3, 1, 2 };
static int P_DATA[4] = { 0, 0, 1, 1 };
static int Q_DATA[4] = { 1, 0, 0, 1 };

static int oldquad = 0;
float *P_ptr, *Q_ptr;
int *new_ptr, *Pdata_ptr, *Qdata_ptr;
int quad_diff, test;

P_ptr = &P_RCV[0];
Q_ptr = &Q_RCV[0];
Pdata_ptr = &P_DATA[0];
Qdata_ptr = &Q_DATA[0];
new_ptr = &quad[0];

if( PQ[0] < 0 )
    {
    P_ptr += 2;
    Q_ptr += 2;
    new_ptr += 2;
    }
```

```
if( PQ[1] < 0 )
   {
   P_ptr += 1;
   Q_ptr += 1;
   new_ptr += 1;
   }

quad_diff = *new_ptr - oldquad;
oldquad = *new_ptr;

if( quad_diff < 0 )
   quad_diff += 4;

for( test = quad_diff; test > 0; test-- )
   {
   Pdata_ptr += 1;
   Qdata_ptr += 1;
   }

baud[0] = *Pdata_ptr;
baud[1] = *Qdata_ptr;


SLICEPQ[0] = *P_ptr;
SLICEPQ[1] = *Q_ptr;
}
/*****************************************************************/
```

The operation of the QPSK demodulator/slicer is the inverse of the modulator. The incoming signal components PQ[0] and PQ[1] are typically corrupted (by noise, phase shift, channel distortion) versions of the transmitted pair. These components are first sliced according to the rules discussed earlier in this chapter and the sliced values are saved and a pointer is established to indicate the quadrant of the received signal. The sliced values will not be of direct use just yet; they will be used in relation to carrier-phase recovery and adaptive equalization in further developments of the program.

The differential decoding of the received quadrant is accomplished by the following code:

```
quad_diff = *new_ptr - oldquad;
oldquad = *new_ptr;
if( quad_diff < 0 )
        quad_diff += 4;
```

The value of quad_diff is then translated back into bits according to the QPSK differential coding rule.

4A.3   16-QAM MODULATOR-DEMODULATOR

Recall that the quadbits in 16-QAM (fig. 4.2-3) are divided into two phase bits, baud[0] and baud[1], and two amplitude bits, baud[2] and baud[3]. The phase bits are differentially encoded in exactly the same manner as the phase bits in QPSK and the amplitude bits are encoded in a manner that makes them invariant to $90°$ rotation.

```
/********************************************************************/
void SIXTEEN_QAM_MODULATOR( void )
{
static int quadrant = 0;
int test;
static int XMIT_QUAD[4] = {1, 0, 2, 3};
static float P_XMIT[16] = {.316, .948, .316, .948, -.316, -.316, -.948, -.948,
                    -.316, -.948, -.316, -.948, .316, .316, .948, .948 };
static float Q_XMIT[16] = {.316, .316, .948, .948, .316, .948, .316, .948,
                    -.316, -.316, -.948, -.948, -.316, -.948, -.316, -.948 };
float *P_ptr, *Q_ptr;
int *QUAD_ptr;
P_ptr = &P_XMIT[0];
Q_ptr = &Q_XMIT[0];
QUAD_ptr = &XMIT_QUAD[0];

if( baud[0] == 1 )
        QUAD_ptr += 2;
if( baud[1] == 1 )
        QUAD_ptr += 1;
quadrant += *QUAD_ptr;
quadrant %= 4;

for( test = quadrant; test > 0; test-- )
        {
        P_ptr += 4;
        Q_ptr += 4;
        }
if( baud[2] == 1 )
        {
        P_ptr += 2;
        Q_ptr += 2;
        }
if( baud[3] == 1 )
        {
        P_ptr += 1;
        Q_ptr += 1;
        }
PQ[0] = *P_ptr;
PQ[1] = *Q_ptr;
}
/********************************************************************/
```

The code for determining the transmitted quadrant is the same as that of the QPSK case. This quadrant is, in turn, used to advance a pointer into the two arrays P_XMIT and Q_XMIT that together hold the coordinates of the 16 points in the signal set. These points are arranged according to quadrant and the pointer ends up at the beginning of the a particular group.

The pointer is then advanced through the group of points in the established quadrant in a manner that is keyed to the amplitude bits baud[3] and baud[4]. As in the QPSK case, the selected pair is placed in PQ[0] and PQ[1].

The 16-QAM demodulator/slicer operates as well in a manner that is similar to the QPSK case but somewhat more involved. The arrays are carefully established so that the process of slicing simultaneously advances pointers that determine not only the received point but the received quadrant as well. Note that knowledge of the quadrant is required for decoding the amplitude bits. The quadrant information is then differentially decoded into the phase bits.

```
/******************************************************************/
void SIXTEEN_QAM_SLICER( void )
{
static float P_RCVR[16] = {.948, .948, .316, .316, .948, .948, .316, .316,
                -.948, -.948, -.316, -.316, -.948, -.948, -.316, -.316 };
static float Q_RCVR[16] = {.948, .316, .948, .316, -.948, -.316, -.948,
                -.316, .948, .316, .948, .316, -.948, -.316, -.948, -.316 };

static int quad[4] = { 0, 3, 1, 2 };
static int BAUD0[4] = { 0, 0, 1, 1 };
static int BAUD1[4] = { 1, 0, 0, 1 };
static int BAUD2[16] = { 1, 0, 1, 0, 1, 1, 0, 0, 1, 1, 0, 0, 1, 0, 1, 0 };
static int BAUD3[16] = { 1, 1, 0, 0, 1, 0, 1, 0, 1, 0, 1, 0, 1, 1, 0, 0 };

static int oldquad = 0;
float *P_ptr, *Q_ptr;
int *new_ptr, *BAUD0_ptr, *BAUD1_ptr, *BAUD2_ptr, *BAUD3_ptr;
int quad_diff, test;

P_ptr = &P_RCVR[0];
Q_ptr = &Q_RCVR[0];
BAUD0_ptr = &BAUD0[0];
BAUD1_ptr = &BAUD1[0];
BAUD2_ptr = &BAUD2[0];
BAUD3_ptr = &BAUD3[0];

new_ptr = &quad[0];

if( PQ[0] < 0 )
   {
   P_ptr += 8;
   Q_ptr += 8;
   BAUD2_ptr += 8;
   BAUD3_ptr += 8;
   new_ptr += 2;
   }
if( PQ[1] < 0 )
   {
   P_ptr += 4;
   Q_ptr += 4;
   BAUD2_ptr += 4;
   BAUD3_ptr += 4;
   new_ptr += 1;
   }
```

```
if( fabs( PQ[0] ) < QAMTHRESH )
   {
   P_ptr += 2;
   Q_ptr += 2;
   BAUD2_ptr += 2;
   BAUD3_ptr += 2;
   }
if( fabs( PQ[1] ) < QAMTHRESH )
   {
   P_ptr += 1;
   Q_ptr += 1;
   BAUD2_ptr += 1;
   BAUD3_ptr += 1;
   }
quad_diff = *new_ptr - oldquad;
oldquad = *new_ptr;

if( quad_diff < 0 )
   quad_diff += 4;

for( test = quad_diff; test > 0; test-- )
   {
   BAUD0_ptr += 1;
   BAUD1_ptr += 1;
   }

baud[0] = *BAUD0_ptr;
baud[1] = *BAUD1_ptr;
baud[2] = *BAUD2_ptr;
baud[3] = *BAUD3_ptr;


SLICEPQ[0] = *P_ptr;
SLICEPQ[1] = *Q_ptr;
}
/*******************************************************************/
```

4A.4 CARRIER PHASE IMPAIRMENTS

The process of installing carrier phase impairments simply involves establishing the values of the phase offset, the carrier frequency offset, and both the frequency and amplitude of the carrier phase jitter. These quantities are then added together.

The transmitted signal is then rotated by the carrier phase distortion at each symbol interval. The important structural issue, one which characterizes the program as a whole, is that the rotation function accepts as input the pair PQ[0] and PQ[1], operates on that pair, and then provides new values of that pair as output. In turn, the new pair PQ[0] and PQ[1] serve as input to the next block.

Note that derotation by the carrier phase estimate $\hat{\theta}$ is accomplished in much the same way. This block is deactivated in SNRTEST1 and 2 but will subsequently be used in Chapter 6 on carrier phase recovery.

```
/*******************************************************************/
float phase_distortion( float phase_off, float freq_off,
                          float freq_jit, float amp_jit)
{
  static float result1, result2, result3, result4;
  result4 += (2*PI) * (freq_jit/f_baud);
  result3 = (PI/180) * amp_jit * sin( result4 );
  result2 += (2*PI) * (freq_off/f_baud);
  result1 = phase_off + result2 + result3;
  return result1;
}
/*******************************************************************/
void ROTATE( float theta )
{
  float temp_p, temp_q;
  temp_p = PQ[0]*cos( theta ) - PQ[1]*sin( theta );
  temp_q = PQ[1]*cos( theta ) + PQ[0]*sin( theta );
  PQ[0] = temp_p;
  PQ[1] = temp_q;
}
/*******************************************************************/
void DEROTATE( float thetahat )
{
  float temp_p, temp_q;
  temp_p = PQ[0]*cos( thetahat ) + PQ[1]*sin( thetahat );
  temp_q = PQ[1]*cos( thetahat ) - PQ[0]*sin( thetahat );
  PQ[0] = temp_p;
  PQ[1] = temp_q;
}
/*******************************************************************/
```

4A.5 ADDITIVE NOISE

The code for generating the in-phase and quadrature components of the noise is very direct. The point worth noting is that the input variables PQ[0] and PQ[1] are modified and then serve as output.

```
/*******************************************************************/
/*  This computation for generating a Gaussian random variable  */
/*  having variance sigsq from a uniform distribution is based on */
/*  Alan R. Miller, "Basic Programs for Scientists and Engineers", */
/*  SYBEX,1981, pg. 26                            */
void noise( void )
{
 int n;
 float temp_p, temp_q;
 temp_p = 0.0;
 temp_q = 0.0;
```

```
for( n=0; n<24; n++ )
        temp_p += ( (float) rand() )/32768.;
for( n=0; n<24; n++ )
        temp_q += ( (float) rand() )/32768.;
temp_p = (temp_p - 12.)*sqrt(sigsq/2);
temp_q = (temp_q - 12.)*sqrt(sigsq/2);
PQ[0] += temp_p;
PQ[1] += temp_q;
}
/*******************************************************************/
```

There are two variations of the program, SNRTEST1 and SNRTEST2. The first of these automatically steps the signal-to-noise ratio from 25 db to 15 db in 1 db intervals in order to reproduce the error-rate curves for the different modulation systems. The second allows choice in the SNR and displays the resulting signal space.

Since we should not yet be using the carrier-phase recovery system, those parameters should be maintained at all 0.

The listings are available from the website http://www.wiley.com/college/kurzweil

CHAPTER 4 PROBLEMS

4.1   Prove eq. 4.1-2.

4.2   **(a)** Referring to fig. 4.1-1, show that if the phase of the demodulating carrier is properly chosen, in this case $\theta = 0$, the upper leg of the receiver rejects the Sin carrier and the lower leg rejects the Cos carrier. Why is it important that $\omega_c = n\omega_s$.

   (b) Find the output of each leg of the receiver for $\theta \neq 0$.

   (c) Referring to fig. 4.1-3a, show that $\theta \neq 0$ corresponds to a rotation of the four-point constellation by an angle θ.

4.3   Demonstrate that the energy per transmitted symbol in eq. 4.1-4 is equal to E_s.

4.4   In fig. 4.1-2, show that the noise output power of the receiver filter is $P_n = \eta/2$.

4.5   **(a)** Show that the average energy per transmitted symbol in the 16-QAM constellation of fig. 4.1-3b is E_s.

   (b) Show that there is a subset of the 16-QAM constellation that is equivalent to a rotated version of the QPSK constellation of fig. 4.1-3a.

4.6   Find the scaling required to make the average energy per transmitted symbol in the 32-QAM signal set of fig. 4.1-3d be equal to E_s. Compute the average probability of error for this signal set as a function of ρ_M.

4.7   **(a)** Find the value of ρ_M required to an error probability of 10^{-6} for the QPSK signal of fig. 4.1-3a. For that value of ρ_M find the average probability of error for the other signal sets in fig. 4.1-3.

   (b) Conversely, find the value of ρ_M required to give an error probability of 10^{-6} for each of the signal sets in fig. 4.1-3.

4.8   Construct a finite-state machine (Moore) model of the differential coding rule of eq. 4.2-3a. Using Markov process techniques, confirm that this differential coding has no effect on the power spectral density of the transmitted signal.

4.9   **(a)** Complete the differential coding table of fig. 4.2-2.

   (b) Repeat (a) for fig. 4.2-3.

4.10 Find a state machine representation of the differential encoding rules for QPSK and 8-PSK.

4.11 A data sequence consists of periodic repetitions of the pattern 00011011. Using the differential coding techniques that were discussed in the text, find the first 10 transmitted symbols for (a) QPSK, (b) 8-PSK, (c) 16-QAM.

4.12 **(a)** For the V.29 constellation shown in fig. 4.3-4, find the appropriate scaling of the points so that the average energy per transmitted symbol is E_s.

(b) Find an expression for the average probability of symbol error in terms of ρ_M. (Hint: Divide the plane into rectangles rotated 45° to the horizontal.)

4.13 Prove the following properties of the Hilbert Transform:
   (a) If $x(t) = x(-t)$, then $\hat{x}(t) = -\hat{x}(-t)$;

   (b)
$$\int_{-\infty}^{\infty} x^2(t)\, dt = \int_{-\infty}^{\infty} \hat{x}^2(t)\, dt$$

   (c)
$$\int_{-\infty}^{\infty} x(t)\hat{x}(t)\, dt = 0$$

4.14 A function $f(t)$ has a Fourier Transform $F(\omega)$ shown below:

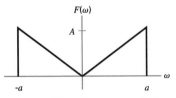

Find the Hilbert Transform of $f(t)$.

4.15 A lowpass signal $m(t)$ has a Fourier Transform $M(\omega)$ and $f(t) = m(t)\mathrm{Cos}(\omega_0 t)$ has the Fourier Transform shown below. If $f(t)$ is the input to a Hilbert Transform circuit, find an expression for the output. Describe what would happen if the carrier frequency were too low.

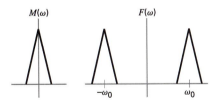

4.16 With $f(t)$ and $F(\omega)$ as given in Problem 4-14, show that
$$s(t) = f(t)\,\mathrm{Cos}(\omega_c t) \pm \hat{f}(t)\,\mathrm{Sin}(\omega_c t)$$
is a SSB (single sideband) signal.

4.17 **(a)** In a Hilbert Transform receiver for a 2400-baud, 16-QAM signal we wish to implement the Hilbert Transform with a transversal filter. If the receiver is to operate with an error rate of 10^{-5} and if the residual mse of the transversal is to be counted as additional noise, how many taps should it have so that it does not materially affect the error rate? Repeat for an error rate of 10^{-6}.

(b) Repeat a for an 8-PSK system.

4.18 Find the lookup table that implements the free-running demodulator for the V.29 modem.

4.19 The Hilbert Transform of narrowband noise $n(t)$ is $\hat{n}(t)$. Show that the cross-correlation of $n(t)$ and $\hat{n}(t)$ is:

(a) $R_{n\hat{n}}(\tau) = -\hat{R}_n(\tau)$

(b) $R_{\hat{n}n}(\tau) = \hat{R}_n(\tau)$

4.20   The power spectral density of a narrowband noise process $n(t)$ is $P_n(\omega)$ as shown below:

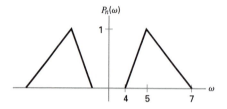

(a) Find the power spectral densities of $n_I(t)$ and $n_Q(t)$.

(b) Find the cross-spectral density of $n_I(t)$ and $n_Q(t)$.

4.21   Confirm the 3 db performance difference between coherent MPSK and DPSK in eq. 4.8-2.

4.22   Using the program SNRTEST2, observe the various signal constellations in the presence of noise and with a phase offset of 30°. Sketch your observations.

4.23   Using the program SNRTEST1, compare the simulation results of error rate vs. SNR for the various signal constellations with the theoretical results. Be careful to use enough symbols in your simulation to give a reasonable result for the expected error rate.

REFERENCES

[A&D62]   E. Arthurs and H. Dym, "On the optimum detection of digital signals in the presence of white Gaussian noise—a geometric interpretation and a study of three basic data transmission systems," *IRE Trans. on Communication Systems*, CS-10, no. 4, Dec. 1962, pp. 336–372.

[Amo80]   F. Amoroso, "The bandwidth of digital data signals," *IEEE Comm. Magazine*, Nov. 1980, pp. 13–24.

[Aus83]   M.C. Austin et al., "QPSK, staggered QPSK, and MSK—a Comparative Evaluation," *IEEE Trans. Communications*, **Com-31**, No. 2, Feb. 1983, pp. 171–182.

[Bin66]   L. Bin, "Decision feedback detection of minimum shift keying," *IEEE Trans. Communications*, **Com-44**, No. 9, Sept. 1966, pp. 1073–1076.

[Bin90]   J.A.C. Bingham, "Multicarrier modulation for data transmission: an idea whose time has come," *IEEE Comm. Magazine*, May 1990.

[Bin00]   John A.C. Bingham, "ADSL, VDSL, and Multicarrier Modems" Wiley-Interscience, 2000.

[CTC91]   J. Chow, J. Tu, and J. Cioffi, "A Discrete Multitone Transceiver System for HDSL Applications," *IEEE Journal on Selected Areas in Communications*, Vol. 9, No. 6, August 1991.

[Cha66]   R.W. Chang, "High-speed multichannel data transmission with bandlimited orthogonal signals," Bell System Technical J., **45**, 1966, pp. 1775–1796.

[Cio97]   J. Cioffi, "Asymmetric digital subscriber lines," in J.D. Gibson (ed.), *The Communications Handbook*, CRC Press, 1997, pp. 450–479.

[Din80]   N. Dinn, "Digital radio: its time has come," *IEEE Comm. Magazine*, **18**, No. 11, Nov. 1980, pp. 6–12.

[For96]   G.D. Forney, L. Brown, M.V. Eyuboglu, & J. Moran, "The V.34 high-speed modem standard," *IEEE Comm. Magazine*, Dec. 1996, pp. 28–33.

[G&H75]   R.D. Gitlin and E. Ho, "The performance of staggered quadrature amplitude modulation in the presence of phase jitter," *IEEE Trans. Communications*, **Com-23**, No. 3, March 1975, pp. 348–352.

[Har82]   F. Harris, "The discrete Fourier transform applied to time domain signal processing," *IEEE Comm. Magazine*, May 1982.

[Hir80]   Hirosaki, "An analysis of automatic equalizers for orthogonally multiplexed QAM systems using the discrete Fourier transform," *IEEE Trans. Communications*, **Com-28**, No. 7, Jan. 1980.

[Hir81]   B. Hirosaki, "An orthogonally multiplexed QAM system using the discrete Fourier transform," *IEEE Trans. Communications*, **COM-29**, No. 7, July 1981.

[Hir86]   B. Hirosaki, "Advanced groupband data modem using orthogonally multiplexed QAM technique," *IEEE Trans. Communications*, **COM-34**, No. 6, June 1986.

[Kur77]   H. Kurematsu, et al., "The QAM2G-10R digital radio equipment using a partial response system," *Fujitsu Scientific and Technical Journal*, **12**, No. 2, June 1977.

[L&P93]   H. Leib and S. Pasupathy, "Error control properties of minimum shift keying," *IEEE Communications Magazine*, January 1993, pp. 52–61.

[Max96]   Kim Maxwell, "Asymmetric Digital Subscriber Line: Interim Technology for the Next Forty Years," *IEEE Communications Magazine*, October 1996, pp. 100–106.

[Mer97]   P. Mermelstein, "The IS-54 digital cellular standard" in J.D. Gibson (ed.), *The Communications Handbook*, CRC Press, 1997, pp. 1246–1256.

[Nog86]   T. Noguchi et al., "Modulation techniques for microwave digital Radio," *IEEE Communications Magazine*, October 1986, **24**, No. 10, pp. 21–30.

[O&S89]   A. Oppenheim and R. Schafer, *Discrete Time Signal Processing*, Prentice-Hall, 1989.

[P&R80]   A. Peled and A. Ruiz, "Frequency domain data transmission using reduced computational complexity algorithms," *International Conference on Acoustics, Speech, and Signal Processing*, April 1980, pp. 964–967.

[P&S81]   V. Prabhu, and J. Salz, "On the performance of phase shift keying systems," *Bell System Technical J.* **60**, No. 10, Dec. 1981, pp. 2307–2341.

[Pas79]   S. Pasupathy, "Minimum shift keying: a spectrally efficient modulation," *IEEE Communications Magazine*, July 1979, pp. 14–22.

[SHL95]   M. Simon, S. Hinedi, and W. Lindsey, *Digital Communication Techniques*, Prentice-Hall, 1995, Chapter 7.

[Sal67]   B. Saltzberg, "Performance of an Efficient Parallel Data Transmission System," *IEEE Trans. Communications*, **COM-15**, No. 6, December 1967.

[Stu97]   G. Stuber, "Modulation methods," in Gibson (ed.), *The Communications Handbook*, CRC Press, 1997, pp. 1246–1256.

[T&H86]   D. Taylor and P. Hartmann, "Telecommunications by microwave digital radio," *IEEE Communications Magazine*, **vol. 24**, No. 8, August 1986, pp. 11–16.

[W&E71]   S. Weinstein and P. Ebert, "Data transmission by frequency-division multiplexing using the discrete Fourier transform," *IEEE Trans. Communications*, **COM-19**, No. 5, October 1971.

[Yam81]   H. Yamamoto, "Advanced 16-QAM techniques for digital microwave radio," *IEEE Communications Magazine*, **19**, No. 5, May 1981, pp. 36–45.

[Yam87]   H. Yamamoto, et al., "Future trends in microwave digital radio," *IEEE Communications Magazine*, **25**, No. 2, February 1987, pp. 40–52.

CHAPTER 5

MAXIMUM LIKELIHOOD SIGNAL DETECTION AND SOME APPLICATIONS

We have, without introducing the name, used maximum likelihood signal detection in Chapters 3 and 4 to decide, given the noisy received signal, which is the most likely transmitted signal. This was done both for baseband and for QAM signaling methods. In this chapter we want to formalize and generalize that approach so that we can extend it in a variety of ways.

From a theoretical point of view, we want to introduce the idea of thinking about signals and noise as points in a function space that is analogous to a Euclidean vector space. This will allow us to characterize and generalize maximum likelihood detection in order to be able to use it in other kinds of modulation techniques. Examples of this will be the development of CPM (continuous phase modulation) in this chapter and multidimensional signaling in subsequent chapters.

The maximum likelihood technique also underlies the basic theories of carrier phase and timing phase recovery. These issues will be discussed in detail in Chapter 6.

Finally, maximum likelihood detection is central to the analytic techniques necessary to address more complex issues such as channel capacity, trellis codes, and the Viterbi Algorithm.

5.1 VECTOR SPACES AND FUNCTION SPACES

Let us begin with a brief review of ordinary two-dimensional Euclidean vector space as shown in fig. 5.1-1. In fig. 5.1-1a we see a pair of vectors $\mathbf{w}_1$ and $\mathbf{w}_2$ that are not colinear and therefore can form a basis for all vectors in the space. That means that any vector in the space can be represented as a linear combination of these two basis vectors. Two such vectors are

$$\mathbf{x} = \alpha_1 \mathbf{w}_1 + \alpha_2 \mathbf{w}_2$$
$$\mathbf{y} = \beta_1 \mathbf{w}_1 + \beta_2 \mathbf{w}_2$$

(5.1-1a)

where the inner product or dot product of these two vectors is denoted as

$$\langle \mathbf{x}, \mathbf{y} \rangle = \|\mathbf{x}\| \cdot \|\mathbf{y}\| \cos \theta = \mathbf{x} \cdot \mathbf{y}$$

(5.1-1b)

The fact that $\mathbf{w}_1$ and $\mathbf{w}_2$ are not orthogonal, or perpendicular to each other, makes the process of analysis and manipulation of these vectors not as easy as it could be.

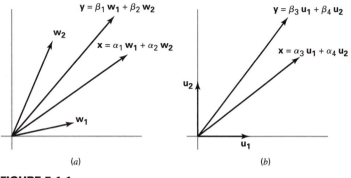

FIGURE 5.1-1

In fig. 5.1-1b the basis vectors $\mathbf{u}_1$ and $\mathbf{u}_2$ are perpendicular or *orthogonal*. In addition, they each have unit length, making them *orthonormal*. The great advantage of having orthogonal or orthonormal basis vectors is the ease with wish we can manipulate them and break them down into their various components.

Recall that if the basis vectors are orthonormal, then the dot product of two vectors

$$\mathbf{x} = \alpha_3 \mathbf{u}_1 + \alpha_4 \mathbf{u}_2$$
$$\mathbf{y} = \beta_3 \mathbf{u}_1 + \beta_4 \mathbf{u}_2 \tag{5.1-1c}$$

is given by

$$\langle \mathbf{x}, \mathbf{y} \rangle = (\alpha_3 \mathbf{u}_1 + \alpha_4 \mathbf{u}_2) \cdot (\beta_3 \mathbf{u}_1 + \beta_4 \mathbf{u}_2) = \alpha_3 \beta_3 + \alpha_4 \beta_4 \tag{5.1-2}$$

since, for orthonormal basis vectors,

$$\langle \mathbf{u}_i, \mathbf{u}_k \rangle = \mathbf{u}_i \cdot \mathbf{u}_k = \begin{cases} 0 & \text{if } i \neq k \\ 1 & \text{if } i = k \end{cases} \tag{5.1-3}$$

All of this is easily generalized to three-dimensional and higher-dimensional Euclidean vector spaces.

Using a set of N orthonormal basis vectors, we can represent the N-dimensional vector

$$\mathbf{v} = \sum_{n=1}^{N} \alpha_n \mathbf{u}_n \tag{5.1-4a}$$

in matrix form as a column vector

$$\mathbf{v} = \begin{bmatrix} \alpha_1 \\ \alpha_2 \\ \cdot \\ \alpha_N \end{bmatrix} \tag{5.1-4b}$$

where the dot or inner product of two vectors is

$$\langle \mathbf{v}_1, \mathbf{v}_2 \rangle = \mathbf{v}_1^T \mathbf{v}_2 = [\alpha_1 \quad \alpha_2 \quad \cdot \quad \alpha_N] \begin{bmatrix} \beta_1 \\ \beta_2 \\ \cdot \\ \beta_N \end{bmatrix} = \sum_{n=1}^{N} \alpha_n \beta_n \tag{5.1-4c}$$

According to fig. 5.1-1, we have two representations of the same vector $\mathbf{x}$:

$$\mathbf{x} = \alpha_1 \mathbf{w}_1 + \alpha_2 \mathbf{w}_2 \tag{5.1-5a}$$

$$\mathbf{x} = \alpha_3 \mathbf{u}_1 + \alpha_4 \mathbf{u}_2 \tag{5.1-5b}$$

with two different sets of basis functions. By the very nature of basis functions, one set $\mathbf{w}$ can be represented as a linear combination of the other set $\mathbf{u}$. In two dimensions,

$$[\mathbf{w}_1 \quad \mathbf{w}_2] = [\mathbf{u}_1 \quad \mathbf{u}_2]\begin{bmatrix} a_{11} & a_{12} \\ a_{21} & a_{22} \end{bmatrix} = [\mathbf{u}_1 \quad \mathbf{u}_2]A \tag{5.1-6a}$$

and

$$\mathbf{x} = [\mathbf{u}_1 \quad \mathbf{u}_2]\begin{bmatrix} \alpha_3 \\ \alpha_4 \end{bmatrix} = [\mathbf{w}_1 \quad \mathbf{w}_2]A^{-1}\begin{bmatrix} \alpha_3 \\ \alpha_4 \end{bmatrix} = [\mathbf{w}_1 \quad \mathbf{w}_2]\begin{bmatrix} \alpha_1 \\ \alpha_2 \end{bmatrix} \tag{5.1-6b}$$

so that

$$\begin{bmatrix} \alpha_1 \\ \alpha_2 \end{bmatrix} = A^{-1}\begin{bmatrix} \alpha_3 \\ \alpha_4 \end{bmatrix} \tag{5.1-6c}$$

Fact 1

If both pairs of basis vectors $(\mathbf{u}_1, \mathbf{u}_2)$ and $(\mathbf{w}_1, \mathbf{w}_2)$ are orthonormal, then one can be generated from the other by pure rotation (without changing the length of any of the vectors or changing the angles between them). Consequently, it follows that the transformational matrix A is given by

$$A = \begin{bmatrix} \cos\theta & \sin\theta \\ -\sin\theta & \cos\theta \end{bmatrix} \tag{5.1-7a}$$

the rotational matrix. The inverse of this matrix is

$$A^{-1} = \begin{bmatrix} \cos\theta & -\sin\theta \\ \sin\theta & \cos\theta \end{bmatrix} \tag{5.1-7b}$$

where

$$A^T = A^{-1} \quad \text{and} \quad \det(A) = |A| = 1 \tag{5.1-7c}$$

Although the relationship of eq. 5.17c was established for othornormal vectors in a two-dimensional Euclidean space, it extends to multidimensional Euclidean spaces as well. The multidimensional equivalent of a two-dimensional rotational matrix is called a *unitary matrix* where $U^T = U^{-1}$.

Fact 2

Let us say that a linear transformation of a vector $\mathbf{x}$ to a vector $\mathbf{y}$ using the set of basis vectors $(\mathbf{w})$ is denoted by

$$\mathbf{y}^{(w)} = B\,\mathbf{x}^{(w)} \tag{5.1-8a}$$

or:

$$\begin{bmatrix} \beta_1 \\ \beta_2 \end{bmatrix} = B \begin{bmatrix} \alpha_1 \\ \alpha_2 \end{bmatrix} \tag{5.1-8b}$$

It follows from eq. 5.1-6c that if we want to express the same transformation of the vector using the basis vectors (**u**),

$$\left\{ A^{-1} \begin{bmatrix} \beta_3 \\ \beta_4 \end{bmatrix} = B A^{-1} \begin{bmatrix} \alpha_3 \\ \alpha_4 \end{bmatrix} \right\} \Rightarrow \left\{ \begin{bmatrix} \beta_3 \\ \beta_4 \end{bmatrix} = A B A^{-1} \begin{bmatrix} \alpha_3 \\ \alpha_4 \end{bmatrix} \right\} \Rightarrow y^{(u)} = A B A^{-1} x^{(u)}$$

$$\tag{5.1-8c}$$

and if A is a *rotational* or *unitary* transformation then $A^T = A^{-1}$.

The Gram-Schmidt Orthonormalization Procedure

The issue here is how to construct a set of orthonormal basis vectors (**u**) from a set of basis vectors (**w**) that are not orthogonal. (Recall from fig. 5.1-1 that a set of vectors need not be orthogonal in order to be linearly independent and therefore form a basis.) The procedure is both straightforward and recursive. It will be illustrated for a three-dimensional space.

We will start by making

$$\mathbf{u}_1 = \frac{\mathbf{w}_1}{\|\mathbf{w}_1\|} \tag{5.1-9a}$$

which is a unit vector in the direction of $\mathbf{w}_1$. As an intermediate step, we will define

$$\mathbf{c}_2 = \mathbf{w}_2 - \langle \mathbf{w}_2, \mathbf{u}_1 \rangle \mathbf{u}_1 \tag{5.1-9b}$$

which creates a vector that is orthogonal to $\mathbf{u}_1$, and we cause that vector to have unit length by defining

$$\mathbf{u}_2 = \frac{\mathbf{c}_2}{\|\mathbf{c}_2\|} \tag{5.1-9c}$$

The next intermediate step is to define

$$\mathbf{c}_3 = \mathbf{w}_3 - \langle \mathbf{w}_3, \mathbf{u} \rangle \mathbf{u}_1 - \langle \mathbf{w}_3, \mathbf{u}_2 \rangle \mathbf{u}_2 \tag{5.1-9d}$$

and

$$\mathbf{u}_3 = \frac{\mathbf{c}_3}{\|\mathbf{c}_3\|} \tag{5.1-9e}$$

and so on.

The Schwarz Inequality

We have already examined the Schwarz Inequality in Chapter 3 in connection with the derivation of the matched filter. This was done in relation to functions. We want to repeat that derivation in relation to vectors.

Let **x** and **y** be vectors. Then

$$\langle \mathbf{x}, \mathbf{y} \rangle \leq \|\mathbf{x}\| \cdot \|\mathbf{y}\| \tag{5.1-10a}$$

with equality only if $\mathbf{x} = \alpha \mathbf{y}$ where α is a constant (perhaps complex).

Proof:

Consider

$$\|\mathbf{x} + \alpha\mathbf{y}\|^2 = \langle \mathbf{x} + \alpha\mathbf{y}, \mathbf{x} + \alpha\mathbf{y} \rangle = \|\mathbf{x}\|^2 + \alpha^* \langle \mathbf{x}, \mathbf{y} \rangle + \alpha \langle \mathbf{x}, \mathbf{y} \rangle^* + |\alpha|^2 \|\mathbf{y}\|^2 \geq 0 \tag{5.1-10b}$$

Since α is an arbitrary constant, we may make it equal to

$$\alpha = -\frac{\langle \mathbf{x}, \mathbf{y} \rangle}{\|\mathbf{y}\|^2} \tag{5.1-10c}$$

and substituting into eq. 5.10c,

$$\|\mathbf{x} + \alpha\mathbf{y}\|^2 = \|\mathbf{x}\|^2 - \frac{|\langle \mathbf{x}, \mathbf{y} \rangle|^2}{\|\mathbf{y}\|^2} \geq 0 \tag{5.1-10d}$$

with equality only when $\alpha = 0$.

end of the proof

By analogy with vectors, we are going to define an *inner product on functions* as:

$$\langle f_i(t), f_j(t) \rangle = \int_0^{T_s} f_i(t) f_j^*(t) \, dt \tag{5.1-11}$$

where each of these functions $f_i(t)$ and $f_j(t)$ may be written as the linear combination of orthonormal basis functions that are defined as

$$\langle \phi_n(t), \phi_m(t) \rangle = \int_0^{T_s} \phi_n(t) \phi_m^*(t) \, dt = \begin{cases} 0 & \text{if } n \neq m \\ 1 & \text{if } n = m \end{cases} \tag{5.1-12}$$

where $n = 1, \cdots, N$ and N is the dimension of the function space.

Accordingly, the function $f(t)$ can be represented as

$$f(t) = \sum_{n=1}^{N} f_n \phi_n(t) \tag{5.1-13}$$

It is important to make the following two observations:

1. The idea of an inner product is a generalization of operations that typically are quite different from each other. The dot product of two vectors is different from the integral of the product of two functions; yet both satisfy the conditions of being inner products.
2. The integral of the product of two functions is over a specific range; in this instance the range is T_s, the symbol interval.

According to Parseval's Theorem, we may extend eq. (5.1-11)

$$\langle f_i(t), f_j(t) \rangle = \int_0^{T_s} f_i(t) f_j^*(t) \, dt = \frac{1}{2\pi} \int_{-\infty}^{\infty} F_i(\omega) F_j^*(\omega) \, d\omega \tag{5.1-14}$$

where $f_i(t)$, $f_j(t)$ are time-limited. It follows that functions that are disjoint in the frequency domain are orthogonal in the time domain.

EXAMPLE 1

The set of functions $\phi_n(t) = \frac{1}{\sqrt{T_s}}e^{jn\omega_s t}$, $n = 0, \pm1, \pm2, \cdots$, are orthonormal over $[t_1, t_1 + T_s]$ where $T_s = 2\pi/\omega_s$. This can be seen by forming

$$\langle\phi_n(t), \phi_m(t)\rangle = \int_{t_1}^{t_1+T_s} \phi_n(t)\phi_m^*(t)\,dt = \frac{1}{T_s}\int_{t_1}^{t_1+T_s} e^{jn\omega_s t}e^{-jm\omega_s t}\,dt$$

$$= \begin{cases} \frac{1}{T_s}\int_{t_1}^{t_1+T_s} dt = 1, & \text{if } n = m \\ \frac{1}{T_s}\int_{t_1}^{t_1+T_s} e^{j(n-m)\omega_s t}\,dt = 0, & \text{if } n \neq m \end{cases} \tag{5.1-15}$$

∎

EXAMPLE 2

A variant of the set of orthogonal complex exponentials is the set of orthogonal trigonometric functions. The following collection of trigonometric identities is well known:

$$\int_{t_1}^{t_1+T_s} \text{Sin}(n\omega_s t)\text{Sin}(m\omega_s t)\,dt = \int_{t_1}^{t_1+T_s} \text{Cos}(n\omega_s t)\text{Cos}(m\omega_s t)\,dt = \begin{cases} 0, & \text{if } n \neq m \\ \frac{T_s}{2}, & \text{if } n = m \end{cases} \tag{5.1-16a}$$

$$\int_{t_1}^{t_1+T_s} \text{Sin}(n\omega_s t)\text{Cos}(m\omega_s t)\,dt = 0, \quad \text{for all } n, m \tag{5.1-16b}$$

With the understanding that orthogonal functions (or vectors) can be made orthonormal simply by scaling them, we can say that trigonometric functions having frequencies that are integer multiples of some fundamental frequency are orthogonal over the fundamental period and that Sin and Cos always have zero inner product. ∎

EXAMPLE 3

Define the set of functions

$$p_n(t) = \begin{cases} 1 & \text{for } nT_s \leq t \leq (n+1)T_s \\ 0 & \text{otherwise} \end{cases} \tag{5.1-17}$$

As shown in fig. 5.1-2, these functions, by not overlapping, are orthogonal in time. Although this orthogonal set appears trivial, it does form the basis for *pulse position modulation*. ∎

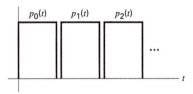

FIGURE 5.1-2

```
1  1  1  1  1  1  1  1  1  1  1  1  1  1  1  1
1  1  1  1  1  1  1  1 -1 -1 -1 -1 -1 -1 -1 -1
1  1  1  1 -1 -1 -1 -1 -1 -1 -1 -1  1  1  1  1
1  1  1  1 -1 -1 -1 -1  1  1  1  1 -1 -1 -1 -1
1  1 -1 -1 -1 -1  1  1  1  1 -1 -1 -1 -1  1  1
1  1 -1 -1 -1 -1  1  1 -1 -1  1  1  1  1 -1 -1
1  1 -1 -1  1  1 -1 -1 -1 -1  1  1 -1 -1  1  1
1  1 -1 -1  1  1 -1 -1  1  1 -1 -1  1  1 -1 -1
1 -1 -1  1  1 -1 -1  1  1 -1 -1  1  1 -1 -1  1
1 -1 -1  1  1 -1 -1  1 -1  1  1 -1 -1  1  1 -1
1 -1 -1  1 -1  1  1 -1  1 -1 -1  1 -1  1  1 -1
1 -1  1 -1 -1  1 -1  1  1 -1  1 -1 -1  1 -1  1
1 -1  1 -1 -1  1 -1  1 -1  1 -1  1  1 -1  1 -1
1 -1  1 -1  1 -1  1 -1 -1  1 -1  1 -1  1 -1  1
1 -1  1 -1  1 -1  1 -1  1 -1  1 -1  1 -1  1 -1
```

First 16 Walsh functions

FIGURE 5.1-3

EXAMPLE 4

The *Walsh Functions* [Wal23], [Cor73], are a set of orthogonal digital waveforms that have application in digital transmission. The first 16 of these functions are shown in fig. 5.1-3 as 16-bit binary numbers. As analog waveforms, the inner product is formed by integrating the product of two of the functions. As digital vectors, the inner product is formed by an AND operation followed by mod-2 addition of the elements of the result. ∎

Using a set of orthonormal functions as a set of basis functions, we have represented a given function $f(t)$ as the linear combination

$$f(t) = \sum_{n=1}^{N} f_n \phi_n(t) \tag{5.1-13a}$$

over the time interval $(0, T_s)$. We can find the value of f_n by taking the inner product

$$\langle f(t), \phi_n(t) \rangle = \int_0^{T_s} \left(\sum_m f_m \phi_m(t) \right) \phi_n^*(t)\, dt = \sum_m f_n \int_0^{T_s} \phi_m(t) \phi_n^*(t)\, dt = f_n \tag{5.1-13b}$$

which is exactly the same process of finding a Fourier Series coefficient.

Similarly, the inner product of two functions $f(t)$ and $g(t)$ is

$$\langle f(t), g(t) \rangle = \int_0^{T_s} \left(\sum_m f_m \phi_m(t) \right) \left(\sum_n g_n^* \phi_n^*(t) \right) dt$$

$$= \sum_n \sum_m f_m g_n^* \int_0^{T_s} \phi_m(t) \phi_n^*(t)\, dt = \sum_n f_n g_n^* \tag{5.1-14}$$

The energy contained in a finite time signal $f(t)$ can therefore be written as the inner product of that signal with itself:

$$E = \langle f(t), f(t) \rangle \int_0^{T_s} f(t) f^*(t)\, dt = \int_0^{T_s} |f(t)|^2\, dt = \sum_n |f_n|^2 \qquad (5.1\text{-}15)$$

In vector space terms, the signal energy is equal to the square of the magnitude of the vector.

This is very similar to Fourier Series except that the Fourier Series basis functions $\{e^{jn\omega_s t}\}$ are orthogonal but not orthonormal and the result would have to be scaled accordingly.

Since the inner product ideas for both vector space and function space give the same results, it follows that the functional notation of eq. 5.1-13a has the vector equivalent:

$$\mathbf{f} = \begin{bmatrix} f_1 \\ f_2 \\ \cdot \\ f_N \end{bmatrix} \qquad (5.1\text{-}16)$$

where we may transport all Euclidean vector operations over to function space.

Fact 3

Let $f(t) = \sum_{n=1}^{N} f_n \phi_n(t)$ be approximated by $f_{N_1}(t) = \sum_{n=1}^{N_1} f_n \phi_n(t)$ where $N_1 < N$. Define the error as

$$\varepsilon_{N_1} = f(t) - f_{N_1}(t) = \sum_{n=1}^{N} f_n \phi_n(t) - \sum_{n=1}^{N_1} f_n \phi_n(t). \qquad (5.1\text{-}17a)$$

It follows that the inner product

$$\langle \varepsilon_{N_1}, f_{N_1}(t) \rangle = 0 \qquad (5.1\text{-}17b)$$

so that the error is always orthogonal to the partial sum. This fact will end up being quite important in a subsequent discussion of adaptive equalizers.

5.2 EIGENVALUES AND EIGENVECTORS

We have seen that a linear transformation of a vector $\mathbf{x}$ can be represented by the matrix equation

$$\mathbf{y} = A\mathbf{x} \qquad (5.2\text{-}1a)$$

where $\mathbf{x}$ and $\mathbf{y}$ are, in general, not colinear. It is of particular interest to examine the set of vectors $\mathbf{u}$ and corresponding set of scalars λ for which

$$A\mathbf{u} = \lambda\mathbf{u} \qquad (5.2\text{-}1b)$$

where the result of the linear transformation of u is the same vector $\mathbf{u}$ multiplied by a scalar λ. This transformation can also be written as a homogeneous system of linear equations

$$[A - \lambda I]\mathbf{u} = 0 \qquad (5.2\text{-}1c)$$

which has a nontrivial solution only if

$$\det[A - \lambda I] = g(\lambda) = 0 \qquad (5.2\text{-}1d)$$

where $g(\lambda)$ is called the characteristic equation of A.

The values of λ that satisfy eq. 5.2-1d are called *eigenvalues* and the corresponding vectors **u** are called *eigenvectors*.

EXAMPLE

The matrix A is defined as $A = \begin{bmatrix} 3 & 4 \\ 2 & 1 \end{bmatrix}$. The characteristic equation of A is given by

$$g(\lambda) = \det \begin{vmatrix} 3 - \lambda & 4 \\ 2 & 1 - \lambda \end{vmatrix} = \lambda^2 - 4\lambda - 5 = 0$$

so that the eigenvalues are $\lambda_1 = 5$ and $\lambda_2 = -1$. In order to find the corresponding eigenvectors, we form the following two equations based on eq. 5.2-1b:

$$A\mathbf{u}_1 = \lambda_1 \mathbf{u}_1 \qquad \text{and} \qquad A\mathbf{u}_2 = \lambda_2 \mathbf{u}.$$

We shall find the two eigenvectors by allowing

$$\mathbf{u}_1 = \begin{bmatrix} a_1 \\ b_1 \end{bmatrix} \qquad \text{and} \qquad \mathbf{u}_2 = \begin{bmatrix} a_2 \\ b_2 \end{bmatrix}$$

and expanding

$$\begin{bmatrix} 3a_1 + 4b_1 \\ 2a_1 + b_1 \end{bmatrix} = 5 \begin{bmatrix} a_1 \\ b_1 \end{bmatrix} \qquad \text{and} \qquad \begin{bmatrix} 3a_2 + 4b_2 \\ 2a_2 + b_2 \end{bmatrix} = - \begin{bmatrix} a_2 \\ b_2 \end{bmatrix}$$

yielding

$$a_1 - 2b_1 = 0 \qquad \text{and} \qquad a_2 + b_2 = 0$$

so that we can choose

$$\mathbf{u}_1 = \begin{bmatrix} 2 \\ 1 \end{bmatrix} \qquad \text{and} \qquad \mathbf{u}_2 = \begin{bmatrix} 1 \\ -1 \end{bmatrix}$$

∎

There are a number of issues to be noted with regard to the previous example. This is a simple example chosen to illustrate the basic features of eigenvalues and eigenvectors. Life is made more complicated when we have complex and/or multiple eigenvalues, which certainly may arise from the characteristic equation. Clearly, as the order of the matrix A gets larger, the amount of computation increases. For an operationally complete and easily accessible general discussion of the subject, the interested reader can consult [Gup76]. For computational assistance in finding eigenvalues and eigenvectors, the reader may use the appropriate functions in MATLAB, MATHCAD, or Mathematica.

In this discussion, we are interested in a particular form of the matrix A corresponding to the autocorrelation matrix of a random vector X. We have seen in Chapter 2 that such a matrix will always have the form

$$A = \begin{bmatrix} a_1 & a_2 + jb_2 & a_3 + jb_3 & a_4 + jb_4 \\ a_2 - jb_2 & a_1 & a_2 + jb_2 & a_3 + jb_3 \\ a_3 - jb_3 & a_2 - jb_2 & a_1 & a_2 + jb_2 \\ a_4 + jb_4 & a_3 - jb_3 & a_2 - jb_2 & a_1 \end{bmatrix} \qquad (5.2\text{-}2)$$

where in addition to the matrix being equal to its conjugate transpose (making it Hermitian), the elements of the both the major and minor diagonals are equal to each other making it Toeplitz. At the moment we are only interested in the Hermetian property of the autocorrelation matrix.

Fact 4

It is well established that the eigenvalues of a Hermitian matrix are always real-valued and that the eigenvectors are orthogonal (and, by scaling, can be made orthonormal).

EXAMPLE

We will define a correlation matrix A and use MATLAB to evaluate the eigenvalues and eigenvectors.

A = [5, 3, 2, 1; 3, 5, 3, 2; 2, 3, 5, 3; 1, 2, 3, 5]

A =

   5 3 2 1
   3 5 3 2
   2 3 5 3
   1 2 3 5

[V,D] = eig(A)

V =

   0.2706   -0.5468   -0.6533   0.4483
  -0.6533   0.4483   -0.2706   0.5468
   0.6533   0.4483   0.2706   0.5468
  -0.2706   -0.5468   0.6533   0.4483

D =

   1.5858   0   0   0
     0   1.9010   0   0
     0   0   4.4142   0
     0   0   0   12.0990

where the diagonal elements of D are the eigenvalues of A and the columns of V are the eigenvectors associated with those eigenvalues. It can be easily confirmed that these eigenvectors constitute an orthonormal set of basis vectors for the four-dimensional Euclidean space. ■

Fact 5

Consider a Hermetian matrix A such as the correlation matrix of a random vector X. The matrix of eigenvectors Ω of the matrix A has the property that its inverse equals its transpose

$$\Omega^{-1} = \Omega^T \tag{5.2-3}$$

A matrix having this property has previously been defined as being *unitary*.

EXAMPLE

Consider the matrix A given in the previous example. The matrix of eigenvectors is denoted as V. Using MATLAB:

inv(V)

ans =

0.2706	-0.6533	0.6533	-0.2706
-0.5468	0.4483	0.4483	-0.5468
-0.6533	-0.2706	0.2706	0.6533
0.4483	0.5468	0.5468	0.4483

V'

ans =

0.2706	-0.6533	0.6533	-0.2706
-0.5468	0.4483	0.4483	-0.5468
-0.6533	-0.2706	0.2706	0.6533
0.4483	0.5468	0.5468	0.4483

confirming the assertion.   ■

Finally we can generalize eq. 5.2-1b to read

$$A\Omega = \Omega D \tag{5.2-4a}$$

where D is the diagonal matrix of eigenvalues. Using eq. 5.2-3, $\Omega^{-1} = \Omega^T$, we get that for a Hermetian (or correlation) matrix

$$A = \Omega D \Omega^T \tag{5.2-4b}$$

Expanding eq. 5.2-4b, and recognizing the eigenvectors to be column vectors, we get

$$A = [\mathbf{u}_1 \quad \mathbf{u}_2 \quad \cdots \quad \mathbf{u}_N] \begin{bmatrix} \lambda_1 & 0 & \cdot\cdot & 0 \\ 0 & \lambda_2 & \cdot\cdot & 0 \\ \cdot\cdot & \cdot\cdot & \cdot\cdot & \cdot\cdot \\ 0 & \cdot\cdot & \cdot\cdot & \lambda_N \end{bmatrix} \begin{bmatrix} \mathbf{u}_1^T \\ \mathbf{u}_2^T \\ \cdot\cdot \\ \mathbf{u}_3^T \end{bmatrix}$$

$$= [\lambda_1\mathbf{u}_1 \quad \lambda_2\mathbf{u}_2 \quad \cdots \quad \lambda_N\mathbf{u}_N] \begin{bmatrix} \mathbf{u}_1^T \\ \mathbf{u}_2^T \\ \cdot\cdot \\ \mathbf{u}_3^T \end{bmatrix}$$

$$= \sum_{n=1}^{N} \lambda_n \mathbf{u}_n \mathbf{u}_n^T \tag{5.2-4c}$$

so that the covariance matrix can be expanded as a sum of matrices $\mathbf{u}_n \mathbf{u}_n^T$ weighted by the corresponding eigenvalues λ_n. Eq. 5.2-4c is one version of Mercer's Theorem.

Finally, recall that X is the random vector having covariance matrix A. Let us define $\mathbf{Y} = \Omega^{\mathbf{T}}\mathbf{X}$ so that the random variables in Y are orthogonal. According to eq. 5.2-3, $\mathbf{X} = \Omega\mathbf{Y}$ so that

$$X = \Omega Y = [\mathbf{u}_1 \mathbf{u}_2 \cdots \mathbf{u}_N] \begin{bmatrix} Y_1 \\ Y_2 \\ \cdot \\ Y_N \end{bmatrix} = \sum_{n=1}^{N} Y_n \mathbf{u}_n \qquad (5.2\text{-}4\text{d})$$

This final result is of particular importance. It shows how any random vector X having a covariance matrix A can be expanded as a linear combination of orthonormal basis vectors and uncorrelated coefficients Y_n.

We shall extend these ideas of eigenvalues and eigenvectors to Euclidean function spaces subsequently.

5.3 REPRESENTATION OF SIGNALS AND ADDITIVE WHITE GAUSSIAN NOISE

Based on the the analysis of QAM-type systems that was done in Chapter 4 and, in particular, fig. 4.1-3, it is relatively straightforward to develop the idea of a signal as a point in signal space.

Given a set of orthonormal basis functions $\{\phi_n(t)\}, n = 1, 2, \ldots, N$ each signal in a set of transmitted signals may be represented as

$$s_m(t) = \sum_{n=1}^{N} s_{mn}(t)\phi_n(t); \quad m = 1, 2, \ldots, M \qquad (5.3\text{-}1)$$

For example, in 16-QAM, we have $N = 2$, corresponding to the two basis functions

$$\phi_1(t) = \sqrt{\frac{2}{T_s}}\mathrm{Cos}(\omega_s t) \qquad \phi_2(t) = \sqrt{\frac{2}{T_s}}\mathrm{Sin}(\omega_s t) \qquad (5.3\text{-}2\text{a})$$

and $M = 16$, corresponding to the number of points in the signal set. The values of the coefficients range over

$$s_{mn} = \pm\sqrt{\frac{E_s}{10}}, \pm 3\sqrt{\frac{E_s}{10}} \qquad (5.3\text{-}2\text{b})$$

so that a particular point in this signal set might look like

$$s_7(t) = -\sqrt{\frac{E_s}{10}}\phi_1(t) + 3\sqrt{\frac{E_s}{10}}\phi_2(t) \qquad (5.3\text{-}2\text{c})$$

and another might look like

$$s_{10}(t) = +3\sqrt{\frac{E_s}{10}}\phi_1(t) + 3\sqrt{\frac{E_s}{10}}\phi_2(t). \qquad (5.3\text{-}2\text{d})$$

The vector representations of these signals are

$$\mathbf{s}_7 = \sqrt{\frac{E_s}{10}}\begin{bmatrix} -1 \\ +3 \end{bmatrix}, \qquad \mathbf{s}_{10} = \sqrt{\frac{E_s}{10}}\begin{bmatrix} +3 \\ +3 \end{bmatrix} \qquad (5.3\text{-}2\text{e})$$

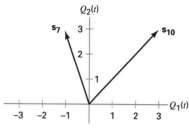

FIGURE 5.3-1

and they are shown in fig. 5.3-1. Note that the magnitude-squared of the vector, which is inner product of the vector with itself, is equal to the energy contained in that signal so that

$$|\mathbf{S}_7|^2 = \langle \mathbf{S}_7, \mathbf{S}_7 \rangle = \mathbf{S}_7^T \mathbf{S}_7 = \frac{E_s}{10}[-1 \quad 3]\begin{bmatrix} -1 \\ 3 \end{bmatrix} = E_s \qquad (5.3\text{-}2\text{f})$$

and

$$|\mathbf{S}_{10}|^2 = \langle \mathbf{S}_{10}, \mathbf{S}_{10} \rangle = \mathbf{S}_{10}^T \mathbf{S}_{10} = \frac{E_s}{10}[3 \quad 3]\begin{bmatrix} 3 \\ 3 \end{bmatrix} = 1.8 E_s \qquad (5.3\text{-}2\text{g})$$

For purposes of making decisions at the receiver, we will want to know the distance between the two points described by these vectors. The distance squared is given by:

$$d_{7,10}^2 = |\mathbf{S}_7 - \mathbf{S}_{10}|^2 = \frac{E_s}{10}[-4 \quad 0]\begin{bmatrix} -4 \\ 0 \end{bmatrix} = 1.6 E_s \qquad (5.3\text{-}2\text{h})$$

According to the Law of Cosines (fig. 5.3-2), the angle between the two vectors is given by

$$a^2 = b^2 + c^2 - 2bc\,\text{Cos}(A) \qquad (5.3\text{-}3\text{a})$$

In signal-space terms, $\text{Cos}(A)$ is the *correlation* between the vectors b and c. This idea of the correlation between two vectors or signal points is not confined to vectors in a two-dimensional space. In an n-dimensional Euclidean space any two vectors form a two-dimensional subspace or plane in which the Law of Cosines holds. It follows that the *correlation* between any two vectors in signal space is given by

$$\rho_{mn} = \frac{|\mathbf{S}_n|^2 + |\mathbf{S}_m|^2 - |\mathbf{S}_n - \mathbf{S}_m|^2}{2|\mathbf{S}_n||\mathbf{S}_m|} \qquad (5.3\text{-}3\text{b})$$

This representation is easily understood in relation to the QAM-type systems that we have analyzed in Chapter 4. But it also must be emphasized that that the possibilities

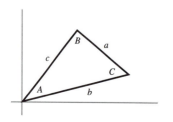

FIGURE 5.3-2

for signal spaces having orthonormal basis functions extend well beyond the simple two-dimensional space characterized by Cos and Sin. In relation to Chapter 9, Trellis Coding, we will deal with communication systems having basis functions that are of dimension higher than two.

In order to continue the discussion, we will now discuss the representation of noise in a Euclidean function space. We shall assume that we are dealing with zero-mean AWGN (additive white Gaussian noise) having a power spectral density

$$P_n(\omega) = \eta/2. \tag{5.3-4}$$

The noise is represented in much the same way that we represent the signal, as a linear combination of the orthonormal basis functions

$$n(t) = \sum_{m=1}^{N} n_m \phi_m(t) + n'(t) \tag{5.3-5a}$$

where $n'(t)$ includes any components of noise that lie outside the range of the basis functions. It follows that:

$$n_m = \int_0^{T_s} n(t)\phi_m^*(t)\,dt \tag{5.3-5b}$$

We want to find the variance of each component of the noise multiplying one of the orthonormal basis functions. In order to do this, we form:

$$E[n_i n_j] = E\left[\int_0^{T_s} n(t)\phi_i^*(t)\,dt \int_0^{T_s} n(u)\phi_j(u)\,du\right]$$

$$= \int_0^{T_s}\int_0^{T_s} E[n(t)n(u)]\phi_i^*(t)\phi_j(u)\,dt\,du$$

$$= \int_0^{T_s}\int_0^{T_s} \bar{R}n(t,u)\phi_i^*(t)\phi_j(u)\,dt\,du$$

$$= \int_0^{T_s}\int_0^{T_s} \frac{\eta}{2}\delta(t-u)\phi_i^*(t)\phi_j(u)\,dt\,du$$

$$= \frac{\eta}{2}\int_0^{T_s} \phi_i^*(t)\phi_j(t)\,dt \tag{5.3-6a}$$

so that

$$E[n_i n_j] = \begin{cases} \sigma_i^2 = \frac{\eta}{2}; & \text{for } i = j \\ 0; & \text{for } i \neq j \end{cases} \tag{5.3-6b}$$

This result is equivalent to the characterization of bandpass noise in Chapter 4. The noise components associated with the different basis functions all have the same power and the noise components in the different dimensions are all uncorrelated.

It should be noted that the whiteness of the noise makes it possible to decompose the noise into *uncorrelated* components with respect to any set of orthonormal basis functions. We shall see in a subsequent section that if the noise is Gaussian but not white, there will be only one set, depending on the autocorrelation function of the noise, of orthonormal basis functions that will allow decomposition with uncorrelated components.

5.4   THE CORRELATION RECEIVER IN THE PRESENCE OF ADDITIVE WHITE GAUSSIAN NOISE

If the signal $s_j(t)$ is transmitted and corrupted with AWGN $n(t)$ as described above, then the received signal is $x_j(t)$, where

$$x_j(t) = \sum_i (s_{ij} + n_i)\phi_i(t) + n'(t) = \sum_i x_{ij}\phi_i(t) + n'(t). \qquad (5.4\text{-}1)$$

In order to obtain the components x_{ij} from the received signal, we may expand the correlation receiver of fig. 3.6-1 to the more general structure of fig. 5.4-1. The received signal $x(t)$ consists of the transmitted signal, which is unknown to the receiver, plus noise. The correlation receiver computes the components of the received signal in the directions of the orthonormal basis functions. Notice that the 'extra' noise component $n'(t)$ is eliminated because it is out of the space defined by the basis functions. The output of the receiver can be thought of as a vector in signal space

$$\mathbf{X} = \begin{bmatrix} x_1 \\ x_2 \\ . \\ x_N \end{bmatrix}. \qquad (5.4\text{-}2)$$

Using QPSK as an example, the relationship of the received signal vector to the possible transmitted signal vectors is shown in fig. 5.4-2. The problem at the receiver is,

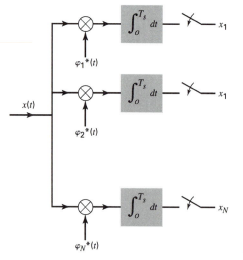

FIGURE 5.4-1

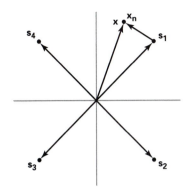

FIGURE 5.4-2

given the received vector, to decide with minimum probability of error which of the possible signals was sent. Looked at a slightly different way, we want to compute, for all values of i,

$$P[s_i(t) \text{ sent}/x(t) \text{ received}] \tag{5.4-3}$$

and choose that transmitted signal that maximizes the conditional probability of eq. 5.4-3. This is called the *maximum a posteriori probability*.

 Please note that we are assuming that each transmitted symbol and each received symbol can be treated separately and independently. This is certainly true at this point in our development, but it will not be true when the data is encoded or when the channel is distorted. In those instances the system has memory and, rather than symbol-by-symbol estimation, we must use maximum likelihood sequence estimation, MLSE. At that point we will have to revisit eq. 5.4-3.

 So our rule is choose $s_i(t)$ if

$$P[s_i(t) \text{ sent}/x(t) \text{ received}] \geq P[s_j(t) \text{ sent}/x(t) \text{ received}] \quad \text{for } i \neq j. \tag{5.4-4a}$$

This equation can be modified using Bayes' Rule to read

$$\frac{P[s_i(t)]P[x(t)/s_i(t)]}{P[x(t)]} \geq \frac{P[s_j(t)]P[x(t)/s_j(t)]}{P[x(t)]} \quad \text{for } i \neq j \tag{5.4-4b}$$

where $P[x(t)]$, the unconditional probability of the received signal, may be eliminated from the inequality. Since the noise is a continuous, rather than a discrete, process, it makes more sense to rewrite the remainder of eq. 5.4-4b in terms of probability densities. In doing so we may now use the results of our development of signal space techniques for the representation of signals and noise. The main point here is that because we are able to represent both signal and noise in terms of their components in relation to a set of orthonormal basis functions, we may use vector representations instead of time-function representations. Therefore

$$x(t) \to \mathbf{x} \quad \text{and} \quad s_i(t) \to \mathbf{s}_i \tag{5.4-4c}$$

and so the decision rule can be represented in vector terms:

$$P[\mathbf{s}_i]f_{\mathbf{x}/\mathbf{s}_i}[\mathbf{x}/\mathbf{s}_i] \geq P[\mathbf{s}_j]f_{\mathbf{x}/\mathbf{s}_j}[\mathbf{x}/\mathbf{s}_j] \quad \text{for } i \neq j \tag{5.4-4d}$$

Finally, it should be recognized that it is typical that all of the transmitted signals are equally likely so that the decision rule becomes: choose $\mathbf{s}_i$ if

$$f_{\mathbf{x}/\mathbf{s}_i}[\mathbf{x}/\mathbf{s}_i] \geq f_{\mathbf{x}/\mathbf{s}_j}[\mathbf{x}/\mathbf{s}_j] \quad \text{for } i \neq j \tag{5.4-4e}$$

The joint conditional probability of eq. 5.4-4e, which is for a single transmitted and received symbol, is known as a *likelihood function* and the process of deciding on which signal was transmitted, i.e.- the decision rule, is called *maximum likelihood*.

Let us recall that from eq. 5.4-1 the noise can be represented as

$$n(t) = x(t) - s_i(t) = \sum_{n=1}^{N} n_n \phi_n(t) = \sum_{n=1}^{N} (x_n - s_{ni}) \phi_n(t) \qquad (5.4\text{-}5)$$

where according to eq. 5.3-6b

$$E[n_i n_j] = \begin{cases} \sigma_i^2 = \dfrac{\eta}{2}; & \text{for } i = j \\ 0; & \text{for } i \neq j \end{cases}$$

the different components of the noise are uncorrelated. Since, in addition, the noise is Gaussian, eq. 5.4-4e can be rewritten as

$$(\pi \eta)^{-(N/2)} \exp\left[-\frac{1}{\eta} \sum_{n=1}^{N} (x_n - s_{ni})^2 \right] \geq (\pi \eta)^{-(N/2)} \exp\left[-\frac{1}{\eta} \sum_{n=1}^{N} (x_n - s_{nj})^2 \right] \qquad \text{for } i \neq j$$

$$(5.4\text{-}6a)$$

Since $\log(x)$ is monotonically related to x, it is possible to simplify eq. 5.4-6a considerably by cancelation of common terms and taking the natural log of both sides. If, in addition, we eliminate the minus sign in the exponent by reversing the inequality, the decision rule becomes: *choose* s_i *if*

$$-\frac{1}{\eta} \sum_{n=1}^{N} (x_n - s_{ni})^2 \leq -\frac{1}{\eta} \sum_{n=1}^{N} (x_n - s_{nj})^2 \qquad \text{for } i \neq j \qquad (5.4\text{-}6b)$$

In a Euclidean space, each of the terms in eq. 5.4-6b is the square of the distance from the signal vector to the received vector. *Our decision rule then reduces to choosing the signal point that is geometrically closest to the received point.* This is, of course, entirely consistent with the decision rule that was developed in QAM-type systems.

In structuring the decision boundaries in the signal space it is often useful to have alternative ways of determining the Euclidean distance between signal points. One way is based on the energy of and correlation between signal points, as shown in eq. 5.3-3b:

$$\rho_{mn} = \frac{|\mathbf{S_n}|^2 + |\mathbf{S_m}|^2 - |\mathbf{S_n} - \mathbf{S_m}|^2}{2|\mathbf{S_n}||\mathbf{S_m}|} \qquad (5.3\text{-}3b)$$

which can be rewritten as

$$d_{mn}^2 = E_m + E_n - 2\rho_{mn} \sqrt{E_m E_n} \qquad (5.4\text{-}7)$$

From our work in Chapter 4, we can use this result to compute the conditional probability of transmitting $\mathbf{S_n}$ and deciding $\mathbf{S_m}$ as:

$$P(\mathbf{S_m}/\mathbf{S_n}) = Q\left[\frac{d_{mn}/2}{\sqrt{\eta/2}} \right] = Q\left[\frac{d_{mn}}{\sqrt{2\eta}} \right] \qquad (5.4\text{-}8)$$

The decision rule may be implemented in a number of different ways. In relation to QAM-type systems we developed a decision device called a slicer. Another way of making

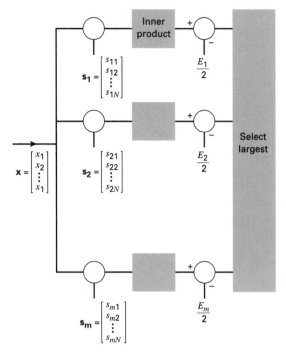

FIGURE 5.4-3

the decision, based on processing the received signal as a vector, is shown in fig. 5.4-3. This structure, known as a *vector receiver*, operates on the following principle:

$$-\frac{1}{\eta}\sum_{n=1}^{N}(x_n - s_{ni})^2 \leq -\frac{1}{\eta}\sum_{n=1}^{N}(x_n - s_{nj})^2 \qquad \text{for } i \neq j$$

$$\Rightarrow \sum_{n=1}^{N}\left(x_n^2 - 2x_n s_{ni} + s_{ni}^2\right) \leq \sum_{n=1}^{N}\left(x_n^2 - 2x_n s_{nj} + s_{nj}^2\right)$$

$$\Rightarrow \sum_{n=1}^{N}\left(-2x_n s_{ni} + s_{ni}^2\right) \leq \sum_{n=1}^{N}\left(-2x_n s_{nj} + s_{nj}^2\right)$$

$$\Rightarrow \left[\sum_{n=1}^{N}x_n s_{ni} - \frac{E_i}{2}\right] \geq \left[\sum_{n=1}^{N}x_n s_{nj} - \frac{E_j}{2}\right] \qquad (5.4\text{-}9)$$

where E_i and E_j are the energies contained in the respective signal vectors and $\sum_{n=1}^{N}x_n s_{ni}$ is the inner product of the received signal vector and one of the transmitted signal vectors. The object is to choose the largest of the results in order to identify the maximum likelihood transmitted signal.

In Chapter 4 we showed a number of different modulation or signaling techniques which were based on QAM (Quadrature Amplitude Modulation) including QPSK, 8-PSK, and 16-QAM. We will reconsider these signaling approaches and develop a number of other approaches from the more general point of view of signal space. This will help to achieve a broader view of modulation.

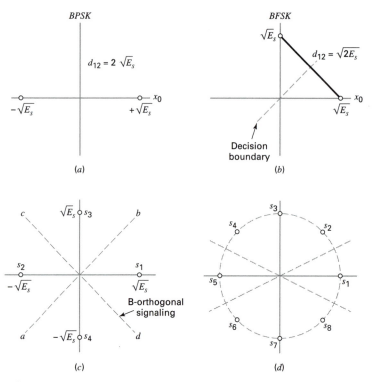

FIGURE 5.4-4

EXAMPLE 1   *Binary Phase-Shift Keying (BPSK) and Antipodal Signaling*

In this system we have two possible equally likely transmitted signals

$$s_1(t), s_2(t) = \pm\sqrt{E_s}\sqrt{2/T_s}\,\mathrm{Cos}(\omega_c t) = \pm\sqrt{E_s}\,\phi(t); \qquad 0 \le t \le T_s \qquad (5.4\text{-}10a)$$

which are shown, in their signal-space version, in fig. 5.4-4a. The point to be noted here is that while the designation BPSK makes sense only in relation to the basis function

$$\phi(t) = \sqrt{2/T_s}\,\mathrm{Cos}(\omega_c t)$$

the signal-space representation will hold true for any $\phi(t)$. This signal-space generalization is called *antipodal signaling*.

It is clear from this figure that the distance between these two points is $d_{12} = 2\sqrt{E_s}$ and the correlation coefficient is given by eq. 5.3-3b

$$\rho_{12} = \frac{|\mathbf{S}_1|^2 + |\mathbf{S}_2|^2 - |\mathbf{S}_1 - \mathbf{S}_2|^2}{2|\mathbf{S}_1||\mathbf{S}_2|} = \frac{E_s + E_s - 4E_s}{2E_s} = -1 \qquad (5.4\text{-}10b)$$

which makes sense because the signal vectors are in opposite directions.

According to eq. 5.4-8,

$$P(\mathbf{S}_2/\mathbf{S}_1) = P(\mathbf{S}_1/\mathbf{S}_2) = Q\left[\frac{d_{12}/2}{\sqrt{\eta/2}}\right] = Q\left[\frac{d_{12}}{\sqrt{2\eta}}\right]$$

$$= Q\left[\frac{2\sqrt{E_s}}{\sqrt{2\eta}}\right] = Q\left[\sqrt{\frac{2E_s}{\eta}}\right] = Q[\sqrt{\rho_M}] \qquad (5.4\text{-}10c)$$

In order to find the probability of error in this system, we note that

$$P_e = P(\mathbf{S}_2/\mathbf{S}_1)P(\mathbf{S}_1) + P(\mathbf{S}_1/\mathbf{S}_2)P(\mathbf{S}_2)$$

$$= \frac{1}{2}Q[\sqrt{\rho_M}] + \frac{1}{2}Q[\sqrt{\rho_M}] = Q[\sqrt{\rho_M}] \qquad (5.4\text{-}10d)$$

which is, of course, the same as the matched filter performance for binary signaling. ∎

EXAMPLE 2   Binary Frequency-Shift Keying (BFSK) or Orthogonal Signaling

In this instance we will go from the general to the specific. Let $\phi_1(t)$ and $\phi_2(t)$ be orthonormal basis functions on the interval $[0, T_s]$. The signaling is binary so we are sending either $s_1(t) = \sqrt{E_s}\phi_1(t)$ or $s_2(t) = \sqrt{E_s}\phi_2(t)$, as shown in fig. 5.4-4b.

In vector terms, the two signals may be represented as

$$\mathbf{S}_1 = \sqrt{E_s}\begin{bmatrix} 1 \\ 0 \end{bmatrix} \quad \text{and} \quad \mathbf{S}_2 = \sqrt{E_s}\begin{bmatrix} 0 \\ 1 \end{bmatrix} \qquad (5.4\text{-}11a)$$

Computing the correlation coefficient of these two signals, we get

$$\rho_{12} = \frac{|\mathbf{S}_1|^2 + |\mathbf{S}_2|^2 - |\mathbf{S}_1 - \mathbf{S}_2|^2}{2|\mathbf{S}_1||\mathbf{S}_2|} = \frac{|\mathbf{S}_1|^2 + |\mathbf{S}_2|^2 - (|\mathbf{S}_1|^2 + |\mathbf{S}_2|^2 - 2\langle\mathbf{S}_1, \mathbf{S}_2\rangle)}{2|\mathbf{S}_1||\mathbf{S}_2|}$$

$$= \frac{\langle\mathbf{S}_1, \mathbf{S}_2\rangle}{|\mathbf{S}_1||\mathbf{S}_2|} = 0 \qquad (5.4\text{-}11b)$$

since the vectors $\mathbf{S}_1$ and $\mathbf{S}_2$ are clearly orthogonal.

Both from eq. 5.4-7 and the geometry of the figure we can see that

$$d_{12}^2 = E_1 + E_2 - 2\rho_{12}\sqrt{E_1 E_2} \Rightarrow d_{12}^2 = 2E_s \Rightarrow \frac{d_{12}}{2} = \sqrt{\frac{E_s}{2}} \qquad (5.4\text{-}11c)$$

According to eq. 5.4-8,

$$P(\mathbf{S}_1/\mathbf{S}_2) = P(\mathbf{S}_2/\mathbf{S}_1) = Q\left[\frac{d_{12}/2}{\sqrt{\eta/2}}\right] = Q\left[\frac{d_{12}}{\sqrt{2\eta}}\right]$$

$$= Q\left[\frac{\sqrt{2E_s}}{\sqrt{2\eta}}\right] = Q\left[\sqrt{\frac{E_s}{\eta}}\right] = Q[\sqrt{\rho_M/2}] \quad (5.4\text{-}11d)$$

Following eq. 5.4-11d, the probability of error for this system is

$$P_e = P(\mathbf{S}_2/\mathbf{S}_1)P(\mathbf{S}_1) + P(\mathbf{S}_1/\mathbf{S}_2)P(\mathbf{S}_2)$$

$$= \frac{1}{2}Q[\sqrt{\rho_M/2}] + \frac{1}{2}Q[\sqrt{\rho_M/2}] = Q[\sqrt{\rho_M/2}] \qquad (5.4\text{-}11e)$$

Note that there is a 3 db performance difference between antipodal and orthogonal signaling.

Binary Frequency-Shift Keying is, under certain circumstances, an example of orthogonal signaling. In BFSK we transmit one of two possible signals:

$$s_1(t) = \sqrt{E_s}\sqrt{2/T_s}\,\text{Cos}\left[\left(\omega_c + \frac{\Delta\omega}{2}\right)t\right] \quad 0 \le t \le T_s$$

$$\qquad\qquad\qquad\qquad\qquad\qquad\qquad\qquad\qquad\qquad\qquad (5.4\text{-}12a)$$

$$s_2(t) = \sqrt{E_s}\sqrt{2/T_s}\,\text{Cos}\left[\left(\omega_c - \frac{\Delta\omega}{2}\right)t\right] \quad 0 \le t \le T_s$$

where $f_s = 1/T_s$ is the symbol rate (in this case bit rate).

In order to compute the correlation coefficent for these equal energy signals, we can use eq. 5.4-11b:

$$\rho_{12} = \frac{\langle \mathbf{S}_1, \mathbf{S}_2 \rangle}{|\mathbf{S}_1||\mathbf{S}_2|} = \frac{2}{T_s} \int_0^{T_s} \text{Cos}\left[\left(\omega_c + \frac{\Delta\omega}{2}\right)t\right]\text{Cos}\left[\left(\omega_c - \frac{\Delta\omega}{2}\right)t\right]dt$$

$$= \frac{1}{T_s} \int_0^{T_s} [\text{Cos}(2\omega_c t) + \text{Cos}(\Delta\omega t)]\, dt = \frac{\text{Sin}(\Delta\omega T_s)}{(\Delta\omega T_s)} \tag{5.4-12b}$$

assuming that the first part of the integral goes to zero. This will be exactly true if the carrier is an integer multiple of the symbol frequency. So the first requirement for BFSK to be an exact example of orthogonal signaling is for $\omega_c = N\omega_s$.

The second requirement is that the two signals be uncorrelated. Clearly, these two signals have a correlation that depends on the frequency difference between them. There are a collection of frequencies for which these signals are uncorrelated and therefore orthogonal. They correspond to

$$\Delta\omega T_s = n\pi \quad \text{or} \quad \frac{\Delta f}{f_s} = \frac{n}{2}; \quad n \neq 0 \tag{5.4-12c}$$

in which case $\rho_{12} = 0$. In short, the two frequencies must differ by an integer multiple of the symbol frequency. Under these circumstances, the signal set may be represented as two orthonormal vectors as shown in fig. 5.4-4b.

Another way of writing eq. 5.4-12c will lead to a consideration of CPM (Continuous-Phase Modulation) later in this chapter. Let

$$h = \frac{\Delta f}{f_s} \tag{5.4-12d}$$

where h is the index of modulation. Note that $h = 1/2$ corresponds to $n = 1$ in eq. 5.4-12c, the smallest frequency difference for which the two signals are orthogonal. It will be left to the Problems to demonstrate that the signal corresponding to $h = 1/2$ is the same as MSK signaling described in Chapter 4. By defining h as we have, we leave the door open for signaling that is not necessarily based on orthogonal functions.

A variation of the receiver structure of fig. 5.4-3 that gives the same results is shown in fig. 5.4-5. If the two signals $s_1(t)$ and $s_2(t)$ are uncorrelated and $s_1(t)$ is received with noise, then the output of the upper leg is the sum of a signal having amplitude $\sqrt{E_s}$ and noise having a variance $\sigma^2 = \eta/2$ and the lower leg has only noise with the same variance. By subtracting the two we will get an output having an amplitude $\pm\sqrt{E_s}$, depending on which signal was sent, and a noise having twice the variance of the original $2\sigma^2 = \eta$. This gives us a decision problem that yields the same result as eq. 5.4-11c.

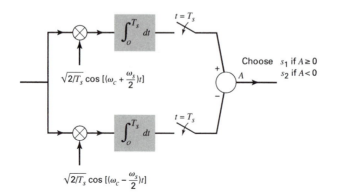

FIGURE 5.4-5

We have seen in fig. 5.1-2 that signals can also be orthogonal because they do not overlap in time. Therefore we can consider the two signals $p_0(t)$ and $p_1(t)$ from that figure as having exactly the same vector representation as eq. 5.4-12a. Consequently a system based on those signals will have the same performance as the coherent BFSK system.

The idea of orthogonal signaling can be expanded to any higher number of dimensions. Finding the probability of error for orthogonal signaling for dimension higher than two is a more complex task that we address in the next section. ∎

EXAMPLE 3 *Bi-Orthogonal Signaling*

The simplest example of bi-orthogonal signaling is defined by the signal vectors

$$\mathbf{S}_1 = \sqrt{E_s}\begin{bmatrix}0\\1\end{bmatrix} \quad \mathbf{S}_2 = \sqrt{E_s}\begin{bmatrix}0\\-1\end{bmatrix} \quad \mathbf{S}_3 = \sqrt{E_s}\begin{bmatrix}1\\0\end{bmatrix} \quad \mathbf{S}_4 = \sqrt{E_s}\begin{bmatrix}-1\\0\end{bmatrix} \qquad \text{(5.4-13a)}$$

and is shown in fig. 5.4-4c. By examination, it is clear that this signal space combines orthogonal signaling with antipodal signaling. This may be achieved both in the time domain and in the frequency domain.

In the time domain we may transmit $\pm p_0(t)$, $\pm p_1(t)$ from fig. 5.1-2. In the frequency domain we may transmit $\pm\mathrm{Sin}(n\omega_s t)$, $\pm\mathrm{Sin}(m\omega_s t)$ as an extension of FSK. We may also transmit $\pm\mathrm{Sin}(\omega_c t)$, $\pm\mathrm{Cos}(\omega_c t)$, which we have previously characterized as a QAM-type signal. Although these are very different modulation schemes, they are, from the point of view of signal space, equivalent. It is also clear that a rotation of the bi-orthogonal signal set of fig. 5.4-4c by $45°$ yields the QPSK signal set.

We want to find the probability of error. This has already been done for QPSK in Chapter 4 but this time we will use a somewhat different approach.

The general expression for probability of error as the sum of joint probabilities can be simplified as a result of the signal vectors being equally probable and the signal set being symmetric:

$$P_e = \sum_{n=1}^{4} P(\text{error}/\mathbf{S}_n)P(\mathbf{S}_n) = P(\text{error}/\mathbf{S}_1) = P(\mathbf{S}_2 \text{ or } \mathbf{S}_3 \text{ or } \mathbf{S}_4/\mathbf{S}_1) \qquad \text{(5.4-13b)}$$

In order to find an expression for this quantity, we must break it down into its component parts. Note that the probability of crossing the line a-b in fig. 5.4-4c is the same the situation in orthogonal signaling, fig. 5.4-4b. This also corresponds to deciding either $\mathbf{S}_2$ or $\mathbf{S}_3$ so that

$$P(\mathbf{S}_2 \text{ or } \mathbf{S}_3/\mathbf{S}_1) = Q(\sqrt{\rho_M/2}) \qquad \text{(5.4-13c)}$$

Similarly, the probability of crossing the line c-d corresponds to deciding $\mathbf{S}_3$ or $\mathbf{S}_4$, so that

$$P(\mathbf{S}_3 \text{ or } \mathbf{S}_4/\mathbf{S}_1) = Q(\sqrt{\rho_M/2}) \qquad \text{(5.4-13d)}$$

The basic axioms of probability theory tell us that

$$P(\mathbf{S}_2 \text{ or } \mathbf{S}_3 \text{ or } \mathbf{S}_4/\mathbf{S}_1) = P(\mathbf{S}_2 \text{ or } \mathbf{S}_3/\mathbf{S}_1) + P(\mathbf{S}_3 \text{ or } \mathbf{S}_4/\mathbf{S}_1) - P(\mathbf{S}_3/\mathbf{S}_1) \qquad \text{(5.4-13e)}$$

since each point should only be counted once. Now $\mathbf{S}_1$ and $\mathbf{S}_3$ are an antipodal pair so that

$$P(\mathbf{S}_3/\mathbf{S}_1) = Q(\sqrt{\rho_M}) \qquad \text{(5.4-14f)}$$

Consequently,

$$P_e = \{2Q[\sqrt{\rho_M/2}] - Q[\sqrt{\rho_M}]\} < 2Q[\sqrt{\rho_M/2}] \qquad \text{(5.4-14g)}$$

which, of course, corresponds to the result for QPSK that we obtained in Chapter 4. ∎

Our example of bi-orthogonal signaling was in two dimensions. The basic idea can be expanded to any number of dimensions. Computation of the probability of error for multidimensional signal sets will be done in the next section.

EXAMPLE 4 *Simplex Signaling*

Simplex signaling is a variant of orthogonal signaling that acheives the same probability of error as orthogonal signaling with a smaller SNR. The actual computation of the probability of error will be put off until the next section. We will focus here on the construction and properties of the simplex signal set.

Consider, for example, the case of three-dimensional orthogonal signaling ($M = 3$) where

$$
\begin{aligned}
s_1(t) &= \sqrt{E_s}\varphi_1(t) \Rightarrow \mathbf{s}_1 = \sqrt{E_s}\varphi_1 \\
s_2(t) &= \sqrt{E_s}\varphi_2(t) \Rightarrow \mathbf{s}_2 = \sqrt{E_s}\varphi_2 \\
s_3(t) &= \sqrt{E_s}\varphi_3(t) \Rightarrow \mathbf{s}_3 = \sqrt{E_s}\varphi_3
\end{aligned}
\tag{5.4-15a}
$$

The average of this signal set is

$$
\bar{S} = \frac{1}{M}\sum_{i=1}^{M}\mathbf{s}_i = \frac{\sqrt{E_s}}{3}(\varphi_1 + \varphi_2 + \varphi_3)
\tag{5.4-15b}
$$

and we will construct a new signal set, the simplex signal set, by subtracting the average from each of the signals in eq. 5.4-15a:

$$
\begin{aligned}
\mathbf{f}_1 &= \mathbf{s}_1 - \bar{S} = \sqrt{E_s}\left(\frac{2}{3}\varphi_1 - \frac{1}{3}\varphi_2 - \frac{1}{3}\varphi_3\right) \\
\mathbf{f}_2 &= \mathbf{s}_2 - \bar{S} = \sqrt{E_s}\left(-\frac{1}{3}\varphi_1 + \frac{2}{3}\varphi_2 - \frac{1}{3}\varphi_3\right) \\
\mathbf{f}_3 &= \mathbf{s}_3 - \bar{S} = \sqrt{E_s}\left(-\frac{1}{3}\varphi_1 - \frac{1}{3}\varphi_2 + \frac{2}{3}\varphi_3\right)
\end{aligned}
\tag{5.4-15c}
$$

This simplex signal set has a number of important properties:

1. Each of the signals has equal energy

$$
E_f = \frac{2}{3}E_s \text{ or, in general, } \frac{M-1}{M}E_s
\tag{5.4-15d}
$$

2. The correlation of any two of the signals is

$$
\rho_{mn} = \frac{\langle \mathbf{f}_m, \mathbf{f}_n \rangle}{|\mathbf{f}_m||\mathbf{f}_n|} = \begin{cases} 1, & \text{if } m = n \\ -\frac{1}{M-1}, & \text{if } m \neq n \end{cases}
\tag{5.4-15e}
$$

3. The dimensionality N of the simplex signal set is equal to $N = M - 1$, where M is the dimensionality of the original M orthogonal signals.

   In our example of $M = 3$, it is easy to establish that $\mathbf{f}_3 = -\mathbf{f}_1 - \mathbf{f}_2$. Since the space that we started with was three-dimensional and the vectors $\mathbf{f}_1$ and $\mathbf{f}_2$ form a two-dimensional subspace, in this case a plane, and $\mathbf{f}_3$ is a linear combination of them, then the dimensionality has been reduced to two.

   In the Problems confimation of this property will be extended to higher dimensions.

4. The distance between any two different signal vectors in the simplex set is $\sqrt{2E_s}$. This can easily be seen in our example for $M = 3$. The square of the distance between $\mathbf{f}_1$ and $\mathbf{f}_2$ is

$$d_{12}^2 = \|\mathbf{f}_1 - \mathbf{f}_2\|^2 \left\| \begin{bmatrix} 2/3 \\ -(1/3) \\ -(1/3) \end{bmatrix} \sqrt{E_s} - \begin{bmatrix} -(1/3) \\ 2/3 \\ -(1/3) \end{bmatrix} \sqrt{E_s} \right\|^2 = \left\| \begin{bmatrix} 1 \\ -1 \\ 0 \end{bmatrix} \sqrt{E_s} \right\|^2 = 2E_s$$

This is exactly the same as the distance between signal vectors in orthogonal signaling. Consequently the probability of error for orthogonal signaling of dimension M will be the same as that for simplex signaling having dimension $N = M - 1$. As indicated previously, computation of that probability of error will be undertaken in the next section. ■

EXAMPLE 5   *Phase Shift Keying*

We have, in Chapter 4, discussed QPSK, which we have seen is equivalent to bi-orthogonal signaling, and 8-PSK. In this example we want to generalize that idea to M-PSK and do the performance analysis in signal-space terms. The signal space representation of 8-PSK shown in fig. 5.4-4d serves to indicate the salient features of all PSK schemes. The signals are equally distributed around a circle, each one of them having the same energy. The actual rotational orientation of the signals is of no consequence to the performance. For the 8-PSK system

$$\mathbf{S}_1 = \sqrt{E_s} \begin{bmatrix} .924 \\ .383 \end{bmatrix} \quad \mathbf{S}_2 = \sqrt{E_s} \begin{bmatrix} .383 \\ .924 \end{bmatrix} \quad \mathbf{S}_3 = \sqrt{E_s} \begin{bmatrix} -.383 \\ .924 \end{bmatrix}$$

$$\mathbf{S}_4 = \sqrt{E_s} \begin{bmatrix} -.924 \\ .383 \end{bmatrix} \quad \mathbf{S}_5 = \sqrt{E_s} \begin{bmatrix} -.924 \\ -.383 \end{bmatrix} \quad \mathbf{S}_6 = \sqrt{E_s} \begin{bmatrix} -.383 \\ -.924 \end{bmatrix}$$

$$\mathbf{S}_7 = \sqrt{E_s} \begin{bmatrix} .383 \\ -.924 \end{bmatrix} \quad \mathbf{S}_8 = \sqrt{E_s} \begin{bmatrix} .924 \\ -.383 \end{bmatrix} \qquad \text{(5.4-16a)}$$

Since each of the signals has the same energy E_s, eq. 5.4-7 reduces to

$$d_{mn}^2 = 2E_s(1 - \rho_{mn}). \qquad \text{(5.4-16b)}$$

where ρ_{mn}, the correlation coefficient, is the Cos of the angle between the two signals. In 8-PSK these angles are 45°, 90°, 135°, 180°. Again using $\mathbf{S}_1$ as a reference signal, we can see that

$$d_{12}^2 = d_{18}^2 = 2E_s(1 - 1/\sqrt{2}) = 2E_s(.293)$$

$$d_{13}^2 = d_{17}^2 = 2E_s(1 - 0) = 2E_s$$

$$d_{14}^2 = d_{16}^2 = 2E_s(1 + 1/\sqrt{2}) = 2E_s(1.707)$$

$$d_{15}^2 = 2E_s(1 - (-1)) = 4E_s$$

(5.4-17c)

We will follow the approach for finding the probability of error that was developed in relation to QPSK in the previous example. Again, since the signal vectors are equally probable and the signal set is symmetric:

$$P_e = \sum_{n=1}^{8} P(\text{error}/\mathbf{S}_n)P(\mathbf{S}_n) = P(\text{error}/\mathbf{S}_1)$$

$$= P(\mathbf{S}_2 \text{ or } \mathbf{S}_3 \text{ or } \mathbf{S}_4 \text{ or } \mathbf{S}_5 \text{ or } \mathbf{S}_6 \text{ or } \mathbf{S}_7 \text{ or } \mathbf{S}_8/\mathbf{S}_1) \qquad (5.4\text{-}17\text{d})$$

Again, in order to find an expression for this quantity, we must break it down into its component parts. Note that the probability of crossing the line a-b in fig. 5.4-4d is the same as deciding either $\mathbf{S}_2$ or $\mathbf{S}_3$ or $\mathbf{S}_4$ or $\mathbf{S}_5$, so that

$$P(\mathbf{S}_2 \text{ or } \mathbf{S}_3 \text{ or } \mathbf{S}_4 \text{ or } \mathbf{S}_5/\mathbf{S}_1) = Q\left[\frac{d_{12}}{\sqrt{2\eta}}\right] = Q(.383\sqrt{\rho_M}) \qquad (5.4\text{-}17\text{e})$$

Similarly, the probability of crossing the line c-d corresponds to deciding $\mathbf{S}_5$ or $\mathbf{S}_6$ or $\mathbf{S}_7$ or $\mathbf{S}_8$, so that

$$P(\mathbf{S}_5 \text{ or } \mathbf{S}_6 \text{ or } \mathbf{S}_7 \text{ or } \mathbf{S}_8/\mathbf{S}_1) = Q\left[\frac{d_{18}}{\sqrt{2\eta}}\right] = Q(.383\sqrt{\rho_M}) \qquad (5.4\text{-}17\text{f})$$

Using eq. 5.4-17d, the basic axioms of probability theory tell us that

$$P_e = P(\mathbf{S}_2 \text{ or } \mathbf{S}_3 \text{ or } \mathbf{S}_4 \text{ or } \mathbf{S}_5 \text{ or } \mathbf{S}_6 \text{ or } \mathbf{S}_7 \text{ or } \mathbf{S}_8/\mathbf{S}_1)$$

$$= P(\mathbf{S}_2 \text{ or } \mathbf{S}_3 \text{ or } \mathbf{S}_4 \text{ or } \mathbf{S}_5/\mathbf{S}_1) + P(\mathbf{S}_5 \text{ or } \mathbf{S}_6 \text{ or } \mathbf{S}_7 \text{ or } \mathbf{S}_8/\mathbf{S}_1) - P(\mathbf{S}_5/\mathbf{S}_1)$$

$$< 2Q(.383\sqrt{\rho_M}) \qquad (5.4\text{-}17\text{g})$$

since each point should only be counted once. Since it is not a simple matter to compute the probability involving the isolated signal point $\mathbf{S}_5$, we are left with the upper bound for 8-PSK: $P_e < 2Q(.383\sqrt{\rho_M})$. For even modestly large values of ρ_M, corresponding to $P_e < 10^{-3}$, this is a very tight bound.

Examples of other PSK systems are left to the Problems. ∎

5.5 PROBABILITY OF ERROR FOR M-ARY ORTHOGONAL AND SIMPLEX SIGNALING

The binary FSK system discussed in the previous section can also be characterized as binary or 2-ary orthogonal signaling because the transmitted signals are, in fact, orthogonal to each other, being simply the result of multiplying orthonormal basis functions by a real number. Since the transmitted signals are assumed to be equally likely, it makes sense to make the further assumption that all of the signal vectors have equal length, i.e., have the same energy.

We want to generalize this to a system in which we transmit one of M such equal-energy orthogonal systems. The transmitted signal set should be of the form

$$s_m(t) = \sqrt{E_s}\phi_m(t); \qquad m = 1, 2, \ldots, M \qquad (5.5\text{-}1\text{a})$$

where the $\phi_m(t)$ constitute an orthonormal signal set. The structure of these signals is particularly simple: each signal vector contains only one nonzero component:

$$\mathbf{s}_1 = \sqrt{E_s}\begin{bmatrix} 1 \\ 0 \\ \cdot \\ 0 \end{bmatrix}, \quad \mathbf{s}_2 = \sqrt{E_s}\begin{bmatrix} 0 \\ 1 \\ \cdot \\ 0 \end{bmatrix}, \quad \ldots \quad \mathbf{s}_M = \sqrt{E_s}\begin{bmatrix} 0 \\ 0 \\ \cdot \\ 1 \end{bmatrix} \qquad (5.5\text{-}1\text{b})$$

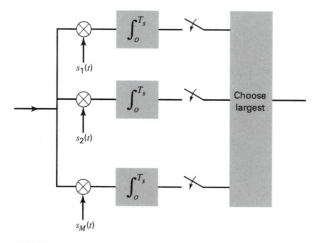

FIGURE 5.5-1

so that the optimum receiver structure, a combination of fig. 5.4-1 and fig. 5.4-5, is as shown in fig. 5.5-1. Examples of such sets of signals which can be constructed for any value of M include:

1. *Pulse Position Modulation*

$$s_m(t) = \begin{cases} \sqrt{E_s}; & \text{for } mT_s \leq t \leq (m+1)T_s \quad m = 1, 2, \ldots, M \\ 0; & \text{otherwise} \end{cases} \qquad (5.5\text{-}2a)$$

■

2. *M-ary Frequency-Shift Keying*

$$s_m(t) = \sqrt{E_s}\sqrt{2/T_s}\mathrm{Cos}[(n\omega_s + m\omega_s)t] \quad 0 \leq t \leq T_s; \quad m = 1, 2, \ldots, M \qquad (5.5\text{-}2b)$$

where $n\omega_s = \omega_c$, the carrier frequency. ■

3. *Sampling Functions*

$$s_m(t) = \sqrt{E_s}\mathrm{Sa}\left(\pi\frac{t - mT_s}{T_s}\right); \quad m = 1, 2, \ldots, M \qquad (5.5\text{-}2c)$$

■

4. *Walsh Functions*

The Walsh functions were described in fig. 5.1-3. They are used for orthogonal signaling in the IS-95 cellular standard. A set of 64 Walsh functions, each 64 bits long, is chosen from the entire set of 64-bit Walsh functions to be maximally far apart from each other in signal space. Six bits are used to choose one of these 64 functions for transmission. The receiver is based on the structure of fig. 5.5-1. The received 64-bit sequence, which may include incorrect bits as a result of noise, is correlated with each of the possible transmitted 64-bit Walsh functions. The largest result is chosen as the most likely transmitted signal. ■

The problem that we wish to address is that of finding the probability of error for such a system. Corresponding to the transmission of a signal vector $\mathbf{s}_m$, the output of the receiver is a vector $\mathbf{z}$:

$$
\mathbf{s}_m = \begin{bmatrix} 0 \\ 0 \\ 0 \\ \sqrt{E_s} \\ 0 \\ 0 \end{bmatrix} \qquad \mathbf{Z} = \begin{bmatrix} z_1 \\ z_2 \\ \cdot \\ z_m \\ \cdot \\ z_M \end{bmatrix} = \begin{bmatrix} n_1 \\ n_2 \\ \cdot \\ \sqrt{E_s} + n_m \\ \cdot \\ n_M \end{bmatrix} \tag{5.5-3}
$$

where all of the noise components are independent gaussian random variables each having variance $\sigma^2 = \eta/2$.

Let us define $P[C/\mathbf{s}_m, n_m]$ as the probability of a correct decision given that $\mathbf{s}_m$ was sent and that the noise accompanying that signal is n_m. Then

$$
\begin{aligned}
P[C/\mathbf{s}_m, n_m] &= \Pr[\text{all } n_i \leq (\sqrt{E_s} + n_m); \quad i \neq m] \\
&= \prod_{\substack{i=1 \\ i \neq m}}^{M} \Pr[n_i \leq (\sqrt{E_s} + n_m)] \\
&= \left[\frac{1}{\sqrt{\pi \eta}} \int_{-\infty}^{\sqrt{E_s}+n_m} \exp(-u^2/\eta)\, du \right]^{M-1}
\end{aligned} \tag{5.5-4a}
$$

We now want to eliminate the noise term n_m by noting that

$$
\begin{aligned}
P[C/\mathbf{s}_m] &= \int_{-\infty}^{\infty} P[C/\mathbf{s}_m, n_m] p(n_m)\, dn_m \\
&= \int_{-\infty}^{\infty} \left[\frac{1}{\sqrt{\pi \eta}} \int_{-\infty}^{\sqrt{E_s}+n_m} \exp(-u^2/\eta)\, du \right]^{M-1} \left[\frac{1}{\sqrt{\pi \eta}} e^{-n_m^2/\eta} \right] dn_m
\end{aligned} \tag{5.5-4b}
$$

Making the substitutions $x^2 = 2u^2/\eta$ and $y = n_m \sqrt{2/\eta}$, we get:

$$
\begin{aligned}
P[C/\mathbf{s}_m] &= \frac{1}{\sqrt{2\pi}} \int_{-\infty}^{\infty} \left\{ \exp(-y^2/2) \left[\frac{1}{\sqrt{2\pi}} \int_{-\infty}^{y+\sqrt{2E_s/\eta}} \exp(-x^2/2)\, dx \right]^{M-1} \right\} dy \\
&= \frac{1}{\sqrt{2\pi}} \int_{-\infty}^{\infty} \{\exp(-y^2/2)[1 - Q(y + \sqrt{\rho_M})]^{M-1}\}\, dy \\
&= \frac{1}{\sqrt{2\pi}} \int_{-\infty}^{\infty} \{\exp(-(y - \sqrt{\rho_M})^2/2)[1 - Q(y)]^{M-1}\}\, dy
\end{aligned} \tag{5.5-4c}
$$

The probability of being correct is:

$$
P[C] = \sum_{m=1}^{M} P[C/\mathbf{s}_m] P(\mathbf{s}_m) = P[C/\mathbf{s}_m]. \tag{5.5-4d}
$$

since each of the transmitted signals $\mathbf{s}_m$ is equally likely. Consequently

$$P[\text{error}] = 1 - P[C]$$

$$= 1 - \frac{1}{\sqrt{2\pi}} \int\limits_{-\infty}^{\infty} \{\exp(-(y - \sqrt{\rho_M})^2/2)[1 - Q(y)]^{M-1}\} \, dy \quad (5.5\text{-}4e)$$

which cannot be solved analytically.

Consequently, we will upper-bound the probability of error using the Union Bound, which says that

$$P\left[\bigcup_n A_n\right] \leq \sum_n P(A_n). \quad (5.5\text{-}5a)$$

In this instance $P(A_n)$ is the probability of error for orthogonal signaling with two signals:

$$P_e(\mathbf{S_1}, \mathbf{S_2}) = Q[\sqrt{\rho_M/2}] \quad (5.5\text{-}5b)$$

Consequently, for the M-ary system:

$$P[\text{error}] \leq (M - 1)Q[\sqrt{\rho_M/2}] \quad (5.5\text{-}5c)$$

Using the bound for the Q function shown in Appendix 2A,

$$Q[x] \leq e^{-(x^2/2)}$$

we can say that

$$P[\text{error}] \leq (M - 1)Q[\sqrt{\rho_M/2}] \leq M Q[\sqrt{\rho_M/2}] \leq M \exp[-\rho_M/4]. \quad (5.5\text{-}5d)$$

Now let us note that $\log_2 M$ is the number bits of information in a transmitted symbol and $\log_2 M/T_s = R$, the rate of information transmission in bits per second. As well, $E_s/T_s = P_s$ is the average power per transmitted symbol. Consequently, we can say that

$$P[\text{error}] \leq \exp\left\{-T_s\left[\frac{P_s}{2\eta} - R \ln 2\right]\right\} \quad (5.5\text{-}5e)$$

Note that as long as

$$R \leq \frac{1}{\ln 2} \frac{P_s}{2\eta} \quad (5.5\text{-}5f)$$

then the upper bound on the probability of making an error goes to zero as T_s gets large. Although tighter upper bounds are possible, this bound gives a crude example of Shannon's Channel Capacity Theorem. We will give a more general discussion of channel capacity in Chapter 8.

As indicated in the previous section, simplex signaling has the same probability of error as orthogonal signaling because in both cases the distance between signal points is $\sqrt{2E_s}$. The distinction between the two is that for orthogonal signaling we have E_s as the energy per transmitted symbol and in simplex signaling we have $E_s \frac{M-1}{M}$. The consequence in eq. 5.5-4e is that $\rho_M \rightarrow \rho_M(1 - \frac{1}{M})$. For $M = 2$ the difference between simplex signaling and orthogonal signaling is 3 db. As M increases, that difference goes to zero.

The error probability for bi-orthogonal signaling can be computed in a manner similar to that for orthogonal signaling.

Let us define $P[C/\mathbf{s}_m, n_m]$ as the probability of a correct decision given that $\mathbf{s}_m$ was sent and that the noise accompanying that signal is n_m. Since $\mathbf{s}_m$ may be either $\pm\sqrt{E_s}$, we require that the noise in the other dimensions not exceed $\pm\sqrt{E_s} + n_m$ in either direction in order to avoid an error. Therefore, using eq. 5.5-4a:

$$P[C/s_m, n_m] = \overline{\Pr[\text{all } n_i \leq |\sqrt{E_s} + n_m|; \quad i \neq m]}$$

$$= \prod_{\substack{i=1 \\ i \neq m}}^{M} \Pr[n_i \leq |\sqrt{E_s} + n_m|]$$

$$= \left[\frac{1}{\sqrt{\pi\eta}} \int_{-(\sqrt{E_s}+n_m)}^{\sqrt{E_s}+n_m} \exp(-u^2/\eta)\, du \right]^{M-1} \tag{5.5-6a}$$

Following the argument of eq. 5.5-4b,

$$P[C/s_m] = \int_{-\infty}^{\infty} P[C/s_m, n_m] p(n_m)\, dn_m$$

$$= \int_{-\infty}^{\infty} \left[\frac{1}{\sqrt{\pi\eta}} \int_{-(\sqrt{E_s}+n_m)}^{\sqrt{E_s}+n_m} \exp(-u^2/\eta)\, du \right]^{M-1} \left[\frac{1}{\sqrt{\pi\eta}} e^{-(n_m^2/\eta)} \right] dn_m \tag{5.5-6b}$$

Again, making the substitutions $x^2 = 2u^2/\eta$ and $y = n_m\sqrt{2/\eta}$, we get:

$$P[C/s_m] = \frac{1}{\sqrt{2\pi}} \int_{-\infty}^{\infty} \left\{ \exp(-y^2/2) \left[\frac{1}{\sqrt{2\pi}} \int_{-(y+\sqrt{2E_s/\eta})}^{y+\sqrt{2E_s/\eta}} \exp(-x^2/2)\, dx \right]^{M-1} \right\} dy$$

$$= \frac{1}{\sqrt{2\pi}} \int_{-\infty}^{\infty} \left\{ \exp(-y^2/2)[1 - 2Q(y + \sqrt{\rho_M})]^{M-1} \right\} dy$$

$$= \frac{1}{\sqrt{2\pi}} \int_{-\infty}^{\infty} \left\{ \exp(-(y - \sqrt{\rho_M})^2/2)[1 - 2Q(y)]^{M-1} \right\} dy \tag{5.5-6c}$$

Note that the difference in the result between eq. 5.5-4c for orthogonal signaling and eq. 5.5-6c for bi-orthogonal signaling is the difference between $Q(y)$ and $2Q(y)$.

Since all the messages are equally likely, we can finally say that the probability of error is given by:

$$P[\text{error}] = 1 - P[C] = 1 - \frac{1}{\sqrt{2\pi}} \int_{-\infty}^{\infty} \left\{ \exp(-(y - \sqrt{\rho_M})^2/2)[1 - 2Q(y)]^{M-1} \right\} dy \tag{5.5-6d}$$

which is, again, not analytically tractable.

In order to upper-bound eq. 5.5-6d, we recall that the probability of error for two-dimensional bi-orthogonal signaling is given by eq. 5.4-14g as:

$$P_e = \{2Q[\sqrt{\rho_M/2}] - Q[\sqrt{\rho_M}]\} < 2Q[\sqrt{\rho_M/2}]$$

and, continuing to parallel the argument for orthogonal signaling from eq. 5.5-5c:

$$P[\text{error}] \leq (M - 1)2Q[\sqrt{\rho_M/2}]. \tag{5.5-7a}$$

and

$$P[\text{error}] \le (M - 1)2Q[\sqrt{\rho_M/2}] \le 2MQ[\sqrt{\rho_M/2}] \le 2M \exp[-\rho_M/4]. \qquad (5.5\text{-}7b)$$

Consequently, the upper bound of eq. 5.5-5e still holds except that the information rate is now $R = \log_2(2M)/T_s$.

5.6 THE MATCHED FILTER RECEIVER IN THE PRESENCE OF NON-WHITE GAUSSIAN NOISE

The extension of the discussion to the optimum detection of signals in non-white Gaussian noise is useful not only in its own right, but also because it gives greater insight into the properties of the matched filter and into the techniques of the signal-space approach. A detailed discussion of this issue may be found in Simon, Hinedi, and Lindsay [SHL95].

The idea is (fig. 5.6-1) that the transmitted waveform is $s(t)$ and the noise, rather than being white, has a power spectral density $P_n(\omega)$. The approach is to insert a "whitening filter" at the input of the receiver having a transfer funtion $H_1(\omega)$ that satisfies

$$H_1(\omega) = \frac{1}{\sqrt{P_n(\omega)}} \qquad (5.6\text{-}1a)$$

The output noise of this filter is white but the transmitted signal $s(t)$ has been distorted by passage through $H_1(\omega)$, resulting in a new signal $g(t)$. According to matched-filter theory, we should now insert a second filter $H_2(\omega)$ that is matched to the signal $g(t)$. This filter will have the transfer function

$$H_2(\omega) = \frac{S^*(\omega)e^{-j\omega T}}{\sqrt{P_n(\omega)}} \qquad (5.6\text{-}1b)$$

where $s(t) \Leftrightarrow S(\omega)$ and T is an arbitrary delay. Consequently $r(t)$, the signal output of this matched filter, will be

$$R(\omega) = S(\omega)H_1(\omega)H_2(\omega) = \frac{S(\omega)S^*(\omega)e^{-j\omega T}}{P_n(\omega)}. \qquad (5.6\text{-}1c)$$

The probability of error for this system will be found from Chapter 3, eq. 3.4-2:

$$\rho = \frac{s_0^2(t_m)}{n_0^2(t)} = \frac{\left|\frac{1}{2\pi}\int_{-\infty}^{\infty} S(\omega)H(\omega)e^{j\omega t_m}\,d\omega\right|^2}{\frac{1}{2\pi}\int_{-\infty}^{\infty} P_n(\omega)|H(\omega)|^2\,d\omega} \qquad (3.4\text{-}2)$$

where $P_e = Q(\rho)$ and $H(\omega) = H_1(\omega)H_2(\omega)$. In order to use the Schwartz Inequality, let

$$G_1(\omega) = S(\omega)\sqrt{P_n(\omega)} \qquad \text{and} \qquad G_2(\omega) = \frac{S(\omega)e^{-j\omega T}}{\sqrt{P_n(\omega)}} \qquad (5.6\text{-}1d)$$

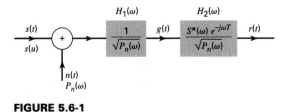

FIGURE 5.6-1

so that

$$\rho = \frac{\left|\frac{1}{2\pi}\int_{-\infty}^{\infty} G_1(\omega)G_2(\omega)\,d\omega\right|^2}{\frac{1}{2\pi}\int_{-\infty}^{\infty}|G_1(\omega)|^2\,d\omega} \leq \frac{\frac{1}{2\pi}\int_{-\infty}^{\infty}|G_1(\omega)|^2\,d\omega\int_{-\infty}^{\infty}|G_2(\omega)|^2\,d\omega}{\frac{1}{2\pi}\int_{-\infty}^{\infty}|G_1(\omega)|^2\,d\omega} \qquad (5.6\text{-}1e)$$

with equality if $G_1(\omega) = cG_2^*(\omega)$. Under this condition of equality,

$$\rho_{\text{MAX}} = \frac{1}{2\pi}\int_{-\infty}^{\infty} \frac{|S(\omega)|^2}{P_n(\omega)}\,d\omega \qquad (5.6\text{-}1f)$$

Note that if the noise is white with $P_n(\omega) = \eta/2$ and the energy in the transmitted signal is equal to E, then eq. 5.6-1f reduces to $\rho_{\text{MAX}} = 2E_s/\eta$, the result for the matched filter in the presence of white Gaussian noise.

There are two significant problems of implementation. The first is the choice of the transmitted signal $s(t)$. The second is the processing time at the receiver. In the presence of white noise the performance of the matched filter system did not depend at all on the shape of the transmitted signal $s(t)$, only on the energy E_s of that signal. This is clearly not the case in eq. 5.6-1f, where the power spectral density of the noise influences the choice of $s(t)$ under the constraint that the energy in $s(t)$ be equal to E_s.

Also in the presence of white noise, we saw the advantage to choosing a transmitted signal that avoided the problem of ISI (intersymbol interference) either by confining $s(t)$ to the symbol interval T_s or by limiting the signal spectrum to meet the Nyquist Criterion. Neither of these options is strictly available to us. If we choose $s(t)$ to be a time-limited pulse, the impulse response of the matched filter plus whitening filter is the inverse Fourier Transform

$$f(t) = \frac{1}{2\pi}\int_{-\infty}^{\infty} \frac{S^*(\omega)e^{-j\omega T}}{P_n(\omega)}\,d\omega \qquad (5.6\text{-}2)$$

Even the simplest non-white noise for which $P_n(\omega) = \sigma^2 \frac{1}{\alpha^2 + \omega^2}$ results in an impulse response that has infinite duration. To actually do the convolution between the received pulse $s(t)$ and the matched filter would result in an infinitely long response, guaranteeing ISI. Alternatively, the matched filter impulse response may be truncated in order to guarantee zero ISI. This loss of processing time will reduce the value of the signal component of the output at the sampling time.

Another approach to addressing this problem of optimum detection of data signals in non-white noise is given in the next section on the Karhunen-Loeve expansion.

5.7 REPRESENTATION OF NON-WHITE GAUSSIAN NOISE: THE KARHUNEN-LOEVE EXPANSION

In the previous section we examined the behavior of the correlation receiver in the presence of white Gaussian noise. We saw that the analysis of this receiver structure was made relatively straightforward because we were able to decompose the noise into a sum of uncorrelated random variables with the same set of basis functions that define the signal. As a consequence, we were able to make all of the performance computations in a simple Euclidean vector space. We would like to do the same thing when the noise is non-white. The difficulty is demonstrated in eq. 5.3-6a. White noise has an autocorrelation function $\bar{R}_n(\tau) = \sigma^2\delta(\tau)$ so that its components are uncorrelated with respect to any orthogonal set of basis functions. This is clearly not the case with non-white noise, where there is only one set of orthogonal basis functions for which the noise is decomposed into uncorrelated random variables. This set depends on the autocorrelation function of the noise. The object is to find that set of basis functions. In order to use signal-space techniques, we will then have to express the signal in terms of these new basis functions. The expansion of the noise in this manner is called the Karhunen-Loeve (K-L) Expansion. Mathematically, we want an expansion of a noise-random process $X(t)$ in the form

$$X(t) = \sum_{n=1}^{\infty} X_n \phi_n(t); \qquad 0 \le t \le T_s \tag{5.7-1a}$$

where

$$X_n = \int_0^{T_s} X(t) \phi_n^*(t)\, dt \tag{5.7-1b}$$

the coefficients X_k are orthogonal random variables, and the basis functions are orthonormal, satisfying

$$\int_0^{T_s} \phi_m(t) \phi_n^*(t)\, dt = \delta_{mn} = \begin{cases} 1; & \text{if } m = n \\ 0; & \text{if } m \ne n \end{cases} \tag{5.7-1c}$$

With this expansion, we can express the autocorrelation function of the not necessarily stationary random process as

$$R_x(t, s) = E\lfloor x(t)x^*(s) \rfloor = E\left[\sum_n X_n \phi_n(t) \sum_m X_m^* \phi_m^*(s) \right]$$

$$= \sum_m \sum_n E[X_n X_m] \phi_n(t) \phi_m^*(s) = \sum_n \lambda_n \phi_n(t) \phi_n^*(s) \tag{5.7-2a}$$

since the random variables X_n are assumed to be orthogonal and

$$E[X_n X_m] = \begin{cases} \lambda_n; & m = n \\ 0; & m \ne n \end{cases} \tag{5.7-2b}$$

Consequently we may write

$$\int_{t_1}^{t_1+T_s} R_x(t, s) \phi_k(s)\, ds = \int_{t_1}^{t_1+T_s} \left[\sum_n \lambda_n \phi_n(t) \phi_n^*(s) \right] \phi_k(s)\, ds$$

$$= \sum_n \lambda_n \phi_n(t) \int_{t_1}^{t_1+T_s} \phi_n^*(s) \phi_k(s)\, ds = \lambda_n \phi_n(t) \tag{5.7-2c}$$

or

$$\int_{t_1}^{t_1+T_s} R_x(t, s) \phi_k(s)\, ds = \lambda_n \phi_n(t) \tag{5.7-2d}$$

To recapitulate, in order to decompose a random process $X(t)$ according to eq. 5.7-1b, we require the orthonormal set of basis functions that satisfy eq. 5.4-6c. This equation is the signal-space equivalent of the formulation of the eigenvalue problem in Euclidean vector space

$$A\mathbf{u}_n = \lambda_n \mathbf{u}_n$$

In eq. 5.7-2d the reference is to the eigenvalues and eigenfunctions of the kernel $R_x(t, s)$.

We can also see that eq. 5.7-2a gives a signal-space version of Mercer's Theorem, which, for wide-sense stationary processes where $s = t + \tau$, is

$$R_x(\tau) = \sum_n \lambda_n \phi_n(t) \phi_n^*(t + \tau) \tag{5.7-3}$$

EXAMPLE 6

We want to find a K-L expansion for the random process $X(t)$ having autocorrelation function $R_x(\tau) = R_x(t_2 - t_1) = \sigma^2 e^{-\alpha|\tau|}$ over the time interval $[-T, T]$.

This problem is worked out in great detail in Davenport and Root [D&R58] and in Leon-Garcia [L-G94] and the interested reader is referred to those sources. The results are as follows:

$$X(t) = \sum_n X_n^c [A_n \text{Cos}(a_n t)] + \sum_n X_n^s [B_n \text{Sin}(b_n t)] \tag{5.7-4}$$

where X_n^c and X_n^s are uncorrelated, zero-mean, Gaussian random variables having variance λ_n.

This expression for $X(t)$ has similarities to a Fourier Series. It is a summation of an orthonormal set of basis functions, Cos and Sin, multiplied by random variables as coefficients. The crucial differences is that the frequencies are not harmonically related as would be the case in a Fourier Series.

According to Leon-Garcia, the values of a_n and b_n are solutions of

$$\text{Tan}(a_n T) = \alpha/a_n \qquad \text{Tan}(b_n T) = -b_n/T \tag{5.7-5a}$$

which exist in each 2π interval as shown in fig. 5.7-1, and

$$A_n^2 = \cfrac{1}{T - \cfrac{\text{Sin}(2a_n T)}{2a_n}} \qquad B_n^2 = \cfrac{1}{T + \cfrac{\text{Sin}(2b_n T)}{2b_n}} \tag{5.7-5b}$$

The decomposition of the noise according to these basis functions will lead to orthogonal noise components in signal space and a similar decomposition of the transmitted signals will lead to signal vectors in the same space. The vector receiver of fig. 5.4-3 is then an appropriate structure. ■

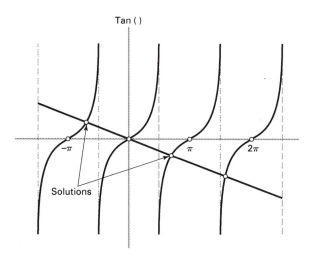

FIGURE 5.7-1

5.8 MAXIMUM LIKELIHOOD SEQUENCE ESTIMATION (MLSE)

Until this point we have used the ideas of maximum likelihood estimation on individual received signal points that have been corrupted by noise. Central to these techniques has been that successive signal points and successive noise samples are uncorrelated, that is, that there is no memory in the system. However, we have seen that a communication system can have memory if the channel generates ISI (intersymbol interference) or if the data is encoded. Note that partial response signaling can be understood either as ISI or as data encoding. The channel also has memory if the noise is not white, but we have dealt with that issue in the previous section on the Karhunen-Loeve Expansion.

We have made the point that either ISI or data encoding can be understood as a Markov process taking the form of a finite state machine having a random binary input. We have also seen that maximum likelihood symbol-by-symbol detection on a channel having ISI will degrade performance because the ISI will lower the threshold against noise. As an alternative, we should consider making decisions by examining an entire sequence of received data, attempting to fit our decisions to the underlying structure of the finite state machine. We shall begin by reviewing how ISI and partial response may represented by a finite state machine.

5.8.1 Representing an ISI Channel as a Finite State Machine

Central to the idea of MLSE is the representation of a channel as a finite state machine or, equivalently, as shown in Chapter 3, a Markov process. We shall proceed by example.

1. Nyquist Signaling

Consider the causal sampled impulse response shown in fig. 5.8-1a as a transversal filter. This impulse response may also be written as the vector

$$F = [f_0 \quad f_1 \quad f_2] = [1 \quad -0.3 \quad +0.2] \tag{5.8-1}$$

Using the same state representation of eq. 3.13-1a and the state transition table of eq. 3.13-1b, where the entries in the table are (next state/current output), the values of the current

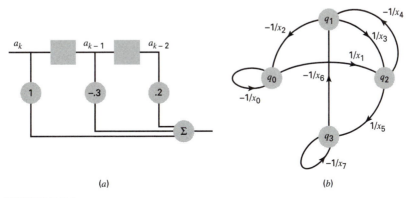

(a)   (b)

FIGURE 5.8-1

output are computed as:

$$
\begin{aligned}
x_0 &= -f_0 - f_1 - f_2 = -1 + 0.3 - 0.2 = -0.9 \\
x_1 &= +f_0 - f_1 - f_2 = +1 + 0.3 - 0.2 = +1.1 \\
x_2 &= -f_0 - f_1 + f_2 = -1 + 0.3 + 0.2 = -0.5 \\
x_3 &= +f_0 - f_1 + f_2 = +1 + 0.3 + 0.2 = +1.5 \\
x_4 &= -f_0 + f_1 - f_2 = -1 - 0.3 - 0.2 = -1.5 \\
x_5 &= +f_0 + f_1 - f_2 = +1 - 0.3 - 0.2 = +0.5 \\
x_6 &= -f_0 + f_1 + f_2 = -1 - 0.3 + 0.2 = -1.1 \\
x_7 &= +f_0 + f_1 + f_2 = +1 - 0.3 + 0.2 = +0.9
\end{aligned}
\tag{5.8-2}
$$

where the state machine is

	$a_k = -1$	$a_k = +1$
q_0	q_0/x_0	q_2/x_1
q_1	q_0/x_2	q_2/x_3
q_2	q_1/x_4	q_3/x_5
q_3	q_1/x_6	q_3/x_7

(5.8-3)

The resulting machine is shown in fig. 5.8-1b. ∎

2. *Partial Response Signaling*

We have seen in Chapter 3 that a partial response encoder may be represented by an ISI model. The Problems for that chapter asked for Moore-machine representations for various PRS systems in order to find their power spectral densities. We shall now find Mealy-machine models of these systems because that model is the one that is appropriate to MLSE.

A model for Class 1 Partial Response (Duobinary) is shown in fig. 5.8-2a. We have seen that the transfer function of this system is

$$
G(z) = 1 + z^{-1}
\tag{5.8-4a}
$$

Since there is only one memory element, the Mealy machine model, fig. 5.8-2b, has only two states: $q_0 = a_{k-1} = -1, q_1 = a_{k-1} = +1$. The state transition/output table

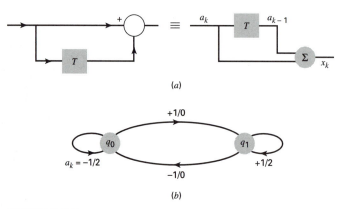

FIGURE 5.8-2

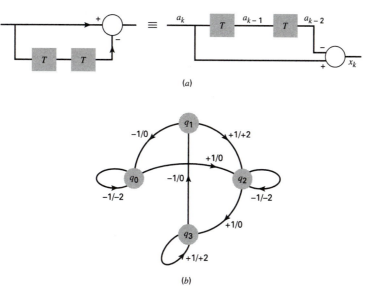

(a)

(b)

FIGURE 5.8-3

is given below:

	$a_k = -1$	$a_k = +1$
q_0	$q_0/x_0 = -2$	$q_1/x_1 = 0$
q_1	$q_0/x_1 = 0$	$q_1/x_0 = +2$

$$(5.8\text{-}4b)$$

A model for Class 4-Partial Response (Modified Duobinary) is shown in fig. 5.8-3a. The transfer function for this system is

$$G(z) = 1 - z^{-2} \qquad (5.8\text{-}5a)$$

Since the memory is equal to 2, the definition of the four states in eq. 3.13-1b and the state transition/output table is

	$a_k = -1$	$a_k = +1$
q_0	$q_0/x_0 = -2$	$q_2/x_1 = 0$
q_1	$q_0/x_1 = 0$	$q_2/x_2 = +2$
q_2	$q_1/x_0 = -2$	$q_3/x_1 = 0$
q_3	$q_1/x_1 = 0$	$q_3/x_2 = +2$

$$(5.8\text{-}6)$$

The state machine is shown in fig. 5.5-3b.

In a partial response system that has additional ISI caused by channel distortion, the values of the output are modified. For example, the channel impulse response vector for a distorted Class-4 system could be

$$F = [1 \quad -0.3 \quad -0.8 \quad 0.1] \qquad (5.8\text{-}7)$$

Knowledge of the sampled impulse response is sufficient information to construct a Mealy-machine model and that is all that we require to proceed.

In order to anticipate any confusion, the Mealy machine for a partial response channel is valid whether or not partial response precoding has been used at the transmitter. Recall that the precoding does not change the power spectral density of the signal and therefore does not change the Markov model. All that the precoding does is prevent decision errors from propagating. ■

5.8.2 Signal Detection for Channels With Memory

We will now continue the discusssion of maximum likelihood detection of signals. The entire input sequence of data to a system characterized by a finite-state machine is

$$a(t) = \sum_{k=0}^{N} a_k \delta(t - kT_s) \tag{5.8-8a}$$

where N is the length of the sequence, and the corresponding output sequence is denoted

$$y(t) = \sum_{k=0}^{N} y_k \delta(t - kT_s) = \sum_{k=0}^{N} (x_k + n_k) \delta(t - kT_s) \tag{5.8-8b}$$

where x_k is the output of the finite state machine and n_k is a sample of a white Gaussian noise process with $P_n(\omega) = \eta/2 = \sigma^2$. Our detection problem is as follows: the received sequence is $y(t)$ and the list of all possible transmitted sequences is denoted $\{a_n(t)\}$; find the particular transmitted data sequence a_n such that

$$P[a_n(t) \text{ sent}/y(t) \text{ received}] > P[a_m(t) \text{ sent}/y(t) \text{ received}] \tag{5.8-9a}$$

for all $m \neq n$. As previously noted, this formulation of the problem is called *maximizing the a posteriori probability*. Through the use of Bayes' Rule, this condition can be rewritten as

$$\frac{P[a_n(t) \text{ sent}]}{P[y(t) \text{ received}]} P[y(t) \text{ received}/a_n(t) \text{ sent}]$$

$$> \frac{P[a_m(t) \text{ sent}]}{P[y(t) \text{ received}]} P[y(t) \text{ received}/a_m(t) \text{ sent}] \tag{5.8-9b}$$

and, if we assume that all data sequences are equally likely,

$$P[y(t) \text{ received}/a_n(t) \text{ sent}] > P[y(t) \text{ received}/a_m(t) \text{ sent}] \tag{5.8-9c}$$

Since the noise is Gaussian, we can write

$$P[y(t) \text{ received}/a_n(t) \text{ sent}] = \left[\frac{1}{\sigma\sqrt{2\pi}}\right]^N \exp\left\{-\frac{1}{2\sigma^2} \sum_{k=0}^{N} [y_k - x_k]^2\right\} \tag{5.8-9d}$$

where

$$x_k = \sum_{j=0}^{v} f_j a_{k-j} \tag{5.8-9e}$$

and where v is the length of the ISI vector F. Maximizing this conditional probability is the same as minimizing the value of

$$D_N = \sum_{k=0}^{N} \left[y_k - \sum_{j=0}^{v} f_j a_{k-j}\right]^2 \tag{5.8-9f}$$

which can be understood as the square of the Euclidean distance between the noisy output and the noiseless output over the entire sequence.

The basic point is that since each value of x_k depends on a sequence of data values rather than on a single one, minimizing D_N requires taking that sequence into account. In some abstract manner, accomplishing this requires that we receive and store the entire received sequence before operating on it in order to find the most likely transmitted data sequence. This is both unrealistic and, fortunately, unnecessary. We can instead find the data points using iteration.

Let $M < N$ be the length of a partial sequence of data and v be the length of the impulse response vector. It follows that

$$D_M = \sum_{k=0}^{M} \left[y_k - \sum_{j=0}^{v} f_j a_{k-j} \right]^2 \qquad (5.8\text{-}10a)$$

and

$$\Delta D = D_M - D_{M-1} = \sum_{k=0}^{M} \left[y_k - \sum_{j=0}^{v} f_j a_{k-j} \right]^2 - \sum_{k=0}^{M-1} \left[y_k - \sum_{j=0}^{v} f_j a_{k-j} \right]^2$$

$$= \left[y_M - \sum_{j=0}^{v} f_j a_{M-j} \right]^2$$

$$= y_M^2 - 2 y_M \sum_{j=0}^{v} f_j a_{M-j} + \sum_{i=0}^{v} f_i a_{M-i} \sum_{j=0}^{v} f_j a_{M-j} \qquad (5.8\text{-}10b)$$

We shall return to this expression at the end of the next section on the Viterbi algorithm in order to decrease the computational complexity of the algorithm.

If we have chosen the first $(M-1)$ transmitted symbols so that D_{M-1} is the smallest cumulative distance up to that point, then minimizing ΔD with respect to the data symbol a_M will also minimize D_M. The difficulty involved in accomplishing this is that at some point in the process a wrong choice may be made, in which case it is necessary to have a way of getting the recursion back on track. This is accomplished by remembering a collection of appropriate paths that go back L symbols, so that while we are processing a_M we are detecting a_{M-L}. This approach is called MLSE, maximum likelihood sequence estimation.

The Viterbi Algorithm is a quite straightforward way of implementing MLSE. Detailed analyses of the probability of error for MLSE and the Viterbi Algorithm may be found in For [72], For [73] and Proakis [Pro89].

5.8.3   The Viterbi Algorithm

The Viterbi Algorithm will be explained through an extended example based on the ISI model of fig. 5.8-1a. Let us assume that the sequential machine of fig. 5.8-1b starts in state q_0 and that the sequence of output observations is:

$$y_1 = -0.8; y_2 = -0.3; -1.2; +1.3; -1.4; +0.7; +0.9; -0.4; y_9 = +1.0; y_{10} = 1.6 \qquad (5.8\text{-}11a)$$

Referring to fig. 5.8-4a, and corresponding to the first output $y_1 = -0.8$, we compute the metrics D_{10}^2 and D_{12}^2, recognizing that q_0 can lead only to q_0 or q_2.

$$D_{10}^2 = (y_1 - x_0)^2 = (-0.8 + 0.9)^2 = 0.01$$
$$D_{12}^2 = (y_1 - x_1)^2 = (-0.8 - 1.1)^2 = 3.61 \qquad (5.8\text{-}11b)$$

Similarly, in fig. 5.8-4b, we compute the metrics D_{20}^2, D_{21}^2, D_{22}^2, D_{23}^2 corresponding to the

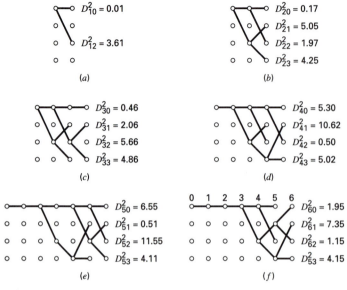

FIGURE 5.8-4

input $y_2 = -0.3$ where we now have a path going to each state.

$$D_{20}^2 = D_{10}^2 + (y_2 - x_0)^2 = 0.01 + (-0.3 + 0.9)^2 = 0.37$$
$$D_{21}^2 = D_{12}^2 + (y_2 - x_4)^2 = 3.61 + (-0.3 + 1.5)^2 = 5.05$$
$$D_{22}^2 = D_{10}^2 + (y_2 - x_1)^2 = 0.01 + (-0.3 - 1.1)^2 = 1.97 \qquad \text{(5.8-11c)}$$
$$D_{23}^2 = D_{12}^2 + (y_2 - x_5)^2 = 3.61 + (-0.3 - 0.5)^2 = 4.25$$

At Stage 3, fig. 5.8-4c, we begin to see the features of the Viterbi Algorithm. At this stage we have two alternative paths to each state. We will compute both cumulative distances to each path, discard the path corresponding to the larger distance, and keep the path corresponding to the smaller distance as a *survivor*.

Define D_{kij}^2 as the square of the distance to state q_i with an immediate predecessor of state q_j at the kth stage of the algorithm. Keep in mind that the structure of the sequential machine does not necessarily allow a transition from each state to every other state. We shall compute these quantities for each of the states q_i.

$$D_{300}^2 = D_{20}^2 + (y_3 - x_0)^2 = 0.37 + (-1.2 + 0.9)^2 = 0.46$$
$$D_{301}^2 = D_{21}^2 + (y_3 - x_2)^2 = 5.05 + (-1.2 + 0.5)^2 = 5.54$$
$$D_{312}^2 = D_{22}^2 + (y_3 - x_4)^2 = 1.97 + (-1.2 + 1.5)^2 = 2.06$$
$$D_{313}^2 = D_{23}^2 + (y_3 - x_6)^2 = 4.25 + (-1.2 + 1.1)^2 = 4.26$$
$$D_{320}^2 = D_{20}^2 + (y_3 - x_1)^2 = 0.37 + (-1.2 - 1.1)^2 = 5.66 \qquad \text{(5.8-11d)}$$
$$D_{321}^2 = D_{21}^2 + (y_3 - x_3)^2 = 5.05 + (-1.2 - 1.5)^2 = 12.34$$
$$D_{332}^2 = D_{22}^2 + (y_3 - x_5)^2 = 1.97 + (-1.2 - 0.5)^2 = 4.86$$
$$D_{333}^2 = D_{23}^2 + (y_3 - x_7)^2 = 4.25 + (-1.2 - 0.9)^2 = 8.66$$

The Viterbi Algorithm now tells us to choose the shortest path to each final state in the trellis so, for example, we will choose the path characterized by D_{300}^2 rather than D_{301}^2 and relabel that choice as D_{30}^2. These new distances and their corresponding paths, the survivors, are shown in fig. 5.8-4c and are listed below.

$$
\begin{aligned}
\text{path 0000} && D_{30}^2 &= 0.46 \\
\text{path 0021} && D_{31}^2 &= 2.06 \\
\text{path 0002} && D_{32}^2 &= 5.66 \\
\text{path 0023} && D_{33}^2 &= 4.86
\end{aligned}
\tag{5.8-11e}
$$

Before making a decision at the left end of the path, we will continue to establish a buffer of length 5 in order to allow the survivor paths to converge at the left end. The length of this buffer L is typically set to two to three times the memory of the channel.

To extend the computation to $k = 4$, we can use the same template.

$$
\begin{aligned}
D_{400}^2 &= D_{30}^2 + (y_4 - x_0)^2 = 0.46 + (+1.3 + 0.9)^2 = 4.87 \\
D_{401}^2 &= D_{31}^2 + (y_4 - x_2)^2 = 2.06 + (+1.3 + 0.5)^2 = 5.30 \\
D_{412}^2 &= D_{32}^2 + (y_4 - x_4)^2 = 5.66 + (+1.3 + 1.5)^2 = 13.50 \\
D_{413}^2 &= D_{33}^2 + (y_4 - x_6)^2 = 4.86 + (+1.3 + 1.1)^2 = 10.62 \\
D_{420}^2 &= D_{30}^2 + (y_4 - x_1)^2 = 0.46 + (+1.3 - 1.1)^2 = 0.50 \\
D_{421}^2 &= D_{31}^2 + (y_4 - x_3)^2 = 2.06 + (+1.3 - 1.5)^2 = 2.10 \\
D_{432}^2 &= D_{32}^2 + (y_4 - x_5)^2 = 5.66 + (+1.3 - 0.5)^2 = 6.30 \\
D_{433}^2 &= D_{33}^2 + (y_4 - x_7)^2 = 4.86 + (+1.3 - 0.9)^2 = 5.02
\end{aligned}
\tag{5.8-11f}
$$

so that the surviviors and cumulative distances are

$$
\begin{aligned}
\text{path 00000} && D_{40}^2 &= 5.30 \\
\text{path 00031} && D_{41}^2 &= 10.62 \\
\text{path 00002} && D_{42}^2 &= 0.50 \\
\text{path 00233} && D_{43}^2 &= 5.02
\end{aligned}
\tag{5.8-11g}
$$

as shown in fig. 5.8-4d.

One more iteration will fill the buffer. The next piece of data is $y_5 = -1.4$ so that the computation of distances is:

$$
\begin{aligned}
D_{500}^2 &= D_{40}^2 + (y_5 - x_0)^2 = 5.30 + (-1.4 + 0.9)^2 = 5.55 \\
D_{501}^2 &= D_{41}^2 + (y_5 - x_2)^2 = 10.62 + (-1.4 + 0.5)^2 = 11.43 \\
D_{512}^2 &= D_{42}^2 + (y_5 - x_4)^2 = 0.50 + (-1.4 + 1.5)^2 = 0.51 \\
D_{513}^2 &= D_{43}^2 + (y_5 - x_6)^2 = 5.02 + (-1.4 + 1.1)^2 = 5.11 \\
D_{520}^2 &= D_{40}^2 + (y_5 - x_1)^2 = 5.30 + (-1.4 - 1.1)^2 = 11.55 \\
D_{521}^2 &= D_{41}^2 + (y_5 - x_3)^2 = 10.62 + (-1.4 - 1.5)^2 = 19.03 \\
D_{532}^2 &= D_{42}^2 + (y_5 - x_5)^2 = 0.50 + (-1.4 - 0.5)^2 = 4.11 \\
D_{533}^2 &= D_{43}^2 + (y_5 - x_7)^2 = 5.02 + (-1.4 - 0.9)^2 = 10.31
\end{aligned}
\tag{5.8-11h}
$$

so the survivors and cumulative distances are

$$
\begin{array}{lll}
\text{path } 000000 & D_{50}^2 = 5.55 \\
\text{path } 000021 & D_{51}^2 = 0.51 \\
\text{path } 000002 & D_{52}^2 = 11.55 \\
\text{path } 000023 & D_{53}^2 = 4.11
\end{array}
\tag{5.8-11i}
$$

as shown in fig. 5.8-4e.

We have now filled the buffer and after the next piece of input data is processed we will take our first output from the left end (after the common starting state q_0). Notice that two of the survivor paths are following one trajectory emerging from $k = 1$ and two are following another. The longer the buffer, the more likely that such a branching will resolve itself without error. In this example the buffer length has been established as $L = 5$ so that after the next input, an output will emerge from the left end of the buffer.

Let's look at one last piece of data, $y_6 = +0.7$. Computing the distances:

$$
\begin{aligned}
D_{600}^2 &= D_{50}^2 + (y_6 - x_0)^2 = 5.55 + (+0.7 + 0.9)^2 = 8.11 \\
D_{601}^2 &= D_{51}^2 + (y_6 - x_2)^2 = 0.51 + (+0.7 + 0.5)^2 = 1.95 \\
D_{612}^2 &= D_{52}^2 + (y_6 - x_4)^2 = 11.55 + (+0.7 + 1.5)^2 = 16.39 \\
D_{613}^2 &= D_{53}^2 + (y_6 - x_6)^2 = 4.11 + (+0.7 + 1.1)^2 = 7.35 \\
D_{620}^2 &= D_{50}^2 + (y_6 - x_1)^2 = 5.55 + (+0.7 - 1.1)^2 = 5.71 \\
D_{621}^2 &= D_{51}^2 + (y_6 - x_3)^2 = 0.51 + (+0.7 - 1.5)^2 = 1.15 \\
D_{632}^2 &= D_{52}^2 + (y_6 - x_5)^2 = 11.55 + (+0.7 + 0.5)^2 = 11.59 \\
D_{633}^2 &= D_{53}^2 + (y_6 - x_7)^2 = 4.11 + (+0.7 - 0.9)^2 = 4.15
\end{aligned}
\tag{5.8-11j}
$$

so the survivors are

$$
\begin{array}{lll}
\text{path } \mathbf{0000}210 & D_{60}^2 = 1.95 \\
\text{path } \mathbf{0000}231 & D_{61}^2 = 7.35 \\
\text{path } \mathbf{0000}213 & D_{62}^2 = 1.15 \\
\text{path } \mathbf{0000}233 & D_{53}^2 = 4.15
\end{array}
\tag{5.8-11k}
$$

as shown in fig. 5.8-4f.

The first output is 0, indicating the common path that has evolved after the starting state. Looking down the line, we see that the next three outputs have already converged to common values of 002. We can see from this example that the length of the buffer is quite important. The longer the buffer, the more likely that a single survivor will have emerged before an output is to be determined.

Finally, we have determined that the first state transition is 00, which, from the state table, corresponds to an input of $a_1 = -1$. The input sequence continues to be reconstructed in this manner.

Some exercises in relation to this example will be given in the Problems.

It was previously noted that if the number of data levels is M and the channel memory is v then the number of states is $Q = M^v$. Under these circumstances it will be very likely that, unlike the situation in our example, there will be more than two ways of getting to any particular state. The Viterbi Algorithm states that there should be *one and only one* survivor sequence, the one with the smallest cumulative distance, leading to each state. This means that the computational algorithm need only remember Q sequences, each of which is L symbols long (L is the length of the buffer).

The six figures fig. 5.8-4a to fig. 5.8-4f are the unfolding of the succession of states in the sequential machine under consideration. The set of all possible paths in this unfolding is called a *trellis* because of its visual similarity to a trellis upon which vines are grown. MLSE evolves a single path through the trellis. We shall be using the idea of a trellis in relation to MLSE and the Viterbi Algorithm in subsequent discussions.

Now let us return to eq. 5.8-10b, where we will now call ΔD_{nj} the difference between y_M and the output that corresponds to a transition from state q_j to state q_n if such a transition is possible. Then

$$\Delta D_{nj} = [y_M - x_{nj}]^2 = y_M^2 - 2y_M x_{nj} + x_{nj}^2. \tag{5.8-12}$$

Since y_M^2 is the same for all of the possible transitions, it may be discarded from the computations. Since x_{nj}^2 is specific to a particular transition, it may be determined in advance and stored in relation to that computation. This leaves the single term that must be computed for each new input, $-2y_M x_{nj}$. This can have the effect of simplifying the computation.

The programming of the Viterbi Algorithm is left to the Problems.

5.8.4   Probability of Error for MLSE

We have previously referred a detailed analysis of the probability of error for MLSE and the Viterbi Algorithm to [For73] and [Pro89]. The main result of this analysis is important to state and understand.

This chapter developed the idea (eq. 5.4-8) that in symbol-by-symbol detection the probability of error is

$$P_e = kQ \left(\sqrt{\frac{d_{\min}^2}{2\eta}} \right) \tag{5.8-13}$$

where $d_{\min}$ is the minimum Euclidean distance between signal points and k is an appropriate constant. For MLSE the probability of error is still given by eq. 5.8-13 but $d_{\min}$ now refers to the minimum distance between sequences.

As an illustration, let us refer to the sequential machine of fig. 5.8-1b and eq. 5.8-2 which gives the outputs for that machine. If the machine is in state q_0 and has inputs equal to -1, then it will remain in q_0 with outputs $x_0 = -0.9$. On the other hand, the machine can go through the sequence $q_0 \to q_2 \to q_1 \to q_0$ with outputs $x_1 = 1.1, x_4 = -1.5, x_2 = -0.5$. The square of the distance between these sequences is

$$\begin{aligned} d^2 &= (x_1 - x_0)^2 + (x_4 - x_0)^2 + (x_2 - x_0)^2 \\ &= (1.1 + 0.9)^2 + (-1.5 + 0.9)^2 + (-0.5 + 0.9)^2 = 4.52 \end{aligned} \tag{5.8-14}$$

The idea is that the sequences that we are comparing start and end in the same state. We are looking for the two nontrivial sequences that have the smallest distance between them. This is the quantity that is used in eq. 3.9-13 for MLSE.

It should be emphasized that MLSE is appropriately used whenever the underlying phenomenon can be described by a state machine or, equivalently, by a Markov process. This includes ISI and trellis coding. These issues will be further developed in Chapters 9 and 10.

5.9 NONCOHERENT DETECTION OF BANDPASS SIGNALS

Differential Phase-Shift Keying (DPSK) was examined in Chapter 4 as an example of detection of a bandpass signal without generating a coherent demodulating carrier at the receiver. Using maximum likelihood techniques, we now want to extend that discussion to other signaling methods, particularly to ASK (Amplitude-Shift Keying) and FSK. The usefulness of this approach was indicated in the discussion of DPSK, namely that noncoherent receivers are far easier to implement than coherent receivers, although there is a performance penalty to be paid.

In order to be able to make the results equally applicable to ASK and FSK it will be assumed that the transmitted signal $s(t)$ can have both amplitude and phase information, so that for $j = 1, 2, \ldots, J$

$$
\begin{aligned}
s_j(t) &= \sqrt{\frac{2}{T_s}} x_j(t) \mathrm{Cos}(\omega_0 t - \varphi_j(t) + \theta) \\
&= a_j(t)\left[\sqrt{\frac{2}{T_s}}\mathrm{Cos}(\omega_0 t + \theta)\right] + b_j(t)\left[\sqrt{\frac{2}{T_s}}\mathrm{Sin}(\omega_0 t + \theta)\right] \\
&= a_j(t)\phi_1(t) + b_j(t)\phi_2(t)
\end{aligned}
\tag{5.9-1a}
$$

where $\phi_1(t)$, $\phi_2(t)$ are orthonormal basis functions having an unknown and random phase θ and where:

$$
\begin{aligned}
a_j(t) &= x_j(t)\mathrm{Cos}(\varphi_j(t)) \\
b_j(t) &= x_j(t)\mathrm{Sin}(\varphi_j(t))
\end{aligned}
\tag{5.9-1b}
$$

The signal is corrupted by additive white Gaussian noise, which, as we have seen, can be similarly represented as

$$
n(t) = n_I(t)\phi_1(t) + n_Q(t)\phi_2(t)
\tag{5.9-1c}
$$

where $n(t)$, $n_I(t)$, $n_Q(t)$ each have power spectral density $\eta/2 = \sigma^2$, the variance of the Gaussian density function. The received signal $r(t)$ is simply the sum of the transmitted signal and noise

$$
\begin{aligned}
r(t) = s_j(t) + n(t) &= r_1(t)\phi_1(t) + r_2(t)\phi_2(t) \\
&= [a_j(t) + n_I(t)]\phi_1(t) + [b_j(t) + n_Q(t)]\phi_2(t)
\end{aligned}
\tag{5.9-1d}
$$

The joint probability density of the received signal conditioned on message j being sent and the value of the random angle θ is the same as the noise probability density function

$$
P[r/j, \theta] = \frac{1}{2\pi\sigma^2}\exp\left[-\frac{1}{2\sigma^2}\left(n_I^2 + n_Q^2\right)\right]
\tag{5.9-2a}
$$

which, by Parseval's Theorem, is equal to

$$
P[r/j, \theta] = \frac{1}{2\pi\sigma^2}\exp\left[-\frac{1}{2\sigma^2}\int_0^{T_s}[r(t) - s_j(t)]^2\, dt\right]
\tag{5.9-2b}
$$

Since θ is random, it is appropriate that we use the expected value, with respect to θ, of the conditional probability. If we are to decide, based on the received signal $r(t)$, whether

$s_j(t)$ or $s_k(t)$ was sent, maximum likelihood theory tells us to decide $s_j(t)$ if

$$P_j E_\theta [P_r/j, \theta] > P_k E_\theta [P_r/k, \theta] \tag{5.9-3}$$

so the idea is devise a circuit that will implement the decision operation of eq. 5.9-3.

The integral in eq. 5.9-2b can be expanded as

$$\int_0^{T_s} [r(t) - s_j(t)]^2 \, dt = \int_0^{T_s} \left[r^2(t) - 2r(t)s_j(t) + s_j^2(t) \right] dt$$

$$= \left[\int_0^{T_s} r^2(t) \, dt + E_j \right] - 2 \int_0^{T_s} r(t)s_j(t) \, dt \tag{5.9-4a}$$

where it should be noted that the first part of the integral does not depend on θ but the second part does. It follows that eq. 5.9-2b can be rewritten as

$$P_r/j, \theta = \frac{1}{2\pi\sigma^2} \exp\left[-\frac{1}{2\sigma^2} \left\{ \int_0^{T_s} r^2(t) \, dt + E_j \right\} \right] \exp\left[\frac{1}{\sigma^2} \int_0^{T_s} r(t)s_j(t) \, dt \right] \tag{5.9-4b}$$

Dropping the multiplying constant and recognizing that θ is uniformly distributed over $[0, 2\pi]$, the expectation with respect to θ of eq. 5.9-2b is

$$E_\theta[P_r/j, \theta] = \exp\left[-\frac{1}{2\sigma^2} \left\{ \int_0^{T_s} r^2(t) \, dt + E_j \right\} \right] \left\{ \frac{1}{2\pi} \int_0^{2\pi} \exp\left[\frac{1}{\sigma^2} \int_0^{T_s} r(t)s_j(t) \, dt \right] d\theta \right\} \tag{5.9-4c}$$

Now consider the integral in the second part of eq. 5.9-4c.

$$\int_0^{T_s} r(t)s_j(t) \, dt = \int_0^{T_s} r(t)[a_j(t)\phi_1(t) + b_j(t)\phi_2(t)] \, dt \tag{5.9-4d}$$

which can be implemented by first doing a demodulation and lowpass filtering as shown in fig. 5.9-1. The demodulating signal is incoherent because it comes from a local oscillator

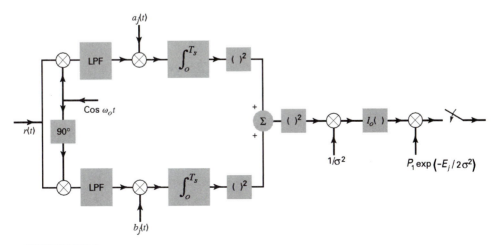

FIGURE 5.9-1

having arbitrary phase. The I and Q results are then correlated with $a_j(t)$ and $b_j(t)$ respectively. The result of this operation is of the form

$$\int_0^{T_s} r(t)s_j(t)\,dt = r_I \mathrm{Cos}\,\theta + r_Q \mathrm{Sin}\,\theta$$

$$= \sqrt{r_{I,j}^2 + r_{Q,j}^2}\,\mathrm{Cos}\left(\tan^{-1}\frac{r_Q}{r_I} + \theta\right)$$

$$= r_{O,j}\mathrm{Cos}\left(\tan^{-1}\frac{r_Q}{r_I} + \theta\right) \qquad (5.9\text{-}4e)$$

Note that in eq. 5.9-4c

$$\frac{1}{2\pi}\int_0^{2\pi}\exp\left[\frac{1}{\sigma^2}\int_0^{T_s} r(t)s_j(t)\,dt\right]d\theta$$

$$= \frac{1}{2\pi}\int_0^{2\pi}\exp\left[\frac{r_{O,j}}{\sigma^2}\mathrm{Cos}\left(\tan^{-1}\frac{r_{Q,j}}{r_{I,j}} + \theta\right)\right]d\theta = I_0\left(\frac{r_{O,j}}{\sigma^2}\right) \qquad (5.9\text{-}4f)$$

where $I_0(\alpha)$ is the 'zero-order modified Bessel function of the first kind'. $I_0(\alpha)$ is a monotonically increasing function for values of $\alpha > 0$ and $I_0(0) = 1$. As shown in fig. 5.9-1, the argument for the Bessel function is obtained by taking the square root of the sum of the squares and dividing the result by σ^2.

We may now rewrite eq. 5.9-4c, noting that it is reasonable to discard the integral involving only $r^2(t)$ because it has no dependence on either j or θ. We may also multiply by P_j in order to have the joint probability that figures into a decision:

$$P_j E_\theta[P_r/j, \theta] = P_j \exp\left[-\frac{E_j}{2\sigma^2}\right]I_0\left(\frac{r_O}{\sigma^2}\right) \qquad (5.9\text{-}5)$$

The circuit of fig. 5.9-1 also shows multiplication by the appropriate constants. This circuit focuses on the computation of the joint probability for a single transmitted signal s_j. Naturally, a communication system will have more than one possible transmitted signal and the idea is that a noncoherent receiver will have such a computation for each of these, as shown in fig. 5.9-2. The sampled outputs from each leg will be compared and

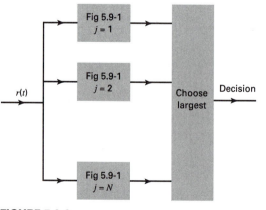

FIGURE 5.9-2

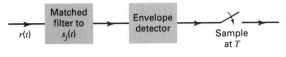

FIGURE 5.9-3

the largest of them chosen. In a system that has equal energy transmissions that are also equally probable it is not necessary to multiply by the constants in order to obtain the proper receiver.

The circuit of fig. 5.9-1 can also be more simply realized with the matched filter circuit of fig. 5.9-3. The proof of this is left to the Problems.

In general, noncoherent receivers perform less well than their coherent counterparts because the process of noncoherent detection involves the comparison of two or more noisy samples. That operation of comparison effectively increases the noise level against which a decision is being made. A very complete discussion of noncoherent receiver structures and their performance is given in [SHL95]. The interested reader may also consult [A&D62].

5.10 CONTINUOUS PHASE MODULATION

We have seen in Chapter 4 that MSK has the property of having phase continuity at the symbol boundaries. We have also seen that MSK is a special case of FSK with the index of modulation $h = 0.5$. It is also a member of the class of signals that are characterized by phase continuity at the symbol boundaries called CPM (Continuous Phase Modulation). A phase modulated carrier has a constant envelope, as distinct from QAM, which can be regarded as a combination of amplitude and phase modulation. Constant envelope signals are advantageous if transmitters are to be driven into nonlinearity. Phase continuity also results in relatively narrowband signals. A few of the key references are [A&S81], [Sun86], [BHMN81], [AAS86].

Following [Z&P85], we shall begin the analysis of this class of signals with the definition of $s(t)$ during the ith bit time as

$$s(t) = \sqrt{\frac{2E_s}{T_s}} \cos \left[\omega_c t + \frac{a_i[t - (i-1)T_s]\pi h}{T_s} + \pi h \sum_{j=1}^{i-1} a_j + \theta_0 \right] \quad (i-1)T_s \leq t \leq iT_s$$

$$(5.10\text{-}1)$$

where a_i is the ith piece of data, which, in principle, can take the values $\pm 1, \pm 3, \ldots, \pm (M-1)$, and M is typically a power of 2. Clearly this signal has continuous phase at the symbol boundaries for any value of h. Note that since phase is always modulo 2π, the accumulated phase will always lie in the region $\pm \pi$. Further, if h, the index of modulation, is the ratio of two integers then the set of possible phases will be finite. As can be seen from fig. 5.10-1, the evolution of the phases at the symbol boundaries forms a trellis and a particular data pattern corresponds to a path through that trellis. The continuity of the phase introduces memory into the system. This will have a major impact on how the signal is detected at the receiver.

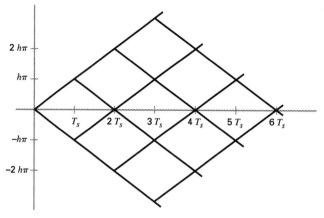

FIGURE 5.10-1

Equation 5.10-1 can be generalized as

$$s(t) = \sqrt{\frac{2E_s}{T_s}} \text{Cos}\left[\omega_c t + a_i \pi h q(t) + \pi h \sum_{j=1}^{i-1} a_i + \theta_0\right] \quad (i-1)T_s \le t \le iT_s$$

$$(5.10\text{-}2a)$$

where

$$q(t) = \int_{-\infty}^{t} g(\lambda - (n-1)T_s)\, d\lambda \qquad (5.10\text{-}2b)$$

and $g(t)$ is a frequency pulse shape function. An even more important generalization is

$$s(t) = \sqrt{\frac{2E_s}{T_s}} \cos[\omega_c t + \varphi_n(t, \alpha_n)] \qquad \text{for all } t \qquad (5.10\text{-}2c)$$

where the subscript n refers to a specific data sequence denoted by α_n.

In order to make eq. 5.10-2a equivalent to eq. 5.10-1, we may define (fig. 5.10-2a)

$$g(t) = \begin{cases} 1/2T_s; & 0 \le t \le T_s \\ 0; & \text{otherwise} \end{cases} \qquad (5.10\text{-}2d)$$

This $g(t)$ occupies one symbol interval and consequently is known as a *full-response* pulse-shape function. Note that $g(t)$ has been scaled so that $q(T_s) = 1/2$. There are many functions

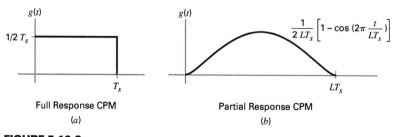

Full Response CPM
(a)

Partial Response CPM
(b)

FIGURE 5.10-2

that meet the full response requirements. We shall consider only $g(t)$ as given in eq. 5.10-2d above.

It is also possible to have $g(t)$ extend over more than one symbol interval. One example among many is the raised cosine pulse extending over L symbol intervals (fig. 5.10-2b):

$$g(t) = \begin{cases} 1/2LT_s[1 - Cos(2\pi t/LT_s)]; & 0 \le t \le LT_s \\ 0; & \text{otherwise} \end{cases} \qquad (5.10\text{-}2e)$$

Note that

$$q(LT_s) = \int_{-\infty}^{LT_s} g(\lambda)\, d\lambda = \frac{1}{2} \qquad (5.10\text{-}2f)$$

Such pulses generate *partial-response* CPM. The reader should be cautioned against attempting to find actual similarities between partial-response CPM and partial-response signaling as discussed in Chapter 3. The same name has been given to two different things so context is everything in making the proper distinctions. The variety of partial response functions and their performance is described in [Sun86], which also provides an extensive list of references. The basic reason for the use of partial-response CPM is that the phase transitions at the symbol boundaries are smoother than the corresponding transitions in the full-response system, leading to power spectral densities that fall off faster outside of the main lobe. This is, of course, advantageous in bandwidth-constrained systems.

We will give a relatively brief discussion of CPM, exclusively dealing with the full response system of eq. 5.10-2d, with the purpose of showing how signal space and maximum likelihood techniques can be used to analyze the performance and conceptualize the receiver structures for this modulation technique.

5.10.1   The Relationship Between FSK and CPM

In understanding the relationship between FSK and CPM we should consider the issue of orthogonality as well as that of phase continuity. We have already seen that the condition of orthogonality is that there are an integer number of half cycles of each one of the transmitted frequencies in a symbol interval T_s, corresponding to the index of modulation equal to $h = 1/2, 1, 3/2, \ldots$ Accordingly, CPM transmission that does not correspond to these values of h does not constitute orthogonal signaling. Orthogonality does not guarantee phase continuity. Figure 5.10-3a shows an FSK signal having three half cycles in one bit period and two half cycles in another. This renders the two signals orthogonal, yet they are clearly not phase-continuous. Figure 5.10-3b shows the same two frequencies, but oriented so that their phases are continuous. It is also the case that orthogonality is not a prerequisite for phase continuity. A continuous-phase FSK signal having any prescribed index of modulation h may easily be generated either by analog or digital means.

The basic issue is that a continuous-phase signal will have a more compact spectrum that one having a discontinuous phase. In the FSK example of Chapter 1 the continuous-

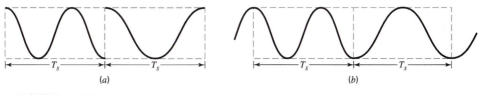

T_s T_s T_s T_s

(a)   (b)

FIGURE 5.10-3

phase signal had a rolloff in magnitude of the order of $1/n^2$ outside the main band, whereas a discontinuous-phase equivalent would have a rolloff of the order of $1/n$. Some examples and more detailed discussion of this issue can be found in [B&D65] and [AAS86]. CPM corresponds to CPFSK (Continuous-Phase FSK) when the frequency pulse shape function $g(t)$ is the full response rectangular pulse of eq. 5.10-2d.

The spectral analysis of CPM is not easily amenable to analytic approaches. Numerical methods and measurement have become the preferred techniques for estimating the bandwidth of these signals. Interested readers should consult [Sun86], [AAS86], and [Z&P85].

5.10.2   Euclidean Distance and Error Probability in CPM

Recall that our discussion of the signal-space approach led to the conclusion that the Euclidean distance between two data signals $s_n(t)$ and $s_m(t)$ in signal space is given by

$$D_{mn}^2 = \int_0^{NT_s} [s_n(t) - s_m(t)]^2 \, dt \qquad (5.10\text{-}3a)$$

where we have chosen NT_s rather than T_s as the upper limit because the CPM signal has memory. Continuing:

$$D_{mn}^2 = \int_0^{NT_s} \left[s_n^2(t) + s_m^2(t) - 2s_n(t)s_m(t) \right] dt$$

$$= 2 \left\{ \int_0^{NT_s} s_n^2(t) \, dt - \int_0^{NT_s} s_n(t)s_m(t) \, dt \right\} \qquad (5.10\text{-}3b)$$

From eq. 5.10-2c, it follows that

$$\int_0^{NT_s} s_n^2(t) \, dt = (NT_s)\left(\frac{1}{2}\right)\left(\frac{2E_s}{T_s}\right) = NE_s \qquad (5.10\text{-}3c)$$

and

$$\int_0^{NT_s} s_n(t)s_m(t) \, dt = \frac{2E_s}{T_s} \int_0^{NT_s} \mathrm{Cos}[\omega_c t + \varphi_n(t, \alpha_n)]\mathrm{Cos}[\omega_c t + \varphi_m(t, \alpha_m)] \, dt$$

$$= \frac{E_s}{T_s} \int_0^{NT_s} \mathrm{Cos}[\varphi_n(t, \alpha_n) - \varphi_m(t, \alpha_m)] \, dt$$

$$+ \frac{2E_s}{T_s} \int_0^{NT_s} \mathrm{Cos}[2\omega_c t + \varphi_n(t, \alpha_n) + \varphi_m(t, \alpha_m)] \, dt \qquad (5.10\text{-}3d)$$

where the second, double frequency, integral goes to zero. It therefore follows that

$$D_{mn}^2 = 2NE_s \left\{ 1 - \frac{1}{NT_s} \int_0^{NT_s} \mathrm{Cos}[\varphi_n(t, \alpha_n) - \varphi_m(t, \alpha_m)] \, dt \right\} \qquad (5.10\text{-}3e)$$

Actual computation of this quantity is a complicated task because it depends on the pulse $g(t)$, the value of the index of modulation h, and the number of signal levels in the data sequence. The basic idea for $g(t)$ of eq. 5.10-2d and binary data is shown in fig. 5.10-1. Keeping in mind that we are trying to find the square of the minimum distance between the trajectories corresponding to two different data streams, let us have the two trajectories start from the same point at $t = 0$ but with two different pieces of data, $+1$ and -1. As a consequence the two paths go to $\pm h\pi$. Since the paths can then converge with the next piece of data, $2T_s$ is the duration of the minimum distance path. The minimum distance then becomes

$$D_{\min}^2 = 4E_s \left\{ 1 - \frac{1}{2T_s} \left[\int_0^{2T_s} \text{Cos}[2\pi h(2q(t) - 2q(t - T_s))] \, dt \right] \right\}$$

$$= 4E_s \left\{ 1 - \frac{1}{2T_s} \left[\int_0^{T_s} \text{Cos}\left[\frac{2\pi ht}{T_s} \right] dt + \int_{T_s}^{2T_s} \text{Cos}\left[\frac{2\pi h(2 - t)}{T_s} \right] dt \right] \right\}$$

$$= 4E_s \left\{ 1 - \frac{\text{Sin}(2\pi h)}{2\pi h} \right\} \tag{5.10-4}$$

For $h < 1/2$ the minimum distance increases as h increases. As h continues to increase, the computation of the minimum distance becomes more complicated because the phase difference must be measured modulo 2π and the actual minimum distance will be less. Considerable effort has been expended in finding exact values for this minimum distance and the interested reader is referred to [Rim91] and [E&L91]. Consequently, the expression of eq. 5.10-4 is generally an upper bound to the actual minimum distance. Note that with a signaling alphabet larger than the binary, the minimum distance would be greater.

The probability of error in this system is, according to signal space techniques, given by eq. 5.8-13:

$$P_e = kQ\left(\sqrt{\frac{d_{\min}^2}{2\eta}} \right)$$

In those circumstances where the minimum distance is an upper bound, then the error is a lower bound.

5.10.3   The Optimum Receiver for CPM

Having determined the Euclidean distance for a CPM signal, we now want to point the interested reader in the direction of optimum receiver structures. The essential issue to keep in mind is that CPM signals have memory and their detection must therefore involve examination of a sequence of received data points rather than bit-by-bit detection. There are two such approaches that are consistent with the discussion of this chapter.

First, Osborne and Luntz [O&L741] use a classic MLSE approach in which they store all possible sequences of length N and make sequence comparison. They make a decision on bit a_i only after observing the data at the ith interval up until the $(i + N)$th interval. Although N should theoretically be infinite, practically it should be "large enough." They correlate the received signal over NT_s with each of the possible N-symbol waveforms and choose the most likely. This makes for a very cumbersome receiver structure even for moderate N. Consequently much attention has been expended on finding suboptimum versions of the receiver that require less extensive computation.

As was pointed out in the previous section, the fact that a CPM signal has memory and can be detected using maximum likelihood techniques means that it is possible to apply the Viterbi Algorithm to this problem. Aulin, Rydbeck and Sundberg [A&S81] and [AAS86] discuss this approach in great detail.

Both of these approaches are quite involved and are outside the main thread of this text.

5.10.4   A Brief Word About Multi-h CPM

We have introduced a class of CPM systems, both full response and partial response, that are characterized by a single value of the modulation index $h = k/p$, where k and p are integers. The Euclidean distance between transmitted signals corresponding to different data values has been shown to be dependent on the value of h. It therefore follows that choosing h to maximize that distance will correspondingly minimize the probability of error.

Multi-h CPM defines a sequence of values for h, $[h_1, h_2, \ldots, h_N]$ that are cycled through periodically for successive transmissions. It turns out that the Euclidean distance in multi-h between possible transmitted sequences is larger than for single-h transmission, so that the probability of error is correspondingly less.

The analysis of multi-h CPM is quite complex and is beyond the scope of this text. The interested reader is referred to [BHMN81], [AAS86], [Z&P85].

CHAPTER 5 PROBLEMS

5.1   A three-dimensional Euclidean space has an orthonormal set of basis vectors $\mathbf{u}_X$, $\mathbf{u}_Y$, $\mathbf{u}_Z$, unit vectors along the X, Y, Z axes. Two vectors defined in this space are:

$$\mathbf{s} = \begin{bmatrix} 1 \\ -2 \\ 3 \end{bmatrix} \quad \text{and} \quad \mathbf{t} = \begin{bmatrix} -2 \\ -1 \\ 4 \end{bmatrix}$$

(a) Find the inner product of these two vectors.

(b) Find the cosine of the angle between the two vectors.

(c) Find the distance between the endpoints of the vectors.

5.2   A four-dimensional Euclidean space has an orthonormal set of basis vectors $\mathbf{u}_W$, $\mathbf{u}_X$, $\mathbf{u}_Y$, $\mathbf{u}_Z$, unit vectors along the W, X, Y, Z axes. Two vectors defined in this space are:

$$\mathbf{s} = \begin{bmatrix} 2 \\ -1 \\ -1 \\ 3 \end{bmatrix} \quad \text{and} \quad \mathbf{t} = \begin{bmatrix} -3 \\ -2 \\ -1 \\ 2 \end{bmatrix}$$

Repeat (a), (b), (c) from Problem 5-1.

5.3   Construct a matrix A that transforms the orthonormal rectangular basis functions of Problem 5-1 to an orthonormal set of cylindrical coordinates r, θ, z. Consult a textbook on electromagnetic theory or vector analysis to find the definition of cylindrical coordinates.

Show that this is a unitary transformation.

5.4   Repeat Problem 3 for spherical coordinates r, θ, φ.

5.5   In Problem 4, let $\theta = 45°$, $\varphi = 30°$. Represent the vector $\mathbf{s}$ of Problem 1 in terms of this new set of basis vectors.

5.6   (a) Using the Gram-Schmidt approach, construct an orthonormal set of basis vectors from

$$\mathbf{r} = \begin{bmatrix} -1 \\ 2 \\ -2 \end{bmatrix} \qquad \mathbf{s} = \begin{bmatrix} 1 \\ -2 \\ 3 \end{bmatrix} \qquad \mathbf{t} = \begin{bmatrix} -2 \\ -1 \\ 4 \end{bmatrix}$$

(b) Assuming that the original basis vectors for part (a) were $\mathbf{u}_X$, $\mathbf{u}_Y$, $\mathbf{u}_Z$, unit vectors along the X, Y, Z axes, find the matrix A that transforms $\mathbf{u}_X$, $\mathbf{u}_Y$, $\mathbf{u}_Z$, to the new basis.

(c) Show that the matrix A is unitary.

5.7   (a) Find the values of A, B, C that make $f_1(t)$ and $f_2(t)$ orthonormal over the interval $t \in [0, \infty]$.

$$f_1(t) = Ae^{-t}u(t) \qquad f_2(t) = Be^{-t}u(t) + Ce^{-2t}u(t)$$

(b) Using eq. 5.1-13b, find the values of a_1 and a_2 to find the series

$$\hat{f}(t) = a_1 f_1(t) + a_2 f_2(t)$$

that approximates the function $f(t) = 3e^{-3t/2}u(t)$.

(c) Find the energy contained in $f(t)$ and in $\hat{f}(t)$ and the percent difference between the two.

5.8   (a) We will first consider the impulse reponse of the brick-wall filter of fig. 3.9-2, which is

$$h_1(t) = \text{Sinc}\left(\pi \frac{t}{T_s}\right)$$

Show that the set of functions $h_1(t - nT_s)$ are orthogonal over the interval $[0, T_s]$. Note that Parseval's Theorem allows us to operate in the frequency domain as well as in the time domain.

(b) Now consider the impulse response of the raised cosine filter of fig. 3.9-3, which is given by

$$h_2(t) = \text{Sinc}\left(\pi \frac{t}{T_s}\right) \frac{\text{Cos}\left(\alpha\pi \frac{t}{T_s}\right)}{1 - \left(2\alpha \frac{t}{T_s}\right)^2}$$

and the set of functions $h_2(t - nT_s)$. Find the inner product, as a function of α, for the functions corresponding to $n = 0, 1, 2$.

5.9   A zero-mean random process has a time-average autocorrelation function $\bar{R}_x(\tau) = e^{-2|\tau|}$ that is sampled at $t = 1, 2, 3, 4, 5$.

(a) Find the resulting covariance matrix.

(b) Using an appropriate computer tool, find the eigenvalues and the matrix of eigenvectors. Show that the matrix of eigenvectors is unitary.

(c) For this example, confirm eq. 5.2-4c, Mercer's Theorem.

5.10   The signal set in a 16-QAM system is defined by the set of vectors

$$\mathbf{s} = \frac{1}{\sqrt{10}} \begin{bmatrix} a \\ b \end{bmatrix} \qquad \text{where} \quad a, b = \pm 1, \pm 3.$$

Find the distance and the correlation between

$$\mathbf{s}_1 = \frac{1}{\sqrt{10}} \begin{bmatrix} 1 \\ -3 \end{bmatrix} \qquad \text{and} \qquad \mathbf{s}_2 = \frac{1}{\sqrt{10}} \begin{bmatrix} -1 \\ 1 \end{bmatrix}$$

5.11   Using any one of the signal points in an 8-PSK system as the reference, find the distance and correlation between the reference point and all other points in the space. Assume that the energy in each signal point is normalized to unity.

5.12   The functions $u_1(t)$ and $u_2(t)$ are used as basis functions for data transmission according to eq. 5.3-1 so that the transmitted signal is

$$s(t) = a_1 u_1(t) + a_2 u_2(t)$$

where $a_1, a_2 = \pm 1$.

(a) Demonstrate that $u_1(t)$ and $u_2(t)$ are orthogonal.

(b) In a matched filter receiver, show the output of each of the filters to each of the basis functions. How do these outputs correspond to the orthogonality of these basis functions?

(c) Determine the correlation of and the distance between the basis functions in signal space.

(d) Compare the signal space characterization of this system to QPSK.

(e) How should the system be modified in order to make it congruent, in a signal-space manner, to FSK?

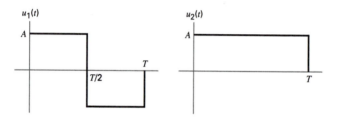

5.13   Consider the two basis functions shown below

(a) Determine the correlation of and the distance between the basis functions in signal space. Show these functions as points in signal space.

(b) If these functions are used for binary transmission in the presence of noise having p.s.d. equal to $\eta/2$, then find the probability of error.

(c) Use the Gram-Schmidt method to find an orthonormal set of basis vectors from these two.

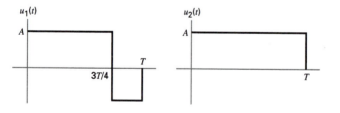

5.14   We have seen in eq. 5.4-12b that for BFSK the correlation of the two signal vectors is

$$\rho_{12} = \frac{\mathrm{Sin}(\Delta\omega T_s)}{(\Delta\omega T_s)}$$

(a) As $\Delta\omega$ increases from zero, sketch the trajectory of the basis vector $\phi_2(t)$ under the assumption that $\phi_1(t)$ remains oriented along the horizontal positive axis. Find the value of $\Delta\omega$ that corresponds to the minimum value of ρ_{12} and determine the position of that vector. Find the probability of error for this circumstance.

(b) Devise an optimal receiver structure for this circumstance.

5.15   Devise an optimum receiver structure for the orthogonal signaling system based on transmitting either $p_0(t)$ or $p_1(t)$ from fig. 5.1-2. Compute the probability of error.

5.16   Find the signal vectors that are produced by rotating those of eq. 5.4-13a by 45°.

5.17 Consider the expression for the probability of error in bi-orthogonal signaling

$$P_e = 2Q\left[\sqrt{\rho_M/2}\right] + Q\left[\sqrt{\rho_M}\right]$$

and let $Q[\sqrt{\rho_M}] = 10^{-6}$. Find P_e and discuss which terms are dominant. Repeat for $Q[\sqrt{\rho_M}] = 10^{-3}$.

5.18 Show that when $h = 1/2$ in eq. 5.4-12d, this corresponds to MSK signaling.

5.19 Show that the simplex signal set corresponding to $M = 4$ is three dimensional.

5.20 Find upper bounds to the probability of error in 16-PSK and 32-PSK systems.

5.21 Write a program in that implements the Viterbi Algorithm for the example in eq. 5.8-6. Let the buffer length L be entered as input variable L. Structure the program so that it may be used for any four-state Mealy machine by entering the outputs for the state transition/output table as input variables as well. A clever programmer will use pointers to lookup tables to implement the state machine. Use this program to verify the results of the example.

Create an experiment that illustrates the performance of this VA receiver for a noisy signal as the buffer length L increases.

Compare the optimal VA performance with binary signaling on a distortionless channel. (Note that the channel taps should be normalized so that the sum of their squares equals unity.)

5.22 It has been asserted in the text that the use of the Viterbi Algorithm in the receiver of a partial response system will recover the 2.1 db of performance lost by partial response in comparison to simple binary signaling. This can be illustrated by modifying the program developed in Problem 5-18 to have partial response transmission and a VA based on a state machine representation of that transmission. Devise an experiment that checks this assertion for Class-1 partial response.

5.23 This problem will demonstrate that MLSE can be used on noncausal channel models. Allow the channel impulse response vector to be $F = [-.2 \ 1 \ -.3 \ +.2]$. Find the Mealy machine model of this system and use the Viterbi Algorithm as the receiver. Expand the program to accommodate eight states in order to do this problem easily. Test the performance in relation to ideal binary signaling.

5.24 **(a)** Find a Mealy-machine representation of the channel described by $F = [1 \ -0.3]$ where we have four possible inputs $a_K = \pm 1, \pm 3$, normalized to having an average energy per transmitted symbol equal to unity.

(b) Modify your VA program to accommodate this circumstance.

(c) Use the program to compare the performance of the VA to that of ideal four-level baseband signaling.

5.25 Prove that the circuit of fig. 5.9-1 can be realized with the circuit of fig. 5.9-3.

REFERENCES

[A&D62]   E. Arthurs, and H. Dym, "On the optimum detection of digital signals in the presence of white Gaussian noise – a geometric interpretation and a study of three basic data transmission systems," *IRE Transactions on Communication Systems*, **CS-10**, No. 4, Dec. 1962, pp. 336–372.

[A&S81]   T. Aulin and C.-E. Sundberg, "Continuous phase modulation—part I: full response signaling," and "Continuous phase modulation—part II: partial response signaling," *IEEE Trans. Communications*, **COM-29**, No. 3, March 1981, pp. 196–225.

[A&S82]   T. Aulin and C.-E. Sundberg, "On the minimum Euclidean distance for a class of signal space codes," *IEEE Trans. on Information Theory*, **IT-28**, No. 1, Jan. 1982, pp. 43–55.

[AAS86]   J.B. Anderson, T. Aulin, and C.-E. Sundberg, *Digital Phase Modulation*, Plenum, 1986.

[B&D65]   W.R. Bennett and J.R. Davey, *Data Transmission*, McGraw-Hill, 1965.

[BHMN81]   V.K. Bhargava, D. Haccoun, R. Matyas, and P. Nuspl, *Digital Communications by Satellite*, Wiley-Interscience, 1981.

[Corr73]   M.S. Corrington, "Solution of differential and integral equations with Walsh functions," *IEEE Transactions on Circuit Theory*, **CT-20**, No. 5, Sept. 1971, pp. 470–476.

[D&R58]   W. Davenport and W. Root, *Random Signals and Noise*, McGraw-Hill, 1958.

[E&L91]   N. Ekanayake and R. Liyanapathirana, "On the exact formula for the minimum squared Euclidean distance of CPFSK," *IEEE Trans. on Communications*, **COM-42**, No. 11, Sept. 1991, pp. 2917–2918.

[For72]   G.D. Forney, "Maximum likelihood sequence estimation of digital sequences in the presence of intersymbol interference," *IEEE Trans. Information Theory*, **IT-18**, May 1972, pp. 363–378.

[For73]   G.D. Forney, "The viterbi algorithm," *Proc. IEEE*, **61**, March 1973, pp. 268–278.

[Gup76]   S.C. Gupta, "Transform and state variable methods in linear systems," John Wiley and Sons, 1976.

[L-G94]   A. Leon-Garcia, *Probability and Random Processes for Electrical Engineering*, Addison-Wesley, 1994.

[O&L74]   W. Osborne and M. Luntz, "Coherent and noncoherent detection of CPFSK," *IEEE Trans. on Communications*, **COM-22**, No. 8, Aug. 1974, pp. 1023–1036.

[Pro89]   J. Proakis, *Digital Communications*, 2nd ed., McGraw-Hill, 1989.

[Rim91]   B. Rimoldi, "Exact formula for the minimum squared Euclidean distance of CPFSK," *IEEE Trans. on Communications*, **COM-39**, No. 9, Sept. 1991, pp. 1280–1282.

[SHL95]   M.K. Simon, S.M. Hinedi, and W.C. Lindsay, *Digital Communication Techniques*, Prentice Hall, 1995.

[Sun86]   C.-E. Sundberg, "Continuous phase modulation," *IEEE Communications Magazine*, **24**, No. 4, April 1986, pp. 25–38.

[Wal23]   J.L. Walsh, "A closed set of normal orthogonal functions," *Amer. J. Math.*, **45**, pp. 5–24, 1923.

[Z&P85]   R. Ziemer and R. Peterson, *Digital Communications and Spread Spectrum Systems*, Macmillan, 1985.

CARRIER PHASE AND TIMING RECOVERY

In Chapters 3 and 4 we addressed some of the basic ideas of baseband and passband QAM type systems, including their signal structure and their theoretical performance in the presence of noise. It was understood that the received signal was being sampled at the proper instant (the instant of sampling is called the timing or sampling phase) and, in the case of bandpass systems, that the frequency and phase of the demodulating carrier was correct so that the constellation of the received signal was properly oriented in relation to the slicer.

We examined square-law timing recovery for baseband signals and forward-acting carrier-phase recovery for bandpass signals in Chapter 2 and discovered that, even if the data signal is noiseless, its statistical properties give rise to jitter in the recovered timing phase or carrier phase. This jitter makes its way through the system to appear as an additional component of noise that degrades system performance. In carrier systems the phase of the received signal will differ from that of the transmitted signal by

$$\theta(t) = \theta_0 + \Delta\omega t + \theta_j \text{Cos}(\omega_j t) \tag{6.0-1}$$

where θ_0 is a constant phase offset that comes from the distance between transmitter and receiver, $\Delta\omega$ is a frequency offset that represents the difference in frequency between the transmitter and receiver oscillators, and the third component is carrier-phase jitter, typically resulting from power-line frequencies and their harmonics phase modulating their way into the system through nonlinearities. In the United States we are concerned with a power-line frequency of 60 hz and its harmonic of 120 hz; in Europe the power line frequency is 50 hz. A carrier-phase recovery system has the job of learning $\theta(t)$ and compensating for it.

In this chapter we are more interested in *decision-directed* methods of carrier and timing-phase recovery because such methods avoid the data dependency of the forward-acting methods. In order to analyze these methods it is first necessary develop some basic ideas about phase-locked loops.

6.1 THE CLASSIC PHASE-LOCKED LOOP

A block diagram of the classic analog phase-locked loop (PLL) is shown in fig. 6.1-1. Both the input signal and the output of the VCO (voltage-controlled oscillator) are considered to be sinusoids:

$$x_i(t) = A_i \text{Cos}(\omega_c t + \theta_i(t))$$
$$x_0(t) = A_0 \text{Sin}(\omega_c t + \theta_0(t)) \tag{6.1-1}$$

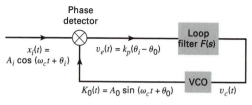

Phase detector

$x_i(t) = A_i \cos(\omega_c t + \theta_i)$

$v_e(t) = k_p(\theta_i - \theta_0)$

Loop filter $F(s)$

VCO

$K_0(t) = A_0 \sin(\omega_c t + \theta_0)$ $v_c(t)$

FIGURE 6.1-1

In some PLLs, these signals may be hard-limited and appear more as square waves than sinusoids. Since the function of the phase detector is to determine the phase difference

$$\theta_e(t) = \theta_i(t) - \theta_0(t) \tag{6.1-2}$$

between the two signals, the important issue is not whether the signals of eq. 6.1-1 are sinusoidal or square, but rather that the phase detector is appropriate to the signals. Note that, as defined, the individual phases and the phase difference may vary with time.

The loop is considered to be 'locked' or in the steady state when the phase difference has settled to zero. The phase detector operates in such a manner as to make it important that one of the signals be represented as a Cos and the other as a Sin.

The transfer characteristic of an ideal phase detector is shown in fig. 6.1-2. Central to the understanding of this phase detector being ideal is that we can measure phase only modulo 2π. Consequently even the ideal phase detector is inherently nonlinear. We should also note that the output of the phase detector is a voltage

$$v_e(t) = k_p \theta_e(t) \tag{6.1-3}$$

An example of a real phase detector is shown in fig. 6.1-3a. The two signals $x_i(t)$ and $x_0(t)$ are squared up (amplified and hard-limited) and applied to an exclusive-OR gate followed by a lowpass filter. It can be seen that when the two inputs are exactly 90 degrees out of phase, Cos and Sin, the output of the exclusive-OR gate is symmetric so that the average value (dc component) is zero. The lowpass filter output is therefore equal to zero. As the phase difference between the two waveforms deviates from 90 degrees, the average value of the XOR gate output waveform increases linearly with the phase difference deviation. As shown in fig. 6.1-3b, the XOR phase detector differs from the ideal when the phase difference exceeds $\pm\pi/2$.

The VCO (voltage-controlled oscillator) is an oscillator whose frequency is linearly controlled, within a given range, by an input voltage. There are many different kinds of circuits that can accomplish this task. The VCO is biased with a dc voltage that causes the

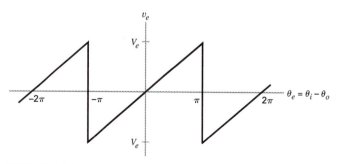

FIGURE 6.1-2

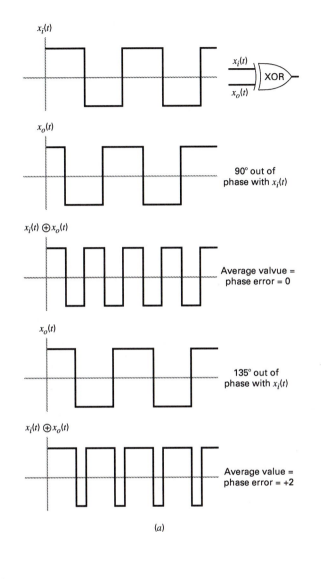

(a)

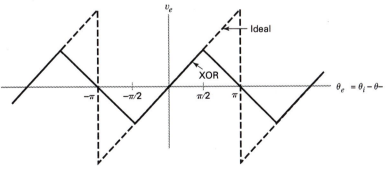

(b)

FIGURE 6.1-3

output frequency nominally to correspond to the anticipated frequency of the input signal. The oscillator frequency may then be varied up or down by externally applied incremental voltages. Referring to fig. 6.1-1, the transfer characteristic of the VCO is

$$\nabla f(t) = k_f v_0(t) \tag{6.1-4}$$

Since phase is the integral of frequency, it follows that the incremental output phase of the VCO is given by

$$\theta_0(t) = \int_{-\infty}^{t} k_f v_0(\tau) \, d\tau \tag{6.1-5}$$

or, in the Laplace Transform domain,

$$\Theta_0(s) = k_f \frac{V_0(s)}{s} \tag{6.1-6}$$

Dynamically, the VCO acts as an integrator, a function which, in a digital PLL, is performed by accumulation. Finally, the PLL contains a loop filter $F(s)$, as shown in fig. 6.1-1.

For the remainder of this analysis it will be assumed that the phase difference of eq. 6.1-2 is sufficiently small that the loop is operating entirely within the linear range of the phase detector. Under these conditions, the PLL can be represented as the simple feedback system of fig. 6.1-4. Note that the input signal and the VCO output are represented not as sinusoids, but as the Laplace Transforms of their phases.

With a bit of algebra, the three basic transfer functions of this simple control system can be easily derived:

The Closed-Loop Transfer Function:

$$\frac{\Theta_0(s)}{\Theta_i(s)} = \frac{k_p k_f F(s)}{s + k_p k_f F(s)} = H(s) \tag{6.1-7a}$$

The Error-Transfer Function:

$$\frac{\Theta_e(s)}{\Theta_i(s)} = \frac{\Theta_i(s) - \Theta_0(s)}{\Theta_i(s)} = \frac{s}{s + k_p k_f F(s)} = 1 - H(s) \tag{6.1-7b}$$

The Loop Filter Output:

$$\frac{V_0(s)}{\Theta_i(s)} = \frac{s}{k_f} \frac{k_p k_f F(s)}{s + k_p k_f F(s)} = \frac{s}{k_f} H(s) \tag{6.1-7c}$$

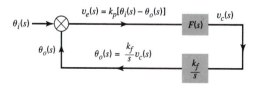

FIGURE 6.1-4

The *order of the PLL* is the order of the denominator of $H(s)$, the number of poles in the closed loop transfer function.

We shall do a few important examples in order to develop an understanding of the linear dynamics of this loop. In order to do these examples, recall that:

Laplace Transform of a Step in Phase:

$$\theta(t) = \nabla\phi u(t) \Rightarrow \Theta(s) = \nabla\phi/s \qquad (6.1\text{-}8a)$$

Laplace Transform of a Step in Frequency (a ramp in phase):

$$\theta(t) = \nabla\omega r(t) \Rightarrow \Theta(s) = \nabla\omega/s^2 \qquad (6.1\text{-}8b)$$

The Final Value Theorem:

$$\lim f(t)|_{t\to\infty} = \lim s F(s)|_{s\to 0} \qquad (6.1\text{-}8c)$$

We will examine the response of the loop for three versions of $F(s)$, the loop filter. The first is $F(s) = 1$, no loop filter at all. (Note that proper design will assure that the lowpass filter associated with the phase detector has a sufficiently large bandwidth as closely to approximate the condition of no loop filter at all.) The second is $F(s)$ simply equal to a lowpass RC circuit and the final one of fig. 6.1-5 is an active RC circuit.

$F(s) = 1$ **(a first-order loop).** From eq. 6.1-7 and eq. 6.1-8a, we have

$$\frac{\Theta_0(s)}{\Theta_i(s)} = \frac{k_f k_p}{s + k_f k_p} = H(s) \qquad (6.1\text{-}9a)$$

$$\Theta_e(s) = \frac{s}{s + k_f k_p}\Theta_i(s) = \frac{s}{s + 1/\tau}\Theta_i(s) \qquad (6.1\text{-}9b)$$

If we consider $H(s)$ to be a filter, then it is readily seen that the equivalent one-sided noise bandwidth of this filter is

$$B_L = \frac{k_f k_p}{4} \text{ hz} \qquad (6.1\text{-}9c)$$

For a step input of phase $\Theta_i(s) = \nabla\phi/s$,

$$\Theta_e(s) = \frac{s}{s + k_p k_f}\Theta_i(s) = \frac{s}{s + 1/\tau}\frac{\nabla\phi}{s}$$

$$\Rightarrow \theta_e(t) = \nabla\phi e^{-t/\tau}u(t) \qquad (6.1\text{-}9d)$$

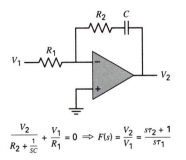

$$\frac{V_2}{R_2 + \frac{1}{sc}} + \frac{V_1}{R_1} = 0 \Rightarrow F(s) = \frac{V_2}{V_1} = \frac{s\tau_2 + 1}{s\tau_1}$$

FIGURE 6.1-5

and similarly

$$\theta_0(t) = \nabla\phi\left[1 - e^{-t/\tau}\right]u(t) \tag{6.1-9e}$$

Note that in the steady state the phase error goes to zero and the output phase equals the input phase.

When the input is a step in frequency or ramp in phase $\Theta_i(s) = \nabla\omega/s^2$,

$$\Theta_e(s) = \frac{s}{s + k_p k_f}\Theta_i(s) = \frac{s}{s + 1/\tau}\frac{\nabla\omega}{s^2}$$

$$\Rightarrow \theta_e(t) = \nabla\omega \cdot \tau\left[1 - e^{-t/\tau}\right]u(t) \tag{6.1-10a}$$

and

$$\theta_0(t) = \nabla\omega \cdot (t - \tau)u(t) + \nabla\omega \cdot \tau e^{-t/\tau}u(t) \tag{6.1-10b}$$

In this instance the steady-state phase error is not equal to zero, but equals $\theta_e(\infty) = \nabla\omega \cdot \tau$, and the steady-state output phase correspondingly lags behind the input phase. To clarify, there must be a steady-state phase error in this system in order to get the VCO to have a steady-state change in frequency to match the change in the input frequency.

This is a major problem because it limits the ability of the loop to track frequency offsets. As the frequency offset gets large, the steady-state error pushes the phase detector toward its nonlinear region and nonlinear oscillations called cycle-slipping. In order to counteract this problem, it is possible to decrease the value of τ by increasing the phase detector and VCO gains. This speeds up the transient response of the loop and also increases the noise bandwidth.

$F(s) = 1/(1 + s\,RC)$ **(a second-order loop).** From eq. 6.1-7b,

$$\Theta_e(s) = \frac{s}{s + \dfrac{k_p k_f}{1 + sRC}}\Theta_i(s) = \frac{s(s + 2d\omega_n)}{s^2 + 2d\omega_n s + \omega_n^2}\Theta_i(s) \tag{6.1-11a}$$

where $\omega_n^2 = \dfrac{k_p k_f}{RC}$ and $d = \dfrac{\omega_n}{2k_p k_f}$. Similarly, from eq. 6.1-7a

$$\Theta_0(s) = \frac{\omega_n^2}{s^2 + 2d\omega_n s + \omega_n^2}\Theta_i(s) \tag{6.1-11b}$$

and again the equivalent one-sided noise bandwidth of the loop is $B_L = (k_p k_f/4)$ hz.

Using the final value theorem of Laplace Transforms, we can determine the steady state values of the phase error when the input is either a step in phase or a step in frequency.

$$\text{Phase Step:} \quad \theta_e(\infty) = s\frac{s(s + 2d\omega_n)}{s^2 + 2d\omega_n s + \omega_n^2}\frac{\nabla\phi}{s}\bigg|_{s=0} = 0 \tag{6.1-11c}$$

$$\text{Frequency Step:} \quad \theta_e(\infty) = s\frac{s(s + 2d\omega_n)}{s^2 + 2d\omega_n s + \omega_n^2}\frac{\nabla\omega}{s^2}\bigg|_{s=0} = \frac{2d}{\omega_n}\nabla\omega = \frac{\nabla\omega}{k_p k_f} \tag{6.1-11d}$$

Note that the steady-state error for the frequency step is the same as for the first-order loop. The difference between the first- and second-order loop is that the transient response of the second-order loop is faster than that of the first, is oscillatory, and has as an extra control parameter the filter time constant RC.

$F(s) = \dfrac{s\tau_2 + 1}{s\tau_1}$ **(second-order loop, fig. 6.1-5).** We can address the issue of response to frequency offset with an active filter:

$$\Theta_e(s) = \frac{s^2}{s^2 + 2d\omega_n s + \omega_n^2}\Theta_i(s) \tag{6.1-12a}$$

where $\tau_1 = R_1 C; \tau_1 = R_2 C; \omega_n = \sqrt{\dfrac{k_p k_f}{\tau_1}}; d = \dfrac{\tau_2 \omega_n}{2}$; and

$$\Theta_0(s) = \frac{s(s + 2d\omega_n)}{s^2 + 2d\omega_n s + \omega_n^2}\Theta_i(s) \tag{6.1-12b}$$

For this loop, the equivalent one-sided noise bandwidth is $B_L = \frac{1}{2}\omega_n(d + \frac{1}{4d})$ hz. With this active filter, the steady-state phase error with a step in frequency as the input is

$$\theta_e(\infty) = s\frac{s^2}{s^2 + 2d\omega_n s + \omega_n^2}\frac{\nabla\omega}{s^2}\bigg|_{s=0} = 0 \tag{6.1-12c}$$

This remarkable result comes from the fact that $F(s)$ has a pole at the origin, a consequence of having an ideal op-amp in the circuit. Since op-amps are not ideal, the steady-state phase error will not actually be zero but will be quite small. We shall see, however, that it is easy to place a pole at the origin (actually at $z = 1$) in a sampled-data PLL.

6.2 NOISE IN THE CLASSIC PHASE-LOCKED LOOP

The analysis of noise in PLLs is an enormously complicated issue, well beyond the scope of this text. Those who are interested may consult the references [Vit66], [Lin72], [L&S73]. Only the simplest results, based upon the phase detector operating linearly, will be presented here.

We shall assume that the input to the PLL consists of carrier with no input signal

$$x(t) = A\cos(\omega_c t) \tag{6.2-1a}$$

and narrowband noise

$$n(t) = n_c(t)\cos(\omega_c t) - n_s(t)\sin(\omega_s t) \tag{6.2-1b}$$

having power spectral density $\eta/2$. If we measure this noise power in the loop bandwidth, we get the input noise power equal to

$$\sigma_n^2 = \left(\frac{\eta}{2}\right)(2B_L) = \eta B_L \tag{6.2-1c}$$

At the loop input, the signal-to-noise ratio is defined as

$$\text{SNR} = \frac{A^2/2}{\eta B_L} = \frac{A^2}{2\eta B_L} \tag{6.2-1d}$$

The input signal plus noise can then be written as

$$x(t) + n(t) = A \, \mathrm{Cos}(\omega_c t) + n_c(t) \, \mathrm{Cos}(\omega_c t) - n_s(t) \, \mathrm{Sin}(\omega_c t)$$

$$= \sqrt{A^2 + n_c^2(t)} \, \mathrm{Cos} \left[\omega_c t + \tan^{-1} \left(\frac{n_s(t)}{A + n_c(t)} \right) \right]$$

$$= A \, \mathrm{Cos} \left[\omega_c t + \frac{n_s(t)}{A} \right] \tag{6.2-2a}$$

for reasonably large SNR. We can then think of phase noise

$$\theta_p(t) = n_s(t)/A \tag{6.2-2b}$$

entering the linearized loop of fig. 6.1-4. The variance or average power of the resulting noise at the output of the VCO is the power spectral density of this input phase noise multiplied by the noise bandwidth of the loop, or

$$\overline{\theta_0^2} = \frac{\eta}{2A^2} 2B_L = \frac{2}{\mathrm{SNR}} \tag{6.2-2c}$$

It should be reiterated that these results come from a limited, linear model of the PLL and simply do not reflect the complexity of the more general analysis. For such an anlysis see the references cited above.

6.3 MAXIMUM-LIKELIHOOD CARRIER-PHASE ESTIMATION

The idea of maximum likelihood detection was introduced in Chapter 5 and led to the idea of using Euclidean distance in analyzing signal sets and received data. Although the mathematics tended to be a bit elevated, the slicer, which is the practical implementation of the idea, is quite direct and intuitive. As we shall see, the maximum likelihood approach to carrier-phase estimation will involve some mathematical complexity but will lead to some very straightforward ideas and implementation. We will follow [Z&P85].

Let us begin by estimating an unknown parameter β associated with a noisy signal

$$y(t) = x(t, \beta) + n(t) \tag{6.3-1a}$$

where $n(t)$ is additive, white Gaussian noise having a power spectral density $P_n(\omega) = \eta/2$. In terms of an orthonormal signal set $\{\varphi_i(t)\}$ we can represent $y(t)$ as

$$y(t) = \sum_k x_k(\beta) \varphi_k(t) + \sum_k n_k \varphi_k(t) \tag{6.3-1b}$$

where

$$x_k(\beta) = \langle x(t, \beta), \varphi_k(t) \rangle = \int_0^{T_s} x(t, \beta) \varphi_k^*(t) \, dt \tag{6.3-1c}$$

and

$$n_k = \langle n(t), \varphi_k(t) \rangle = \int_0^{T_s} n(t) \varphi_k^*(t) \, dt \tag{6.3-1d}$$

where, since the noise is white, $E[n_k n_j^*] = \frac{\eta}{2}\delta_{kj}$. It follows that we can write

$$y(t) = \sum_k y_k \varphi_k(t) = \sum_k [x_k(\beta) + n_k]\varphi_k(t) \qquad (6.3\text{-}1e)$$

where $y_k = [x_k(\beta) + n_k]$ are components of a vector random variable where $E[y_k/\beta] = x_k(\beta)$. We can then write the joint probability density function

$$P_{Y/\beta} = \lim_{N \to \infty} \left(\frac{1}{\sigma\sqrt{2\pi}}\right)^N \exp\left[-\frac{1}{2\sigma^2}\sum_{k=1}^{N}|y_k - x_k(\beta)|^2\right] \qquad (6.3\text{-}2a)$$

Note that the unconditional probability density function of the random vector Y is

$$P_Y = \lim_{N \to \infty} \left(\frac{1}{\sigma\sqrt{2\pi}}\right)^N \exp\left[-\frac{1}{2\sigma^2}\sum_{k=1}^{N}|y_k|^2\right] \qquad (6.3\text{-}2b)$$

We are interested in finding the value of β that maximizes the conditional probability. To find this value, noting that the y_k does not depend on β, we could differentiate the conditional probability directly or, equivalently, we could form the likelihood ratio

$$\frac{P_{Y/\beta}}{P_Y} = \lim_{N \to \infty} \exp\left[\frac{1}{2\sigma^2}\sum_{k=1}^{N}[y_k^* x_k(\beta) + y_k x_k^*(\beta) - |x_k(\beta)|^2]\right] \qquad (6.3\text{-}2c)$$

and differentiate it with respect to β. Since $\ln(x)$ is monotonically related to x, it makes no difference and is more convenient if we differentiate the log of the likelihood ratio:

$$L(\beta) = \ln\left[\frac{P_{Y/\beta}}{P_Y}\right] = \lim_{N \to \infty}\left[\frac{1}{2\sigma^2}\sum_{k=1}^{N}[y_k^* x_k(\beta) + y_k x_k^*(\beta) - |x_k(\beta)|^2]\right] \qquad (6.3\text{-}3a)$$

According to Parseval's Theorem, if

$$x(t) = \sum_k x_k \varphi_k(t) \qquad \text{and} \qquad y(t) = \sum_k y_k \varphi_k(t) \qquad (6.3\text{-}3b)$$

then

$$\langle x(t), y(t) \rangle = \int_0^{T_s} x(t)y^*(t)\,dt = \sum_k x_k y_k^* \qquad (6.3\text{-}3c)$$

It follows that the log-likelihood ratio can be rewritten as

$$L(\beta) = \frac{1}{2\sigma^2}\int_0^{T_s}[y^*(t)x(t,\beta) + y(t)x^*(t,\beta) - x(t,\beta)x^*(t,\beta)]\,dt \qquad (6.3\text{-}3d)$$

and in order to maximize $L(\beta)$ we must set

$$\frac{\partial L(\beta)}{\partial \beta} = \frac{1}{2\sigma^2}\int_0^{T_s}\left[y^*(t)\frac{\partial x(t,\beta)}{\partial \beta} + y(t)\frac{\partial x^*(t,\beta)}{\partial \beta} - x(t,\beta)\frac{\partial x^*(t,\beta)}{\partial \beta}\right.$$

$$\left. - \frac{\partial x(t,\beta)}{\partial \beta}x^*(t,\beta)\right]dt = 0 \qquad (6.3\text{-}4a)$$

or

$$\frac{\partial L(\beta)}{\partial \beta} = \frac{1}{2\sigma^2} 2\mathrm{Re}\left\{\int_0^{T_s}[y(t) - x(t, \beta)]\frac{\partial x^*(t, \beta)}{\partial \beta}\,dt\right\} = 0 \qquad (6.3\text{-}4b)$$

and, if the functions are real,

$$\frac{\partial L(\beta)}{\partial \beta} = \frac{1}{\sigma^2}\left\{\int_0^{T_s}[y(t) - x(t, \beta)]\frac{\partial x(t, \beta)}{\partial \beta}\,dt\right\} = 0 \qquad (6.3\text{-}4c)$$

Alternatively, we can maximize $L(\beta)$ from the coefficients of the orthogonal function expansion of $y(t)$ and $x(t, \beta)$. Again, if these functions are real:

$$\frac{\partial L(\beta)}{\partial \beta} = \frac{1}{\sigma^2}\lim_{N\to\infty}\sum_{k=1}^{N}[y_k - x_k(\beta)]\frac{\partial x_k(\beta)}{\partial \beta} = 0 \qquad (6.3\text{-}4d)$$

EXAMPLE 1 *Estimating the Phase of a Sinusoid*

We want to find a maximum likelihood estimate the phase of a sinusoid in the presence of noise:

$$y(t) = A\,\mathrm{Cos}(\omega_0 t + \theta) + n(t) \qquad \text{where} \quad \omega_0 = 2\pi/T_s. \qquad (6.3\text{-}5a)$$

The orthonormal functions that will be used in the expansion of of this expression are

$$\varphi_1(t) = \sqrt{\frac{2}{T_s}}\,\mathrm{Cos}(\omega_0 t) \qquad \text{and} \qquad \varphi_2(t) = \sqrt{\frac{2}{T_s}}\,\mathrm{Sin}(\omega_0 t) \qquad (6.3\text{-}5b)$$

so that we may also write $y(t)$ as

$$y(t) = y_1\varphi_1(t) + y_2\varphi_2(t)$$
$$= [x_1 + n_1]\varphi_1(t) + [x_2 + n_2]\varphi_2(t) \qquad (6.3\text{-}5c)$$

where

$$x_1(\theta) = \sqrt{\frac{T_s}{2}}\,A\,\mathrm{Cos}\,\theta \qquad \text{and} \qquad x_2(\theta) = -\sqrt{\frac{T_s}{2}}\,A\,\mathrm{Sin}\,\theta \qquad (6.3\text{-}5d)$$

Since $N = 2$ and $y(t)$ is real, eq. 6.3-2a becomes

$$L(\theta) = \frac{1}{2\pi\sigma^2}[[2y_1 x_1(\theta) - |x_1(\theta)|^2] + [2y_2 x_2(\theta) - |x_2(\theta)|^2]]$$

$$= \frac{1}{2\pi\sigma^2}\left\{\left[2y_1\sqrt{\frac{T_s}{2}}\,A\,\mathrm{Cos}\,\theta - \left|\sqrt{\frac{T_s}{2}}\,A\,\mathrm{Cos}\,\theta\right|^2\right]\right.$$

$$\left. + \left[-2y_2\sqrt{\frac{T_s}{2}}\,A\,\mathrm{Sin}\,\theta - \left|\sqrt{\frac{T_s}{2}}\,A\,\mathrm{Sin}\,\theta\right|^2\right]\right\}$$

$$= \frac{1}{\pi\sigma^2}\sqrt{\frac{T_s}{2}}\,A[y_1\,\mathrm{Cos}\,\theta - y_2\mathrm{Sin}\,\theta] - \frac{1}{2\pi\sigma^2}\frac{T_s A^2}{2} \qquad (6.3\text{-}6a)$$

where

$$y_1 = \int_0^{T_s} y(t)\varphi_1(t)\,dt = \sqrt{\frac{2}{T_s}} \int_0^{T_s} y(t)\,\mathrm{Cos}(\omega_0 t)\,dt$$

$$y_2 = \int_0^{T_s} y(t)\varphi_2(t)\,dt = \sqrt{\frac{2}{T_s}} \int_0^{T_s} y(t)\,\mathrm{Sin}(\omega_0 t)\,dt$$

(6.3-6b)

Substituting, we get

$$L(\theta) = \frac{1}{\pi\sigma^2} A \left[\mathrm{Cos}\,\theta \int_0^{T_s} y(t)\,\mathrm{Cos}(\omega_0 t)\,dt - \mathrm{Sin}\,\theta \int_0^{T_s} y(t)\,\mathrm{Sin}(\omega_0 t)\,dt \right] - \frac{1}{2\pi\sigma^2}\frac{T_s A^2}{2}$$

(6.3-7a)

and differentiating,

$$\frac{\partial L(\theta)}{\partial\theta} = -\frac{1}{\pi\sigma^2} A \left[\mathrm{Sin}\,\hat\theta \int_0^{T_s} y(t)\,\mathrm{Cos}(\omega_0 t)\,dt + \mathrm{Cos}\,\hat\theta \int_0^{T_s} y(t)\,\mathrm{Sin}(\omega_0 t)\,dt \right] = 0$$

(6.3-7b)

We can draw two conclusions from this result.

1. We can directly derive an estimate of θ as

$$\tan\hat\theta = \frac{\int_0^{T_s} y(t)\,\mathrm{Sin}(\omega_0 t)\,dt}{\int_0^{T_s} y(t)\,\mathrm{Cos}(\omega_0 t)\,dt}$$

(6.3-7c)

which is the maximum likelihood estimate. This is shown in fig. 6.3-1a.

2. We can indirectly derive the estimate by generating a sinusoid having the same frequency and an estimate of the phase. This reformulation is given by

$$-\frac{1}{\pi\sigma^2} A \left[\mathrm{Sin}\,\hat\theta \int_0^{T_s} y(t)\,\mathrm{Cos}(\omega_0 t)\,dt + \mathrm{Cos}\,\hat\theta \int_0^{T_s} y(t)\,\mathrm{Sin}(\omega_0 t)\,dt \right]$$

$$= -\frac{2A}{2\pi\sigma^2} \int_0^{T_s} y(t)\,\mathrm{Sin}(\omega_0 t - \hat\theta)\,dt = 0$$

(6.3-7d)

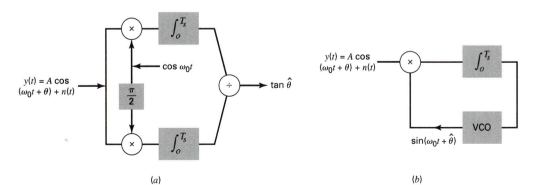

(a)   (b)

FIGURE 6.3-1

which can be interpreted as a phase-locked loop, as shown in fig. 6.3-1b, where the integrate-and-dump operation is a form of lowpass filter.

We can find the variance of θ, due to noise, by invoking the Cramer-Rao lower bound

$$\text{var}(\theta) \geq \left\{ -E\left[\frac{\partial^2 L(\theta)}{\partial \theta^2}\right] \right\}^{-1} \tag{6.3-8a}$$

Differentiating eq. 6.3-7b, we get

$$\frac{\partial^2 L(\theta)}{\partial \theta^2} = -\frac{A}{\pi \sigma^2} \int_0^{T_s} y(t) \cos(\omega_0 t + \theta)\, dt \tag{6.3-8b}$$

and

$$-E\left[\frac{\partial^2 L(\theta)}{\partial \theta^2}\right] = \frac{A}{\pi \sigma^2} \int_0^{T_s} E[y(t)] \cos(\omega_0 t + \theta)\, dt$$

$$= \frac{A}{\pi \sigma^2} \int_0^{T_s} A \cos(\omega_0 t + \theta) \cos(\omega_0 t + \theta)\, dt = \frac{2A^2}{2\pi \sigma^2} \frac{T_s}{2} = \frac{2T_s}{2\pi \sigma^2} P_s$$

$$\tag{6.3-8c}$$

so that

$$\text{var}(\theta) \geq \frac{2\pi \sigma^2}{2T_s} \frac{1}{P_s} = 2\pi \, B_n \frac{\sigma^2}{P_s} \tag{6.3-8d}$$

where $B_n = \frac{1}{2T_s}$ is the one-sided equivalent noise bandwidth, in hz, of the finite-time integrator and P_s is the signal power.

Assertion:

The one-sided noise bandwidth of a finite-time integrator from 0 to T_s is $B_n = 1/2T_s$. ∎

Proof:

The transfer function of a finite-time integrator is

$$H(s) = \frac{1 - e^{-sT_s}}{s} \Rightarrow H(\omega) = \frac{1 - e^{-j\omega T_s}}{j\omega} = T_s \, \text{Sa}\left(\frac{\omega T_s}{2}\right) e^{-j\omega T_s/2}$$

so that the two-sided noise bandwidth is

$$\text{BW} = \frac{\frac{1}{2\pi} \int_{-\infty}^{\infty} |H(\omega)|^2\, d\omega}{|H(0)|^2} = \frac{1}{2\pi} \int_{-\infty}^{\infty} \text{Sa}^2\left(\frac{\omega T_s}{2}\right) d\omega$$

$$= \frac{1}{2\pi} \frac{2}{T_s} \int_{-\infty}^{\infty} \text{Sa}^2(x)\, dx = \frac{1}{2\pi} \frac{2}{T_s} \pi = \frac{1}{T_s}$$

end of the Proof

It follows that

$$\text{var}(\theta) \geq 1/\text{SNR} \tag{6.3-9}$$

This example has shown that maximum likelihood estimation of carrier phase of an umodulated sinusoid leads to an implementation that is consistent with a phase-locked loop. The variance of the estimate is the same of as the noise in the VCO output of a PLL. We will see that the PLL emerges as central to the estimation of modulated carriers as well.

EXAMPLE 2   *Estimating the Demodulating Carrier Phase for BPSK*

In contrast to the sinusoid having fixed amplitude and unknown phase, we now want to consider that same sinusoid carrying a binary data signal

$$y(t) = a_n A \cos(\omega_0 t + \theta) + n(t) \qquad \text{where} \quad \omega_0 = 2\pi/T_s \qquad (6.3\text{-}10a)$$

where

$$P[a_n = +1] = P[a_n = -1] = \frac{1}{2}. \qquad (6.3\text{-}10b)$$

Using the same orthonormal functions as in Example 1, we can write

$$y(t) = y_1 \varphi_1(t) + y_2 \varphi_2(t)$$

$$= a_n A \sqrt{\frac{T_s}{2}} \{[\cos(\theta) + n_1]\varphi_1(t) - [\sin(\theta) + n_2]\varphi_2(t)\} \qquad (6.3\text{-}10c)$$

Since the two values of the data symbol are equally likely, we may work with the simple average of the two density functions corresponding to the two values

$$f_{Y/\theta}(y_1, y_2/\theta) = \frac{1}{2} f_{Y/\theta, a_n=+1} + \frac{1}{2} f_{Y/\theta, a_n=-1} \qquad (6.3\text{-}11a)$$

or

$$f_{Y/\theta}(y_1, y_2/\theta)$$

$$= \frac{1}{2} \frac{1}{2\pi\sigma^2} \left\{ \exp\left[-\frac{1}{2\sigma^2} \left[\left(y_1 - A\sqrt{\frac{T_s}{2}}\cos\theta\right)^2 + \left(y_2 + A\sqrt{\frac{T_s}{2}}\sin\theta\right)^2 \right] \right] \right.$$

$$\left. + \exp\left[-\frac{1}{2\sigma^2} \left[\left(y_1 + A\sqrt{\frac{T_s}{2}}\cos\theta\right)^2 + \left(y_2 - A\sqrt{\frac{T_s}{2}}\sin\theta\right)^2 \right] \right] \right\}$$

$$(6.3\text{-}11b)$$

or

$$f_{Y/\theta}(y_1, y_2/\theta) = K \left\{ \exp\left[-\frac{1}{\sigma^2} A\sqrt{\frac{T_s}{2}} [-y_1 \cos\theta + y_2 \sin\theta] \right] \right.$$

$$\left. - \exp\left[-\frac{1}{\sigma^2} A\sqrt{\frac{T_s}{2}} [y_1 \cos\theta - y_2 \sin\theta] \right] \right\} \qquad (6.3\text{-}11c)$$

or

$$f_{Y/\theta}(y_1, y_2/\theta) = K_1 \cosh\left[\frac{A}{\sigma^2} \sqrt{\frac{T_s}{2}} [y_1 \cos\theta - y_2 \sin\theta] \right] \qquad (6.3\text{-}11d)$$

where K and K_1 are constants.

We will form the maximum likelihood estimate from eq. 6.3-11d. For convenience, we will form $\ln \text{Cosh}(x)$ and differentiate it with repect to θ. This gives us

$$\frac{\partial L(\theta)}{\partial \theta} = \frac{d}{dx}[\ln \cosh(x)]\frac{dx}{d\theta} = \frac{\sinh(x)}{\cosh(x)}\frac{dx}{d\theta} = \tanh(x)\frac{dx}{d\theta} \qquad (6.3\text{-}12a)$$

so that

$$\frac{\partial L(\theta)}{\partial \theta} = \tanh\left(\frac{A}{\sigma^2}\sqrt{\frac{T_s}{2}}[y_1\,\text{Cos}\,\theta - y_2\,\text{Sin}\,\theta]\right)\frac{d\left(\frac{A}{\sigma^2}\sqrt{\frac{T_s}{2}}[y_1\,\text{Cos}\,\theta - y_2\,\text{Sin}\,\theta]\right)}{d\theta} = 0 \qquad (6.3\text{-}12b)$$

or

$$\frac{\partial L(\theta)}{\partial \theta} = \tanh\left(\frac{A}{\sigma^2}\sqrt{\frac{T_s}{2}}[y_1\,\text{Cos}\,\theta - y_2\,\text{Sin}\,\theta]\right)\frac{d\left(\frac{A}{\sigma^2}\sqrt{\frac{T_s}{2}}[y_1\,\text{Cos}\,\theta - y_2\,\text{Sin}\,\theta]\right)}{d\theta} = 0 \qquad (6.3\text{-}12c)$$

or:

$$\frac{\partial L(\theta)}{\partial \theta} = \left\{\tanh\left[\frac{A}{\sigma^2}\sqrt{\frac{T_s}{2}}[y_1\,\text{Cos}\,\theta - y_2\,\text{Sin}\,\theta]\right]\right\}\left(\frac{A}{\sigma^2}\sqrt{\frac{T_s}{2}}[y_1\,\text{Sin}\,\theta + y_2\,\text{Cos}\,\theta]\right) = 0 \qquad (6.3\text{-}12d)$$

and, switching to the time domain,

$$\frac{\partial L(\theta)}{\partial \theta} = \left\{\tanh\left[\frac{A}{\sigma^2}\int_0^{T_s} y(t)\,\text{Cos}(\omega_c t + \theta)\,dt\right]\right\}\left(\frac{A}{\sigma^2}\int_0^{T_s} y(t)\,\text{Sin}(\omega_c t + \theta)\,dt\right) = 0. \qquad (6.3\text{-}12e)$$

This maximum likelihood result may be represented structurally as shown in fig. 6.3-2. It is commonly known as a Costas loop. The operation of tanh in this loop is of considerable interest. Note that

$$\tanh(x) = \frac{e^x - e^{-x}}{e^x + e^{-x}} = \begin{cases} +1 & \text{if } x \gg 0 \\ -1 & \text{if } x \ll 0 \end{cases} \qquad (6.3\text{-}13)$$

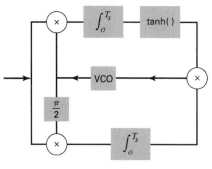

BPSK Costas loop

FIGURE 6.3-2

At high signal-to-noise ratio it follows that the operation of tanh is that of a slicer. To summarize then, the maximum likelihood phase estimate for a BPSK data signal is derived by demodulating the signal by a sine and cosine carrier derived from a common VCO, doing separate integrate and dump operations (lowpass filtering) on the two results, slicing the cosine channel, and multiplying the decided data with the output of the sine channel to drive the VCO.

This structure can be understood as a *decision-directed* carrier recovery because the error in the PLL is being generated by the decided value of the signal. ∎

EXAMPLE 3 *Estimating the Demodulating Carrier Phase for QPSK*

It will be seen that the maximum likelihood analysis for QPSK will carry over to all QAM-type signals and will lead to the general idea of a decision-directed carrier recovery system.

A QPSK signal can be represented as

$$y(t) = a_n A \, \mathrm{Cos}(\omega_0 t + \theta) + b_n A \, \mathrm{Sin}(\omega_0 t + \theta) + n(t) \qquad \text{where} \quad \omega_0 = 2\pi / T_s \qquad (6.3\text{-}14a)$$

and where

$$P[a_n = +1] = P[a_n = -1] = P[b_n = +1] = P[b_n = -1] = \frac{1}{2} \qquad (6.3\text{-}14b)$$

Using the same orthonormal functions as in Example 1, we can write

$$y(t) = y_1 \varphi_1(t) + y_2 \varphi_2(t)$$

$$= A \sqrt{\frac{T_s}{2}} \{ [a_n \, \mathrm{Cos}(\theta) + b_n \, \mathrm{Sin}(\theta) + n_1] \varphi_1(t) - [a_n \, \mathrm{Sin}(\theta) - b_n \, \mathrm{Cos}(\theta) + n_2] \varphi_2(t) \}$$

$$(6.3\text{-}14c)$$

We will also average the conditional probability density over the two sets of data

$$f_{Y/\theta}(y_1, y_2/\theta) = \frac{1}{4} f_{Y/\theta, a_n=+1, b_n=+1} + \frac{1}{4} f_{Y/\theta, a_n=+1, b_n=-1} + \frac{1}{4} f_{Y/\theta, a_n=-1, b_n=+1}$$

$$+ \frac{1}{4} f_{Y/\theta, a_n=-1, b_n=-1} \qquad (6.3\text{-}15a)$$

yielding

$$f_{Y/\theta}(y_1, y_2/\theta) = \frac{1}{4} \frac{1}{2\pi\sigma^2}$$

$$\times \left\{
\begin{aligned}
&\exp\left(-\frac{1}{2\sigma^2}\left[\left(y_1 - A\sqrt{\frac{T_s}{2}}(\mathrm{Cos}\,\theta + \mathrm{Sin}\,\theta)\right)^2 + \left(y_2 + A\sqrt{\frac{T_s}{2}}(\mathrm{Sin}\,\theta - \mathrm{Cos}\,\theta)\right)^2\right]\right) \\
&+\exp\left(-\frac{1}{2\sigma^2}\left[\left(y_1 - A\sqrt{\frac{T_s}{2}}(\mathrm{Cos}\,\theta - \mathrm{Sin}\,\theta)\right)^2 + \left(y_2 + A\sqrt{\frac{T_s}{2}}(\mathrm{Sin}\,\theta + \mathrm{Cos}\,\theta)\right)^2\right]\right) \\
&+\exp\left(-\frac{1}{2\sigma^2}\left[\left(y_1 - A\sqrt{\frac{T_s}{2}}(-\mathrm{Cos}\,\theta + \mathrm{Sin}\,\theta)\right)^2 + \left(y_2 + A\sqrt{\frac{T_s}{2}}(-\mathrm{Sin}\,\theta - \mathrm{Cos}\,\theta)\right)^2\right]\right) \\
&+\exp\left(-\frac{1}{2\sigma^2}\left[\left(y_1 - A\sqrt{\frac{T_s}{2}}(-\mathrm{Cos}\,\theta - \mathrm{Sin}\,\theta)\right)^2 + \left(y_2 + A\sqrt{\frac{T_s}{2}}(-\mathrm{Sin}\,\theta + \mathrm{Cos}\,\theta)\right)^2\right]\right)
\end{aligned}
\right\}$$

$$(6.3\text{-}15b)$$

We can analyze this horrible expression by recognizing that each one of the four exponents is of the form

$$
\left[\left(y_1 - A\sqrt{\frac{T_s}{2}}(a\,\mathrm{Cos}\,\theta + b\,\mathrm{Sin}\,\theta) \right)^2 + \left(y_2 + A\sqrt{\frac{T_s}{2}}(a\,\mathrm{Sin}\,\theta - b\,\mathrm{Cos}\,\theta) \right)^2 \right] \quad (6.3.15c)
$$

which, when expanded, yields

$$
y_1^2 + y_2^2 + A^2\frac{T_s}{2}(a^2 + b^2) - 2A\sqrt{\frac{T_s}{2}}[y_1(a\,\mathrm{Cos}\,\theta + b\,\mathrm{Sin}\,\theta) - y_2(a\,\mathrm{Sin}\,\theta - b\,\mathrm{Cos}\,\theta)]
$$
$$
(6.3\text{-}15d)
$$

The terms not involving θ will drop out when we differentiate, so we may neglect them now. We may also factor out the exponential term involving $2A\sqrt{T_s/2}$ so that, in terms of subsequent differentiation, we need look only at

$$
\sum_{a=\pm 1}\sum_{b=\pm 1} \exp([y_1(a\,\mathrm{Cos}\,\theta + b\,\mathrm{Sin}\,\theta) - y_2(a\,\mathrm{Sin}\,\theta - b\,\mathrm{Cos}\,\theta)])
$$
$$
= \sum_{a=\pm 1}\sum_{b=\pm 1} \exp(a[y_1\,\mathrm{Cos}\,\theta - y_2\,\mathrm{Sin}\,\theta]\exp b[y_1\,\mathrm{Sin}\,\theta + y_2\,\mathrm{Cos}\,\theta]) \quad (6.3\text{-}15e)
$$

which can be written as

$$
\sum_{a=\pm 1}\sum_{b=\pm 1} e^{a\alpha}e^{b\beta} = e^{\alpha}e^{\beta} + e^{\alpha}e^{-\beta} + e^{\alpha}e^{\beta} + e^{-\alpha}e^{-\beta}
$$
$$
= [e^{\alpha} + e^{-\alpha}][e^{\beta} + e^{-\beta}] = 4\,\mathrm{Cosh}(\alpha)\,\mathrm{Cosh}(\beta) \quad (6.3\text{-}15f)
$$

Putting this all together, we can form an appropriate likelihood ratio by taking the natural log of this quantity, so that

$$
L(\theta) = \ln[\cosh(\alpha)] + \ln[\cosh(\beta)] \quad (6.3\text{-}16a)
$$

and

$$
\frac{\partial L(\theta)}{\partial \theta} = \tanh(\alpha)\frac{d\alpha}{d\theta} + \tanh(\beta)\frac{d\beta}{d\theta} = 0 \quad (6.3\text{-}16b)
$$

or

$$
\frac{\partial L(\theta)}{\partial \theta} = -\tanh(y_1\mathrm{Cos}\,\theta - y_2\mathrm{Sin}\,\theta)[y_1\mathrm{Sin}\,\theta + y_2\mathrm{Cos}\,\theta]
$$
$$
+ \tanh(y_1\mathrm{Sin}\,\theta + y_2\mathrm{Cos}\,\theta)[y_1\mathrm{Cos}\,\theta - y_2\mathrm{Sin}\,\theta] = 0 \quad (6.3\text{-}16c)
$$

and going back to the time domain

$$
\frac{\partial L(\theta)}{\partial \theta} = -\tanh\left(\frac{A}{\sigma^2}\int_0^{T_s} y(t)\,\mathrm{Cos}(\omega_c t + \theta)\,dt\right)\left[\frac{A}{\sigma^2}\int_0^{T_s} y(t)\,\mathrm{Sin}(\omega_c t + \theta)\,dt\right]
$$
$$
+ \tanh\left(\frac{A}{\sigma^2}\int_0^{T_s} y(t)\mathrm{Sin}(\omega_c t + \theta)\,dt\right)\left[\frac{A}{\sigma^2}\int_0^{T_s} y(t)\,\mathrm{Cos}(\omega_c t + \theta)\,dt\right] = 0
$$
$$
(6.3\text{-}16d)
$$

which is shown in fig. 6.3-3.

We note that the phase estimate in this example has been based on a single received data point. A filter at the input to the VCO serves the purpose of basing the phase estimate on the past history of the received data as well.

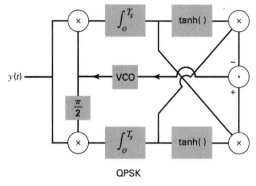

QPSK

FIGURE 6.3-3

Recognizing tanh as a slicer, we see that we have the decision-directed carrier recovery that was promised. A carrier recovery for a more general QAM system would simply involve replacing the two-level slicer in each leg with a slicer appropriate to the signal set. It is useful to characterize the generation of the error signal for the VCO in the following manner: if the received signal point and the decided signal point are each considered to be vectors, then the error signal is the cross-product of these two vectors. ■

Much has been written on carrier phase recovery systems of this type. The interested reader is referred to [L&S73], [Sti71], [L&S71], [Spi77], [L&S71], [L&S72], [Sim79-1, 79-2], [B&M83], [M&M74].

One major conclusion is that although the performance of decision-aided carrier recovery is satisfactory at large values of SNR, the appearance of more frequent decision errors at low SNR tends to degrade the performance seriously. At that point forward-acting carrier recovery reappears as an alternative.

In Chapter 4, fig. 4.6-5, it was observed that a complex digital equalizer would be needed in the receiver in order to compensate for channel distortion. The maximum likelihood carrier-phase recovery receiver for QAM signal sets is easily joined with this equalizer, as shown in fig. 6.3-4 where specific reference is made to QPSK. Note that in this circuit the integrate and dump operation has been replaced by a lowpass filter, a substitution that

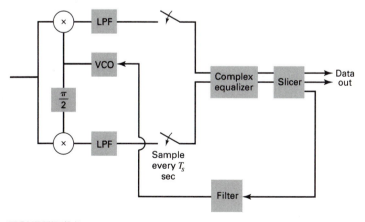

FIGURE 6.3-4

was discussed in Chapter 4, and the sampled signal is applied to a baseband-equivalent complex-valued equalizer. Decisions are made at the equalizer output.

6.4   A DECISION-DIRECTED ADAPTIVE ROTATOR

The difficulty with the approach of fig. 6.3-4 is that the PLL controling the frequency of the demodulating carrier will include the delay caused by the equalizer. Classical control theory informs us that such a delay will cause instability in the loop. In seminal papers, Falconer [Fal76-1], [Fal76-2] has proposed circumventing this problem by splitting the carrier recovery into two parts, a 'free-running' demodulator and an adaptive rotator, and inserting the complex equalizer between them, as shown in fig. 6.4-1. We have already discussed this split in the carrier recovery in Chapter 4 and we will discuss the properties and the design of the adaptive rotator below.

In Falconer's approach, the equalizer is operating on data that has been demodulated but still contains the phase distortion (phase and frequency offset as well as phase jitter) that the carrier-phase recovery system is intended to address. This makes for some interesting dynamics between the equalizer and the rotator, a phenomenon that will be discussed in Chapter 11.

A detailed view of the relationship between the slicer and the adaptive rotater is shown in fig. 6.4-1. Until this point, the slicer has simply been used to make decisions. Now the slicer will also be used to compute errors between the received signal and the decided signal. Two kinds of errors will be computed by the slicer: rectangular errors to drive the adaptive equalizer (to be discussed in Chapter 11) and the angular error between the received signal and the decided signal. In computing the angular error, the slicer is operating as a phase detector for a phase-locked loop. Note that this angular error is filtered and then accumulated or integrated before being applied to the adaptive rotator, thereby completing the loop. In Chapter 11 we will revisit the generation of the error to drive the adaptive equalizer.

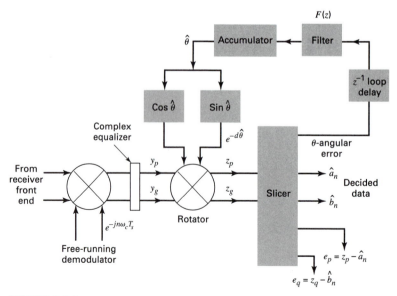

FIGURE 6.4-1

The logic of this structure is very straightforward. If there is no difference at all in the phases of the received carrier and the demodulating carrier, then the only source of an angular difference between the received signal and the decided signal will be noise. The primary purpose of the filter is to smooth or average the noise. The resulting average of zero is then accumulated from sample to sample, resulting in an accumulator output of zero, so that the rotator ends up doing nothing at all, which is exactly what it should be doing under the circumstances. On the other hand, if there is a frequency difference between the received carrier and the demodulating carrier, then the received signal will be rotated by some small amount in each sample with respect to the decided signal. The filter averages this small amount over a number of samples to get rid of the noise and then accumulates the angle so that the rotator always has the received signal in proper alignment with the slicer. The output of the accumulator is the angular estimate $\hat{\theta}$. In order to do the rotation, means must be provided to determine the Sin and Cos of this angle.

The rotation consists of the operation

$$z_I = y_I \, \text{Cos}(\hat{\theta}) + y_Q \, \text{Sin}(\hat{\theta})$$

$$z_Q = y_I \, \text{Sin}(\hat{\theta}) - y_Q \, \text{Cos}(\hat{\theta})$$

The first task is to examine how the slicer is to be used as a phase detector. In fig. 6.4-2 a QPSK signal set is used as an example. The sampled received signal into the slicer is denoted by the pair (z_1, z_Q) indicating the in-phase and quadrature components. The decisions are $(\hat{a}_n, \hat{b}_n)$, which are estimates of the original data.

In order to compute θ, note that

$$\text{Sin}(\theta) = \text{Sin}(\theta + \alpha - \alpha) = \text{Sin}(\theta + \alpha)\text{Cos}(\alpha) - \text{Cos}(\theta + \alpha)\text{Sin}(\alpha)$$

$$= \frac{z_Q}{\sqrt{z_I^2 + z_Q^2}} \cdot \frac{\hat{a}_n}{\sqrt{\hat{a}_n^2 + \hat{b}_n^2}} - \frac{z_I}{\sqrt{z_I^2 + z_Q^2}} \cdot \frac{\hat{b}_n}{\sqrt{\hat{a}_n^2 + \hat{b}_n^2}}$$

$$= \frac{z_Q \hat{a}_n - z_I \hat{b}_n}{\sqrt{z_I^2 + z_Q^2}\sqrt{\hat{a}_n^2 + \hat{b}_n^2}} \tag{6.4-2a}$$

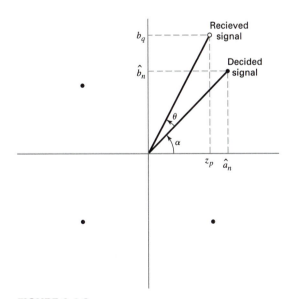

FIGURE 6.4-2

and

$$\text{Cos}(\theta) = \text{Cos}(\theta + \alpha - \alpha) = \text{Cos}(\theta + \alpha)\,\text{Cos}(\alpha) + \text{Sin}(\theta + \alpha)\,\text{Sin}(\alpha)$$

$$= \frac{z_I}{\sqrt{z_I^2 + z_Q^2}} \cdot \frac{\hat{a}_n}{\sqrt{\hat{a}_n^2 + \hat{b}_n^2}} + \frac{z_Q}{\sqrt{z_I^2 + z_Q^2}} \cdot \frac{\hat{b}_n}{\sqrt{\hat{a}_n^2 + \hat{b}_n^2}}$$

$$= \frac{z_I \hat{a}_n + z_Q \hat{b}_n}{\sqrt{z_I^2 + z_Q^2}\sqrt{\hat{a}_n^2 + \hat{b}_n^2}} \tag{6.4-2b}$$

so that

$$\text{Tan}(\theta) = \frac{\text{Sin}(\theta)}{\text{Cos}(\theta)} = \frac{z_Q \hat{a}_n - z_I \hat{b}_n}{z_I \hat{a}_n + z_Q \hat{b}_n} \tag{6.4-2c}$$

Since the computation of the angular error depends on the slicer decision, the PLL is called *decision-directed*. Wrong decisions may send the PLL in the wrong direction for a period of time, but this has not been found to be a major issue at reasonably high SNR.

There are a number of observations to be made about this computation. First, it should be noted that for small angles the Tan is a very close approximation to the angle. Second, this is a true tangent, independent of the magnitudes of the the particular signal points, a very important point in multilevel signal sets. Third, it has already been established that Tan is the maximum-likelihood estimate of the angle.

From a computational point of view, the biggest problem in determining Tan(θ) in this way is the division operation. This operation may not be provided in particular digital signal processors and may therefore have to be manufactured using other operations. There are a number of suboptimal ways of estimating Tan(θ) without using a division that give acceptable results for signal sets such as QPSK and 8-PSK in which each signal has the same magnitude. For signal sets in which the signals have differing amplitudes, such as 16-QAM, eliminating the division can reduce the estimate of Tan(θ) to simply being a value having the same sign. This may seriously degrade performance. Interested readers may consult Bingham [Bin88].

The characteristics of the phase detector are determined by the signal set. As shown in fig. 6.4-3, the phase-detector characteristics for such signal sets as QPSK and 8-PSK are quite straightforward because all the signals have equal energy.

For a 16-QAM signal set the situation is quite different, as shown in fig. 6.4-4a . As the received signal rotates with respect to the slicer, the different points in the signal set cross decision boundaries at different angles. As shown by Simon [Sim74], these different angles are 16.9°, 18.4°, 20.8°, and 32.3°, causing discontinuities in the phase detector in the range ±45°.

We must now compute an average phase-detector characteristic, taking into account that for a given rotation any one of the four points in a quadrant may be the actual received point. The result is shown in fig. 6.4-4b. As an example of the computation, consider the first discontinuity at 16.9°. Just before the decision boundary is crossed at $\theta = 16.9°-$, the angular deviation from each point is the same, 16.9°, so that

$$\overline{\text{Tan}(\theta)} = \frac{1}{4}[4 \cdot \text{Tan}(16.9)] = .304$$

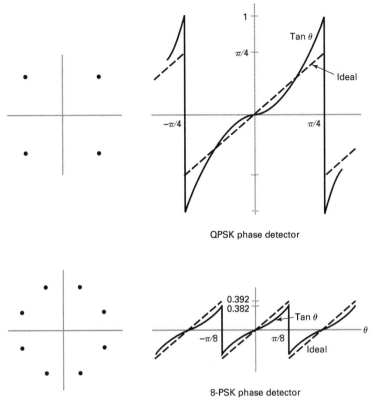

QPSK phase detector

8-PSK phase detector

FIGURE 6.4-3

Just after the decision boundary is crossed, at $\theta = 16.9°+$, the rotated (3,3) point is no longer in the (3,3) decision region but is in the (1,3) decision region, and is now at -19.66 with respect to that point. Consequently, we now have

$$\overline{\mathrm{Tan}(\theta)} = \frac{1}{4}[3 \cdot \mathrm{Tan}(16.9) - \mathrm{Tan}(19.66)] = .138$$

The remaining pattern may be computed in a similar way.

The biggest problem of this phase characteristic is that there is a "false lock" at point at $27°$, so that there is a distinct possibility that the PLL may hang up at this point. Both Simon [Sim79] and Bingham [Bin88] analyze this phenomenon and the reader who wishes more detail is referred to those sources. Simon also shows that as the SNR increases the false lock phenomenon tends to disappear. The reader may also wish to consult [L&C83]. More complex signal sets present even more interesting problems in analyzing the phase detector.

It should be noted that for particularly complex signal sets it may be reasonable to divide the two-dimensional decision space into many small rectangles, and associate with each rectangle a decision, a rectangular error for the equalizer, and an angular error for the carrier phase recovery. This information may then be entered into a lookup table or a ROM and addressed by the digital received data. This approach certainly speeds up the operation of the slicer by bypassing all operations of comparison, decision, and division, reducing the slicer to just a few operations.

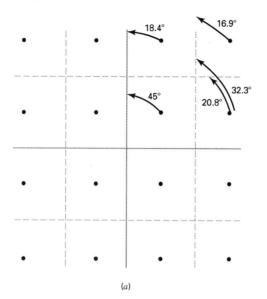

(a)

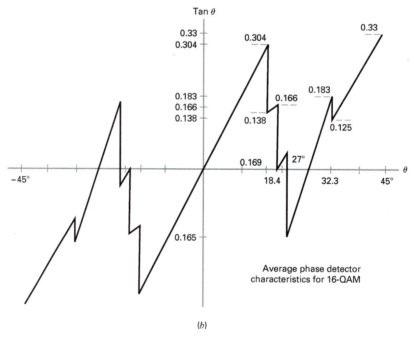

(b)

FIGURE 6.4-4

6.5 A SAMPLED-DATA PLL FOR IMPLEMENTING AN ADAPTIVE ROTATOR

We have seen that the function of the slicer in a QAM system can be expanded to include phase detection. In the following discussion it will be assumed that this phase detector is operating linearly. Referring to Gill and Gupta [G&G72-1], [G&G72-2], it will now be

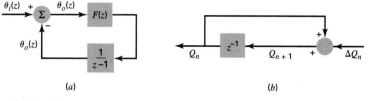

(a)   (b)

FIGURE 6.5-1

demonstrated that the resulting system is equivalent to a sampled-data PLL as shown in fig. 6.5-1a. We define:

$$\theta_i(k) = \text{carrier phase at time } k, \text{ input to the loop}$$

$$\theta_0(k) = \text{output phase of the loop} \tag{6.5-1a}$$

so that with the assumption of a linear phase detector

$$\theta_0(k) = F(z) \sum_{j=0}^{k-1} [\theta_i(j) - \theta_0(j)] \tag{6.5-1b}$$

and rewriting to drop the summation

$$\theta_0(k+1) = F(z)[\theta_i(k) - \theta_0(k)] + \theta_0(k) \tag{6.5-1c}$$

or

$$\theta_0(z) = \frac{F(z)}{z-1}[\theta_i(z) - \theta_0(z)] \tag{6.5-1d}$$

Note that there is a delay of one symbol interval between $\theta_0(k+1)$ and $\theta_0(k)$ so that in implementing the algorithm we will have to guarantee that delay. To demonstrate that $1/(z-1)$ is an integrator, note that in fig. 6.5-1b

$$\nabla\phi_n + \phi_n = \phi_{n+1} \qquad \text{and} \qquad \phi_n = z^{-1}\phi_{n+1} \tag{6.5-2a}$$

so that

$$\phi_n = \frac{1}{z-1}\nabla\phi_n. \tag{6.5-2b}$$

For this loop, the closed-loop response is

$$\theta_0(z) = \frac{\dfrac{F(z)}{(z-1)}}{1 + \dfrac{F(z)}{(z-1)}}\theta_i(z) = H(z)\theta_i(z) \tag{6.5-3a}$$

and the error function (which is the angular error out of the slicer) is

$$\theta_e(z) = [\theta_0(z) - \theta_i(z)] = \frac{1}{1 + \dfrac{F(z)}{(z-1)}}\theta_i(z) \tag{6.5-3b}$$

We are interested in examining the response of this loop to a frequency offset. The frequency offset is represented by $\theta_i(t) = \alpha t$ hz $= 360\alpha t$ degrees/sec. Then

$$\theta_i(z) = 360\alpha T_s \frac{z}{(z-1)^2} \tag{6.5-4}$$

where T_s is the symbol interval. We will examine the response of the loop to a frequency offset when $F(z)$ is a simple first-order filter

$$F(z) = k\frac{z-a}{z-b} \tag{6.5-5a}$$

so that

$$\theta_e(z) = \frac{1}{1 + \frac{F(z)}{(z-1)}}\theta_i(z) = \frac{(z-1)(z-b)}{(z-1)(z-b) + k(z-a)}\theta_i(z)$$

$$= 360\alpha T_s \frac{(z-1)(z-b)}{(z-1)(z-b) + k(z-a)}\frac{z}{(z-1)^2} \tag{6.5-5b}$$

We can find the steady-state value of the phase error using the final value theorem of z-transforms:

$$\theta_e(\infty) = \lim_{z\to 1}[(z-1)\theta_e(z)]$$

$$= \left[360\alpha T_s \frac{z(z-b)}{(z-1)(z-b) + k(z-a)}\right]_{z=1}$$

$$= 360\alpha T_s \frac{(1-b)}{k(1-a)} \tag{6.5-6a}$$

so that we can define

$$\eta = \frac{\theta_e(\infty)}{\alpha} = 360 T_s \frac{(1-b)}{k(1-a)} \text{ degrees of error/hz of offset} \tag{6.5-6b}$$

This is exactly the same kind of result as we had in analyzing the steady-state error to a frequency offset in an analog PLL. In that PLL we had to change the form of the loop filter in order to drive that error to zero and even then we had to ignore stray capacitance. In this sampled-data PLL we can set this steady-state error to zero by the simple expedient of setting the coefficent $b = +1$, which is simple enough to do in a digital system.

The design of the loop filter then involves choosing the values of k and a taking into account:

1. loop stability
2. loop acquisition time
3. noise bandwidth of the loop

With regard to loop stability, note that the denominator of the closed-loop transfer fuction is

$$(z-1)(z-b) + k(z-a) \tag{6.5-7a}$$

and, for $b = +1$

$$(z-1)^2 + k(z-a) = z^2 - (2-k)z + (1-ka). \tag{6.5-7b}$$

Unlike the analog second-order PLL, where stability was assured no matter what the choice of filter or gain parameters, it is possible to choose both k and a so that the poles of the closed-loop system lie outside the unit circle, rendering the system unstable. For this simple system, testing for this condition is simply finding the roots of the quadratic. In order to test for stability in higher-order loops it may be necessary to use the Jury criterion, a sampled-data form of the Routh criterion, or, more easily, the root-finding capability of MATLAB. An excellent reference for the Jury criterion is Gupta [Gup76].

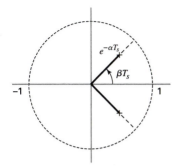

FIGURE 6.5-2

For this second-order loop, the design process is quite straightforward and we shall develop a design methodology. The phase-error transfer function of eq. 6.5-5b is written as:

$$\theta_e(z) = \frac{(z-1)^2}{(z-1)^2 + k(z-a)}\theta_i(z) = \frac{(z-1)^2}{z^2 - (2-k)z + (1-ka)}\theta_i(z)$$

$$= \frac{(z-1)^2}{z^2 - 2e^{-\alpha T_s}\cos(\beta T_s)z + e^{-2\alpha T_s}}\theta_i(z) \qquad (6.5\text{-}8a)$$

It is well known that the z-domain poles (fig. 6.5-2) are at

$$z = e^{-\alpha T_s}e^{\pm j\beta T_s} \qquad (6.5\text{-}8b)$$

The transient is therefore the sampled version of

$$\theta_e(t) = Ae^{-\alpha t}\cos(\beta t + \phi)u(t) \qquad (6.5\text{-}8c)$$

which, in the Laplace Transform domain, has poles at $s = -\alpha \pm j\beta$, giving a denominator of

$$s^2 + 2\alpha s + (\alpha^2 + \beta^2) \qquad (6.5\text{-}8d)$$

which can also be written as

$$s^2 + 2d\omega_n s + \omega_n^2 \qquad (6.5\text{-}8e)$$

where, for a simple RLC circuit,

$$Q = \frac{1}{2d} \quad \text{and} \quad 3 \text{ db BW} = \frac{1}{2\pi}\frac{\omega_n}{Q} = \frac{1}{2\pi}2d\omega_n = \frac{1}{2\pi}2\alpha \text{ hz} \qquad (6.5\text{-}8f)$$

EXAMPLE 1 *A Rapidly Converging Loop*

Let $\alpha = 1/2T_s$ so that when $t = 10T_s$ we have $e^{-\alpha t} = e^{-\alpha\,10T_s} = e^{-5} = .007$. It follows that $e^{-\alpha T_s} = e^{-.5} = 0.607$. For the purposes of this example, we shall assume that $T_s = 1/2400$. We shall arbitrarily choose $\beta T_s = \pi/6$ so that the frequency of oscillation of the transient is $\beta = \frac{1}{T_s}\frac{\pi}{6} = 2400(\pi/6) = 400\pi \Rightarrow 200$ hz. This means that a cycle of oscillation will correspond to 12 samples.

In the z-domain the poles are at

$$z = 0.607e^{\pm j\pi/6} = .526 \pm j.304$$

and the denominator is

$$z^2 - (2)(.607)(.866)z + (.607)^2 = z^2 - 1.052z + .368$$

so that $(2-k) = 1.052$ and $k = .948$ and $(1-ka) = .368$ or $a = .667$. To find the closed-loop bandwidth

$$3 \text{ db BW} = \frac{1}{2\pi}2\alpha \text{ hz} = \frac{1}{2\pi} \cdot 2400 = 382 \text{ hz}$$

Using the MATLAB script shown below, we can show the response of the PLL for a given set of frequency and phase offsets. Note that we establish the one symbol delay in the loop by using, at the beginning of a loop, a value of $\hat{\theta}$ computed at the end of the previous loop.

```
K1  = .948;
a11 = -.667;
b11 = -1;
phase_off = pi/6;
freq_off = 10;

thetahat = 0.0;
Z11 = 0.0;
theta = 0.0;

for  n = 1:100

        theta = theta + phase_off + 2*pi*freq_off/2400;

        alpha = theta - thetahat;
        y(n) = alpha;

        T1  = alpha*K1 - b11*Z11;
        result1 = T1 + a11*Z11;
        Z11 = T1;

        thetahat = thetahat + result1;

end
x = 1:1:100;
plot(x, y)
```

As shown in fig. 6.5-3a, the loop does, in fact, converge rapidly. We can also find the frequency response of this loop, fig. 6.5-3b, demonstrating that rapid convergence corresponds to large bandwidth.

```
A= [1 -1.052 .368];
B = [1 -2 1];
[h, w] = freqz(B, A, 512);
plot(w/pi, abs(h))
```

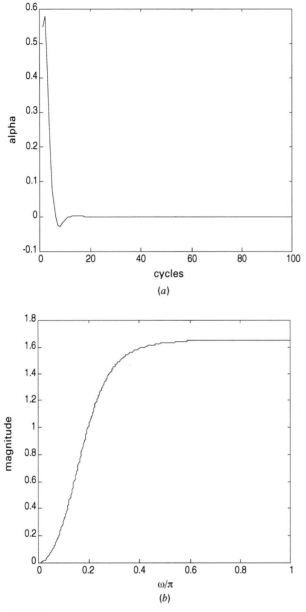

FIGURE 6.5-3

EXAMPLE 2   *A Slowly Converging Loop*

In this example we will make $\alpha = 1/20T_s$ so that when $t = 100T_s$, as opposed to $t = 10T_s$ in the rapidly converging loop, we have $e^{-\alpha t} = e^{-\alpha 10 T_s} = e^{-5} = .007$. It follows that $\alpha T_s = .05$ and $e^{-\alpha T_s} = e^{-.05} = .951$. Again, we will choose $\beta T_s = \pi/6$ so that the frequency of oscillation of the transient is $\beta = \frac{1}{T_s}\frac{\pi}{6} = 2400\frac{\pi}{6} = 400\pi \Rightarrow 200$ hz.

In the z-domain the poles are at

$$z = .951 e^{\pm j\pi/6} = .824 \pm j.476$$

and the denominator is

$$z^2 - (2)(.951)(.866)z + (.951)^2 = z^2 + 1.65z + .907$$

so that $(2-k) = 1.65$ and $k = .35$ and $(1-ka) = .907$ or $a = .266$. To find the closed-loop bandwidth

$$3 \text{ db BW} = \frac{1}{2\pi} 2\alpha \text{ hz} = 38.2 \text{ hz}$$

We can use the same MATLAB script to observe the convergence, fig. 6.5-4a.

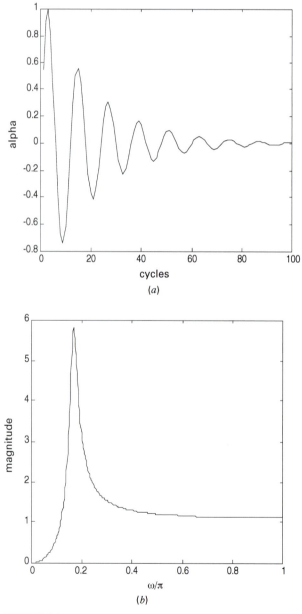

(a)

(b)

FIGURE 6.5-4

K1 = .35;
a11 = −.266;

The slow convergence corresponds to a narrowband filter as shown in fig. 6.5-4b.

```
A= [1  -1.65  .907];
B  = [1  -2  1];
[h, w]  =  freqz(B, A,  512);
plot(w/pi,  abs(h))
```

Clearly, there is an inverse relationship between how fast the loop converges and the loop bandwidth. Since the loop noise depends on the loop bandwidth, it also follows that a fast loop is noisier than a slow loop. ∎

6.6 NOISE AND CARRIER-PHASE JITTER IN SAMPLED-DATA CARRIER PHASE RECOVERY LOOPS

The simple second-order loop described above corrects for both phase offset and frequency offset but does not do an effective job of dealing with carrier-phase jitter. According to Falconer, the amount by which the jitter is attenuated by the loop is given by $|H(e^{j2\pi f_j})|$ where f_j is the frequency of the jitter and $H(z)$ is the closed-loop transfer function. For a second-order loop to suppress the jitter effectively, its closed-loop transfer function would have to extend out to the jitter frequency. Such a large loop bandwidth would increase the phase noise that is associated with the angular output of the slicer.

The noise bandwidth of the loop is computed as

$$B_L = \frac{1}{2\pi} \int_{-\pi}^{\pi} H(e^{-j\omega}) H(e^{j\omega}) \, d\omega \tag{6.6-1a}$$

where

$$\frac{\theta_0(z)}{\theta_i(z)} = H(z) = \frac{k(z+a)}{z^2 + (k-2)z + ka}. \tag{6.6-1b}$$

The noise in the error angle, which is added on a power basis to the noise that comes through with the signal in the absence of a carrier recovery, has a variance

$$\sigma_{\theta_\varepsilon}^2 = 1/\rho \quad \text{where} \quad \rho = \frac{A^2/2}{\eta B_L} = \frac{A^2}{2\eta B_L}. \tag{6.6-1c}$$

Since this noise is rotational, it may be decomposed into two equal rectangular components, each having a variance $\sigma_{\theta_\varepsilon}^2/2$, which may then be added to the in-phase and quadrature components of the noise in order to determine the error rate.

This expression for the angular noise power is the same as for the analog PLL. It is by no means exact and is reasonable only when the closed-loop filter bandwidth is much less than $f_s = 1/T_s$, the symbol frequency of the system. This is certainly the case for the V.22 with $f_s = 600$ hz or the V.32 with $f_s = 2400$ hz.

Two different approaches have been advanced to suppress phase jitter more effectively in the carrier-phase recovery system.

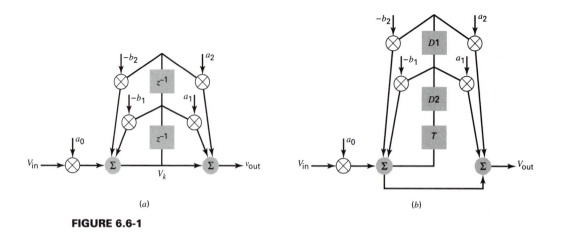

(a)   (b)

FIGURE 6.6-1

Bingham [Bin82] proposes to design a very narrow second-order loop as described above, having a loop bandwidth of perhaps 10 hz, in order to minimize the noise. He then adds two very narrow bandpass sections in parallel centered at the jitter frequency and its second harmonic. These filter sections serve to feed the jitter components back to the phase detector where they may be canceled. The difficulty with this approach is that the resulting PLL is 6th-order and loop stabilty is a major issue. In order to address this issue, Bingham uses a bilinear integrator $\frac{1}{2}\frac{z+1}{z-1}$ instead of the integrator $\frac{1}{z-1}$. Each of the narrow bandpass filters is a biquad section of the form

$$H(z) = a_0 \frac{z^2 + a_1 z + a_2}{z^2 + b_1 z + b_2} \qquad (6.6\text{-}2a)$$

A standard realization of $H(z)$ is shown in fig. 6.6-1a. To demonstrate this, note that

$$V_x = V_{in} - (b_1 z^{-1} + b_2 z^{-2}) V_x$$

or

$$V_{in} = (1 + b_1 z^{-1} + b_2 z^{-2}) V_x \qquad (6.6\text{-}2b)$$

and

$$V_{out} = (1 + a_1 z^{-1} + a_2 z^{-2}) V_x$$

so that $H(z) = \frac{V_{out}}{V_{in}}$. The actual implementation of the biquad in software requires a temporary storage location designated as T as shown in fig. 6.6-1b, because the program shifts pieces of data sequentially rather than together. The two z^{-1} terms are denoted $D1$ and $D2$. Therefore:

$$\begin{aligned} T &= a_0 \cdot V_{in} - b_1 D1 - b_2 D2 \\ V_{out} &= T + a_1 D1 + a_2 D2 \\ D2 &= D1 \\ D1 &= T \end{aligned} \qquad (6.6\text{-}2c)$$

For reference purposes and for purposes of setting up a number of Problems, Bingham's design for the loop filter for a V.22 bis type modem (16QAM, baud rate $f_b = 600$ hz, jitter frequency $f_j = 60$ hz, 120 hz) is given below:

$$F(z) = \frac{0.158z - 0.151}{z - 0.989} + \frac{0.09(z-1)^2}{z^2 - 1.61z + 0.99} + \frac{0.045(z-1)^2}{z^2 - 0.615z + 0.99} \qquad (6.6\text{-}3)$$

Note that this approach would require two separate designs, one for 60 hz power lines and the other for 50 hz power lines.

Gitlin [Git82] has proposed an adaptive predictor to reduce not only sinusoidal carrier phase jitter, but all carrier-phase jitter that has a nonwhite power spectral density. The predictor is in the form of an FIR filter that adapts according to an LMS algorithm. We shall discuss the LMS algorithm in Chapter 11 and return to the adaptation in that context.

Following Gooch and Ready [G&R87], this structure will be developed according to fig. 6.6-2. In fig. 6.6-2a, we show a second-order PLL of the kind discussed in Section 6.5 that compensates for both phase offset and frequency offset. The z^{-1} term emphasizes the delay of one symbol interval that is required between the generation of the phase output

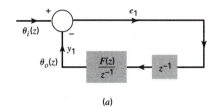

(a)

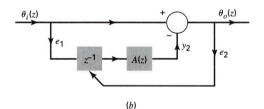

(b)

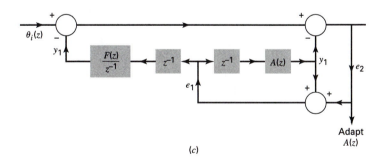

(c)

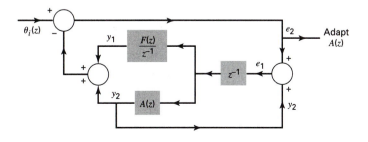

(d)

FIGURE 6.6-2

and its application to the phase detector. It is understood that since the PLL is narrowband, it will not interfere with phase jitter. In fig. 6.6-2b we assume that the input, which has distortion that consists only of phase jitter, is applied to a structure that has been given the name FIR-ALE (finite-impulse response-adaptive phase predictor) or APP (adaptive phase predictor). The transversal filter $A(z)$ will adapt to select the sinusoidal phase jitter as well as nonwhite or correlated components of phase jitter. These selected components are then subtracted from the original signal.

The PLL and the FIR-ALE are connected together in fig. 6.6-2c. Note that the signal $e_1 = e_2 + y_2$ is the same as e_1 in fig. 6.6-2a in the absence of phase jitter in the input. With phase jitter in the input, e_1 contains a replica of that phase jitter that passes through the PLL and is subtracted from the input.

The final version of the system is shown in fig. 6.6-2d. In order to have $A(z)$ adapt properly, it is necessary to first allow the PLL portion to converge so that phase and frequency offset are removed from the phase. When this is accomplished, $A(z)$ may be allowed to adapt.

Cupo and Gitlin [C&G89] have done a full analysis of this structure and have demonstrated that the adaptation coefficient μ is subject to the limits

$$0 < \mu < \frac{2}{L\left[\sum_{m=0}^{J} \frac{A_m^2}{2} + \sigma^2\right]} \tag{6.6-4}$$

where L is the number of taps in $A(z)$, J is the number of sinusoidal jitter components, A_m is the amplitude of the mth component, and σ^2 is the system noise.

Consideration of the performance of this structure, the number of taps required, and the speed of convergence is left to the Problems in Chapter 11.

6.7 SOFTWARE IMPLEMENTATION OF THE ADAPTIVE ROTATOR

We have seen in fig. 6.4-2 and eq. 6.4-2c how the input to the slicer may be used to compute the angular error that will drive the PLL in the carrier-phase recovery system. In order to demonstrate the algorithm more clearly, we shall again turn to the modem simulation program developed in Appendix 4A. An second version of the block diagram of the program, including the equalizer, the slicer, and the adaptive rotator, is shown in fig. 6.7-1.

The phase distortion, including phase offset, frequency offset, and carrier phase jitter rotate the transmitted signal through an angle θ, as indicated in eq. 6.0-1. The operation that is performed is $e^{j\theta_n}$. Noise is then added and the noisy complex baseband received signal (z_I, z_Q) is generated. This signal is applied to the derotate block which performs the operation $e^{-j\hat{\theta}_n}$, where $\hat{\theta}_n$ is an estimate of θ that was computed in the previous symbol interval. The derotated signal (z_{Ir}, z_{Qr}) is then applied to the slicer, which determines the symbol estimates $(\hat{a}_n, \hat{b}_n)$ and the data which is then sent off to the descrambler. The block Error takes both the derotated signal and the symbol estimates and computes

$$\alpha = \tan^{-1}\left[\frac{z_{Qr}\hat{a}_n - z_{Ir}\hat{b}_n}{z_{Ir}\hat{a}_n + z_{Qr}\hat{b}_n}\right] \tag{6.7-1}$$

In turn, α is the input to the carrier filter, which has been described above. The output of the carrier filter is $\hat{\theta}_{n+1}$, which will be used in the next symbol interval.

Some issues relating to the design of the carrier filter will be explored in the Problems. The implementation of the FIR-ALE will be deferred to Chapter 11 in conjunction with the discussion of adaptive equalization. Also in relation to Chapter 11, it should be noted that

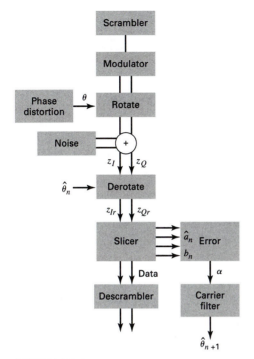

FIGURE 6.7-1

the block Error will subsequently be expanded to include generation of the error signals required for adaptation of the equalizer.

6.8 MAXIMUM LIKELIHOOD TIMING RECOVERY

Maximum likelihood estimation [Sal67], [W&L69] can also be used to find the correct timing phase for sampling a received data signal. The received signal in a baseband PAM system is

$$y(t) = \sum_n a_n h(t - nT_s + \tau) + n(t) \tag{6.8-1}$$

where $h(t)$ is the impulse response of the system, τ is the timing phase, $a_n = \pm 1$ is equally likely binary data, and $n(t)$ is AWGN having power spectral density $\eta/2$.

Following Gitlin and Salz [G&S71], the idea is to minimize the probability of error with respect to τ. The likelihood ratio with τ as the variable comes from eq. 6.3-3d and eq. 6.3-11a, which can be rewritten to read (with certain constants omitted)

$$L(\tau) = E\left\{ \exp\left[-\frac{1}{\eta} \int_0^{NT_s} \left[y(t) - \sum_n a_n h(t - nT_s + \tau) \right]^2 dt \right] \right\} \tag{6.8-2a}$$

where $E\{\ \}$ is the expectation over the data symbols. The upper limit of integration has been chosen to be a multiple of the symbol rate T_s so that the estimate of τ will come from an averaging process. Expanding the square term in the integral, we can see that $y^2(t)$ does not depend on τ and so may be dropped. We can also see that

$$E\left\{\exp\left[-\frac{1}{\eta}\int_0^{NT_s}\left[\sum_n a_n h(t-nT_s+\tau)\right]^2 dt\right]\right\} \qquad (6.8\text{-}2b)$$

is, for large N, simply the energy in the signal, also independent of τ. Consequently we are left with

$$L(\tau) = E\left\{\exp\left[\frac{2}{\eta}\int_0^{NT_s}\sum_n a_n y(t)h(t-nT_s+\tau)\,dt\right]\right\}$$

$$= E\left\{\exp\left[\frac{2}{\eta}\sum_n a_n \int_0^{NT_s} y(t)h(t-nT_s+\tau)\,dt\right]\right\} \qquad (6.8\text{-}2c)$$

where, for sufficiently large N,

$$z_n(\tau) = \int_0^{NT_s} y(t)h(t-nT_s+\tau)\,dt \qquad (6.8\text{-}2d)$$

is the output of a filter matched to $h(t)$, sampled at times $nT_s+\tau$. Consequently, eq. 6.8-2c can be rewritten as

$$L(\tau) = E\left\{\prod_n \exp\left[\frac{2}{\eta}a_n z_n(\tau)\right]\right\}$$

$$= E\left\{\prod_n \frac{\exp\left[\frac{2}{\eta}z_n(\tau)\right]+\exp\left[-\frac{2}{\eta}z_n(\tau)\right]}{2}\right\} = E\left\{\prod_n \text{Cosh}\left[\frac{2}{\eta}z_n(\tau)\right]\right\}$$

$$(6.8\text{-}2e)$$

In order to maximize $L(\tau)$ conveniently, we shall work with the natural log

$$\ln L(\tau) = E\left\{\sum_n \ln\left\{\text{Cosh}\left[\frac{2}{\eta}z_n(\tau)\right]\right\}\right\} \qquad (6.8\text{-}2f)$$

The first thing to note is that $\ln \text{Cosh}(x) \approx x^2/2$ for small values of x (the proof will be left to the Problems) so that, for small values of SNR, we may approach the optimum timing phase by applying the output of the matched filter to a square law device. This method of timing recovery was explored in Chapter 2 and it is of interest to know that it is in relation to maximum likelihood techniques. For larger values of SNR, square-law timing recovery becomes increasingly suboptimal but has the advantage of being very easy to implement.

In a QAM receiver with a Hilbert Transform front end, the two rails give us

$$y(t) = \text{Re}\left\{\sum_n d_n h(t-nT_s+\tau)e^{j(\omega_c t+\theta(t))}\right\}$$

$$(6.8\text{-}2a)$$

$$\hat{y}(t) = \text{Im}\left\{\sum_n d_n h(t-nT_s+\tau)e^{j(\omega_c t+\theta(t))}\right\}$$

where $d_n = a_n + jb_n$. Squaring these two outputs separately, summing the results, and

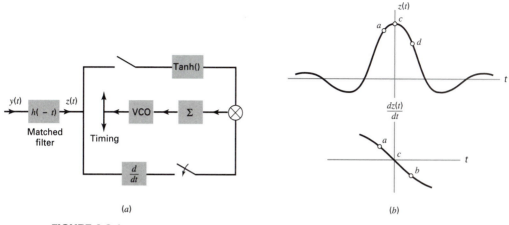

FIGURE 6.8-1

averaging or filtering to remove the double frequency terms yields

$$y^2(t) + \hat{y}^2(t) = \sum_n |d_n|^2 h^2(t - nT_s + \tau) \tag{6.8-2b}$$

removing any dependence on the carrier. In effect, this approach makes square-law timing recovery in bandpass and baseband systems equivalent.

A more rigorous approach to maximum likelihood timing recovery will involve differentiating the log-likelihood function of eq. 6.8-2f so that

$$m(\tau) = \frac{\partial[\ln L(\tau)]}{\partial \tau} = E\left\{\sum_n \frac{\partial}{\partial \tau} \ln\left\{\text{Cosh}\left[\frac{2}{\eta} z_n(\tau)\right]\right\}\right\}$$

$$= E\left\{\sum_n \frac{2}{\eta} \frac{\text{Sinh}\left[\frac{2}{\eta} z_n(\tau)\right]}{\text{Cosh}\left[\frac{2}{\eta} z_n(\tau)\right]} \frac{\partial z_n(\tau)}{\partial \tau}\right\} = E\left\{\sum_n \frac{2}{\eta} \text{Tanh}\left[\frac{2}{\eta} z_n(\tau)\right] \frac{\partial z_n(\tau)}{\partial \tau}\right\} = 0$$

$$\tag{6.8-3}$$

identifies the maximum likelihood estimate. As shown in fig. 6.8-1a, a PLL-like structure will implement this loop. In this implementation it should recalled that for reasonably high SNR the Tanh function is approximated by a slicer and the summation is an averaging filter. Different filter configurations, such as lowpass filters, may also be used. In order to get a qualitative idea of what is happening in this circuit, define $z(t)$ as the output of the matched filter when $h(t)$ is the input; in other words, $z(t) = h(t) * h(-t)$ where $*$ denotes convolution. Consider $z(t)$ and its derivative in fig. 6.8-1b. If the timing is such that $z(t)$ and its derivative are sampled at $t = a$, then the product of the sliced value of $z(t)$ and the derivative is positive, calling upon the VCO to advance the timing phase. If the sampling is at $t = b$ then the product of $z(t)$ and the derivative is negative, calling upon the VCO to retard the timing phase. Clearly, sampling at $t = c$ yields a zero error to the VCO. The object of the algorithm is to set the timing phase at the peak of $z(t)$.

The source of timing jitter is that $h(t)$ extends over multiple symbol intervals, that $h(t)$ comes from a distorted channel so the matched filter output is not Nyquist, and that the data is random. This makes the zero crossing of the derivative into a random variable. This system has been widely used in magnetic recording [Cio90].

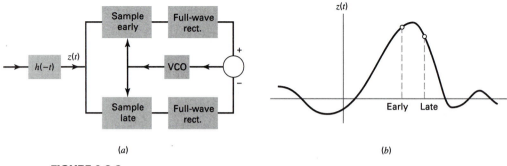

(a) (b)

FIGURE 6.8-2

A suboptimal variant of the system of fig. 6.8-1b is called *early-late timing recovery*, shown in fig. 6.8-2a. It may be obtained by approximating the derivative of the log-likelihood function of eq. 6.8-3 by a difference

$$m(\tau) = \frac{\partial [\ln L(\tau)]}{\partial \tau} = E\left\{ \sum_n \frac{\partial}{\partial \tau} \ln\left\{ \text{Cosh}\left[\frac{2}{\eta} z_n(\tau) \right] \right\} \right\}$$

$$\approx E\left\{ \sum_n \left\{ \ln \text{Cosh}\left[\frac{2}{\eta} z_n(\tau + \varepsilon) \right] - \ln \text{Cosh}\left[\frac{2}{\eta} z_n(\tau - \varepsilon) \right] \right\} \right\} \quad (6.8\text{-}4a)$$

For large values of x, $\text{Cosh}(x) \approx e^x$ and $\ln \text{Cosh}(x) \approx x$ so that

$$m(\tau) = \frac{\partial [\ln L(\tau)]}{\partial \tau} \approx E\left\{ \sum_n \frac{2}{\eta} \{ z_n(\tau + \varepsilon) - z_n(\tau - \varepsilon) \} \right\} \quad (6.8\text{-}4b)$$

As shown in fig. 6.8-2b, the idea is to try to locate the peak of $z(t)$ by locating the 'early' sample and the 'late' sample so that the difference in the sample values is zero. That difference is, of course, an average value over a large number of samples and, in the case of a distorted pulse, the midpoint of the samples may not exactly correspond to the peak. Nevertheless, the system works and has application in satellite communications.

The implementations shown in both fig. 6.8-1a and fig. 6.8-2a operate according to the same pattern: the sampling time is sequentially adjusted according to an algorithm that looks like

$$\tau_{n+1} = \tau_n + \alpha[g(z_n, \tau_n)] \quad (6.8\text{-}4a)$$

where α is a constant step size and $g(z_n, \tau_n)$ has, in each implementation, been defined in such a way that

$$E[g(z_n, \tau_n)] = m(\tau). \quad (6.8\text{-}5b)$$

This algorithm, known as the Robbins-Monro algorithm, will converge to $m(\tau) = 0$, thereby sequentially approching the maximum likelihood estimate. We will be using this approach in a number of timing recovery techniques.

Using this approach on a Nyquist pulse having a 30 db SNR and a bandwidth of 3000 hz, Gitlin and Salz [G&S71] demonstrate effective convergence to the optimum timing phase in 10 symbols.

6.9 BAUD-RATE TIMING RECOVERY

Note that all of the methods of timing recovery that we have discussed until now have operated either directly on the analog received signal or on samples of the received signal taken at least at twice the baud rate.

By way of contrast, the requirements of the ISDN, including sampling the received signal at the baud rate and doing the remaining processing digitally, have given rise to the need for finding the optimum timing phase based on baud-rate samples of the received signal [M&M76], [Abo94].

The first step is to show how we can obtain baud-rate samples of the channel impulse response from the sampled received data. The analog received signal is

$$x(t) = \sum_m d_m h(t - mT_s) \tag{6.9-1a}$$

and sampled at the baud rate is

$$x_n = \sum_m d_m h_{n-m} = d_0 h_n + \sum_{m \neq 0} d_m h_{n-m} \tag{6.9-1b}$$

where the analog and sampled versions of $h(t)$ are shown in fig. 6.9-1.

Let $\hat{d}_0$ be the decided data so that

$$\hat{d}_0 x_n = \hat{d}_0 d_0 h_n + \sum_{m \neq 0} \hat{d}_0 d_m h_{n-m} \tag{6.9-1c}$$

and assuming that the data is binary and uncorrelated and that the decision is correct, we have

$$E[\hat{d}_0 x_n] = h_n \tag{6.9-1d}$$

This gives us a way of estimating the magnitude of the impulse response at the sampling point provided that there is no delay between the data and the decision. Any delay will require compensation.

Other baud rate samples of the impulse response can be estimated in a similar manner. For example,

$$x_{n-1} = d_0 h_{n-1} + \sum_{m \neq 0} d_m h_{n-1-m} \Rightarrow E[\hat{d}_0 x_{n-1}] = h_{n-1} \tag{6.9-2a}$$

and

$$x_{n+1} = d_0 h_{n+1} + \sum_{m \neq 0} d_m h_{n+1-m} \Rightarrow E[\hat{d}_0 x_{n+1}] = h_{n+1} \tag{6.9-2b}$$

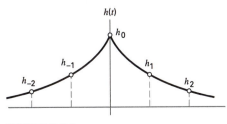

FIGURE 6.9-1

Notice that estimating h_{n-1} involves cross-correlating the present decision with the previous data. This is easily accomplished by shifting the data into a transversal filter. On the other hand, estimating h_{n+1} leads to difficulties because it involves waiting for the next piece of data, a delay that should be avoided.

This problem can be addressed by reformulating eq. 6.9-2b to be

$$x_n = d_{-1} h_{n+1} + \sum_{m \neq -1} d_m h_{n-m} \Rightarrow E[\hat{d}_{-1} x_n] = h_{n+1} \tag{6.9-3a}$$

and

$$x_n = d_{-2} h_{n+2} + \sum_{m \neq -2} d_m h_{n-m} \Rightarrow E[\hat{d}_{-2} x_n] = h_{n+2} \tag{6.9-3b}$$

and so on.

The idea, then, is that by saving both past data and past decisions in shift registers or transversal filters, we have the information that is necessary to give us estimates of the baud-rate samples of the system impulse response $h(t)$.

Now let us recall the Robbins-Monro iterative algorithm (eq. 6.8-5b) in the form

$$\tau(n+1) = \tau(n) - \lambda z_n(\tau) \tag{6.9-4}$$

for sequentially approaching a good timing phase. The object is to construct an appropriate function $z_n(\tau)$. Some examples are in order.

A Nyquist pulse and a distorted version of that pulse are shown in fig. 6.9-2a. In the Nyquist pulse it is clear that the timing function

$$z_n(\tau) = h_1 - h_{-1} \tag{6.9-5a}$$

will be appropriate for identifying the peak of $h(t)$. Since we are constrained to establish the timing phase from the unequalized signal, using the same timing function for the distorted

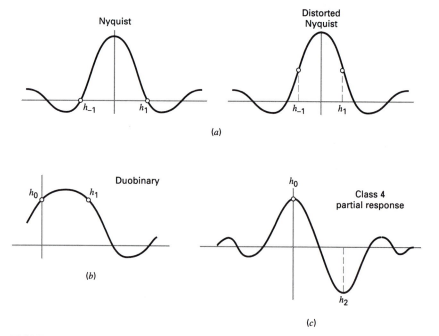

(a)

(b)

(c)

FIGURE 6.9-2

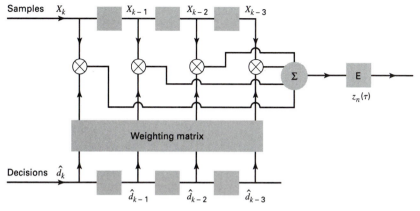

FIGURE 6.9-3

pulse will be workable but will not end up getting us to the peak. It will, however, get us more or less close to the peak.

A duobinary pulse (Class-1 partial response) is shown in fig. 6.9-2b. An appropriate timing function for this system is

$$z_n(\tau) = h_0 - h_1 \tag{6.9-5b}$$

A Class 4 partial response pulse is shown in fig. 6.9-2c. An appropriate timing function for this system is

$$z_n(\tau) = h_0 + h_2 \tag{6.9-5c}$$

Generally, the timing function may be constructed according to the structure shown in fig. 6.9-3. It is important to remember to average or take the expectation of these computed values befor applying them to the Robbins-Monro algorithm. Timing jitter is clearly dependent on the variance of the estimate, which in turn is reduced by increasing the number of terms in the computation of the expectation.

Finally, this approach allows the user very easily to make up timing functions that suit particular needs. These functions may involve as many terms as deemed appropriate.

6.10 TIMING RECOVERY FOR QAM SYSTEMS WITH BAUD-RATE EQUALIZATION

We have seen that systems based on maximum likelihood timing recovery algorithms search for the sampling phase that corresponds to the peak of the output of the receiver filter, the end-to-end channel impulse response. It is fairly straightforward to demonstrate that this is typically not the best sampling phase for QAM systems where the channel equalization is to be performed on baud rate samples of the complex received distorted signal.

This issue was first analyzed by Lyon [Lyon75-1] [Lyon75-2] for analog timing recovery systems and then restated by Godard [God78] for an all-digital receiver. We will follow Godard's approach. The general approach is called BECM (Band Edge Component Maximization).

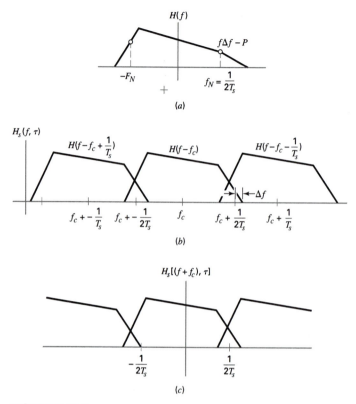

FIGURE 6.10-1

The output of the front end of a Hilbert Transform receiver in a QAM system may be written as

$$y_+(t) = y(t) + j\hat{y}(t) = \sum_n d_n h(t - nT_s)e^{j2\pi f_c t} + v(t) \tag{6.10-1}$$

where $\hat{y}(t)$ is the Hilbert transform of $y(t)$, $h(t)$ is the baseband equivalent complex impulse-response of the system including raised cosine shaping and a distorted channel, and $v(t)$ is complex bandpass noise.

$H(f)$, the Fourier Transform of $h(t)$, is shown in fig. 6.10-1a. The following identities deriving from the Poisson Sum Formulas will be useful:

$$h(t - nT_s) \Leftrightarrow H(f)e^{-jn2\pi f T_s} \tag{6.10-2a}$$

$$h(t - nT_s)e^{j2\pi f_c t} \Leftrightarrow H(f - f_c)e^{-jn2\pi(f - f_c)T_s} \tag{6.10-2b}$$

so that

$$y(t) \Leftrightarrow Y(f) = \sum_n d_n H(f - f_c)e^{-jn2\pi(f - f_c)T_s} \tag{6.10-2c}$$

We will now sample $y(t)$ at the baud rate, every T_s seconds, at an arbitrary timing phase τ. The Fourier Transform of the sampled signal will be found step by step. First:

$$\{y'(t) = y(t + \tau)\} \Leftrightarrow \left\{ Y'(f) = \sum_n d_n H(f - f_c)e^{-jn2\pi(f - f_c)T_s}e^{-j2\pi f\tau} \right\} \tag{6.10-3a}$$

Next, the sequence of samples of $y'(t)$ can be written as

$$y^*(t) = \sum_k y'(kT_s)\delta(t - kT_s) = y'(t)\delta_T(t) \qquad (6.10\text{-}3b)$$

having a Fourier Transform

$$Y^*(f) = Y'(f) \otimes \delta_{f_s}(f) = \frac{1}{T_s}\sum_k Y'\left(f - \frac{k}{T_s}\right)$$

$$= \frac{1}{T_s}\sum_k\sum_n d_n H\left(f - f_c - \frac{k}{T_s}\right)e^{-jn2\pi\left(f - f_c - \frac{k}{T_s}\right)T_s}e^{-j2\pi\left(f - \frac{k}{T_s}\right)\tau}$$

$$= \frac{1}{T_s}\sum_n d_n e^{-jn2\pi(f - f_c)T_s}\sum_k H\left(f - f_c - \frac{k}{T_s}\right)e^{-j2\pi\left(f - \frac{k}{T_s}\right)\tau} \qquad (6.10\text{-}3c)$$

where we define

$$H_s(f, \tau) = \sum_k H\left(f - f_c - \frac{k}{T_s}\right)e^{-j2\pi\left(f - \frac{k}{T_s}\right)\tau} \qquad (6.10\text{-}3d)$$

which is depicted in fig. 6.10-1b. The sampled spectrum $H_s(f, \tau)$ is what the equalizer will be trying to equalize. Note that there will be a demodulation before the equalizer so that the actual signal that gets to the complex baseband equalizer of fig. 4.6-5, $H_s[(f + f_c), \tau]$, is as shown in fig. 6.10-1c.

The character of the aliasing that results from baud-rate sampling depends on the timing phase. A poor choice of timing phase, even sampling at the peak of the data pulse, can create dips and even nulls in the sampled spectrum around the Nyquist frequency. Equalizing these dips and nulls results in noise amplification that degrades the performance of the sytem. The object of the Godard algorithm is to settle upon a timing phase that maximizes the energy in the region of aliasing.

The raised cosine shaping in the channel guarantees that the only overlap or aliasing that will occur in fig. 6.10-1b is between adjacent spectra. Note that $H_s(f, \tau)$ clearly depends on the value of τ, although that dependence is not explicitly seen in fig. 6.10-1b. Godard's proposal is to choose the timing phase τ in order to maximize the energy of $H_s(f, \tau)$. This energy is computed as

$$\varepsilon^2(\tau) = T_s\int_{f_c}^{f_c + 1/T_s} |H_s(f, \tau)|^2\, df \qquad (6.10\text{-}4a)$$

Since the overlap is only on adjacent spectra, we can, for the purposes of integration, write eq. 6.10-4a as

$$\varepsilon^2(\tau) = T_s\int_{f_c}^{f_c + 1/T_s}\left|\frac{1}{T_s}H(f - f_c)e^{j2\pi f\tau} + \frac{1}{T_s}H\left(f - f_c - \frac{1}{T_s}\right)e^{j2\pi\left(f - \frac{1}{T_s}\right)\tau}\right|^2 df$$

$$(6.10\text{-}4b)$$

In order to evaluate this integral, consider

$$|a + b|^2 = (a + b)(a^* + b^*) = |a|^2 + |b|^2 + 2\,\text{Re}[ab^*] \qquad (6.10\text{-}4c)$$

In the integral of eq. 6.10-4b, the magnitude square of the individual terms eliminates dependence on τ, making those terms into a simple constant. Therefore:

$$\varepsilon^2(\tau) = \text{const} + 2\,\text{Re}\left[\left[e^{j2\pi\frac{\tau}{T_s}}\right]\int_{f_c}^{f_c+1/T_s} H(f - f_c)H^*\left(f - f_c - \frac{1}{T_s}\right)df\right] \quad (6.10\text{-}4d)$$

Since the object is to find the value of τ that maximizes $\varepsilon^2(\tau)$, the constant can be dropped and the limits of integration changed to encompass only the region of overlap of the aliased spectra. Thus:

$$\varepsilon^2(\tau) = 2\,\text{Re}\left[\int_{f_c+\frac{1}{2T_s}-\nabla f}^{f_c+\frac{1}{2T_s}+\nabla f} H(f - f_c)H^*\left(f - f_c - \frac{1}{T_s}\right)df\left[e^{j2\pi\frac{k}{T_s}}\right]\right] \quad (6.10\text{-}4e)$$

When the integration is done, the result will be of the form

$$\varepsilon^2(\tau) = 2\,\text{Re}\left[Ae^{j\alpha}e^{j2\pi\frac{\tau}{T_s}}\right] = 2A\text{Cos}\left(2\pi\frac{\tau}{T_s} + \alpha\right) \quad (6.10\text{-}4f)$$

where α depends on the channel. Clearly $\varepsilon^2(\tau)$ will be maximized at a sampling time that maximizes the Cos. Note that this maximum occurs every T_s sec. This forms the basis for a Robbins-Monro iterative algorithm of the form

$$\tau(n + 1) = \tau(n) + \lambda\frac{\partial\varepsilon^2(\tau)}{\partial\tau} \quad (6.10\text{-}5)$$

where we can see that the best timing phase corresponds to the derivative $(\partial\varepsilon^2(\tau)/\partial\tau) = 0$. In order to implement the algorithm we must be able to compute the derivative from the received data.

From eqs. 6.10-4e and 6.10-4f, it is easy to see that this derivative is

$$\frac{\partial\varepsilon^2(\tau)}{\partial\tau} = -\frac{4\pi A}{T_s}\text{Sin}\left(2\pi\frac{\tau}{T_s} + \alpha\right)$$

$$= -\frac{4\pi}{T_s}\text{Im}\left[\int_{f_c+\frac{1}{2T_s}-\nabla f}^{f_c+\frac{1}{2T_s}+\nabla f} H(f - f_c)H^*\left(f - f_c - \frac{1}{T_s}\right)df\left[e^{j2\pi\frac{\tau}{T_s}}\right]\right] \quad (6.10\text{-}6)$$

According to fig. 6.10-1b, the limits of integration in eq. 6.10-6 may be changed to $\pm\infty$ and consequently we may invoke Parseval's Theorem to evaluate the integral in the time domain.

Let $b(t) = e^{-\alpha t}u(t)$ be the impulse response of a narrow lowpass filter, so that

$$b(t)e^{j\beta t} = b(t)\text{Cos}(\beta t) + jb(t)\text{Sin}(\beta t) \quad (6.10\text{-}7a)$$

is the complex impulse response of a narrowband bandpass filter. If the input to this filter is a signal $x(t)$, the output is a filtered signal $y(t)$, the real output, and the Hilbert Transform

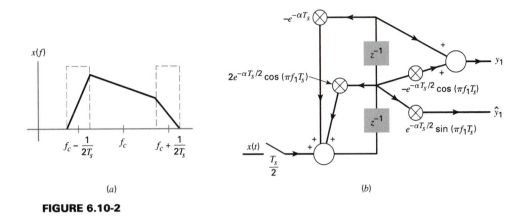

(a) (b)

FIGURE 6.10-2

of $y(t)$ is the imaginary output $\hat{y}t$. Recall the z-transforms

$$e^{-\alpha t}\text{Cos}(\beta t)u(t) \Leftrightarrow \frac{z^2 - ze^{-\alpha T}\text{Cos}(\beta T)}{z^2 - 2ze^{-\alpha T}\text{Cos}(\beta T) + e^{-2\alpha T}} \qquad (6.10\text{-}7b)$$

$$e^{-\alpha t}\text{Sin}(\beta t)u(t) \Leftrightarrow \frac{ze^{-\alpha T}\text{Sin}(\beta T)}{z^2 - 2ze^{-\alpha T}\text{Cos}(\beta T) + e^{-2\alpha T}} \qquad (6.10\text{-}7c)$$

where T is the sampling interval. This kind of filter, shown in fig. 6.10-2a, will be used to extract the components of the input signal $x(t)$ around the frequencies $f_1 = f_c + \frac{1}{2T_s}$ and $f_2 = f_c - \frac{1}{2T_s}$. In order to avoid aliasing in this process, $x(t)$ will have to be sampled every $T_s/2$ sec. We will need two filters of the type shown in fig. 6.10-2b for each of the frequencies f_1 and f_2, a total of four filters. The filters centered at f_1 have transfer functions:

$$F_1(z) = \frac{z^2 - ze^{-\alpha T_s/2}\text{Cos}(\pi f_1 T_s)}{z^2 - 2ze^{-\alpha T_s/2}\text{Cos}(\pi f_1 T_s) + e^{-\alpha T_s}} \qquad (6.10\text{-}8a)$$

$$\hat{F}_1(z) = \frac{ze^{-\alpha T_s/2}\text{Sin}(\pi f_1 T_s)}{z^2 - 2ze^{-\alpha T_s/2}\text{Cos}(\pi f_1 T_s) + e^{-\alpha T_s}} \qquad (6.10\text{-}8b)$$

and those centered at f_2 have the same form with appropriate parameter changes.

The output of the filters centered at f_1 is to be multiplied by the complex conjugate of the one centered at f_2 and the imaginary part extracted. This can be expressed as

$$\text{Im}[(y_1 + j\hat{y}_1)(y_2 - j\hat{y}_2)] = \hat{y}_1 y_2 - \hat{y}_2 y_1 \qquad (6.10\text{-}9)$$

The result is sampled at the symbol rate and, as shown in fig. 6.10-3, applied to the control circuitry of a VCO. The purpose of the averaging loop and the constant μ_2 is to correct for frequency offset between the incoming data clock and the VCO output. The purpose of the constant μ_1 is to find the sampling phase that maximizes the energy. The resulting algorithm for updating timing phase is

$$\tau(n + 1) = \tau(n) - \mu_1 p_n - \mu_2 \sum_k p_k \qquad (11.64)$$

Note that aliasing issues require that the filters in fig. 6.10-3 require inputs every $T_s/2$ sec, even though the equalizer itself will operate on samples every T_s sec. Further discussion of this issue will be reserved until Chapter 11.

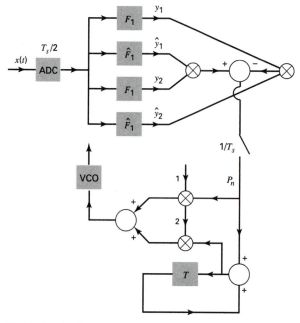

FIGURE 6.10-3

An analysis of the timing jitter that results from this approach is quite daunting and not easily available. However, a simulation approach for this algorithm is suggested in the Problems for Chapter 7.

There will be additional discussion of timing recovery in Chapters 10 and 11.

CHAPTER 6 PROBLEMS

6.1 Redesign the carrier recovery filter of eq. 6.6-3 to correspond to a symbol rate of 2400 hz. Expanding the MATLAB script of Sec. 6.5, Example 1, add carrier-phase jitter at 60 hz and 120 hz and conduct some experiments to show how the filter suppresses this jitter.

This Problem will be continued as Problem 11-12 for an examination of the effectiveness of the PLL-ALE structure discussed in Sec. 6.6.

6.2 Explain what is required to make the maximum likelihood carrier recovery system for offset QPSK operate in the same way as that for standard QPSK.

6.3 In relation to eq. 6.8-2f, show that $\ln \text{Cosh}(x) \approx x^2/2$.

6.4 The received pulse shape in fig. 6.8-1 is

$$f(t) = \begin{cases} \frac{1}{2}\left(1 + \text{Cos}\left(\frac{2\pi t}{T_s}\right)\right) & -\frac{T_s}{2} \le t \le \frac{T_s}{2} \\ 0 & \text{otherwise} \end{cases}$$

and the input waveform is

$$z(t) = \sum_n a_n f(t - \tau - nT_s)$$

where $a_n = \pm 1$ and τ is the sampling phase. Find an expression for the input to the VCO as a function of τ. In so doing, treat the summation at the VCO input as an expectation operator. What can be said about timing jitter in this example?

6.5 Find an expression for the expected value of the input to the VCO in a baud-rate timing recovery system for:

(a) Duobinary

(b) Class-4 Partial Response.

REFERENCES

[Abo94]   T. Aboulnasr et al., "Characterization of a symbol rate timing recovery technique for a 2BQ1 digital receiver," *IEEE Trans. Communications*, **Comm-42**, No. 2/3/4, Feb./March/April, 1994, pp. 1409–1414.

[B&M83]   N. Blachman and S. Mousavinezhad, "Carrier-tracking loop performance for quaternary and binary PSK signals," *IEEE Trans. Aerospace and Electronic Systems*, **AES-19**, No. 2, March 1983, pp. 162–166.

[Bin82]   J.A.C. Bingham, "Design of carrier loop filters to track phase jitter," *Asilomar Conference on Circuits, Systems, and Computers*, Monterey, CA, 1982.

[Bin88]   J.A.C. Bingham, "The theory and practice of modem design," Wiley Interscience, 1988.

[C&G89]   R.L. Cupo and R.D. Gitlin, "Adaptive carrier recovery systems for digital data communications receivers," *IEEE J. Selected Areas in Communications*, **7**, No. 9, Dec. 1989, pp. 1328–1331.

[Cio90]   J. Cioffi et al., "Adaptive equalization in magnetic-disk storage channels," *IEEE Communications Magazine*, Feb. 1990.

[Fal76-1]   D.D. Falconer, "Jointly adaptive equalization and carrier recovery in two-dimensional digital communication systems," *Bell System Technical J.*, **55**, No. 3, March 1976, pp. 317–394.

[Fal76-2]   D.D. Falconer, "Analysis of a gradient algorithm for simultaneous passband equalization and carrier phase recovery," *Bell System Technical J.*, **55**, No. 4, April 1976, pp. 409–4211.

[G&G72-1]   G.S. Gill and S.C. Gupta, "First order discrete phase-locked loop with application to demodulation of angle-modulated carrier," *IEEE Trans. Communications*, **Comm-20**, No. 6, June 1972, pp. 454–462.

[G&G72-2]   G.S. Gill and S.C. Gupta, "On higher order discrete phase-locked loops," *IEEE Trans. Aerospace and Electronic Systems*, **AES-8**, No. 5, Sept. 1972, pp. 615–623.

[G&R87]   R.P. Gooch and M.J. Ready, "An adaptive phase lock loop for phase jitter tracking," *Twenty-First Asilomar Conference on Signals, Systems, and Computers*, Monterey, CA, 1987.

[G&S71]   R.D. Gitlin and J. Salz, "Timing recovery in PAM systems," *Bell System Technical J.*, **50**, No. 5, May-June 1971, pp. 1645–1669.

[Git82]   R.D. Gitlin, "Adaptive phase jitter tracker," U.S. Pat. 4320526, March 16, 1982.

[God78]   D.N. Godard, "Passband timing recovery in an all-digital modem receiver," *IEEE Trans. Communications*, **Comm-26**, No. 5, May 1978, pp. 517–523.

[Gup76]   S.C. Gupta, "Transform and state variable methods in linear systems," John Wiley and Sons, 1976.

[L&S71]   W.C. Lindsay and M.K. Simon, "Data aided tracking loop," *IEEE Trans. Communications*, **Comm-19**, No. 4, April 1971, pp. 157–168.

[L&S72]   W.C. Lindsay and M.K. Simon, "Carrier synchronization and detection of polyphase signals," *IEEE Trans. Communications*, **Comm-20**, No. 2, Feb. 1972, pp. 441–454.

[Lin72]   W.C. Lindsay, *Synchronization systems in Communications*, Prentice-Hall, 1972.

[L&S73]   W.C. Lindsay and M.K. Simon, *Telecommunication Systems Engineering*, Prentice-Hall, 1973.

[L&C83]   A. Leclert and P. Vandamme, "Universal carrier recovery loop for QASK and PSK signal sets," *IEEE Trans. Communications*, **Comm-31**, No. 1, Jan. 1983, pp. 130–136.

[Lyon75-1] D.L. Lyon, "Timing recovery in synchronous equalized data communication," *IEEE Trans. Communications*, **Comm-23**, Feb. 1975, pp. 269–274.

[Lyon75-2] D.L. Lyon, "Envelope derived timing recovery in QAM and SQAM systems," *IEEE Trans. Communications*, **Comm-23**, Nov. 1975, pp. 1327–1331.

[M&M74]   R. Matyas and P. McLane, "Decision-aided tracking loops for channels with phase jitter and intersymbol interference," *IEEE Trans. Communications*, **Comm-22**, No. 8, Aug. 1974, pp. 1014–1023.

[M&M76]   K.H. Mueller and M. Muller, "Timing recovery in digital synchronous data receivers," *IEEE Trans. Communications*, **Comm-24**, No. 5, May 1976, pp. 516–531.

[Sal67]   B.R. Saltzberg, "Timing recovery for synchronous binary data transmission," *Bell System Technical J.*, **46**, No. 3, March 1967, pp. 593–622.

[Sim74]   Marvin K. Simon, "Carrier synchronization and detection of QASK signal sets," *IEEE Trans. Communications*, **Comm-22**, No. 2, Feb. 1974, pp. 98–105.

[Sim79-1] Marvin K. Simon, "On the optimality of the MAP estimation loop for carrier phase tracking BPSK and QPSK signals," *IEEE Trans. Communications*, **Comm-27**, No. 1, Jan. 1979, pp. 158–165.

[Sim79-2] Marvin K. Simon, "False lock performance of quadriphase receivers," *IEEE Trans. Communications*, **Comm-27**, No. 11, Nov. 1979, pp. 1660–1670.

[Sti71]   J.J. Stiffler, *Theory of Synchronous Communications*, Prentice-Hall, 1971.

[Spi77]   J.J. Spilker, Jr., *Digital Communications by Satellite*, Prentice-Hall, 1977.

[Vit 66] A.J. Viterbi, *Principles of Coherent Communications*, McGraw-Hill, 1966.

[W&L69]   P.A. Wintz and E.J. Luecke, "Performance of optimum and suboptimum synchronizers," *IEEE Trans. Communication*, **Comm-17**, No. 3, June 1969, pp. 380–389.

[Z&P85]   R. Ziemer and R. Peterson, "Digital Communications and spread spectrum systems," Macmillan, 1985.

CHANNEL MODELS FOR COMMUNICATION SYSTEMS

The idea of the baseband equivalent of a bandpass communications system was developed in Chapter 4. Part of that model was the baseband equivalent of the channel. In this chapter the baseband equivalent model of some important communications channels will be developed. These include the standard telephone line, the twisted-pair digital subscriber loop, coaxial cable, digital microwave line-of-sight antennas, satellite channels, and fading channels for wireless communications.

The baseband representation of these different channel models will provide a unifying thread in subsequent discussions of equalization. This common representation will also make it straightforward to do computer simulation of these systems.

7.1 A RECAPITULATION OF THE BASEBAND EQUIVALENT CHANNEL MODEL

We established in Chapter 4 that the general form of a bandpass data signal is

$$
\begin{aligned}
s(t) &= s_I(t)\,\mathrm{Cos}(\omega_c t) - s_Q(t)\,\mathrm{Sin}(\omega_c t) \\
&= \mathrm{Re}[(s_I(t) + js_Q(t))(\mathrm{Cos}(\omega_c t) + j\,\mathrm{Sin}(\omega_c t))] \\
&= \mathrm{Re}[s_+(t)] = \mathrm{Re}[s(t) + j\hat{s}(t)] = \mathrm{Re}\left[\hat{s}(t)e^{j\omega_c t}\right]
\end{aligned}
\tag{7.1-1}
$$

and that the impulse response $h(t)$ of the bandpass channel, which is represented by a bandpass filter, has a similar form:

$$
\begin{aligned}
h(t) &= h_I(t)\,\mathrm{Cos}(\omega_c t) - h_Q(t)\,\mathrm{Sin}(\omega_c t) \\
&= \mathrm{Re}[(h_I(t) + jh_Q(t))(\mathrm{Cos}(\omega_c t) + j\,\mathrm{Sin}(\omega_c t))] \\
&= \mathrm{Re}[h_+(t)] = \mathrm{Re}[h(t) + j\hat{h}(t)] = \mathrm{Re}\left[\tilde{h}(t)e^{j\omega_c t}\right]
\end{aligned}
\tag{7.1-2}
$$

and the response $r(t)$ of the channel to an input $s(t)$ is also a bandpass signal:

$$
\begin{aligned}
r(t) &= r_I(t)\,\mathrm{Cos}(\omega_c t) - r_Q(t)\,\mathrm{Sin}(\omega_c t) \\
&= \mathrm{Re}[(r_I(t) + jr_Q(t))(\mathrm{Cos}(\omega_c t) + j\,\mathrm{Sin}(\omega_c t))] \\
&= \mathrm{Re}[r_+(t)] = \mathrm{Re}[r(t) + j\hat{r}(t)] = \mathrm{Re}\left[\tilde{r}(t)e^{j\omega_c t}\right]
\end{aligned}
\tag{7.1-3}
$$

It was established that the time-domain relationship among these quantities corresponds to the complex convolution

$$
r_I(t) + jr_Q(t) = (h_I(t) + jh_Q(t)) \otimes (s_I(t) + js_Q(t))
\tag{7.1-4a}
$$

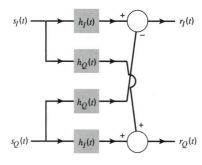

FIGURE 7.1-1

or

$$r_I(t) = h_I(t) \otimes s_I(t) - h_Q(t) \otimes s_Q(t) \qquad (7.1\text{-}4\text{b})$$

and

$$r_Q(t) = h_I(t) \otimes s_Q(t) + h_Q(t) \otimes s_I(t). \qquad (7.1\text{-}4\text{c})$$

This set of relationships forms the basis for the baseband equivalent model of the communication system shown in fig. 7.1-1. This model can be modified for computer simulation by using sampled versions of the time-domain signals, represented as tranversal or FIR filters as shown in fig. 7.1-2.

As we shall see in Chapter 10, Equalization of Distorted Channels, the appropriate sampling rate for the channel model depends on the particular equalization algorithm that will be used. For the purposes of this presentation we can be satisfied with two different channel models: one based on samples taken every T_s seconds and another based on samples taken every $T_s/2$ seconds (where $f_s = 1/T_s$ is the symbol or baud rate).

It is often necessary to obtain the time-domain model of the baseband equivalent channel from frequency-domain data that is given in the passband. The basis for doing this was also established in Chapter 4. The frequency domain representation of a bandpass channel $H(\omega)$ is shown in fig. 4.6-2 where the positive and negative frequency portions are denoted as $H_1(\omega)$ and $H_2(\omega)$ and where the consequence of $h(t)$ being a real impulse reponse is the requirement that

$$H_1(\omega) = H_2^*(-\omega) \qquad (7.1\text{-}5\text{a})$$

or

$$H_1(\omega + \omega_c) = H_2^*(-\omega + \omega_c) \qquad (7.1\text{-}5\text{b})$$

after these responses are shifted to baseband.

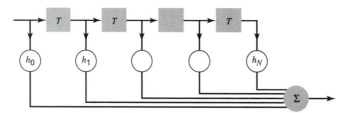

FIGURE 7.1-2

It was demonstrated in Chapter 4 that the Fourier Transforms of $h_I(t)$ and $h_Q(t)$ are given by

$$h_I(t) \Leftrightarrow H_I(\omega) = [H_1(\omega + \omega_c) + H_2(\omega - \omega_c)] \qquad (7.1\text{-}5c)$$

$$h_Q(t) \Leftrightarrow H_Q(\omega) = -j[H_1(\omega + \omega_c) - H_2(\omega - \omega_c)] \qquad (7.1\text{-}5d)$$

where these operations are shown in fig. 4.6-4b.

Note that $H_I(\omega)$ is an even function (fig. 4.6-4c) and $H_Q(\omega)$ is odd (fig. 4.6-4d). According to Fourier Transform theory, this means that the corresponding time functions are respectively even and odd. As we shall see, in real systems where the channel is not strictly bandlimited, the time-domain baseband impulse responses will not be strictly even or odd but will tend in those directions.

In the next section we develop a program that is specifically designed to find the sampled baseband in-phase and quadrature components of the channel impulse response for the telephone line. As shown in fig. 7.1-1, this impulse response can be represented as a complex transversal filter having complex taps

$$h_I(nT) + jh_Q(nT). \qquad (7.1\text{-}6a)$$

where T is the time between samples.

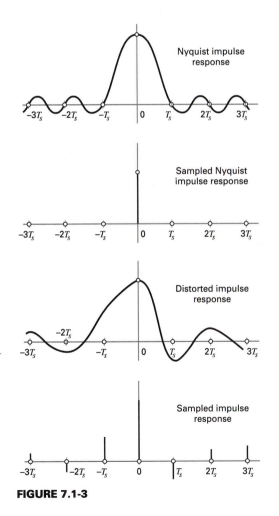

FIGURE 7.1-3

If the channel were ideal and the system were Nyquist, the baseband equivalent of the bandpass channel would look like

$$H = [0 + j0, 0 + j0, 1 + j0, 0 + j0, 0 + j0] \qquad (7.1\text{-}6b)$$

and a distorted channel would look like

$$H = [-.05 + j.02, .15 + j.08, 1 + j.20, -.10 - j.04, .02 - j.01] \qquad (7.1\text{-}6c)$$

where the deviations from the ideal represent the intersysmbol interference terms, as illustrated in fig. 7.1-3.

7.2   THE ORDINARY TELEPHONE LINE

We have used examples of modems operating on the ordinary telephone line in Chapter 4 as illustrations of the use of QAM type signals in data transmission. A detailed analysis of the telephone line is useful in a number of ways. Since the attenuation and delay characteristics of the telephone line have been well established by EIA-496-A, shown in Appendix 7A, we are able to work easily with an important channel. With this data we shall be able to do a detailed example on how to compute sampled versions of $h_I(t)$ and $h_Q(t)$ for the ordinary telephone line. This will not only serve to establish basic principles for the characterization of all bandpass channels, but will also, in subsequent discussions of equalization, enable us to use data that will produce realistic results.

This detailed example is embodied in a program CHAN768 that is available on the website. The sections below will simultaneously develop the theory behind this program and the structure of the code.

7.2.1   Computing a Linear Approximation to the Phase Characteristic of a Bandpass Channel

In characterizing a bandpass channel we shall include both the transmitter and receiver filters, each of which is the square root of raised cosine, into the amplitude characteristic so that the raw channel data for amplitude will be multiplied by a raised cosine as shown in fig. 7.2-1.

The transfer function of the resulting bandpass channel is given by

$$H(\omega) = A(\omega)e^{j\theta(\omega)} \qquad (7.2\text{-}1)$$

and is shown in fig. 7.2-2.

We are first interested in a linear approximation to the phase characteristic $\theta(\omega)$

$$\theta(\omega) \approx \hat{\theta}(\omega) = -(\omega - \omega_c)T_{\text{av}} + \theta_0 \qquad (7.2\text{-}2a)$$

where T_{av} is the average time delay in the filter and θ_0 is the phase offset at the center

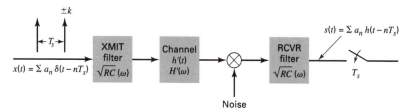

FIGURE 7.2-1

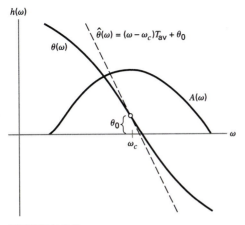

FIGURE 7.2-2

frequency. A somewhat easier approximation to implement is

$$\hat{\hat{\theta}}(\omega) = \hat{\theta}(\omega) + \omega_c T_{av} = \omega T_{av} + \theta_0. \qquad (7.2\text{-}2b)$$

We want to minimize

$$f(T_{av}, \theta_0) = \int_0^\infty A^2(\omega)[\theta(\omega) - \omega T_{av} - \theta_0]^2 \, d\omega \qquad (7.2\text{-}3a)$$

with respect to T_{av} and θ_0. Since we are minimizing the average value (integral) of the square of a function, the operation is termed least-mean-square or LMS. The quantity inside the brackets is multiplied by $A^2(\omega)$ in order to give lesser weight to the phase contributions at frequencies where less power is transmitted through the filter. As indicated above, $A(\omega)$ will also include the full raised cosine shaping that combines the transmitting and receiving filters because our object will be to find a baseband equivalent for the entire bandpass system.

Since we are simultaneously minimizing with respect to two variables, we must take two partial derivatives and set them each equal to zero.

$$\frac{\partial f(T_{av}, \theta_0)}{\partial T_{av}} = 2 \int_0^\infty A^2(\omega)[\theta(\omega) - \omega T_{av} - \theta_0]\omega \, d\omega$$

$$= 2 \int_0^\infty \omega A^2(\omega)\theta(\omega) \, d\omega - 2T_{av} \int_0^\infty \omega^2 A^2(\omega) \, d\omega - 2\theta_0 \int_0^\infty \omega A^2(\omega) \, d\omega = 0$$

$$\qquad (7.2\text{-}3b)$$

and

$$\frac{\partial f(T_{av}, \theta_0)}{\partial \theta_0} = 2 \int_0^\infty A^2(\omega)[\theta(\omega) - \omega T_{av} - \theta_0] \, d\omega$$

$$= 2 \int_0^\infty \omega A^2(\omega)\theta(\omega) \, d\omega - 2T_{av} \int_0^\infty \omega A^2(\omega) \, d\omega - 2\theta_0 \int_0^\infty A^2(\omega) \, d\omega = 0$$

$$\qquad (7.2\text{-}3c)$$

In order to solve for T_{av} and θ_0, it is necessary to compute a number of factors that are shown below in both integral and summation form, anticipating the writing of a program to accomplish these tasks.

$$T_1 = \int_0^\infty \omega A^2(\omega)\theta(\omega)\, d\omega = \sum_1^N A_i^2 \theta_i \omega_i$$

$$T_2 = \int_0^\infty \omega^2 A^2(\omega)\, d\omega = \sum_1^N A_i^2 \omega_i^2$$

$$T_3 = \int_0^\infty \omega A^2(\omega)\, d\omega = \sum_1^N A_i^2 \omega_i$$

$$T_4 = \int_0^\infty A^2(\omega)\theta(\omega)\, d\omega = \sum_1^N A_i^2 \theta_i \qquad \text{(7.2-3d)}$$

$$T_5 = \int_0^\infty \omega A^2(\omega)\, d\omega = \sum_1^N A_i^2 \omega_i$$

$$T_6 = \int_0^\infty A^2(\omega)\, d\omega = \sum_1^N A_i^2$$

It follows that eqs. 7.2-3b and 7.2-3c can be written as

$$T_1 = T_2 T_{av} + T_3 \theta_0$$
$$T_4 = T_5 T_{av} + T_6 \theta_0 \qquad \text{(7.2-3e)}$$

and solving, we find

$$T_{av} = \frac{T_1 T_6 - T_3 T_4}{T_2 T_6 - T_3 T_5} \qquad \text{(7.2-4a)}$$

$$\theta_0 = \frac{T_2 T_4 - T_1 T_5}{T_2 T_6 - T_3 T_5} \qquad \text{(7.2-4b)}$$

These two quantities will be subsequently used in order to establish the proper origin of coordinates and rotational orientation of the in-phase and quadrature components of the impulse response of $h_I(t)$ and $h_Q(t)$.

7.2.2   The Inverse Discrete Fourier Transform

The last step in a considerable amount of data preparation will be that of taking the inverse discrete Fourier Transform of a frequency-domain data array in order to obtain a time-domain data array. The specification of this step at this point will illustrate how the parameters of the data with which we are working actually come into play.

The inverse Fourier Transform is defined as

$$f(t) = \frac{1}{2\pi} \int_{-\infty}^\infty F(\omega)e^{j\omega t}\, d\omega \qquad \text{(7.2-5a)}$$

so that the inverse discrete Fourier Transform is

$$f(kT) = \frac{1}{2\pi} \sum_{n=-(N/2)}^{(N/2)-1} F(n\Omega) e^{jn\Omega kT} \, \Omega \tag{7.2-5b}$$

The basic parameters in this expression are chosen corresponding to the particular problem that is to be solved. The telephone line data, given in Appendix 7A, with which we shall be dealing is given as amplitude and delay every 25 hz, so we set $\Omega = 2\pi \times 25$. Corresponding to modems that have been discussed in Chapter 4, the symbol rate in this problem is $f_s = 2400$ hz. We want eight samples of $f(kT)$ every symbol interval T_s so that we can investigate how changing the timing phase in the receiver affects system performance. This makes $T = 1/19,200$.

Substituting these numbers and multiplying $f(kT)$ by T in order to normalize the peak to unity, we have

$$f(kT) = \frac{25}{19200} \sum_{n=-(N/2)}^{(N/2)-1} F(n\Omega) e^{j2\pi nk/768} = \frac{1}{768} \sum_{n=-(N/2)}^{(N/2)-1} F(n\Omega) e^{j2\pi nk/768} \tag{7.2-5c}$$

Finally, we note that both quantities that we are trying to find, $h_I(t)$ and $h_Q(t)$, are real. According to Fourier Transform theory, if $f(t)$ is real and $F(\omega) = R(\omega) + jX(\omega)$ then

$$f(t) = \frac{1}{2\pi} \int_{-\infty}^{\infty} [R(\omega) \cos(\omega t) - X(\omega) \sin(\omega t)] \, d\omega \tag{7.2-5d}$$

and in the discrete domain

$$f(kT) = \frac{1}{768} \sum_{n=-(N/2)}^{(N/2)-1} \left[R(n\Omega) \cos\left(2\pi \frac{nk}{768}\right) - X(n\Omega) \sin\left(2\pi \frac{nk}{768}\right) \right] \tag{7.2-5e}$$

Taking into account that our inverse transform will have 768 time-domain points, we will make life easier by arranging the frequency-domain data to have the same number of points. This results in a situation where a standard FFT will not work so that the coding will have to be done directly. We also note that in C the index runs from $n = 0$ so that the actual computation will require

$$f(kT) = \frac{1}{768} \sum_{n=0}^{767} \left[R(n\Omega) \cos\left(2\pi \frac{(n-384)k}{768}\right) - X(n\Omega) \sin\left(2\pi \frac{(n-384)k}{768}\right) \right] \tag{7.2-5f}$$

7.2.3 The Structure of the Program CHAN768

Consistent with the data given in Appendix 7A, the telephone-line data consists of 117 samples each of four attenuation characteristics (A, B, C, flat) and six delay characteristics (1, 2, 3, 4, 5, 3 msec) taken at 25 hz intervals. These data files have been placed in a directory PHONE and the user selects one of each to be read into the program. The file 'flat' is a test file corresponding to 0 db attenuation across the band and the file '3 msec' is a test file corresponding to a uniform 3 msec delay across the band. Corresponding to the variations that exist in actual telephone lines, the user may select any combination of attenuation and delay characteristic, hence an A2 or a B3 line. The C3 line is often considered to be a worst-case line.

In the program CHAN768 given on the website, the attenuation is changed from db to absolute value and the delay is integrated to give phase. A 12.5% raised cosine, corresponding to a symbol rate of 2400 hz, is created and multiplied by the attenuation characteristic. This incorporates the transmit and receiver filters, each of which is the square root of raised cosine, into the channel. The resulting data arrays of 117 points each are then centered in the required 768-point array as amp(ω) and phase(ω).

We then compute T_{av} and θ_0 as indicated above. Since the value of T_{av} is the time delay of the channel, the real and imaginary portions of the transfer function are computed as

$$FR(\omega) = \text{amp}(\omega)\,\text{Cos}[\omega(t - T_{av})]$$
$$FI(\omega) = \text{amp}(\omega)\,\text{Sin}[\omega(t - T_{av})]$$

(7.2-6a)

and the resulting data are centered in 768-point arrays that go from $n = 0$ to $n = 767$.

Using eqs. 7.1-5c,d, we compute the real and imaginary parts of both the in-phase and quadrature portions of the channel impulse response. Since these quantities are centered at $n = 384$, they must be subsequently 'demodulated'.

```
for(n=330; n<=438; n++)
    {
        FPR[n] = FR[n] + FR[768-n];
        FPI[n] = FI[n] - FI[768-n];
        FQR[n] = -FI[n] - FI[768-n];
        FQI[n] = FR[n] - FR[768-n];
    }
```

(7.2-6b)

The next step is to use eq. 7.2-5e to take the inverse discrete Fourier Transform of these quantities:

```
for(n=0; n<768; n++)
    {
    for(m=0; m<768; m++)
        {
```
$$HP[n] \mathrel{+}= FPR[m]\,\text{Cos}\!\left(2\pi\frac{n(m-384)}{768}\right) - FPI[m]\,\text{Sin}\!\left(2\pi\frac{n(m-384)}{768}\right);$$
$$HQ[n] \mathrel{+}= FQR[m]\,\text{Cos}\!\left(2\pi\frac{n(m-384)}{768}\right) - FQI[m]\,\text{Sin}\!\left(2\pi\frac{n(m-384)}{768}\right);$$
```
        }
    }
```

(7.2-6c)

This result is rotated by θ_0, the average phase shift through the channel, a quantity that was computed earlier. In practice, the carrier recovery system will remove this constant phase offset and, in order to get the desired result, so do we.

```
for(n=0; n<768; n++)
    {
        HPR[n] = HP[n]  Cosθ0 + HQ[n]  Sinθ0;
        HQR[n] = HQ[n]  Cosθ0 - HP[n]  Sinθ0;
    }
```

(7.2-6d)

The peak value of the in-phase portion of the impulse response occurs at $n = 0$. We normalize the response so that the peak equals unity and then we rearrange the response in the array so that the peak occurs at $n = 384$. Finally, the user is allowed to choose the timing phase and to generate a sampled channel impulse response having samples either every T_s or $T_s/2$ seconds.

Examples of the use of this program are left to the Problems.

Although this program has been written explicitly for finding the in-phase and quadrature components of the impulse response of the ordinary telephone line, modifying the program for a similar computation for any bandpass channel is a straightforward task. Similarly, it is quite direct to modify (and substantially simplify) this program to find the sampled impulse response of a baseband channel. Consequently, in subsequent discussions of other channels we will not undertake the task of actually computing impulse responses from frequency-domain data and leave that to the Problems.

7.3 TWISTED PAIR AND COAXIAL CABLE

The term 'twisted-pair' literally describes the configuration of two typically 24-26-gauge copper wires. Ideally a twisted pair can simply be treated as a transmission line. In actual application there are additional considerations that must be taken into account.

The twisted-pair portion of the ordinary telephone line, that portion linking the individual subscriber to the telephone company central office (fig. 7.3-1) is, as we have noted in Chapters 3 and 4, increasingly being used for data transmission independent of the bandwidth restrictions of ordinary telephone transmission. A high-speed data network delivers data to the computer at the central office, which, in turn, retransmits that data to the subscriber at rates consistent with the bandwidth of the twisted pair. In this context the twisted pair is referred to as the DSL (Digital Subscriber Loop).

The loop itself is completely passive but there typically are impedance discontinuities that arise from the splicing of wires having differing gauges and from the existence of open stubs called 'bridge taps' that are vestiges of previous connections. A central part of the strategy for using this loop for low-cost digital transmission is avoid any attempts to make these lines better by fixing the impairments, and rather to design systems that are able to overcome them.

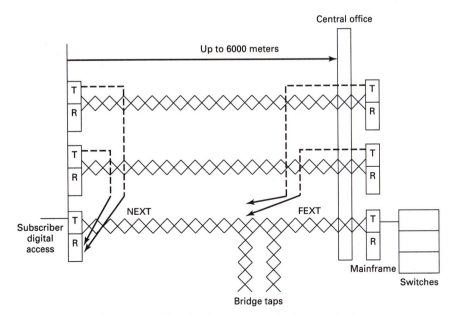

FIGURE 7.3-1 ([GH&W92] fig. 1.2-6 reproduced with permission)

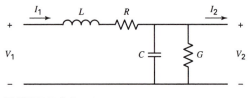

FIGURE 7.3-2

The first use of this subscriber loop for direct transmission of data was the ISDN, described in Chapter 3. This system transmits data at 160 kbs with 2 bits/symbol in a standard four-level baseband signaling configuration. The symbol rate is therefore 80 kbaud. A description of this system may be found in [ABG81] and [GVC84].

The more recent characterization of the DSL defines a bandpass channel by excluding the low frequencies up to around 200 khz and encompassing a higher band into the megahertz range. This division in frequency range roughly corresponds to an analytic division on how the transfer function of the twisted pair should be characterized.

The classical analysis of transmission lines results in a model (fig. 7.3-2) of a section of line of length L consisting of series inductance and resistance and parallel capacitance and conductance. This section is described by a two-port model based on the chain matrix

$$\begin{bmatrix} V_1 \\ I_1 \end{bmatrix} = \begin{bmatrix} A & B \\ C & D \end{bmatrix} \begin{bmatrix} V_2 \\ I_2 \end{bmatrix} = \begin{bmatrix} \cosh(\gamma L) & Z_0 \sinh(\gamma L) \\ \frac{\sinh(\gamma L)}{Z_0} & \cosh(\gamma L) \end{bmatrix} \begin{bmatrix} V_2 \\ I_2 \end{bmatrix} \tag{7.3-1a}$$

where

$$Z_0 = \sqrt{\frac{R + j\omega L}{G + j\omega C}} \tag{7.3-1b}$$

is the characteristic impedance of the line and

$$\gamma = \sqrt{(R + j\omega L)(G + j\omega C)} \tag{7.3-1c}$$

is the propagation constant. Were the line lossless, the characteristic impedance would reduce to $Z_0 = \sqrt{L/C}$ (in this regard, transmission lines in the form of coaxial cables are typically described as 50 ohm or 75 ohm lines) and the the propagation constant to $\gamma = j\omega\sqrt{LC}$, implying zero attenuation during propagation.

The use of this model for the computation of subscriber-loop characteristics is illustrated by Bingham [Bin00].

The distinction between the higher and the lower frequency ranges is based on the relative magnitudes of resistance and reactance in Z_0 and γ. At the lower frequencies both the characteristic impedance and the propagation constant are complex so that there is significant phase distortion as well as amplitude distortion.

At higher frequencies the propagation constant becomes very close to purely imaginary, resulting in the absence of significant phase distortion (the group delay is virtually constant across the band. Similarly, the characteristic impedance of the line becomes purely resistive. However, the increasing significance of the skin effect of the resistance causes the amplitude distortion to become severe. The twisted-pair channel has been modeled as [K&S90]

$$|H_c(f)|^2 = e^{-\alpha\sqrt{f}} \tag{7.3-2a}$$

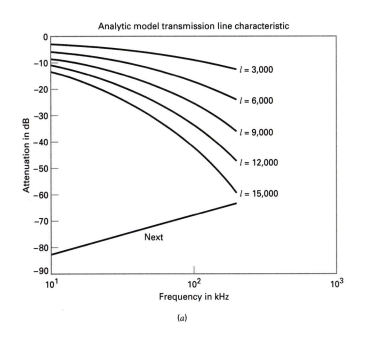

(a)

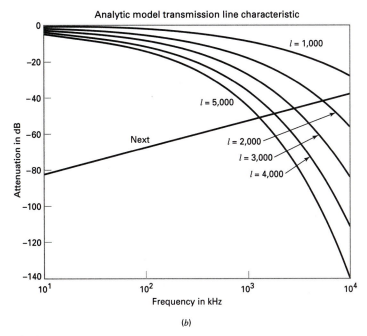

(b)

FIGURE 7.3-3

where f is the frequency in khz and $\alpha = k(\ell/\ell_0)$. These parameters have been defined as

$$\ell = \text{length of channel in ft}$$
$$\ell_0 = \text{reference length of (18,000 ft)} \qquad (7.3\text{-}2b)$$
$$k = 1.158, \text{ a constant}$$

The reference length refers to the longest local loop that is considered for feasible digital transmission. Equation 7.3-2a is plotted in fig. 7.3-3a over the frequency range that is appropriate for ISDN transmission and in fig. 7.3-3b over the frequency range that is used for a variety of DSL transmission modes to be discussed below. Clearly the channel is very dependent on line length and it is this factor that is the first limitation on both the ability to implement ISDN on all local loops and the appropriate data rate for the DSL.

In addition to Gaussian noise, twisted-pair loops suffer from two related impairments, near-end (NEXT) and far-end (FEXT) crosstalk resulting from the bundling of many twisted pairs into a single cable, as shown in fig. 7.3-1. Since this crosstalk arises from the addition of transmitted data signals, it is not surprising that it can be characterized [CGS83] as a cyclostationary random process having a power spectral density that can be used in the standard manner in linear systems.

NEXT arises from transmitters that are operating on the same side of the cable as a receiver and is therefore not attenuated by the transmission line. The characterization of the magnitude-squared of the NEXT transfer function can be quite complex [Bin00], depending on the specific characteristics of the line and the number of lines in the cable. The form of the NEXT is given as

$$|H_{\text{NEXT}}(f)|^2 = K_{\text{NEXT}} f^{3/2} \tag{7.3-3a}$$

where an average-case characterization is

$$10 \log_{10} |H_{\text{NEXT}}(f)|^2 = -75 + 16 \log_{10}(N) + 15 \log_{10}(f) \tag{7.3-3b}$$

where f is in Mhz and N is the number of lines in the cable. This average case (for $N = 25$) is plotted along with the attenuation characteristics in figs. 7.3-3a and 7.3-3b.

The FEXT power generated by the transmitters at the far end of the cable is

$$|H_{\text{FEXT}}(f)|^2 = K_{\text{FEXT}} f^2 \ell \tag{7.3-4a}$$

and it appears at the receiver as

$$K_{\text{FEXT}} f^2 \ell |H_c(f)|^2 \tag{7.3-4b}$$

The idea is that crosstalk from the far end enters the receiver cable for the entire length of the line, explaining its linear dependence on the line length, but is also attenuated by the full length of the line, explaining the multiplication by $|H_c(f)|^2$ in eq. 7.3-4b. The average-case characterization of FEXT is

$$10 \log_{10} |H_{\text{FEXT}}(f)|^2 = -75 + 16 \log_{10}(N) + 20 \log_{10}(f) + 10 \log_{10}(\ell) \tag{7.3-4c}$$

f is in Mhz, N is the number of lines in the cable, and ℓ is in kfeet.

Consequently the total power at the receiver is

$$\{|H_c(f)|^2 + |H_{\text{NEXT}}(f)|^2 + |H_c(f)|^2 |H_{\text{FEXT}}(f)|^2\} P_S(f) \tag{7.3-5}$$

It is left to the Problems to show that it is only at very high frequencies and long line length that FEXT exceeds NEXT as the primary disturbance at the receiver.

In Chapter 8, Channel Capacity and Coding, we shall see that NEXT interference is the fundamental limitation on high-speed data transmission.

Considerable effort has been expended in characterizing realistic local loops (see Cox and Adams [C&A85]). Some examples this characterization are given by Lechleider [Lec89] and shown in fig. 7.3-4. The effect of bridged taps on the attenuation characteristic of the twisted pair in the ISDN frequency range and the impulse response of the line under these conditions are given by Lin and Tzeng [L&T88] and are shown in fig. 7.3-5. It is particularly worth noting that this impulse response, unlike the impulse response of the ordinary telephone line, does not have any leading echoes. As we shall see in Chapter 10,

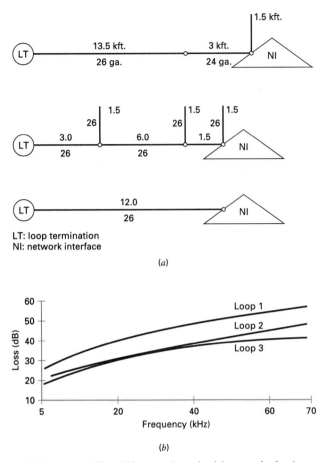

FIGURE 7.3-4   ([Lec89] reproduced with permission)

Equalization of Distorted Channels, this characteristic is of particular importance in the design of ISDN circuitry.

Transmission over the DSL is an emerging technology that is beyond the scope of this discussion. The interested reader is also referred to Maxwell [Max96], Cioffi [Cio97], Bingham [Bin00] and the entire issue of the *IEEE Journal on Selected Areas of Communications*, September 1991. Depending on the length and condition of the line, bit rates exceeding 6 Mbps have been achieved with a 1-Mhz bandwidth.

It should be noted that twisted pair is also used for data transmission quite independently of the local loop in a variety of applications where the line length is relatively short and high-speed transmission is required.

Since both twisted pair and coaxial cable are transmission lines, their analysis is the same although their parameters differ somewhat. The primary difference between twisted pair and coaxial cable is in coaxial cable's shielding, which is more effective at higher frequencies, leaving twisted pair the less expensive and equally effective medium at lower frequencies. Coaxial cable systems may, however, be made more effectively wideband than twisted pair through the use of repeaters. Repeaters, spaced periodically, serve to limit high frequency attenuation. The characteristics of coaxial cable are discussed in some detail in Smith [Smi93]. The problem of the accumulation of timing jitter as a limitation on repeatered coaxial cable has been discussed in Chapter 2.

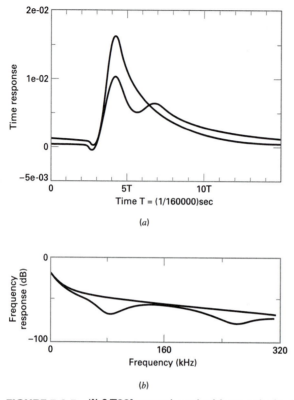

FIGURE 7.3-5 ([L&T88] reproduced with permission)

7.4 MULTIPATH AND RAYLEIGH FADING CHANNELS

A multipath channel is characterized by more than one path between the transmitter antenna and the receiver antenna, as shown in fig. 7.4-1a. The major consequence of multipath is the possibility of the 'fading' of the received signal at the receiver antenna. This phenomenon can be illustrated easily by considering the received signal

$$x(t) = A \, Cos(\omega_c t) + B \, Cos[\omega_c(t + \tau)] \qquad (7.4\text{-}1a)$$

where the second component, the multipath signal, is an attenuated (i.e. $B < A$) and delayed version of the 'main' signal. A small bit of trigonometry yields

$$x(t) = A \left[1 + \left(\frac{B}{A} \right)^2 + 2\frac{B}{A} \, Cos(\omega_c \tau) \right]^{1/2} Cos \left[\omega_c t + \tan^{-1} \frac{\frac{B}{A} \, Sin(\omega_c \tau)}{1 + \frac{B}{A} \, Cos(\omega_c \tau)} \right]$$

$$(7.4\text{-}1b)$$

The magnitude of $x(t)$ is plotted vs. $\omega_c \tau$ (fig. 7.4-1b) and is characterized by alternate reinforcement and cancellation at peaks occurring when $\omega_c \tau$ is a multiple of π. The cancellation is called *fading*. If τ is held constant the fading is clearly frequency-selective. In this sense it can be understood as channel amplitude distortion. It is clear from eq. 7.4-1b that the phase of the received signal is also frequency-dependent. We have already seen that frequency-selective amplitude and phase distortion translate into ISI in the time domain. It is

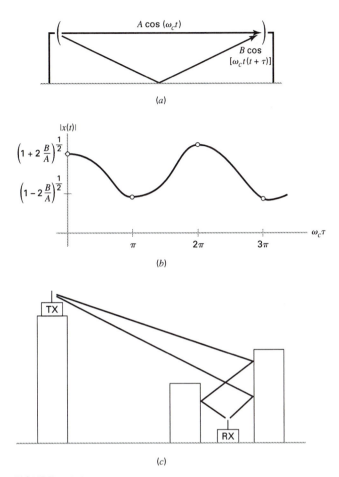

FIGURE 7.4-1

also evident that a constant frequency and a varying delay τ will also result in fading. Fading that arises in this manner is characterized as Doppler-shift. It is most directly associated with the changing distance between antennas in a mobile communications system. The channel in a mobile or cellular communication system, fig. 7.4-1c, can be exceedingly complex. Unlike a line-of-sight system, there is no main ray between transmitter and receiver; the received signal is composed of many reflections of the transmitted signal.

    Referring to fig. 7.4.1a, even if both antennas are fixed, the transmission characteristics of the channel change randomly with time. This time variation of the channel will be manifested as fading characterized as Doppler-shift even without any relative antenna motion. In a digital microwave LOS (line-of-sight) system, changing atmospheric conditions may cause the multipath components to change in amplitude and delay. In a mobile system with a stationary receiver the channel may change as a result of vehicular traffic in the transmission path. In fig. 7.4-1c even a small change in the receiver position can result in significant change in the channel. The characterization of fading will therefore include both time dispersion and frequency dispersion.

    In relation to a transmitted bandlimited data signal, a fade may occur within the signal band or may correspond to a frequency outside of it. The random nature of the physical phenomena that give rise to the multipath will cause the fade to have a time-varying magnitude and to jump in and out of the signal band.

Just as the channels discussed earlier in this chapter were modeled as transversal filters having delays of T_s, the symbol rate, or integer fractions T_s/M, we intend to model the multipath channel in the same way. We shall also discuss some of the characterizations that are unique to fading channels. In order to accomplish this we shall draw upon the work of Bello [Bel63], Kailath [Kal61], Stein [Ste87], Ehrman et al. [Ehr82], Jeruchim et al. [JBS92], and Sklar [Skl97].

A 'discrete multipath' model may be understood by transmitting a bandpass data signal $s(t)$ and receiving

$$y(t) = \sum_n a_n(t)s(t - \tau_n(t)) \tag{7.4-2a}$$

where it is made clear that both the attenuation and the delay associated with the various paths are time-variant. Using the techniques developed in Chapter 4, eq. 7.4-2a can be rewritten to express the same relationship in terms of the complex envelopes of the signals involved, yielding a baseband equivalent

$$\tilde{y}(t) = \sum_n \tilde{a}_n(t)\tilde{s}(t - \tau_n(t)) \tag{7.4-2b}$$

and allowing the definition of the time-varying baseband equivalent impulse response

$$\tilde{h}(\tau;t) = \sum_n \tilde{a}_n(t)\delta(\tau - \tau_n(t)) \tag{7.4-2c}$$

which is to be understood as the response of the system at time t to an impulse applied at time $(t - \tau)$.

A diffuse multipath model consists of many multipath components and may be written as an integral

$$\tilde{y}(t) = \int_{-\infty}^{\infty} \tilde{a}(\tau, t)\tilde{s}(t - \tau) \, d\tau \tag{7.4-2d}$$

Kailath [Kal61] has demonstrated that if the transmitted signal $s(t)$ is bandlimited and consists of digital data transmitted at $f_s = 1/T_s$ symbols/sec then the impulse response of eqs. 7.4-2c and 7.4-2d can be represented as a complex transversal filter having fixed delays and time-varying tap coefficients, as shown in fig. 7.4-2. These delays may be equal to the symbol interval T_s or some rational fraction T_s/M of that delay. The total delay in this transversal filter, corresponding to significantly large taps, is called the *delay spread* of the impulse response. We shall subsequently discuss how to characterize these tap coefficients as random variables in order to develop time domain models for digital microwave and mobile wireless systems.

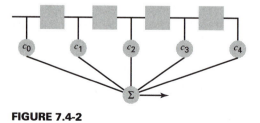

FIGURE 7.4-2

7.4.1 Time-Domain and Frequency-Domain Characterizations of Fading Channels

Bello [63] has defined the *delay cross-power spectral density* of the channel as

$$R_c(\tau_1, \tau_2; t, \Delta t) = \frac{1}{2} E[\tilde{a}^*(\tau_1, t)\tilde{a}(\tau_2, t + \Delta t)] \qquad (7.4\text{-}3a)$$

and has characterized this system as WSSUS (wide-sense stationary uncorrelated scattering). With this characterization, which means that signal variations that arrive with different delays are considered to be uncorrelated, eq. 7.4-3a can be reduced to

$$R_c(\tau_1, \tau_2; \Delta t) = R_c(\tau_1; \Delta t)\delta(\tau_1 - \tau_2)$$
$$= \frac{1}{2} E[\tilde{a}^*(\tau, t)\tilde{a}(\tau, t + \Delta t)] = R_c(\tau; \Delta t) \qquad (7.4\text{-}3b)$$

In this expression τ refers to the delay of the signal resulting from dispersion and Δt refers to the 'Doppler-shift' that results from motion or from time variation in the channel parameters.

Although both forms of distortion may exist simultaneously, it is useful to consider them separately. In order to accomplish this, define the

$$\text{multipath intensity profile} \quad R_I(\tau) = R_c(\tau; \Delta t)|_{\Delta t = k_1} \qquad (7.4\text{-}4a)$$

where the subscript I refers to ISI, and the

$$\text{spaced-time correlation function} \quad R_D(\Delta t) = R_c(\tau; \Delta t)|_{\tau = k_2} \qquad (7.4\text{-}4b)$$

where the subscript D refers to Doppler and where k_1 and k_2 are arbitrary constants. Each of these functions have Fourier Transforms

$$R_I(\tau) \Leftrightarrow S_I(\Delta f) \qquad (7.4\text{-}4c)$$

where $S_I(\Delta f)$ is the *spaced-frequency correlation function*, and

$$R_D(\Delta t) \Leftrightarrow S_D(\nu) \qquad (7.4\text{-}4d)$$

where $S_D(\nu)$ is the *Doppler power spectrum*. Representative pictures of these functions, adapted from Sklar [Skl97], are shown in fig. 7.4-3. Bello has pointed out that there is as essential duality in these functions, i.e.

$$R_I(\tau) \overset{\text{dual}}{\longleftrightarrow} S_D(\nu) \qquad (7.4\text{-}4e)$$

and

$$R_D(\Delta t) \overset{\text{dual}}{\longleftrightarrow} S_I(\Delta f) \qquad (7.4\text{-}4f)$$

We now shall examine the significance of these functions. Let us first consider eq. 7.4-4c. As indicated above, the multipath intensity profile $R_I(\tau)$ (fig. 7.4-3a) can be understood as the impulse response of a time-invariant channel which has a transfer function $S_I(\Delta f)$. Channel distortion, a frequency-domain characterization, and channel dispersion, a time-domain characterization, are therefore simply different ways of looking at the same phenomenon. Consequently, the delay spread T_m indicates the effective duration of the intersysmbol interference. Except for some specialized nomenclature, this ISI can be understood in very much the same way as ISI has been defined for previously discussed channel models. In this model if the delay spread $T_m < T_s$, the symbol duration, then there is no actual ISI but rather there is simply a reduction in amplitude. Such a narrow $R_I(\tau)$ will yield a very wide $S_I(\Delta f)$ (fig. 7.4-3b) referred to as 'flat fading' or 'frequency

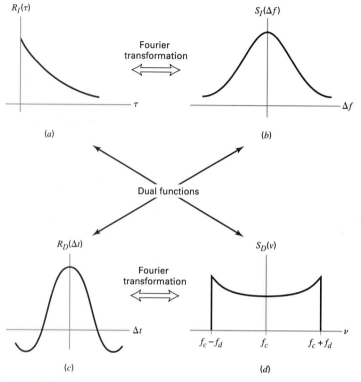

FIGURE 7.4-3

non-selective fading'. On the other hand, a narrow or notched $S_I(\Delta f)$, 'frequency-selective fading', corresponds to a wide $R_I(\tau)$ or channel-induced ISI.

In order to characterize the relationship between $R_I(\tau)$ and $S_I(\Delta f)$ more precisely, consider the example

$$R_1(\tau) = e^{-(\tau/T)}u(\tau) \Leftrightarrow S_I(\Delta f) = \frac{1}{1 + j(\omega/\omega_c)} \tag{7.4-5a}$$

where $f_c = 1/T$ is the 3 db cutoff frequency. The *delay spread* is $T_m = \alpha T$, where α is an appropriate constant that indicates the effective duration of the ISI. The definition of the *coherence bandwidth* of the received signal is the reciprocal of the delay spread:

$$f_0 = 1/T_m \tag{7.4-5b}$$

This simple relationship between time duration and bandwidth is not necessarily suitable for more complex time responses and so the *rms delay spread* of $R_I(\tau)$ is defined as:

$$\sigma_\tau = \sqrt{\overline{\tau^2} - \bar{\tau}^2} \tag{7.4-5c}$$

The *rms coherence bandwidth* is then defined as a constant times the reciprocal of the rms delay spread:

$$f_0 = k/\sigma_\tau \tag{7.4-5d}$$

As Sklar points out, the value of the constant depends on the particular defintion of bandwidth that is used, a definition that varies from application to application.

The *spaced-time correlation function* $R_D(\Delta t)$, eqs. 7.4-4b and 4d, measures the correlation at the receiver of two sinusoids having the same frequency transmitted at different times t_1 and t_2, where $\Delta t = t_2 - t_1$. If the channel has not changed between t_1 and t_2, then $R_D(\Delta t)$ will be a constant and its Fourier Transform, the *Doppler Power Spectrum* $S_D(v)$, will simply be an impulse function.

Clarke [Cla68] has shown that for a dense-scatter channel model with constant velocity of the receiving antenna, the *spaced-time correlation function* $R_D(\Delta t)$ is given by:

$$R_D(\Delta t) = J_0(kV\Delta t) \tag{7.4-6a}$$

where $J_0(\ )$ is the zero-order Bessel function of the first kind, V is the velocity, and $k = 2\pi/\lambda$ is the free-space phase constant. This means that the Doppler shift frequency is $f_D = V/\lambda$. Under these circumstances the corresponding *Doppler Power Spectrum is*

$$S_D(v) = \frac{1}{\pi f_D \sqrt{1 - \left(\frac{v - f_c}{f_D}\right)^2}} \tag{7.4-6b}$$

where f_c is the carrier frequency. These functions are shown in figs. 7.4-3c,d.

The *coherence time* T_0 is the time over which $R_D(\Delta t)$ is essentially constant. Channel dispersion (a time-domain phenomenon that is characterized as ISI) that arises from Doppler shift is measured by the relationship of T_0 to the symbol duration T_s. If $T_s < T_0$, denoted as *slow fading*, then there is no resulting ISI. On the other hand if $T_s > T_0$, denoted as *fast fading*, there will be significant channel dispersion and, consequently, ISI will be manifested.

The relationship between the coherence time and the Doppler shift is

$$T_0 = \beta/f_D \tag{7.4-6c}$$

where β is a constant that depends on the particular channel. This relationship is clearly the dual of eq. 7.4-5d.

7.4.2 The Microwave Line of Sight Channel

Digital microwave transmission involves antennas in a direct transmission path as shown in fig. 7.4-1a. Under many atmospheric conditions that simple and direct transmission is what occurs and the corresponding error rate is very small, of the order of 10^{-10}. Under other atmospheric conditions a multipath situation arises in which the transmitted signal, reflected from various layers of the atmosphere, arrive at the receiving antenna somewhat delayed with respect to the main transmission.

We will follow the characterization of this distortion, the three-ray model, made by Rummler [Rum79], [Rum81], [Rum86] and elaborated by Jeruchim [JBS92], as shown in fig. 7.4-4. The three-ray model assumes that the impulse response of the multipath channel

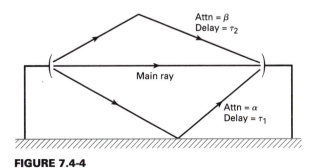

FIGURE 7.4-4

is of the form

$$h(t) = \delta(t) + \alpha(t - \tau_1) + \beta(t - \tau_2) \tag{7.4-7a}$$

where τ_1 and τ_2 are the delays associated with the multipath in relation to the main signal path. It follows that the transfer function is:

$$H(\omega) = 1 + \alpha e^{-j\omega\tau_1} + \beta e^{-j\omega\tau_2} \tag{7.4-7b}$$

in which we assume that the variable ω represents $(\omega - \omega_c)$. The model requires that $\tau_1 \ll 1/B$ where B is the bandwidth of the transmitted signal, so that $e^{-j\omega\tau_1} \approx 1$ and eq. 7.4-7b can be rewritten as:

$$H(\omega) = 1 + \alpha + \beta e^{-j\omega\tau} \tag{7.4-7c}$$

The final form of the model, now basically a two-path model, recalls the basic characteristic of multipath that is being simulated, the existence of a notch in the transfer function. In order to make it explicit, this notch is located at ω_0 so that

$$H(\omega) = a\left[1 - be^{-j(\omega-\omega_0)\tau}\right] \tag{7.4-8a}$$

where $a = 1 + \alpha$ and $b = -\beta/(1 + \alpha)$. A bit of manipulation shows that

$$|H(\omega)|^2 = a^2[1 + b^2 - 2b\cos[(\omega - \omega_0)\tau]] \tag{7.4-8b}$$

which is typically represented in db as

$$A(f) = 10\log_{10}|H(f)|^2 \tag{7.4-8c}$$

The phase of $H(\omega)$ is

$$\varphi(\omega) = \tan^{-1}\left(\frac{b\sin[(\omega - \omega_0)\tau]}{1 - b\cos[(\omega - \omega_0)\tau]}\right) \tag{7.4-8d}$$

and the group delay is the the negative of the deriviative of $\varphi(\omega)$ with respect to ω. It is left to the Problems to demonstrate that

$$D(f) = -\frac{1}{2\pi}\frac{d\varphi(\omega)}{d\omega} = \frac{\tau}{2\pi}\frac{b[b - \cos[2\pi(f - f_0)\tau]]}{1 + b^2 - 2b\cos[2\pi(f - f_0)\tau]} \tag{7.4-8e}$$

It should be understood that the parameters in this model must be appropriately chosen in order to have the model correspond to the characteristics of actual multipath channels and that these choices are based upon considerable experimentation. The first of these is the limitation that $\tau \approx 1/6B$ where B is the channel bandwidth. A bandwidth of 25 Mhz corresponds to $\tau = 6.3$ nsec, an example which is cited in the literature.

A typical plot of $A(f)$ and $D(f)$ vs. $(f - f_0)$ is given in fig. 7.4-5. In this plot we see the periodic notches in amplitude and the changes in group delay corresponding to those notches. The group delay may correspond to a minimum-phase sytstem, $b < 1$, or a non-minimum-phase system, $b > 1$.

For a minimum-phase system we may rewrite eq. 7.4-8a as

$$H_1(f) = a\left[1 - be^{-j2\pi(f-f_0)\tau}\right] \tag{7.4-9a}$$

where

$$A_1 = -20\log_{10}a \qquad B_1 = -20\log_{10}(1 - b) \tag{7.4-9b}$$

For a non-minimum-phase system

$$H_2(f) = ab\left[e^{-j2\pi(f-f_0)\tau} - \frac{1}{b}\right] \tag{7.4-9c}$$

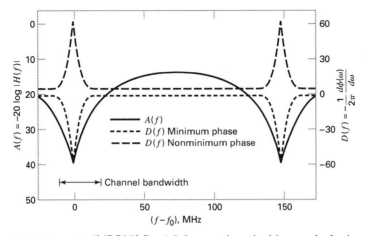

FIGURE 7.4-5 ([JBS92] fig. 4.3-2 reproduced with permission)

where

$$A_2 = -20 \log_{10} ab, \qquad B_2 = -20 \log_{10}\left(1 - \frac{1}{b}\right) \tag{7.4-9d}$$

According to Balaban [Bal85], B_1 and B_2 are exponentially distributed with a mean of 3.8 db and A_1 and A_2 are Gaussian random variables having a standard deviation of 5 db and a mean of

$$m = 24.6\frac{B^4 + 500}{B^4 + 800} \text{ db} \tag{7.4-9e}$$

The location of the notch frequency is also random with the range of $\omega_0\tau$ being $[-\pi, \pi]$ where it is five times more likely that $|\omega_0\tau| < \pi/2$ than $|\omega_0\tau| > \pi/2$.

To sum up, the effect of multipath is to create periodic notches in the channel transfer function. A notch may occur in the transmission band B or outside of it, having different effects on the system impulse response. Since the operation of the medium is statistical, the notch will jump in and out of the transmission band according to the statistical relationships that have been given.

The reader should understand that this presentation is intended simply to convey a sense of the issue. A serious understanding of the phenomenon of multipath will require considerable pursuit of the references.

Our primary task is to show how to represent this channel as a complex baseband impulse response. We can see that a given set of parameters determines the multipath channel in the frequency domain. From that determination we need only follow the approach taken earlier in this chapter to find the time-domain equivalent. This is left to the Problems.

7.4.3 The Rayleigh Fading Channel

The presentation given above gives a deterministic characterization of a channel that is typically quite random. A mobile channel in a changing environment such as an urban center, fig. 7.4-1c, will often change quite suddenly because the many multipath signals will add in random ways. The statistical characterization of this channel is based upon the assumption that the resulting received random signal will be of the form

$$x(t) = A(t)\text{Cos}(\omega_c t) + B(t)\text{Sin}(\omega_c t) \tag{7.4-10a}$$

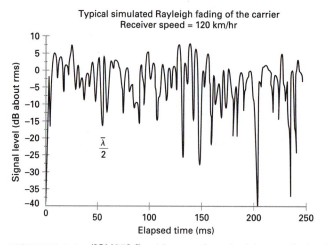

FIGURE 7.4-6 ([Skl97] fig. 10 reproduced with permission)

where ω_c is the carrier frequency and $A(t)$ and $B(t)$ are zero-mean Gaussian random variables having average power or variance σ^2. This expression can also be written as

$$x(t) = r(t) \operatorname{Cos} [\omega_c t + \varphi(t)] \qquad (7.4\text{-}10b)$$

where

$$r(t) = \sqrt{A^2(t) + B^2(t)} \qquad \text{and} \qquad \varphi(t) = \tan^{-1}\left(\frac{B(t)}{A(t)}\right) \qquad (7.4\text{-}10c)$$

It is well known that $r(t)$ will have the Rayleigh probability density function

$$p(r) = \begin{cases} \dfrac{r}{\sigma^2} \exp\left(-\dfrac{r}{2\sigma^2}\right); & 0 \le r \le \infty \\ 0; & 0 \le r \end{cases} \qquad (7.4\text{-}10d)$$

and that the phase will be uniformly distributed. As a consequence, the multipath channel is known as the *Rayleigh Fading Channel*. The magnitude characteristic of such a channel as a function of time is shown in fig. 7.4-6.

We now want to return to the original intention of representing this random channel as a complex transversal filter consistent with the model developed at the beginning of this chapter. Considerable research has been devoted to this topic, much of which has been recapitulated by Rappaport [Rap96, Chapter 4]. In particular, we shall refer to the model developed by Clarke [Cla68], Smith [Smi75], and Gans [Gan72].

The development of the channel model is divided into two parts, the channel distortion caused by Doppler shift followed by the effects of frequency selective fading.

The Doppler shift model is shown in fig. 7.4-7. The Doppler Power Spectrum for a single Doppler shift f_D of eq. 7.4-6b is the key equation for this model. The amplitude spectrum $\sqrt{S_D(\nu)}$ is specified with N points (typically N is a power of 2). The frequency spacing of these points is $\Delta f = 2f_D/(N - 1)$. Note that the results of taking an IFFT (Inverse Fast Fourier Transform) of these samples will be N points spaced $\Delta t = 1/\Delta f$ seconds apart. Each of the frequency components has a random amplitude according to a Gaussian probability density function. These are constructed by generating independent complex Gaussian samples for the $N/2$ positive frequencies and generating the negative frequency components as the complex conjugates of their positive counterparts.

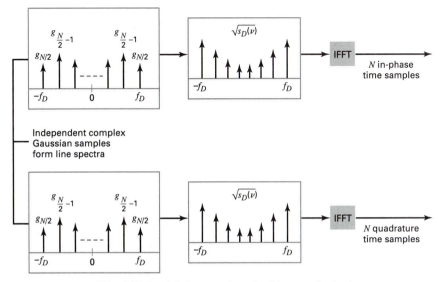

FIGURE 7.4-7 ([Rap96] fig. 4.2-3 reproduced with permission)

The two rails of fig. 7.4-7 produce the real and imaginary parts of a sequence of complex time-domain samples of the channel response. The magnitude of these samples will be Rayleigh distributed and their phases will be uniformly distributed.

A line-of-sight component may be added to this model by adding a strong, deterministic frequency component in the center of randomly generated spectrum.

Frequency-selective fading or multipath may be added to this model according to fig. 7.4-8. The outputs of several of these Doppler simulators may be used to control the channel taps of an FIR structure having non-uniform delays.

There are a few more steps that must be taken in order to have this channel model in the form of a T_s or T_s/m spaced transversal filter in a form suitable for simulation. First, the

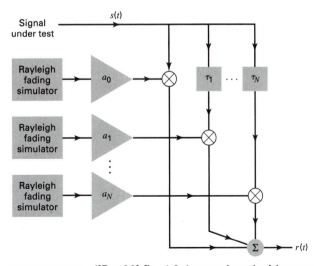

FIGURE 7.4-8 ([Rap96] fig. 4.2-4 reproduced with permission)

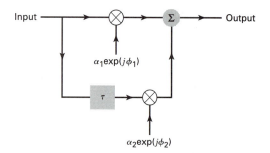

FIGURE 7.4-9   ([Rap96] fig. 4.2-5 reproduced with permission)

output $r(t)$ in fig. 7.4-8 must be decimated into samples having the correct interval. Next, if the transmitter and receiver filters form a raised cosine, the decimated samples must be translated into the frequency domain and be modified by the raised cosine template. These samples may then be translated back into the time domain to form the channel model.

A particularly useful version of this approach is the two-ray model of multipath distortion shown in fig. 7.4-9. The two complex coefficients are independent outputs of the Rayleigh Fading Simulators of fig. 7.4-7. The value of τ may be varied to produce a variety of multipath models.

The development of a program to generate these models is an appropriate student assignment and is addressed in the Problems.

Again, the reader is cautioned that there is a vast body of research on Rayleigh fading channels. This discussion is intended only as the barest introduction to the issue.

CHAPTER 7 PROBLEMS

7.1   Using the program CHAN768, find baud rate and $T_s/2$ channel models for:

(a) the C3 telephone line

(b) the C-MS3 telephone line (this is a uniform 3 msec delay)

Compare the ISI in these two channels having the same amplitude distortion but one of them with severe delay distortion and the other with no delay distortion.

7.2   Consider the amplitude files NOTCHL and NOTCHC, associated with the program CHAN768, which place a 20 db dip on the left and in the center of an otherwise distortionless raised cosine channel. Use these files, together with the uniform delay MS3 characteristic, to generate baud-rate and $T_s/2$ time domain models of a typical Rayleigh fading channel.

Compare the ISI of this channel with that of the typical telephone line.

7.3   The impulse response of a Rayleigh fading channel is often represented as a Gaussian pulse

$$p(t) = \frac{1}{\sigma\sqrt{2\pi}} \exp(-t^2/2\sigma^2)$$

where σ is the rms delay spread. Compare the mean-square values of the ISI contributed by this pulse when σ is the following fractional values of the baud rate T_s: a. $\sigma = 0.01T_s$, b. $\sigma = 0.1T_s$, c. $\sigma = T_s$.

Draw some conclusions about the relationship of delay spread to the need for equalization of the channel.

7.4   The object of this problem is to find the impulse response of a twisted pair used for baseband transmission. Assuming that the delay distortion is uniform at the frequencies in question, the transfer function can be found from eq. 7.3-2a. This transfer function will be multiplied by an appropriate baseband raised-cosine characteristic. Use the FFT function of Matlab to find the impulse response. Estimate the channel sampled impulse response for a 100 khz symbol rate, and for wire lengths of 1/4, 1/2, and 3/4 of the reference length.

7.5 We now want to find the impulse response of twisted pair used as a bandpass channel having a center frequency of 600 khz, a symbol rate of 1 Mhz, and a 20% raised cosine characteristic. Either by reformulating CHAN768 or by using Matlab, find the sampled impulse response of the channel for wire lengths of 1/4, 1/2, and 3/4 of the reference length.

7.6 Develop a program that simulates the Rayleigh Fading channel model of fig. 7.4-7.

7.7 Make a plot of eq. 7.3-5 similar to fig. 7.3-3 to show the effect of FEXT on the DSL.

7.8 Prove eq. 7.4-8e.

REFERENCES

[ABG81]   S.V. Ahmed, P.P. Bohn, and N.L. Gottfried, "A tutorial on two-wire digital transmission in the loop plant," *IEEE Trans. Communications*, **Com-29**, No. 11, Nov. 1981, pp. 1554–1564.

[Bal85]   P. Balaban, "Statistical models for amplitude and delay of selective fading," *ATT Tech. J.*, **64**, No. 10, 1985, pp. 2525–2550.

[Bel63]   P. Bello, "Characterization of randomly time-variant linear channels," *IEEE Trans. Communications*, **CS-11**, Dec. 1963, pp. 360–393.

[Bin00]   J.A.C. Bingham, "*ADSL, VDSL, and Multicarrier Modulation,*" Wiley Interscience, 2000.

[C&A85]   S.A. Cox and P.F. Adams, "An analysis of digital transmission techniques for the local network," *British Telecom Tech. J.*, **3**, No. 3, July 1985, pp. 73–75.

[CGS83]   J.C Campbell, A.J. Gibbs, and B.M. Smith, "The cyclostationary nature of crosstalk interference from digital signals in multipair cable — Part I and Part II," *IEEE Trans. Communications*, **Com-31**, May 1983, pp. 629–649.

[Cio97]   J.M. Cioffi, "Asymmetric digital subscriber lines," in J.D. Gibson (ed.), *The Communications Handbook*, CRC Press, 1997, pp. 450–479.

[Cla68]   R.H. Clarke, "A statistical theory of mobile radio reception," *Bell System Technical J.*, **47**, No. 6, July-Aug. 1968, pp. 957–1000.

[Ehr82]   L. Ehrman, "Real time software simulation of the HF radio channel," *IEEE Trans. Communications*, **Com-30**, No. 8, Aug. 1982, pp. 1809–1818.

[Gan72]   M.J. Gans, "A power spectral theory of propagation in the mobile radio environment," *IEEE Trans. Vehicular Technology*, **VT-21**, Feb. 1972, pp. 27–38.

[GH&W92] R.D. Gitlin, J.F. Huyes, and S.B. Weinstein, "*Data Communications Principles,*" Plenum, 1992.

[GVC84]   P.J. van Gerwen, N.A.M. Verhoeckx, and T.A.C.M. Claasen, "Design considerations for a 144 kbit/s digital transmission unit for the local telephone network," *IEEE J. Selected Areas of Communication*, **SAC-2**, No. 2, March 1984, pp. 314–323.

[JBS92]   M. Jeruchim, P. Balaban, and S. Shanmugam, *Simulation of Communication Systems*, Plenum, 1992.

[K&S90]   I. Kalet and S. Shamai (Schitz), "On the capacity of a twisted-wire pair: Gaussian model," *IEEE Trans. Communications*, **38**, No. 3, March 1990, pp. 379–383.

[Kal61]   T. Kailath, "Channel characterization: time-variant dispersive channels," in E.J. Baghdady (ed.), *Lectures on Communication System Theory*, McGraw-Hill, 1961.

[L&T88]   N.S. Lin and C.P. Tzeng, "Full-duplex data over local loops," *IEEE Communications Magazine*, **26**, No. 2, Feb. 1988, pp. 31–42.

[Lec89]   J.W. Lechleider, "Line codes for digital subscriber lines," *IEEE Communications Magazine*, **27**, No. 9, Sept. 1989, pp. 25–32.

[Max96]   Kim Maxwell, "Asymmetric digital subscriber line: interim technology for the next forty years," *IEEE Communications Magazine*, Oct. 1996, pp. 100–106.

[Rap96]   T.S. Rappaport, *Wireless Communications*, Prentice-Hall, 1996.

[Rum79]   W.D. Rummler, "A new selective fading model: application to propagation data," *Bell System Technical J.*, **58**, No. 5, May-June 1979, pp. 1037–1071.

[Rum81]   W.D. Rummler, "More on the multipath fading channel model," *IEEE Trans. Communications*, **Com-29**, No. 3, March 1981, pp. 346–352.

[Rum86]   W.D. Rummler et al., "Multipath fading channel models for microwave digital radio," *IEEE Communications Magazine*, **24**, No. 11, Nov. 1986, pp. 30–42.

[Skl97]   B. Sklar, "Rayleigh fading channels in mobile digital communication systems, Part I: characterization; Part II: mitigation," *IEEE Communications Magazine*, **35**, No. 9, Sept. 1997, pp. 136–155.

[Smi75]   J.I. Smith, "A computer generated multipath fading simulation for mobile radio," *IEEE Trans. Vehicular Technology*, **VT-24**, Aug. 1975, pp. 39–40.

[Smi93]   David Smith, *Digital Transmission Systems*, 2nd ed., Van Nostrand, 1993.

[Ste87]   S. Stein, "Fading channel issues in system engineering," *IEEE J. Selected Areas of Communication*, **SAC-5**, No. 2, Feb. 1987, pp. 68–89.

ATTENUATION AND DELAY CHARACTERISTICS OF THE ORDINARY TELEPHONE LINE

EIA-496-A

Attenuation Distortion Curves (dB)

Freq(Hz)	A	B	C	Freq(Hz)	A	B	C	Freq(Hz)	A	B	C
300	0.66	2.60	4.60	1275	0.00	0.40	1.02	2250	0.26	2.93	7.43
325	0.62	2.02	3.64	1300	0.00	0.45	1.13	2275	0.28	3.03	7.62
350	0.58	1.56	3.83	1325	0.00	0.50	1.23	2300	0.30	3.12	7.80
375	0.55	1.19	2.16	1350	0.00	0.55	1.34	2325	0.32	3.21	7.95
400	0.51	0.90	1.60	1375	0.00	0.60	1.47	2350	0.34	3.31	8.12
425	0.47	0.74	1.18	1400	0.00	0.66	1.60	2375	0.36	3.40	8.31
450	0.44	0.58	0.82	1425	0.00	0.72	1.76	2400	0.38	3.50	8.50
475	0.41	0.44	0.53	1450	0.00	0.78	1.91	2425	0.40	3.61	8.71
500	0.38	0.30	0.30	1475	0.00	0.84	2.05	2450	0.43	3.71	8.92
525	0.36	0.16	0.13	1500	0.00	0.90	2.20	2475	0.45	3.81	9.14
550	0.33	0.04	−0.01	1525	0.00	0.97	2.34	2500	0.48	3.90	9.35
575	0.31	−0.04	−0.12	1550	0.00	1.04	2.48	2525	0.51	3.98	9.58
600	0.28	−0.09	−0.20	1575	0.00	1.10	2.64	2550	0.55	4.08	9.80
625	0.25	−0.11	−0.25	1600	0.00	1.16	2.80	2575	0.58	4.18	10.00
650	0.22	−0.12	−0.29	1625	0.00	1.21	2.98	2600	0.62	4.30	10.20
675	0.20	−0.13	−0.33	1650	0.00	1.27	3.15	2625	0.66	4.44	10.38
700	0.18	−0.14	−0.35	1675	0.01	1.33	3.33	2650	0.71	4.57	10.57
725	0.17	−0.14	−0.36	1700	0.01	1.40	3.50	2675	0.76	4.71	10.78
750	0.15	−0.14	−0.37	1725	0.02	1.48	3.67	2700	0.83	4.85	11.00
775	0.13	−0.14	−0.36	1750	0.02	1.56	3.84	2725	0.91	4.99	11.25
800	0.11	−0.13	−0.35	1775	0.03	1.63	4.02	2750	1.00	5.13	11.49
825	0.09	−0.12	−0.33	1800	0.04	1.70	4.20	2775	1.09	5.29	11.74
850	0.07	−0.11	−0.30	1825	0.05	1.77	4.40	2800	1.18	5.45	12.00
875	0.05	−0.09	−0.26	1850	0.06	1.83	4.58	2825	1.29	5.62	12.27
900	0.04	−0.07	−0.20	1875	0.07	1.89	4.74	2850	1.40	5.81	12.54
925	0.02	−0.06	−0.17	1900	0.08	1.95	4.90	2875	1.51	6.00	12.81
950	0.01	−0.05	−0.12	1925	0.09	2.00	5.03	2900	1.63	6.20	13.10
975	0.00	−0.03	−0.06	1950	0.10	2.06	5.17	2925	1.75	6.42	13.38
1000	0.00	0.00	0.00	1975	0.11	2.13	5.33	2950	1.89	6.65	13.69
1025	0.00	0.03	0.07	2000	0.12	2.20	5.50	2975	2.05	6.89	14.03
1050	0.00	0.06	0.15	2025	0.14	2.29	5.69	3000	2.23	7.15	14.40

(Continued)

(Continued)

Freq(Hz)	A	B	C	Freq(Hz)	A	B	C	Freq(Hz)	A	B	C
1075	0.00	0.09	0.22	2050	0.15	2.37	5.88	3025	2.44	7.42	14.83
1100	0.00	0.12	0.30	2075	0.17	2.44	6.06	3050	2.67	7.72	15.25
1125	0.00	0.15	0.37	2100	0.18	2.50	6.25	3075	2.91	8.04	15.67
1150	0.00	0.19	0.45	2125	0.19	2.55	6.42	3100	3.18	8.40	16.10
1175	0.00	0.23	0.54	2150	0.20	2.60	6.61	3125	3.43	8.78	16.50
1200	0.00	0.27	0.65	2175	0.22	2.67	6.80	3150	3.76	9.20	16.97
1225	0.00	0.31	0.78	2200	0.23	2.75	7.00	3175	4.14	9.66	17.50
1250	0.00	0.36	0.90	2225	0.25	2.84	7.22	3200	4.58	10.15	18.10

Tolerances are −.25 dB on the negative side (less attenuation) and +.25 dB or +10% of dB reading on the positive side (more attenuation), whichever is greater. Attenuation must be monotonically increasing from 300 hz to 200 hz and 3200 hz to 3500 hz.

Envelope Delay Distortion

Freq(Hz)	1	2	3	4	5	Freq(Hz)	1	2	3	4	5
300	0.00	3.76	2.79	7.79	7.82	1450	0.00	0.04	−0.03	0.12	0.15
325	0.00	3.28	2.40	7.15	7.18	1475	0.00	0.03	−0.03	0.10	0.13
350	0.00	2.87	2.08	6.54	6.57	1500	0.00	0.03	−0.03	0.09	0.12
375	0.00	2.51	1.80	5.97	6.00	1525	0.00	0.03	−0.03	0.07	0.10
400	0.00	2.20	1.58	5.44	5.47	1550	0.00	0.02	−0.03	0.05	0.09
425	0.00	1.94	1.41	4.91	4.94	1575	0.00	0.02	−0.03	0.04	0.07
450	0.00	1.72	1.27	4.44	4.47	1600	0.00	0.01	−0.02	0.03	0.06
475	0.00	1.52	1.13	4.03	4.06	1625	0.00	0.01	−0.02	0.02	0.04
500	0.00	1.36	1.02	3.68	3.71	1650	0.00	0.01	−0.01	0.01	0.03
525	0.00	1.23	0.92	3.38	3.41	1675	0.00	0.00	−0.01	0.00	0.01
550	0.00	1.11	0.82	3.11	3.14	1700	0.00	0.00	0.00	0.00	0.00
575	0.00	1.00	0.74	2.86	2.89	1725	0.00	0.00	0.01	0.00	−0.01
600	0.00	0.91	0.66	2.62	2.65	1750	0.00	0.00	0.02	0.00	−0.02
625	0.00	0.83	0.59	2.42	2.45	1775	0.00	0.00	0.03	0.00	−0.03
650	0.00	0.76	0.52	2.22	2.25	1800	0.00	0.00	0.04	0.01	−0.04
675	0.00	0.70	0.47	2.04	2.07	1825	0.00	0.00	0.04	0.01	−0.04
700	0.00	0.64	0.42	1.87	1.90	1850	0.00	0.01	0.05	0.02	−0.05
725	0.00	0.58	0.37	1.72	1.75	1875	0.00	0.01	0.06	0.03	−0.05
750	0.00	0.54	0.34	1.58	1.61	1900	0.00	0.02	0.06	0.05	−0.05
775	0.00	0.49	0.30	1.45	1.48	1925	0.00	0.02	0.09	0.06	−0.05
800	0.00	0.46	0.28	1.34	1.37	1950	0.00	0.03	0.11	0.08	−0.05
825	0.00	0.42	0.25	1.24	1.27	1975	0.00	0.04	0.13	0.10	−0.05
850	0.00	0.39	0.23	1.16	1.19	2000	0.00	0.04	0.15	0.12	−0.05
875	0.00	0.36	0.21	1.08	1.11	2025	0.00	0.05	0.18	0.15	−0.04
900	0.00	0.34	0.19	1.00	1.03	2050	0.00	0.06	0.20	0.17	−0.04
925	0.00	0.31	0.17	0.93	0.96	2075	0.00	0.07	0.23	0.20	−0.04
950	0.00	0.28	0.14	0.86	0.89	2100	0.00	0.08	0.25	0.22	−0.03
975	0.00	0.26	0.13	0.80	0.83	2125	0.00	0.09	0.28	0.25	−0.03
1000	0.00	0.24	0.11	0.75	0.78	2150	0.00	0.10	0.30	0.27	−0.03
1025	0.00	0.22	0.09	0.68	0.71	2175	0.00	0.11	0.33	0.30	−0.02

(Continued)

(Continued)

Freq(Hz)	1	2	3	4	5	Freq(Hz)	1	2	3	4	5
1050	0.00	0.20	0.07	0.62	0.65	2200	0.00	0.12	0.35	0.32	−0.02
1075	0.00	0.18	0.05	0.58	0.59	2225	0.00	0.12	0.38	0.35	−0.01
1100	0.00	0.16	0.03	0.51	0.54	2250	0.00	0.13	0.40	0.37	−0.01
1125	0.00	0.15	0.02	0.47	0.50	2275	0.00	0.14	0.43	0.40	0.00
1150	0.00	0.13	0.01	0.42	0.45	2300	0.00	0.16	0.46	0.43	0.00
1175	0.00	0.12	0.00	0.38	0.41	2325	0.00	0.17	0.49	0.46	0.01
1200	0.00	0.11	−0.01	0.35	0.38	2350	0.00	0.18	0.52	0.49	0.02
1225	0.00	0.10	−0.02	0.32	0.35	2375	0.00	0.19	0.55	0.52	0.02
1250	0.00	0.09	−0.02	0.29	0.32	2400	0.00	0.20	0.59	0.56	0.02
1275	0.00	0.08	−0.02	0.26	0.29	2425	0.00	0.22	0.62	0.59	0.03
1300	0.00	0.07	−0.03	0.24	0.27	2450	0.00	0.23	0.66	0.63	0.03
1325	0.00	0.08	−0.03	0.21	0.24	2475	0.00	0.25	0.71	0.68	0.03
1350	0.00	0.08	−0.03	0.19	0.22	2500	0.00	0.27	0.76	0.73	0.04
1375	0.00	0.05	−0.03	0.17	0.20	2525	0.00	0.29	0.81	0.78	0.05
1400	0.00	0.05	−0.03	0.15	0.18	2550	0.00	0.31	0.86	0.83	0.06
1425	0.00	0.04	−0.03	0.14	0.16	2575	0.00	0.33	0.92	0.89	0.07
2600	0.00	0.36	0.98	0.95	0.08	2925	0.00	0.83	2.08	2.05	0.42
2625	0.00	0.39	1.04	1.01	0.10	2950	0.00	0.88	2.21	2.18	0.46
2650	0.00	0.42	1.10	1.07	0.12	2975	0.00	0.94	2.35	2.32	0.51
2675	0.00	0.44	1.17	1.14	0.14	3000	0.00	1.01	2.49	2.46	0.55
2700	0.00	0.47	1.25	1.22	0.16	3025	0.00	1.06	2.63	2.60	0.60
2725	0.00	0.50	1.34	1.31	0.19	3050	0.00	1.15	2.80	2.77	0.65
2750	0.00	0.53	1.43	1.40	0.21	3075	0.00	1.23	2.99	2.96	0.71
2775	0.00	0.56	1.51	1.48	0.24	3100	0.00	1.32	3.20	3.17	0.78
2800	0.00	0.60	1.58	1.56	0.26	3125	0.00	1.40	3.43	3.40	0.84
2825	0.00	0.64	1.66	1.63	0.29	3150	0.00	1.51	3.68	3.65	0.93
2850	0.00	0.68	1.75	1.72	0.32	3175	0.00	1.64	3.95	3.92	1.03
2875	0.00	0.72	1.84	1.81	0.35	3200	0.00	1.78	4.24	4.21	1.14
2900	0.00	0.77	1.95	1.92	0.38						

Tolerances are −.05 ms on the negative side and +.05 ms or 10% on the positive side, whichever is greater. Delay must be monotonically increasing from 300 hz to 200 hz and 3200 hz to 3500 hz.

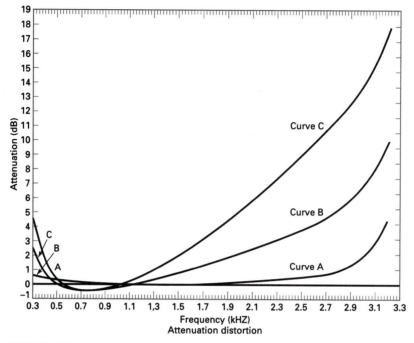

FIGURE 7A-1

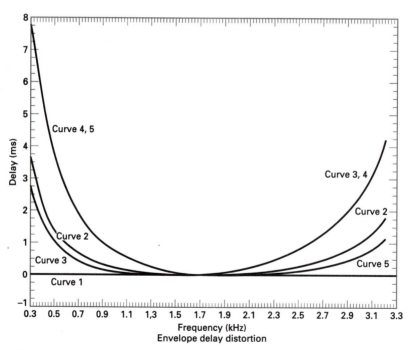

FIGURE 7A-2

CHAPTER 8

CHANNEL CAPACITY AND CODING

8.1 INTRODUCTION

The issue of Channel Capacity in communications is a fairly subtle one. Just what channel capacity is and how it is achieved can be the subject of considerable misunderstanding. As is well known, the ideas of Channel Capacity were developed by Claude Shannon [Sha48], [S&W49] and subsequently articulated by many authors. In a brief paper, Massey [Mas84] has discussed the significance of Shannon's work to communications. In this chapter we will begin by presenting an elementary and heuristic approach to this issue and then go on to a more sophisticated theoretical development.

Channel capacity has to do with how fast data can be transmitted on a channel *without making errors*. Until now we have been concerned with the probability of error associated with transmission. Now we must take that probability of error, associated with a given signal-to-noise ratio, and see how much further we can reduce it by encoding the data. In other words, the idea of channel capacity is completely tied up with the idea of data encoding. In this development it will take us a while to get to this point and the reader is urged to be patient.

First let us remind ourselves of the similarities and distinctions between the ideal and the bandlimited binary data transmission systems shown in fig. 8.1-1. The ideal system of fig. 8.1-1a, which is ideal because the channel has infinite bandwidth, can sustain transmission at any symbol rate. The infinite bandwidth means that there will be no distortion of the transmitted signal. For a given transmitted energy level E_s, a given noise power spectral density $\eta/2$, and a matched filter receiver, the probability of error is $Q(2E_s/\eta)$ independent of the rate of transmission $f_s = 1/T_s$. We will see, however, that the capacity of this system is not infinite. The "real" system of fig. 8.1-1b is real because of its bandlimited channel and consequently does not allow data transmission at any speed. It allows data transmission only at twice the Nyquist frequency and must have transmit and receive filters that simultaneously satisfy the Nyquist Criterion and form a matched filter system. We have seen that this real system, shown in fig. 8.1-1b as having raised cosine filtering, and the ideal system operating with the same signal-to-noise ratio will have the same probability of error when they are transmitting at the same symbol rate.

For a real system having a fixed symbol rate and fixed transmitter power, increasing the bit rate means going to a more complex modulation scheme having multiple bits per symbol, leading to an increased probability of error. Up to a certain point, this increase in error probability can be compensated for by encoding the data and thereby providing the redundancy that allows error-correcting capability. Past that point, coding cannot compensate and the error probability just increases. That critical point is called Channel Capacity and, from the point of view of redundancy, is measured in bits/symbol. The number of symbols

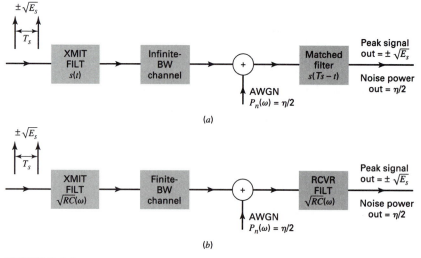

FIGURE 8.1-1

per second is determined by the channel's Nyquist bandwidth and leads to the measurement of capacity in bits/sec.

8.2   CHANNEL MODELS

Central to the discussion of Channel Capacity is the idea of a channel model that is subject to probabilistic manipulation. Such models are just a short step from the signal-space models that we have been using to compute probability of error. A collection of such models is shown in fig. 8.2-1. In fig. 8.2-1a we see the signal space for standard binary transmission where $p = Q(\sqrt{\rho_m})$ is the probability of crossing a decision boundary. In the channel model we have transmitted symbols x_0 and x_1 and received symbols y_0 and y_1 which are, of course, simply the decided versions of the transmitted symbols. The meaning of the connections are conditional probabilities:

$$P(y_0/x_0) = P(y_1/x_1) = (1 - p)$$
$$P(y_1/x_0) = P(y_0/x_1) = p$$

(8.2-1)

For reasons that should be self-evident, this channel is called the BSC or binary symmetric channel.

Similar channel models are shown in other parts of fig. 8.2-1. A channel based on QPSK is shown in fig. 8.2-1b with a different value of $p = Q(\sqrt{\rho_M/2})$ and conditional probabilities appropriate to the modulation scheme.

It is important to note that the definition of a channel in this context is not simply the physical medium of transmission but also includes the modulation scheme that is used. The SN changes the value of p but does not change the structure of the channel.

The models for 8-PSK and 16 QAM shown in figs. 8.2-1c and 8.2-1d are somewhat more complex and are shown in a simplified form that assumes that an error involves moving only to an adjacent point. Since the probability of ending up at remote points is quite small, this an acceptable approximation.

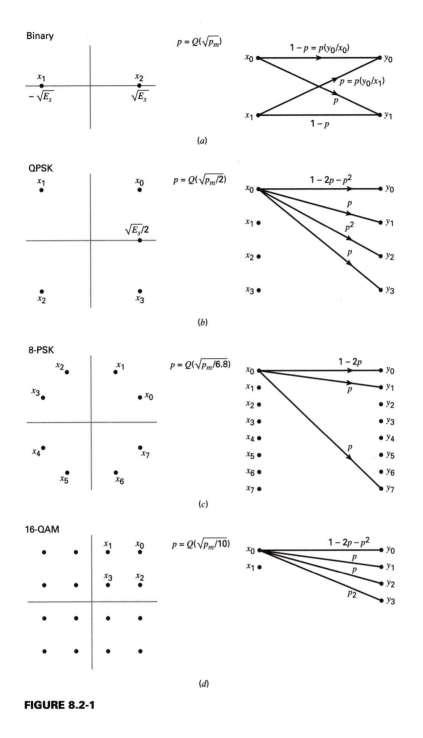

FIGURE 8.2-1

8.3   CHANNEL CODING FOR ERROR REDUCTION

We have seen that binary baseband transmission can be modeled as the BSC (binary symmetric channel) having a symbol probability of error equal to p, where $p \ll 1$. The object of channel coding is to reduce that probability without doing anything to the physical channel.

FIGURE 8.3-1

As shown in fig. 8.3-1, the BSC is preceded by a channel encoder and followed by a channel decoder. Binary symbols are transmitted on the channel at the symbol rate $f_s = 2f_N$, twice the Nyquist frequency, as a consequence of the bandlimited character of the system.

It is now necessary to make a sharp distinction between the output of the data source (the input of the channel encoder), which consists of actual binary data, occurring at a rate of f_b bits/second, and the binary symbols transmitted on the channel (the output of the channel encoder) at the rate f_s. As a simple example, consider the coding scheme that has the channel encoder repeat the data bit three times:

Data Bit	Channel Symbol
0	000
1	111

(8.3-1)

This means that the data rate from the source is one third of the symbol rate on the channel. Coding, the introduction of redundancy into the transmitted signal, has resulted in the slowing down of the data from the source. A basic parameter of information theory is the information rate

$$R = \frac{f_b}{f_s} \text{ information bits/channel symbol}$$

(8.3-2)

In the example of eq. 8.3-1, $R = 1/3$.

In order to examine how this encoding scheme may be used to detect and correct errors, assume that the source digit is zero and that the transmitted channel sequence is 000. The received sequence may be without error or may contain up to three errors. For the BSC each error has probability p and the errors occur independently. The possible received sequences together with their probabilities of occurrence are shown in Table 8.1. Since $p \ll 1$, it follows that a larger number of errors is less likely than a smaller number. Reversing the situation by transmitting 111, it is easily seen that a table complementary to Table 8.1 arises.

The channel decoding problem can be characterized as: given a received sequence, which transmitted sequence (or codeword) should we conclude was sent and what is the probability of error (i.e., of choosing the wrong codeword)?

TABLE 8.1.

Transmitted Sequence	Received Sequence	Number of Errors	Probability of Occurrence
000	000	0	$(1-q)^3$
	001	1	$q(1-q)^2$
	010	1	$q(1-q)^2$
	100	1	$q(1-q)^2$
	011	2	$q^2(1-q)$
	101	2	$q^2(1-q)$
	110	2	$q^2(1-q)$
	111	3	q^3

Since the source digits 0,1 are equally probable and independent and the channel is symmetric, it follows that the decoding rule should be symmetric as well and it should be devised to minimize the probability of error. These criteria lead to the following decoding rule:

a. if the received sequence is 000, 001, 010, 100 then conclude that the transmitted sequence is 000;

b. if the received sequence is 111, 110, 101, 011 then conclude that the transmitted sequence is 111.

This approach is called *maximum likelihood decoding.*

In this particular example a single error of transmission does not lead to a decoding error but two or three such errors do. It follows that the probability of error at the decoder output is the probability of two or three errors or

$$P_e = 3p^2(1 - p) + p^3 \tag{8.3-3}$$

As an example, if $p = 10^{-2}$ then the error probability in the coded system is $P_e \cong 3 \times 10^{-4}$, a considerable improvement. This code can be said to correct single errors.

To decrease the probability of decoding error still further, the same coding scheme can be extended to

$$\begin{aligned} 0 &\to 00000 = c_1 \\ 1 &\to 11111 = c_2 \end{aligned} \tag{8.3-4}$$

with $R = 1/5$. The same method of analysis as was used above shows that this code can correct all patterns of both single and double errors of transmission and the probability of decoding error is

$$P_e = 10q^3(1 - q)^2 + 5q^4(1 - q) + q^5 \tag{8.3-5}$$

so that with $q = 10^{-2}$, $P_e \cong 10^{-5}$.

Although extension of this method will lead to arbitrarily low probability of error, it will also require a value of R that gets very small.

8.4 THE NOISY CHANNEL CODING THEOREM AND CHANNEL CAPACITY—A FIRST LOOK

Shannon's Noisy Channel Coding Theorem states that it is possible to find channel coding schemes that make the probability of decoding error arbitrarily small without making R go to zero as long as $R < C$, the channel capacity. The key to accomplishing this is to have the channel encoder process K data bits at a time (as opposed to a single data bit in the coding schemes of eqs. 8.3-2 and 8.3-4) and deliver N channel symbols where $R = K/N$ bits/symbol. Note that we continue to assume that the channel symbol rate is f_s symbols per second.

In order to understand how this is accomplished, let us return to the coding rule of eq. 8.3-4 in order to establish some notation. Using the maximum likelihood approach developed above, the decoding rule for this code is easily constructed as shown in Table 8.2.

The Hamming distance, or simply distance, between two sequences is defined as the number of places in which they differ. Table 8.2 is arranged so that the channel sequences

TABLE 8.2

	$c_1 = 00000$	$c_2 = 11111$
	00001	11110
	00010	11101
$d=1$	00100	11011
	01000	10111
	10000	01111
	00011	11100
	00101	11010
	01001	10110
	10001	01110
$d=2$	00110	11001
	01010	10101
	10010	01101
	01100	10011
	10100	01011
	11000	00111

having distance $d = 1$ and $d = 2$ from the two codewords c_1 and c_2 are associated with those codewords. As a consequence, this code is able to correct all patterns of one and two errors but any pattern of three or more errors will result in a decoding error. The two codewords are distance $d = 5$ from each other.

We may think of Table 8.2 as describing two non-intersecting decoding spheres centered around the codewords c_1 and c_2 consisting of all of the channel sequences having distance $d = 1, 2$ from the respective codewords. When the distance D between the codewords is odd, the error-correcting capability of the code is given by:

$$D = 2e + 1 \qquad (8.4\text{-}1)$$

Having established the terminology of codeword, distance and decoding sphere, we shall now demonstrate how the channel encoder can process K source bits at a time and transmit N corresponding channel symbols. The channel encoder is characterized by the matrix operation

$$
[\cdot \ \cdot 0 \ \cdot 1 \ \cdot]
\underset{\longleftarrow \ K \ \longrightarrow}{}
\overset{\overset{N\text{-}K}{\longleftrightarrow}}{
\begin{bmatrix}
1 & 0 & 0 & \cdot & & \cdot & 0 & & \\
0 & 1 & & \cdot & 0 & & & & \\
& & & & \cdot & & & P & \\
& & & & & \cdot & & & \\
& & & & 1 & 0 & & & \\
& & & & 0 & 1 & & & \\
\end{bmatrix}
}
\qquad (8.4\text{-}2)
$$

$$\underset{\longleftarrow \qquad N \qquad \longrightarrow}{}$$

The K source bits form a vector that multiplies a $K \times N$ generator matrix, producing a vector of N channel symbols. Modulo 2 arithmetic is used in the matrix multiplication.

The generator matrix is composed of a K-dimensional identity matrix and $K \times (N-K)$ parity check matrix P of zeros and ones. The identity matrix guarantees that the codewords corresponding to the 2^K different sets of source bits are linearly independent. The P matrix may be chosen at random. Each of the 2^K codewords consists of the K source bits followed by $(N-K)$ check digits. Codes of this form are called (N, K) block codes.

It should be noted that the generator matrices of actual codes (e.g., cyclic codes, BCH codes, etc.) have algebraic structure that allows both the encoding and decoding to be mechanized in very elegant ways. However these generator matrices are always reducible, through linear combinations on their rows, to a matrix having the form of eq. 8.4-2.

Clearly, the codewords form a subset containing 2^K of the 2^N possible sequences having length N. As a result of transmission errors, any of the 2^N sequences may appear at the decoder. In order to accomplish the decoding, we must establish decoding spheres around each of the codewords. Optimizing the error-correcting capability of the code means making the codewords mutually as far as possible from other in order to increase as much as possible the common radius e of the decoding spheres. Corresponding to the dimensionality of the matrix P, there are $2^{K(N-K)}$ (N, K) codes which will have widely varying error-correcting capabilities.

The geometry of the situation makes it clear that that will be some code in which the minimum distance between any two codewords is maximized, thereby maximizing the error-correcting capability of codes having the specific values of N, K. In order to proceed, it is not necessary to find this code, only to know that it exists.

Since there are 2^K codewords it would appear at first glance that each of the decoding spheres ought to contain $2^{(N-K)}$ sequences. It has been shown, however, that this is not the case except in a few rare instances. Typically, the actual number of sequences in a decoding sphere is

$$\sum_{j=1}^{e} \binom{N}{j} < 2^{N-K} \tag{8.4-3}$$

where e is the maximum error-correcting capability of an (N, K) code. This is exactly analogous to the sphere-packing problem where there are always spaces between the spheres, even when they are touching. A strictly sphere decoding approach does not assign the sequences in these spaces to any of the codewords. Our analysis will proceed on this basis. Subsequently we will modify the results.

Let $d_{\min}$ be defined as the maximum possible value for the minimum distance for a particular (N, K) code. Recall that the error-correcting capability of the code is given by $d = 2e + 1$. There are two bounds on the value of $d_{\min}$ that are of consequence to this argument.

Let the value of q_r be defined by

$$h_2(p_r) = 1 - R \tag{8.4-4a}$$

where

$$h_2(x) = -x \log_2(x) - (1 - x) \log_2(1 - x) \tag{8.4-4b}$$

as shown in fig. 8.4-1, then

Gilbert Bound:
For most codes selected at random (by randomly selecting the P matrix) and for sufficiently large N,

$$d_{\min} \geq N p_r \tag{8.4-5a}$$

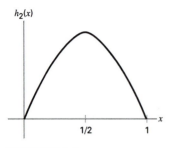

FIGURE 8.4-1

Elias Bound:
For every code and for sufficiently large N,

$$d_{\min} \leq 2Np_r(1 - p_r) \qquad (8.4\text{-}5b)$$

These bounds are shown in fig. 8.4-2.

Having figured out how many errors a code can correct, let us turn to an examination of how many transmission errors can be made on a channel. For a codeword of length N and a BSC having probability of error q, the probability of making j errors is

$$P[j \text{ errors}] = \binom{N}{j} q^j (1 - q)^{N-j} = P_j \qquad (8.4\text{-}6a)$$

the expected number of errors is

$$E[\#\text{errors}] = Np \qquad (8.4\text{-}6b)$$

and the variance of the number of errors is

$$\text{var}[\#\text{errors}] = \sqrt{p(1 - p)/N} \qquad (8.4\text{-}6c)$$

This distribution is shown in fig. 8.4-3. Note that as N increases the variance of the number of errors goes to zero or, equivalently, the probability that the number of errors is different from Np goes to zero. Another way of saying this is

$$P[\#\text{errors} > Np] \xrightarrow[N \to \infty]{} 1$$
$$P[\#\text{errors} < Np] \xrightarrow[N \to \infty]{} 0 \qquad (8.4\text{-}6d)$$

The question then becomes: does the error-correcting capability of the code exceed Np or, equivalently, does $p_r/2$ exceed p? If $p_r/2 > p$ then the code can correct more than the

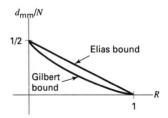

FIGURE 8.4-2

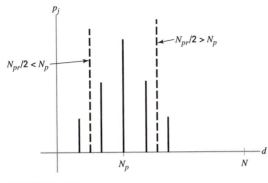

FIGURE 8.4-3

expected number of errors and

$$P[\text{decoding errors}] \xrightarrow[N \to \infty]{} 0 \qquad (8.4\text{-}7a)$$

but if $p_r/2 < p$

$$P[\text{decoding errors}] \xrightarrow[N \to \infty]{} 1 \qquad (8.4\text{-}7b)$$

Recalling from eq. 8.4-4a that $h_2(p_r) = 1 - R$, we may define

$$h_2(2p) = 1 - C_D \qquad (8.4\text{-}8a)$$

so that

$$\{p_r/2 > p\} \Rightarrow \{R < C_D\}$$
$$\{p_r/2 < p\} \Rightarrow \{R > C_D\} \qquad (8.4\text{-}8b)$$

where C_D is the sphere decoding channel capacity. All of this leads to the following proposition:

PROPOSITION

Encoded data is transmitted at a rate R bits/symbol with a block length N. It is decoded according to the minimum distance rule using strict sphere decoding. If $R < C_D$ it is possible, by increasing N, to find codes that reduce the probability of decoding error to an arbitrarily small value. If $R > C_D$ then increasing N guarantees that the probability of a decoding error goes to unity. ∎

This proposition is a crude version of Shannon's Noisy Channel Coding Theorem. It is crude because the decoding rule based on non-intersecting spheres fails to assign the large numbers of sequences lying in the interstices of these spheres to any of the codewords. Were these sequences properly assigned, the code would be able to correct most error patterns resulting from somewhat more than e errors. This would make the actual channel capacity C somewhat greater than C_D. In fact, as we shall subsequently see, the actual channel capacity of a BSC is given by

$$h_2(p) = 1 - C \qquad (8.4\text{-}8c)$$

We shall be more precise about channel capacity and about the behavior of the probability of error at rates below capacity in subsequent sections.

8.5 INFORMATION, ENTROPY, AND CHANNEL CAPACITY

In the subsequent discussion of Channel Capacity we will be doing some mathematical manipulations involving probabilities on a general channel model. In order to make these manipulations more understandable we refer to the example of Section 2.1, fig. 2.1-1 in which each transmission can be one of three different symbols. More detailed discussions can be found in [Ash65], [Bla87], [Bla90], [Gal68], [Laf90], [Wil96], and [C&T91]. In this model the successive transmissions are independent. Note that the number of transmissions or symbols per second does not figure into the channel model. That rate is determined by the Nyquist frequency of the system.

We are interested in the amount of information that is being transmitted. In order to get that value we must first have a mathematical definition of information: the information contained in a single symbol x_i is given by

$$I(x_i) = -\log_2 P(x_i) \quad \text{bits} \tag{8.5-1a}$$

An important reason for adopting this as the measure of information is the requirement that if two successive symbols are independent then the values of information will add. Note that

$$
\begin{aligned}
I(x_i x_j) = -\log_2 P(x_i x_j) &= -\log_2[P(x_i)P(x_j)] \\
&= -\log_2[P(x_i)] - \log_2[P(x_j)] \\
&= I(x_i) + I(x_j) \quad \text{bits}
\end{aligned}
\tag{8.5-1b}
$$

so that the logarithmic measure of information meets this requirement.

Since the expression for the information contained in a symbol is a function of the probability of that symbol, it follows that information is a random variable. Therefore the average information of the source is written as the expectation of a random variable:

$$I(X) = \sum_i P(x_i)I(x_i) = -\sum_i P(x_i)\log_2 P(x_i) = H(X) \tag{8.5-1c}$$

where $I(X) = H(X)$ is called the entropy of the source and is measured in bits/symbol.

EXAMPLE 1

In the system of fig. 2.1-1,

$$I(X) = -[.2\log_2(.2) + .3\log_2(.3) + .5\log_2(.5)] = 1.49 \text{ bits/symbol}$$

A little experimentation leads to the result that $I(X)$ is maximized when the probabilities of each of the three symbols are equal (and equal to 1/3). Under these conditions $I(X) = H(X) = 1.59$ bits/symbol. ∎

In general it can be shown that the information or entropy of a source is maximized if all the symbols are equally likely. The idea here is that information is the resolution of uncertainty and uncertainty is maximized with equally likely symbols.

The *conditional information* of a symbol is defined as

$$I(x_i/y_i) = -\log_2 P(x_i/y_i) \tag{8.5-2a}$$

and the *average conditional information* is

$$H(X/y_i) = \sum_i P(x_i/y_i)I(x_i/y_i) \tag{8.5-2b}$$

and further averaged over all the outputs we obtain the *equivocation*

$$H(X/Y) = \sum_j P(y_j)H(X/y_j)$$
$$= -\sum_i \sum_j P(y_j)P(x_i/y_j)\log_2 P(x_i/y_j) \tag{8.5-2c}$$

The definition of *mutual information* is:

$$I(X, Y) = H(X) - H(X/Y)$$
$$= -\sum_i P(x_i)\log_2 P(x_i) + \sum_i \sum_j P(y_j)P(x_i/y_j)\log_2 P(x_i/y_j)$$
$$= -\sum_i \sum_j P(x_i, y_j)\log_2 P(x_i) + \sum_i \sum_j P(x_i, y_j)\log_2 P(x_i/y_j)$$
$$= \sum_i \sum_j P(x_i, y_j)\log_2 \frac{P(x_i/y_j)}{P(x_i)P(y_j)}$$
$$= \sum_i \sum_j P(x_i, y_j)\log_2 \left(\frac{P(y_j/x_i)}{P(y_j)}\right)$$
$$= \sum_i \sum_j P(x_i)P(y_j/x_i)\log_2 \left(\frac{P(y_j/x_i)}{\sum_i P(x_i)P(y_j/x_i)}\right) \tag{8.5-3a}$$

Although each of the forms of mutual information has its uses, the final form is computationally most direct because all of the quantities come directly from the definition of the channel.

It also turns out to be quite important that $I(X, Y) = I(Y, X)$. This is easily shown.

$$I(X, Y) = H(X) - H(X/Y)$$
$$= \sum_i \sum_j P(x_i, y_j)\log_2 \frac{P(x_i, y_j)}{P(x_i)P(y_j)}$$
$$= \sum_i \sum_j P(x_i, y_j)\log_2 \left(\frac{P(x_i/y_j)}{P(x_i)}\right)$$
$$= \sum_i \sum_j P(y_j)P(x_i/y_j)\log_2 \left(\frac{P(x_i/y_j)}{P(x_i)}\right)$$
$$= H(Y) - H(Y/X) = I(Y, X) \tag{8.5-3b}$$

Shannon's great leap was to identify $H(X)$ as the average information provided by the source, $H(X/Y)$ as the information lost due to the channel noise, and $I(X, Y)$ as the information available to the receiver. Since the channel is given, the only variable in this expression is the set of probabilities on the input symbols. The definition of channel capacity then becomes

$$C = \max_{P(x_i)} I(X, Y) \quad \text{bits/symbol} \tag{8.5-3c}$$

This definition is not intuitively obvious. Interested readers are encouraged to consult Shannon's original paper [Sha48] as well as the references cited above.

Equation 8.5-3c is a constrained maximization problem since

$$
\begin{aligned}
&1. \ P(x_i) \geq 0, \quad \forall i \\
&2. \ \sum_i P(x_i) = 1
\end{aligned}
\tag{8.5-3d}
$$

One of the basic outcomes of the theory is that if the channel is symmetric, as are all of the channels that we have considered, then $I(X, Y)$ is maximized when all of the input probabilities are equal.

With this, we can find the expressions for channel capacity for each of the channels that we have considered.

EXAMPLE 2   Binary: $p = Q(\sqrt{\rho_m})$

Let us note that if, in a symmetric channel, the inputs are equally likely then so are the outputs. Then

$$
C = I(X, Y) = \sum_i \sum_j P(x_i)P(y_j/x_i) \log_2 \left(\frac{P(y_j/x_i)}{P(y_j)} \right)
$$

so that

$$
C = 2\left[\frac{1}{2}p \log_2 \frac{p}{1/2} + \frac{1}{2}(1-p) \log_2 \frac{(1-p)}{1/2} \right]
$$

$$
= p \log_2 2p + (1-p) \log_2 2(1-p) \quad \text{bits/symbol}
$$

and with a bit of manipulation

$$
C = (p + 1 - p) + p \log_2 p + (1-p) \log_2(1-p)
$$

$$
= 1 - h_2(p) \quad \text{bits/symbol}
\tag{8.5-4a}
$$

Note that since $\lim(x \log_2 x) \underset{x \to 0}{\longrightarrow} 0$ for this channel as p approaches 0, corresponding to a very high signal-to-noise ratio, the channel capacity C approaches 1 bit/symbol.

This is an crucial and subtle point. Until now we have thought about the binary channel carrying one bit of information for each transmitted symbol. Now we must see that, if there is any noise in the system, a transmission of a large number N of symbols will inevitably mean that fewer than N will be received correctly because of decision errors. Therefore we have received fewer bits of information than we transmitted. In order to achieve error-free transmission it will be necessary to encode the data in a manner described in Section 8.3. This will reduce the rate of actual data transmission although the symbol rate on the channel will remain the same. As long as there is any noise on the channel it will not be possible to have error-free communication at one bit per symbol.   ∎

EXAMPLE 3   QPSK: $p = Q(\sqrt{\rho_m/2})$

By symmetry,

$$
C = 4 \sum_{j=0}^{3} \frac{1}{4} P(y_j/x_0) \log_2 \left[\frac{P(y_j/x_0)}{1/4} \right]
$$

$$
= 2p \log_2(4p) + p^2 \log_2(4p^2) + (1 - 2p - p^2) \log_2[4(1 - 2p - p^2)] \tag{8.5-4b}
$$

∎

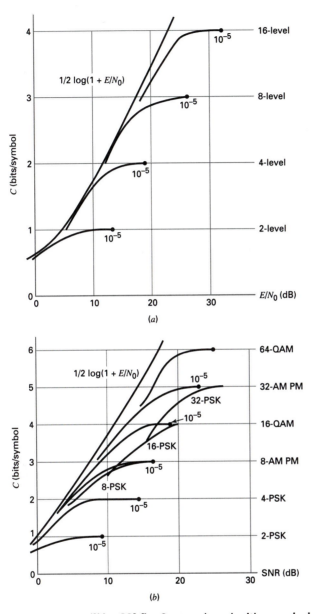

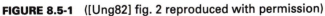

FIGURE 8.5-1 ([Ung82] fig. 2 reproduced with permission)

Clearly as the SNR becomes greater and p becomes smaller, the value of C approaches 2 bits/symbol.

EXAMPLE 4 *8-PSK:* $p = Q(\sqrt{\rho_m/6.8})$

$$C = 8 \sum_{j=0}^{7} \frac{1}{8} P(y_j/x_0) \log_2 \left[\frac{P(y_j/x_0)}{1/8} \right]$$

$$= 2p \log_2(8p) + (1 - 2p) \log_2[8(1 - 2p)] \qquad (8.5\text{-}4c)$$

so that as $p \to 0$, $C \to 3$ bits/symbol. ∎

EXAMPLE 5 *16-QAM:* $p = Q(\sqrt{\rho_m/10})$

$$C = 16 \sum_{j=0}^{15} \frac{1}{16} P(y_j/x_0) \log_2 \left[\frac{P(y_j/x_0)}{1/16} \right]$$

$$= 2p \log_2(16p) + p^2 \log_2(16p^2) + (1 - 2p - p^2) \log_2[16(1 - 2p - p^2)] \quad (8.5\text{-}4d)$$

Again, as $p \to 0$, the value of $C \to 4$ bits/symbol. ∎

The curves of Channel Capacity vs. ρ_m for the four modulation schemes that we have considered are shown in fig. 8.5-1. Along with them is shown the capacity of what is called a general waveform channel, which will be derived in the next section.

$$C = \log_2(1 + \rho_M) \quad \text{bits/symbol} \quad (8.5\text{-}5)$$

The implication here is that if we use, for example, a 16-QAM modulation scheme then the most that we ever can hope to attain is 4 bits/symbol no matter how low we make the noise. On the other hand, it is implied that there are other modulation schemes, which are not specified, that will allow the capacity of the channel to increase. In fact, as the SNR increases so does the channel capacity, because even if we are transmitting many bits per symbol there will be very few errors.

8.6 THE CAPACITY OF AN IDEAL BANDLIMITED CHANNEL HAVING AWGN

We have discovered that we can increase the capacity of a channel by constructing increasingly complex waveforms to transmit on a bandlimited channel. The capacity of a 16-QAM channel exceeds that of an 8-PSK channel, and so on. We now want to find the capacity of a bandlimited channel in which there is no direct reference to a particular modulation scheme.

Recall that in the discrete case we computed the entropy of a discrete random variable

$$X = (x_1, x_2, \ldots, x_N) \quad (8.6\text{-}1a)$$

having a probability distribution

$$P(x_1), P(x_2), \ldots, P(x_N) \quad (8.6\text{-}1b)$$

as

$$I(X) = \sum_i P(x_i) I(x_i) = - \sum_i P(x_i) \log_2 P(x_i) = H(X). \quad (8.6\text{-}1c)$$

If the random variable is continuous, eq. 8.6-1c is modified to

$$H(X) = - \int_{-\infty}^{\infty} p(x) \log_2[p(x)] \, dx \quad (8.6\text{-}1d)$$

and, similarly

$$
I(X, Y) = H(X) - H(X/Y)
$$
$$
= -\int_{-\infty}^{\infty} p(x) \log_2[p(x)]\, dx + \int_{-\infty}^{\infty}\int_{-\infty}^{\infty} p(x/y) \log_2[p(x/y)]\, dx\, dy
$$
$$
= H(Y) - H(Y/X)
$$
$$
= -\int_{-\infty}^{\infty} p(y) \log_2[p(y)]\, dy + \int_{-\infty}^{\infty}\int_{-\infty}^{\infty} p(y/x) \log_2[p(y/x)]\, dx\, dy \quad \text{(8.6-1e)}
$$

In the waveform channel the transmitted signal is $x(t)$ and the received signal is

$$
y(t) = x(t) + n(t) \tag{8.6-2}
$$

where $n(t)$ is additive white Gaussian noise having a power spectral density $\eta/2$.

We shall first find the capacity of an ideal lowpass channel having unity gain in the band $[-W, W]$ hz and AWGN having an average power or variance σ^2. It is assumed that the noise is limited to the same bandwidth as the signal. According to the sampling theorem, such bandlimited functions can be decomposed into a series of orthonormal functions

$$
x(t) = \sum_{i=1}^{N} x_i \phi_i(t)
$$
$$
y(t) = \sum_{i=1}^{N} y_i \phi_i(t) \tag{8.6-3a}
$$
$$
n(t) = \sum_{i=1}^{N} n_i \phi_i(t)
$$

where

$$
\phi_i(t) = \text{Sinc}\left(\pi \frac{t - iT_s}{T_s} \right) \tag{8.6-3b}
$$

and $T_s = 1/2W$ or $2WT_s = 1$. The coefficients in these expansions are samples of the respective waveforms taken every T_s seconds. As a consequence of the orthogonality of the basis functions, the successive samples of both signal and noise in each expansion are mutually independent.

We want to evaluate the mutual information $I(X, Y) = H(Y) - H(Y/X)$ for this channel. To begin, we will work on $H(Y)$. Because we have sampled the waveform at the Nyquist frequency, we are able to compute $H(Y)$ based on the samples rather than on the continuous waveform, so that:

$$
H(Y) = -\int_{-\infty}^{\infty} p(y) \log_2[p(y)]\, dy
$$
$$
= -\int_{-\infty}^{\infty} \cdots \int_{-\infty}^{\infty} P(y_1, \ldots, y_N) \log_2[P(y_1, \ldots, y_N)]\, dy_1 \cdots dy_N \quad \text{(8.6-3c)}
$$

and since these samples are independent,

$$H(Y) = -\int_{-\infty}^{\infty} \cdots \int_{-\infty}^{\infty} \prod_{i=1}^{N} P(y_i) \log_2 \left[\prod_{j=1}^{N} P(y_j) \right] dy_1 \cdots dy_N$$

$$= \int_{-\infty}^{\infty} \cdots \int_{-\infty}^{\infty} \prod_{i=1}^{N} P(y_i) \sum_{j=1}^{N} \log_2 [P(y_j)] \, dy_1 \cdots dy_N \qquad (8.6\text{-}3d)$$

In order to see how this integral simplifies, let's do an example with $N = 3$.

$$-\int_{-\infty}^{\infty} \int_{-\infty}^{\infty} \int_{-\infty}^{\infty} \prod_{i=1}^{3} P(y_i) \sum_{j=1}^{3} \log_2 [P(y_j)] \, dy_1 \cdots dy_3$$

$$= -\int_{-\infty}^{\infty} \int_{-\infty}^{\infty} \int_{-\infty}^{\infty} P(y_1) P(y_2) P(y_3) \{ \log_2 [P(y_1)] + \log_2 [P(y_2)] + \log_2 [P(y_3)] \} \, dy_1 \, dy_2 \, dy_3$$

$$(8.6\text{-}3e)$$

One of these integrals looks like

$$-\int_{-\infty}^{\infty} \int_{-\infty}^{\infty} \int_{-\infty}^{\infty} \prod_{i=1}^{3} P(y_i) \log_2 [P(y_i)] \, dy_1 \cdots dy_3$$

$$= \int_{-\infty}^{\infty} P(y_1) \{ \log_2 [P(y_1)] \} \, dy_1 \int_{-\infty}^{\infty} P(y_2) \, dy_2 \int_{-\infty}^{\infty} P(y_3) \, dy_3$$

$$= \int_{-\infty}^{\infty} P(y_1) \{ \log_2 [P(y_1)] \} \, dy_1 \qquad (8.6\text{-}3f)$$

so it follows that

$$\int_{-\infty}^{\infty} p(y) \log_2 [p(y)] \, dy = N \int_{-\infty}^{\infty} P(y_i) \{ \log_2 [P(y_i)] \} \, dy_1 \qquad (8.6\text{-}4a)$$

Similarly,

$$\int_{-\infty}^{\infty} \int_{-\infty}^{\infty} p(y/x) \log_2 [p(y/x)] \, dx \, dy = N \int_{-\infty}^{\infty} \cdots \int_{-\infty}^{\infty} p(y_i/x_i) \log_2 [p(y_i/x_i)] \, dx_i \, dy_i$$

$$(8.6\text{-}4b)$$

Consequently, the average mutual information is

$$\frac{1}{N} I(X, Y) = -\int_{-\infty}^{\infty} P(y_i) \{ \log_2 [P(y_i)] \} \, dy_i + \int_{-\infty}^{\infty} \int_{-\infty}^{\infty} p(y_i/x_i) \log_2 [p(y_i/x_i)] \, dx_i \, dy_i$$

$$(8.6\text{-}4c)$$

Now since $y_i = x_i + n_i$ where

$$p(n_i) = \frac{1}{\sigma \sqrt{2\pi}} e^{-n_i^2 / 2\sigma^2} \qquad (8.6\text{-}4d)$$

the second integral in eq. 8.6-4c simplifies to

$$\int_{-\infty}^{\infty}\int_{-\infty}^{\infty} p(y_i/x_i)\log_2[p(y_i/x_i)]\,dx_i\,dy_i = \int_{-\infty}^{\infty} p(n_i)\log_2[p(n_i)]\,dn_i$$

$$= \int_{-\infty}^{\infty} \frac{1}{\sigma\sqrt{2\pi}}e^{-n_i^2/2\sigma^2}\left[-\log_2\sigma\sqrt{2\pi} - \frac{n_i^2}{2\sigma^2}\log_2 e\right]$$

$$= -\log_2\sigma\sqrt{2\pi} - \frac{\sigma^2}{2\sigma^2}\log_2 e = -\frac{1}{2}\log_2 e - \frac{1}{2}\log_2 2\pi\sigma^2 \qquad (8.6\text{-}4e)$$

which is a constant.

It follows then that in order to find the capacity we should maximize the average mutual information with respect to the output probability

$$C = \max_{P(y_i)} \frac{1}{N}I(X, Y) = -\int_{-\infty}^{\infty} P(y_i)\{\log_2[P(y_i)]\}\,dy_i + \text{constant} \qquad (8.6\text{-}5a)$$

subject to the constraints that since $P(y_i)$ must be a valid probability density function and since the variance or average power in $y(t)$ must be the sum of the powers of the transmitted signal and the noise:

$$1. \quad \int_{-\infty}^{\infty} P(y_i)\,dy_i = 1 \qquad (8.6\text{-}5b)$$

$$2. \quad \int_{-\infty}^{\infty} y_i^2 P(y_i)\,dy_i = P_{av} + \sigma^2. \qquad (8.6\text{-}5c)$$

In order to do the maximization, we form the Lagrangian

$$-\int_{-\infty}^{\infty} P(y_i)\{\log_2[P(y_i)]\}\,dy_i + a_1 \int_{-\infty}^{\infty} P(y_i)\,dy_i + a_2 \int_{-\infty}^{\infty} y_i^2 P(y_i)\,dy_i$$

$$= \int_{-\infty}^{\infty} P(y_i)\{-\log_2[P(y_i)] + a_1 + a_2 y_i^2\}\,dy_1 \qquad (8.6\text{-}5d)$$

which is maximized when

$$\log_2 p(y_i) = a_1 + a_2 y_i^2 \qquad (8.6\text{-}5e)$$

or

$$p(y_i) = Ae^{-a_2 y_i^2} \qquad (8.6\text{-}5f)$$

But in order for $p(y_i)$ to be a proper density function it must have a variance

$$\sigma_y^2 = P_{av} + \sigma^2 \qquad (8.6\text{-}5g)$$

so that

$$p(y_i) = \frac{1}{\sqrt{2\pi(P_{av} + \sigma^2)}}e^{-y_i^2/2(P_{av} + \sigma^2)} \qquad (8.6\text{-}5h)$$

It follows from eq. 8.6-5a that

$$C = H(Y) - H(n) = \left[\frac{1}{2} + \frac{1}{2}\log_2[2\pi(P_{av} + \sigma^2)]\right] - \left[\frac{1}{2} + \frac{1}{2}\log_2[2\pi\sigma^2]\right]$$

$$= \frac{1}{2}\log_2\left[\frac{P_{av} + \sigma^2}{\sigma^2}\right] = \frac{1}{2}\log_2\left[1 + \frac{P_{av}}{\sigma^2}\right] \qquad (8.6\text{-}6a)$$

but $P_{av} = E_{av}/T_s$ and $\sigma^2 = \frac{\eta}{2} \cdot (2W)$ where $2WT_s = 1$ so that

$$C = \frac{1}{2}\log_2\left(1 + \frac{E_{av}}{(\eta/2)2WT_s}\right) = \frac{1}{2}\log_2(1 + \rho_M) \quad \text{bits/symbol} \qquad (8.6\text{-}6b)$$

For a QAM system where we have two carriers occupying the same bandwidth, the capacity is doubled:

$$C = \log_2(1 + \rho_M) \quad \text{bits/symbol} \qquad (8.6\text{-}6c)$$

The curves for the baseband and the passband systems are given in fig. 8.5-1.

8.7   THE CAPACITY OF AN ARBITRARY WAVEFORM CHANNEL SUBJECT TO A POWER CONSTRAINT

Our intention is to find the capacity of an arbitrary channel where the transmitter is operating under a maximum power limitation. The channel may be represented as the transfer function $H(\omega)$ of an arbitrary linear system and the noise, assumed to be stationary and Gaussian, has an arbitrary power spectral density $N(\omega)$.

8.7.1   The Capacity of Parallel Channels Subject to an Overall Power Constraint

The first step in determining the capacity of an arbitrary waveform channel subject to a power constraint is to break the arbitrary channel into N parallel channels, each of which is assumed to be ideal. This approach, which in practical terms is implemented by multicarrier transmission, is illustrated in fig. 8.7-1a. As shown, the noise power in each of these subchannels is assumed to be different, these variances denoted respectively as σ_n^2, $n = 1, \ldots N$. Although this may initially seem counterintuitive, the reader is urged to be patient.

The total transmitter power is P and the power in each subchannel P_n is to be allocated in a way that maximizes the capacity subject to the constraint that $P = \sum_{n=1}^{N} P_n$. It is well known [Bla87] that the optimal way to allocate this power is to have the total power in

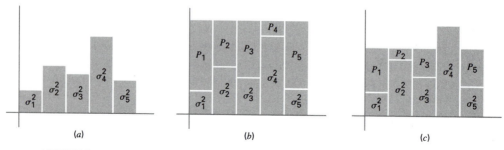

(a)   (b)   (c)

FIGURE 8.7-1

each subchannel, signal power plus noise power, be the same, as shown in fig. 8.7-1b. This allocation of transmitter power is called "water pouring" and the analogy extends to fig. 8.7-1c where a subchannel having an excessive noise level is allocated no power at all. The overall power level is given by the parameter B as shown.

The resulting channel capacity, a function of the power P, is given from eq. 8.6-6a as

$$C(P) = \sum_{n=1}^{N} \frac{1}{2} \log_2 \left(1 + \frac{P_n}{\sigma_n^2} \right)$$

$$= \sum_{n=1}^{N} \frac{1}{2} \max \left[0, \log_2 \left(\frac{B}{\sigma_n^2} \right) \right] \qquad (8.7\text{-}1a)$$

since $B = P_n + \sigma_n^2$ and the contribution to the capacity will be zero if there is no power allocated to that subchannel. The value of B is determined parametrically by the additional constraint

$$P = \sum_{n=1}^{N} \max[0, B - \sigma_n^2] \qquad (8.7\text{-}1b)$$

This does not lend itself to a closed-form analytic technique for finding B; rather, numerical techniques are more appropriate.

8.7.2 The Capacity of an Arbitrary Channel Subject to a Power Constraint

In characterizing the noise power in the different subchannels in fig. 8.7-1a as being different, we were, in effect, looking at fig. 8.7-2a in a somewhat different way. Our actual problem is that of a transmitted signal $x(t)$ applied to a channel $H_1(f)$ and then added to Gaussian noise having power spectral density $N(f)$. The simple transformation of fig. 8.7-2b allows us to compare the noise power spectral density $n'(t)$ directly against the power spectral density of the signal $x(t)$. Even if $n(t)$ is white, $n'(t)$ will vary across the band as

$$N'(f) = \frac{N(f)}{|H_1(f)|^2} \qquad (8.7\text{-}2)$$

Accordingly, the summations of eqs. 8.7-1a and 8.7-1b can be rewritten as integrals:

$$C(P) = \frac{1}{2} \int_{-\infty}^{\infty} \max \left[0, \log_2 \left(\frac{|H_1(f)|^2 B}{N(f)} \right) \right] df \qquad (8.7\text{-}3a)$$

$$P = \int_{-\infty}^{\infty} \max \left[0, \left(B - \frac{N(f)}{|H_1(f)|^2} \right) \right] df \qquad (8.7\text{-}3b)$$

Again, these integrals require a numerical solution.

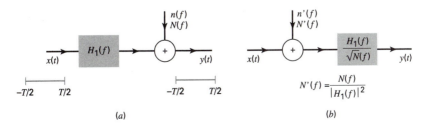

FIGURE 8.7-2

EXAMPLE 6

A more or less 'practical' example of this computation is that of a lowpass channel having the transfer function of an nth-order Butterworth filter where

$$|H_1(f)|^2 = \frac{1}{1 + (f/f_0)^{2n}} \tag{8.7-4a}$$

and the noise is white with $P_n(\omega) = \eta/2$. In this instance

$$C(P) = \frac{1}{2} \int_{-W}^{W} \log_2 \left(\frac{B}{\frac{\eta}{2}[1 + (f/f_0)^{2n}]} \right) df \tag{8.7-4b}$$

and

$$P = \int_{-W}^{W} \left\{ B - \frac{\eta}{2}[1 + (f/f_0)^{2n}] \right\} df \tag{8.7-4c}$$

where W is the solution to

$$\left\{ B - \frac{\eta}{2}[1 + (f/f_0)^{2n}] \right\} = 0 \tag{8.7-4}$$

This result leads to the perhaps counterintuitive conclusion that as the total available transmitter power increases it makes sense to use more of the available bandwidth for signal transmission, even if the channel attenuation in that region continues to increase. ∎

Considerable attention has been devoted in the literature ([Gal68], [Wyn66], [SLP62], [Sle76]) to a rigorous proof of eqs. 8.7-3a and 8.7-3b, deriving them from the capacity of parallel channels, eqs. 8.7-1a and 8.7-1b, through the use of the Karhunen–Loeve expansion, which was discussed in Chapter 5. According to this expansion, the nonwhite noise process $n'(t)$ in fig. 8.7-2b can be expanded over the finite time interval $[-T/2, T/2]$ in a series of orthogonal functions

$$n'(t) = \sum_{i=1}^{\infty} n_i' \varphi_i(t) \tag{8.7-5a}$$

where the noise coefficients n_i' are independent Gaussian random variables having variance λ_i. The input signal $s(t)$ may also be expanded with the same set of orthogonal functions

$$x(t) = \sum_{i=1}^{\infty} x_i \varphi_i(t) \tag{8.7-5b}$$

The orthogonal basis functions fortunately do not have to be found in order to proceed with the analysis.

As a consequence of the expansion, the different components of signal plus noise $(x_i + n_i')$ may be considered to be on orthogonal or parallel channels so that eqs. 8.7-1a and 8.7-1b become

$$C(P) = \sum_{i=1}^{N} \frac{1}{2} \max \left[0, \log_2 \left(\frac{B}{\lambda_i^2} \right) \right] \tag{8.7-6a}$$

$$P = \sum_{n=1}^{N} \max \left[0, B - \lambda_i^2 \right] \tag{8.7-6b}$$

In order to get to eqs. 8.7-3a and 8.7-3b, we want to show the relationship of the power spectral density to the eigenvalues of a linear system. The references demonstrate a variety of ways of developing this relationship. We will outline only one of these.

Let us start by imagining the output of fig. 8.7-2b to be sampled and an $M \times M$ correlation matrix formed from those samples. We have seen that this matrix will always be Toeplitz. As M gets very large, an Mth-order Toeplitz matrix becomes a *circulant matrix* in the limit.

A circulant matrix has the form

$$C = \begin{bmatrix} x_0 & x_1 & x_2 & x_3 & x_4 \\ x_4 & x_0 & x_1 & x_2 & x_3 \\ x_3 & x_4 & x_0 & x_1 & x_2 \\ x_2 & x_3 & x_4 & x_0 & x_1 \\ x_1 & x_2 & x_3 & x_4 & x_0 \end{bmatrix} \qquad (8.7\text{-}7a)$$

where, as can be seen, each row is a cyclic shift of the row above. The resulting matrix must always be square. Let

$$X = [x_0 \quad x_1 \quad x_2 \quad x_3 \quad x_4] \qquad (8.7\text{-}7b)$$

be the first row of this matrix and let Y be the DFT (Discrete Fourier Transform) of X. It follows that

$$Y = [y_0 \quad y_1 \quad y_2 \quad y_3 \quad y_4] = XF \qquad (8.7\text{-}7c)$$

where $\omega = e^{j2\pi/5}$ and

$$F = \begin{bmatrix} 1 & 1 & 1 & 1 & 1 \\ 1 & \omega^{-1} & \omega^{-2} & \omega^{-3} & \omega^{-4} \\ 1 & \omega^{-2} & \omega^{-4} & \omega^{-6} & \omega^{-8} \\ 1 & \omega^{-3} & \omega^{-6} & \omega^{-9} & \omega^{-12} \\ 1 & \omega^{-4} & \omega^{-8} & \omega^{-12} & \omega^{-16} \end{bmatrix} \qquad (8.7\text{-}7d)$$

Clark [Cla85] demonstrates first that the inverse DFT matrix is given by

$$F^{-1} = \frac{1}{5} \begin{bmatrix} 1 & 1 & 1 & 1 & 1 \\ 1 & \omega & \omega^2 & \omega^3 & \omega^4 \\ 1 & \omega^2 & \omega^4 & \omega^6 & \omega^8 \\ 1 & \omega^3 & \omega^6 & \omega^9 & \omega^{12} \\ 1 & \omega^4 & \omega^8 & \omega^{12} & \omega^{16} \end{bmatrix} \qquad (8.7\text{-}7e)$$

and then shows that for any circulant matrix

$$F^{-1}CF = D = \begin{bmatrix} y_0 & 0 & 0 & 0 & 0 \\ 0 & y_1 & 0 & 0 & 0 \\ 0 & 0 & y_2 & 0 & 0 \\ 0 & 0 & 0 & y_3 & 0 \\ 0 & 0 & 0 & 0 & y_4 \end{bmatrix} \qquad (8.7\text{-}7f)$$

indicating that the components of Y, the DFT of the vector X, are the eigenvalues of the circulant matrix C based on X.

The correlation matrix of a channel, a Toeplitz matrix, consists of samples of the channel autocorrelation function $R(\tau)$ which has Fourier Transform $P(\omega)$. The correlation matrix asymptotically becomes becomes a circulant matrix ([Gra72], [Mak81]) as its order becomes greater. As this happens the DFT of the sampled autocorrelation function becomes the Fourier Transform of the continuous autocorrelation function and consequently the eigenvalues of the correlation matrix become the magnitude of $P(\omega)$. This may be expressed mathematically first with the finite-order eigenvalue relationship

$$Ru = \lambda u \qquad (8.7\text{-}8a)$$

of the Mth-order matrix correlation matrix R. As $M \to \infty$ the Fourier Transform of both sides yields

$$P(f)\mathbf{U}(f) = \lambda \mathbf{U}(f) \qquad (8.7\text{-}8b)$$

where $P(f)$ is the magnitude-squared of the spectrum of signal plus noise. The only way that eq. 8.7-8b can be satisfied is if $\mathbf{U}(f) = \delta(f - f_c)$, the Fourier Transform of a sinusoid, and λ_c is the magnitude of $P(f_c)$.

In the case of fig. 8.72b,

$$P(f) = \lambda = \frac{N(f)}{|H_1(f)|^2} \qquad (8.7\text{-}9)$$

which justifies eqs. 8.7-3a and 8.7-3b.

8.8 THE CAPACITY OF SOME COMMON CHANNELS

8.8.1 The Ordinary Telephone Line

A general discussion of the channel capacity of the ordinary telephone line and the means of attaining this capacity is given by Lucky [LSW68].

We shall make an approximation to the capacity of a telephone line by assuming a bandwidth $B = 2400$ hz corresponding to the Nyquist bandwidth of the V.32 modem. We shall also assume that the SNR = 30 db. Note that the typical transmitter power in a telephone is 0 dbm or $P = 1$ mw. This means that $\sigma^2 = 10^{-6} = \eta B$ so that $\eta = 4.2 \times 10^{-10}$. According to eq. 8.6-6b,c the channel capacity is

$$C = B\left[\log_2\left(1 + \frac{P_{av}}{\sigma^2}\right)\right] \quad \text{bits/sec}$$
$$= [2400][\log_2(1 + 1000)] = [2400][9.97] = 23{,}921 \text{ bits/sec} \qquad (8.8\text{-}1)$$

As we saw in Chapter 4, the V.34 operate at speeds of 28.8 kbits/sec up to 33.6 kbits/sec actually uses a Nyquist bandwidth in excess of 2400 hz. The V.90 modem is able to operate at its maximum speed of 56 kbits/sec only if the phone line is terminated in a PCM channel bank.

8.8.2 An Infinite-Bandwidth Ideal Channel

Start with an ideal lowpass channel having bandwidth $[-W, W]$, noise having a psd $\eta/2$, and signal power equal to P. The noise power in that bandwidth is ηW. The channel capacity in bits/sec is then given by

$$C = \frac{2W}{2}\log_2\left(1 + \frac{P}{\eta W}\right) \qquad (8.8\text{-}2a)$$

Letting $P/\eta W = \alpha$ yields

$$C = \frac{P}{\eta\alpha}\log_2(1+\alpha) = \frac{1}{\ln 2}\frac{P}{\eta}\frac{\ln(1+\alpha)}{\alpha} \qquad (8.8\text{-}2b)$$

With P constant and $W \to \infty$ it follows that $\alpha \to 0$ and the infinite bandwidth capacity becomes

$$C_\infty = \frac{1}{\ln 2}\frac{P}{\eta}\lim_{\alpha\to 0}\frac{\ln(1+\alpha)}{\alpha} = \frac{1}{\ln 2}\frac{P}{\eta} \qquad (8.8\text{-}2c)$$

With the same values of P and η as in the previous example,

$$C_\infty = \frac{1}{\ln 2}\frac{P}{\eta} = 1.44\frac{10^{-3}}{4.2 \cdot 10^{-10}} = 3.4 * 10^6 \text{ bits/sec} \qquad (8.8\text{-}2d)$$

Clearly there is advantage to having greater bandwidth, even with an absolute constraint on the transmitted power.

8.8.3 The Twisted Pair Digital Subscriber Loop (DSL)

A model of the twisted pair channel, which was discussed in Chapter 7, is shown in fig. 8.8-1. The designated transmitted signal and the NEXT are shown as having the same power spectral densities $P_s(f)$; both arise from data signals having the same statistics. The two paths have differing transfer functions, the signal transfer function $|H_c(f)|^2$ of eq. 7.3-2a and the NEXT transfer function $|H_{\text{NEXT}}(f)|^2$ of eq. 7.3-3a. These two transfer functions are shown together in fig. 7.3-3a,b. Note both that the $H_c(f)$ decreases with increasing frequency and the NEXT power increases with increasing frequency.

The channel capacity of the DSL will be found from eq. 8.6-6a,b,c by treating the NEXT as nonwhite Gaussian noise. From fig. 8.8-1, we find

$$C_{\text{NEXT}} = \int_{f\in A}\left[\log_2\left(1 + \frac{|H_c(f)|^2 P_s(f)}{|H_x(f)|^2 P_s(f)}\right)\right]df$$

$$= \int_{f\in A}\left[\log_2\left(1 + \frac{|H_c(f)|^2}{|H_x(f)|^2}\right)\right]df \quad \text{bits/sec} \qquad (8.8\text{-}3)$$

where A is the range of integration. Note that since both the signal power and the NEXT power have the same spectral density, the capacity is independent of that density and dependent only upon the ratio of signal-to-NEXT transfer functions. For the analytic model

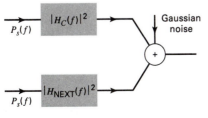

FIGURE 8.8-1

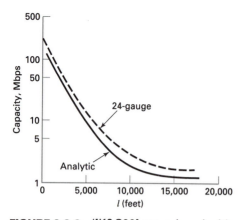

FIGURE 8.8-2 ([K&S90] reproduced with permission)

of the transfer functions referred to above

$$C_{\text{NEXT}} = \int_0^\infty \left[\log_2 \left(1 + \frac{e^{-\alpha\sqrt{f}}}{\beta f^{3/2}} \right) \right] df \quad \text{bits/sec} \tag{8.8-4}$$

Kalet [K&S90] has evaluated this integral numerically for some typical parameters as a function of line length, resulting in the graph of C_{NEXT} vs. line length shown in fig. 8.8-2. Some important numbers that arise from this graph are

Cable Length (ft)	Analytic Model Capacity (Mbps)	24-gauge Model Capacity (Mbps)
600	120.3	176.9
6000	5.95	9.71
18000	1.19	1.56

The differences in channel capacity vs. cable length provide the basis for different classes of service on the DSL. The interested reader is referred to the *IEEE Journal on Selected Areas in Communications*, "High Speed Digital Subscriber Lines," Vol. 9, No. 6, August 1991.

Some additional work on the capacity of twisted pair has been done by Shamai (Schitz) [Sha90] and Lechleider [Lec90].

8.9 EXPONENTIAL ERROR BOUNDS FOR RATES BELOW CAPACITY

In Section 8.4 we took a crude look at the idea of channel capacity from the point of view of channel coding. Based upon that analysis, we can project that the probability of error will be approximated by an exponential bound

$$P_e \approx e^{-N(C-R)} \tag{8.9-1}$$

where N is the block length of the code, R is the rate of information transmission, and C is the channel capacity. The idea is that a given error probability can be achieved by either transmitting at a low enough rate or using a long enough code. This exponential bound is similar to the bound on error probabilty for uncoded orthogonal transmission in eq. 5.5-5e.

Gallager [Gal65] has developed such a bound and, in so doing, has derived Shannon's equation for channel capacity, eq. 8.5-3c, in a different way. This alternative approach will give additional insight into the idea of capacity.

We have already seen that a block code takes K source bits and transmits them as N channel symbols. This means that there are $M = 2^K$ possible transmitted sequences of N channel symbols, which are called codewords x_m, $m = 1 \ldots M$. Any of the 2^N possible channel sequences y may appear at the receiver but those that are different from the codewords indicate that an error caused by channel noise has occurred. Using maximum likelihood decoding, a received sequence y is decoded as x_m if

$$P(y/x_m) > P(y/x_{m'}) \quad \text{for all } m' \neq m \tag{8.9-2a}$$

A decoding error occurs if the application of eq. 8.9-2a does not lead to the correct transmitted codeword. The probability of such a decoding error, conditioned on x_m having been transmitted, can be expressed as:

$$P_{em} = \sum_y P(y/x_m)\varphi_m(y) \tag{8.9-2b}$$

where

$$\varphi_m(y) = \begin{cases} 1 & \text{if } P(y/x_m) \leq P(y/x_{m'}) \text{ for some } m' \neq m \\ 0 & \text{otherwise} \end{cases} \tag{8.9-2c}$$

The key to finding the exponential error bound is to upper-bound $\varphi_m(y)$ as:

$$\varphi_m(y) \leq \left[\frac{\sum_{m' \neq m} P(y/x_{m'})^{1/1+\rho}}{P(y/x_m)^{1/1+\rho}} \right]^\rho ; \quad \rho > 0 \tag{8.9-3a}$$

As Gallager observes, this bound is not intuitively obvious but can be demonstrated to be valid. Since the right-hand side of eq. 8.9-3a is always non-negative, the bound is valid for $\varphi_m(y) = 0$. When $\varphi_m(y) = 1$ some term in the numerator will always be greater than the denominator. Raising the bracketed ratio to the power ρ keeps the right-hand side greater than unity.

Substituting eq. 8.9-3a into eq. 8.9-2b yields

$$P_{em} \leq \sum_y P(y/x_m)^{1/1+\rho} \left[\sum_{m' \neq m} P(y/x_{m'})^{1/1+\rho} \right]^\rho ; \quad \text{for } \rho > 0 \tag{8.9-3b}$$

The bound of eq. 8.9-3b is implicitly for a particular set of codewords. An ensemble of sets of codewords, which is the same as an ensemble of codes, can be generated by choosing each codeword according to the probability measure $P(x)$ so that the probability associated with the codewords $x_1, \ldots, x_M$ is $\prod_{m=1}^M P(x_m)$. Since at least one of these codes will have a probability of error as small as the ensemble average probability of error of eq. 8.9-3b, we have

$$\bar{P}_{em} \leq \sum_y P(y/x_m)^{1/1+\rho} \left[\sum_{m' \neq m} P(y/x_{m'})^{1/1+\rho} \right]^\rho \quad \text{for } \rho > 0 \tag{8.9-3c}$$

Imposing the additional restriction that $1 \geq \rho$, we may simplify eq. 8.9-3c by averaging at the numbered places indicated in eq. 8.9-3d.

$$
\overline{\underset{1}{} \quad \underset{2 \quad 4}{} \quad \underset{3}{}} \\
\sum_y P(y/x_m)^{1/1+\rho} \left[\sum_{m' \neq m} P(y/x_{m'})^{1/1+\rho} \right]^\rho \tag{8.9-3d}
$$

Note that the terms that are being summed are random variables and the average of such a sum is the sum of the averages. Consequently the potion of the averaging under 1 may be removed. Under 2 we have the product of two independent random variables and we note that the average of the product equals the product of the averages. We therefore have

$$
\bar{P}_{em} \leq \sum_y \overline{P(y/x_m)^{1/1+\rho}} \left[\sum_{m' \neq m} \overline{P(y/x_{m'})^{1/1+\rho}} \right]^\rho ; \quad \text{for } 0 \leq \rho \leq 1 \tag{8.9-3e}
$$

We now want use the fact that $\overline{Z^\rho} \leq \overline{Z}^\rho$ (the proof is left to the Problems). With this, it is possible to remove 3 and, again exchanging sum and average, remove 4. We are therefore left with

$$
\bar{P}_{em} \leq \sum_y \overline{P(y/x_m)^{1/1+\rho}} \left[\sum_{m' \neq m} \overline{P(y/x_{m'})^{1/1+\rho}} \right]^\rho ; \quad \text{for } 0 \leq \rho \leq 1 \tag{8.9-3f}
$$

Now note that since the codewords are chosen with probability $P(x)$, we have

$$
\overline{P(y/x_m)^{1/1+\rho}} = \sum_x P(x) P(y/x)^{1/1+\rho} \tag{8.9-3g}
$$

which is independent of m. Consequently eq. 8.9-3g can be used for both the m and the m' terms in eq. 8.9-3f. Finally, the summation over $m \neq m'$ is over $M - 1$ terms and we get

$$
\bar{P}_{em} \leq (M-1)^\rho \sum_y \sum_x P(x) P(y/x)^{1/1+\rho} \left[\sum_x P(x) P(y/x)^{1/1+\rho} \right]^\rho
$$

$$
= (M-1)^\rho \sum_y \left[\sum_x P(x) P(y/x)^{1/1+\rho} \right]^{1+\rho} \quad \text{for } 0 \leq \rho \leq 1 \tag{8.9-4}
$$

If the channel is memoryless, the bound of eq. 8.9-4 can be simplified considerably. Under those circumstances,

$$
P(y/x) = \prod_{n=1}^N P(y_n/x_n) \quad \text{and} \quad P(x) = \prod_{n=1}^N P(x_n) \tag{8.9-5a}
$$

and substituting into eq. 8.9-4 we get

$$
\bar{P}_{em} \leq (M-1)^\rho \sum_y \left[\sum_x \prod_{n=1}^N P(x_n) P(y_n/x_n)^{1/1+\rho} \right]^{1+\rho} \tag{8.9-5b}
$$

The order of multiplication and addition can be exchanged, first within the square brackets

$$
\bar{P}_{em} \leq (M-1)^\rho \sum_y \left[\prod_{n=1}^N \sum_x P(x_n) P(y_n/x_n)^{1/1+\rho} \right]^{1+\rho} \tag{8.9-5c}
$$

and then by taking the product out of the brackets entirely:

$$\bar{P}_{em} \le (M-1)^{\rho} \prod_{n=1}^{N} \sum_{y} \left[\sum_{x} P(x_n)P(y_n/x_n)^{1/1+\rho} \right]^{1+\rho} \tag{8.9-5d}$$

Demonstration of the validity of the operation of exchanging the order of multiplication and addition is left to the Problems.

Further simplification of eq. 8.9-5d can be accomplished first noting that the channel input and output symbols may be represented as x_j, $j = 1, \ldots, J$; $y_k, k = 1, \ldots K$. Then we may denote $P(x_j) = P_j$ and $P(y_k/x_j) = P_{jk}$. The value of $M-1$ may be upperbounded by

$$(M-1) < M = 2^{NR} \tag{8.9-5e}$$

where $R = (K/N)$ source bits per channel symbol. Finally, each of the terms that the product is operating on is identical so the product becomes equivalent to raising a single one of these terms to the Nth power. The result is

$$\bar{P}_{em} \le \exp\left\{ -N\left[-\rho R - \ln \sum_{j=1}^{J} \left(\sum_{k=1}^{K} P_k P_{jk}^{1/1+\rho} \right)^{1+\rho} \right] \right\} \tag{8.9-6a}$$

or

$$\bar{P}_{em} \le e^{-N[-\rho R + E_0(\rho, P)]}; \quad 0 \le \rho \le 1 \tag{8.9-6b}$$

where

$$E_0(\rho, P) = -\ln \sum_{j=1}^{J} \left(\sum_{k=1}^{K} P_k P_{jk}^{1/1+\rho} \right)^{1+\rho} \tag{8.9-6c}$$

As long as $E_0(\rho, P) > \rho R$ the upper bound on the error probability will decrease as N increases. In turn, $E_0(\rho, P)$ can be maximized with respect to P and ρ. The result is

$$E(R) = \max_{P} E(R, P) = \max_{\rho, P}[-\rho R + E_0(\rho, P)] \tag{8.9-6d}$$

and

$$P_e < e^{-NE(R)} \tag{8.9-6e}$$

We shall demonstrate that $E(R) > 0$ for $0 < R < C$, the channel capacity. This is another way of understanding the significance of channel capacity. Part of this development is the construction of a graph of $E(R)$ vs. R. In turn, this requires a closer examination of $E_0(\rho, P)$.

An examination of eq. 8.9-6c reveals the following properties of $E_0(\rho, P)$:

1.	$E_0(\rho, P) = 0$	for $\rho = 0$	(8.9-7a)
2.	$E_0(\rho, P) > 0$	for $\rho > 0$	(8.9-7b)
3.	$\dfrac{\partial E_0(\rho, P)}{\partial \rho} > 0$	for $\rho > 0$	(8.9-7c)

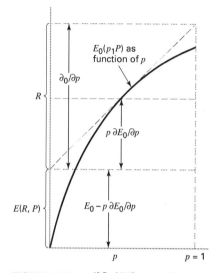

FIGURE 8.9-1   ([Gal65] reproduced with permission)

4.
$$\frac{\partial^2 E_0(\rho, P)}{\partial \rho^2} \leq 0 \qquad \text{for } \rho > 0 \tag{8.9-7d}$$

5.
$$\left.\frac{\partial E_0(\rho, P)}{\partial \rho}\right|_{\rho=0} = \sum_{k=1}^{K}\sum_{j=1}^{J} P_k P_{jk} \ln \frac{P_{jk}}{\sum_{i=1}^{K} P_i P_{ji}} = I(P) \tag{8.9-7e}$$

where $I(P)$ is the average mutual information of the channel. Recall that the channel capacity C was earlier defined in eq. 8.5-3c as the maximum over the input probability vector P_k of the mutual information $I(P)$.

Based upon the characteristics of eq. 8.9-7, a typical graph of $E_0(\rho, P)$ vs. ρ is shown in fig. 8.9-1. Note that $E_0(\rho, P)$ is positive for all $\rho > 0$, is zero at the origin, has a positive and decreasing slope, and has a slope at the origin equal to $I(P)$, the mutual information. Based upon these properties, it becomes fairly direct to accomplish the maximization indicated for $E(R)$ in eq. 8.9-6d:

$$\frac{\partial E(R, P)}{\partial \rho} = -R + \frac{\partial E_0(\rho, P)}{\partial \rho} = 0 \tag{8.9-8a}$$

corresponding to

$$R = \frac{\partial E_0(\rho, P)}{\partial \rho} \tag{8.9-8b}$$

where such a solution must exist in the range

$$\left.\frac{\partial E_0(\rho, P)}{\partial \rho}\right|_{\rho=1} \leq R \leq I(P) \tag{8.9-8c}$$

Within this range of R, the graph of $E(R)$ vs. R can best be expressed through the parametric equations

$$E(R, P) = E_0(\rho, P) - \rho \frac{\partial E_0(\rho, P)}{\partial \rho} \tag{8.9-9a}$$

$$R = \frac{\partial E_0(\rho, P)}{\partial \rho} \tag{8.9-9b}$$

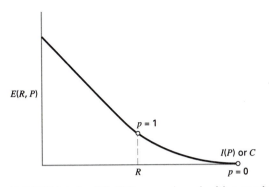

FIGURE 8.9-2   ([Gal65] reproduced with permission)

and noting that

$$\frac{\partial E(R, P)}{\partial R} = \frac{\partial E(R, P)/\partial \rho}{\partial R/\partial \rho} = -\rho \qquad (8.9\text{-}9c)$$

we can see (fig. 8.9-2) that $E(R, P)$ has a zero slope at $R = I(P)$ corresponding to $\rho = 0$. As R decreases to $R = \frac{\partial E_0(\rho, P)}{\partial \rho}\big|_{\rho=1}$, the slope decreases to -1, corresponding to $\rho = 1$. For $R \leq \frac{\partial E_0(\rho, P)}{\partial \rho}\big|_{\rho=1}$, the maximum of eq. 8.9-6d corresponds to $\rho = 1$ so that

$$E(R) = -R + E_0(1, P) \qquad (8.9\text{-}10)$$

Another way of constructing the graph of $E(R)$ vs. R is shown in fig. 8.9-3. Here the bound is considered to the envelope of the set of lines

$$E(R) = \text{lub}[-\rho R + E_0(\rho, P)] \qquad (8.9\text{-}11)$$

for all values of $0 \leq \rho \leq 1$.

Finally, we note that the value of $R = I(P)$ is to be maximized over the input probability distribution P, yielding the definition of Channel Capacity originally proposed by Shannon.

Despite its mathematical complexity, this development therefore has double value. First, it demonstrates that the definition of Channel Capacity can be arrived at by upperbounding the probability of error as well as by Shannon's insight about maximizing mutual information. Second, it reveals the relationship between information rate and acheivable probability of error with appropriate coding. In that regard it can be seen that at rates very

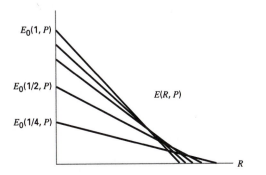

FIGURE 8.9-3   ([Gal65] reproduced with permission)

close to capacity a large value of the code block length N is required to achieve a reasonably small error probability.

The illustration of this bound for the binary symmetric channel will be left to the Problems.

CHAPTER 8 PROBLEMS

8.1 In eq. 8.7-4, let $n = 1$ corresponding to a lowpass RC circuit. Solve the parametric equations and develop a graph of $C(P)$ vs. P.

8.2 **(a)** From eq. 8.9-3e, prove that $\overline{Z^\rho} \le \overline{Z}^\rho$ for $0 \le \rho \le 1$.

(b) Show that the order of multiplication and addition can be exchanged in eq. 8.9-5c,d.

8.3 Find the capacity of channels having the following conditional probability matrices:

$$\begin{bmatrix} 1-\beta & \beta & 0 \\ \beta & 1-\beta & 0 \\ 0 & 0 & 1 \end{bmatrix}, \quad \begin{bmatrix} \dfrac{1-p}{2} & \dfrac{1-p}{2} & \dfrac{p}{2} & \dfrac{p}{2} \\ \dfrac{p}{2} & \dfrac{p}{2} & \dfrac{1-p}{2} & \dfrac{1-p}{2} \end{bmatrix}$$

8.4 Illustrate the bound of eq. 8.9-11 for the binary symmetric channel.

REFERENCES

[Ash65] Robert Ash, *Information Theory*, Wiley, 1965.

[Bla87] Richard Blahut, *Principles and Practice of Information Theory*, Addison-Wesley, 1987.

[Bla90] Richard Blahut, *Digital Transmission of Information*, Addison-Wesley, 1990.

[C&T91] T. Cover and J. Thomas, *Elements of Information Theory*, Wiley, 1991.

[Cla85] A.P. Clark, *Equalizers for Digital Modems*, Halsted, 1985.

[Gal65] Robert G. Gallager, "A simple derivation of the coding theorem and some applications," *IEEE Trans. Information Theory*, **IT-11**, Jan. 1965, pp. 3–18.

[Gal68] Robert G. Gallager, *Information Theory and Reliable Communication*, Wiley, 1968.

[Gra72] R.M. Gray, "On the asymptotic eigenvalue distribution of Toeplitz matrices," *IEEE Trans. Information Theory*, **IT-18**, No. 6, Nov. 1972, pp. 725–730.

[K&S90] I. Kalet and S. Shamai (Schitz), "On the capacity of a twisted-wire pair: Gaussian model," *IEEE Trans. Communications*, **Com. 38**, No. 3, March 1990, pp. 379–383.

[Laf90] Pierre Lafrance, *Fundamental Concepts in Communication*, Prentice Hall, 1990.

[Lec90] J.W. Lechleider, "The capacity of NEXT-impaired subscriber loops," Globecom 1990, Dec. 2-5, 1990, **2**, pp. 1161–1165.

[Lsw68] R.W. Lucky, J. Salz, and E.J. Weldon, "Principles of Data Communication," McGraw-Hill, 1968.

[Mak81] J. Makhoul, "On the eigenvectors of symmetric Toeplitz matrices," *IEEE Trans. Acoustics, Speech, and Signal Processing*, **ASSP-29**, No. 4, Aug. 1981, pp. 868–872.

[Mas84] J.L. Massey, "Information theory: the Copernican system of communications," *IEEE Communications Magazine*, **22**, No. 12, Dec. 1984.

[S&W49] C.E. Shannon and W.W. Weaver, *The Mathematical Theory of Communication*, University of Illinois Press, 1949.

[SLP62]   D. Slepian, H. Landau, and H. Pollak, "Prolate spheroidal wave functions, Fourier analysis and uncertainty I, II, III," *Bell System Technical J.*, Jan. 1961, pp. 43–84; July 1962, pp. 1295–1336.

[Sha48]   C.E. Shannon, "A mathematical theory of communication," *Bell System Technical J.*, **27**, July and Oct. 1948, pp. 379–423 and 623–656.

[Sha90]   S. Shamai (Schitz), "On the capacity of a twisted-wire pair: peak power constraint," *IEEE Trans. Communications*, **Com. 38**, No. 3, March 1990, pp. 368–378.

[Sle76]   D. Slepian, "On bandwidth," *Proc. IEEE*, **64**, No. 3, March 1976.

[Wil96]   Stephen Wilson, *Digital Modulation and Coding*, Prentice Hall, 1990.

[Wyn66]   A.D. Wyner, "The capacity of the band-limited Gaussian channel," *Bell System Technical J.*, March 1966, pp. 359–395.

CHAPTER *9*

TRELLIS CODING AND MULTIDIMENSIONAL SIGNALING

Trellis coding is a method of combining channel coding and signal modulation into a single process for the purpose of reducing the probability of error in the presence of AWGN without increasing the required bandwidth. Since trellis coding is relatively easy to implement it has become widely used in modems for the telephone line and promises to extend the performance of magnetic recording systems and transmission subjected to fading. This chapter will introduce the basic ideas of trellis codes primarily through the use of examples. Readers who wish a more complete mathematical approach to the subject should refer to the referenced literature; in particular, two recent books on the subject are Biglieri et al. [BDMS91] and Schlegel [Sch97].

In order better to appreciate the idea of trellis coding, we shall briefly recapitulate the idea of block coding and introduce the idea of convolutional coding.

9.1 BLOCK CODING AND CONVOLUTIONAL CODING

Consider first a data source producing a data stream of equally likely 1s and 0s at the rate of k bits/sec. As shown in fig. 9.1-1a, the data stream is encoded and the resulting stream of channel symbols, which also can be understood as consisting of 1s and 0s, are applied to the channel at the rate of n symbols/sec where $n > k$. Data transmitted in this way is said to be transmitted at the rate of $R = k/n$ bits/symbol. It should be noted that the channel itself may operate with any of the modulation techniques that have been discussed in Chapters 4 and 5, such as QPSK, 16-QAM, etc. The point to be made is the separation of coding and transmission in this traditional approach to the channel-coding problem.

A block code is formed by gathering up a block of K data bits and using them to generate a block of N channel symbols where $K/N = k/n = R$. The process by which this is accomplished can be understood as multiplying a row vector of K data bits into a $\mathbf{K} \times \mathbf{N}$ generator matrix as shown in fig. 9.1-1b. There are therefore 2^K possible transmitted code-words, each one a block of N channel symbols. Since channel noise creates the possibility of decision error, the decoder will have to contend with the possibility of 2^N length-N possible received channel sequences. The Hamming distance between two channel sequences is the number of places in which they differ.

An ideal decoding rule associates each of the 2^N possible received channel sequences with the nearest one of the 2^K codewords. Consequently, the minimum distance between any two codewords $d_{\min}$ should be maximized in order to have a code that corrects the largest possible number of errors. It is important to note that block coding treats each successive block of K bits entirely separately. In order to mechanize both the encoding and decoding

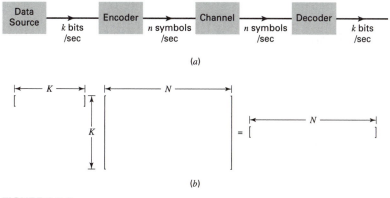

FIGURE 9.1-1

processes, the generator matrix is given an algebraic structure. Examples of such structures are the BCH or Reed-Solomon Codes.

A convolutional encoder works quite differently. As shown in fig. 9.1-2a, a K-bit frame of data is fed into a shift register that holds V such frames. The value of V is known as the *constraint length* of the code. These KV bits are logically combined into an encoded N symbol frame. In the example shown, $K = 3$, $N = 4$, and $V = 5$. The rate of transmission is $R = K/N$ bits per channel symbol. Note that the shift register provides for a memory of V frames of data.

A somewhat more accessible encoder is shown in fig. 9.1-2b. In this example the frame length is $K = 1$, the constraint length is $V = 3$, and the channel sequence has length $N = 2$. Consequently the rate is $R = 1/2$.

This encoder can be understood to be a sequential machine having eight states, corresponding to a shift register of length $KV = 3$, and two inputs. (Note that the sequential machine of fig. 9.1-2a would have $2^K = 8$ possible inputs and $2^{KV} = 2^{15}$ possible states, truly a formidable enterprise.) The definition of a state is the triplet (s_2, s_1, s_0) and, since the output does not directly depend on the input, the result is a Moore machine. The state transition table of this machine is easily found as:

TABLE 9.1-1

	0	1	Output
q_0	q_0	q_4	11
q_1	q_0	q_4	11
q_2	q_1	q_5	01
q_3	q_1	q_5	10
q_4	q_2	q_6	11
q_5	q_2	q_6	00
q_6	q_3	q_7	10
q_7	q_3	q_7	01

Recall that an ISI model for a distorted channel was represented as a sequential machine in Chapter 5 just as this encoder is represented as a sequential machine. Since the sequential machine approach led to the use of a trellis diagram to represent the evolving output and the Viterbi Algorithm for the purpose of detection of the input in the ISI model, we

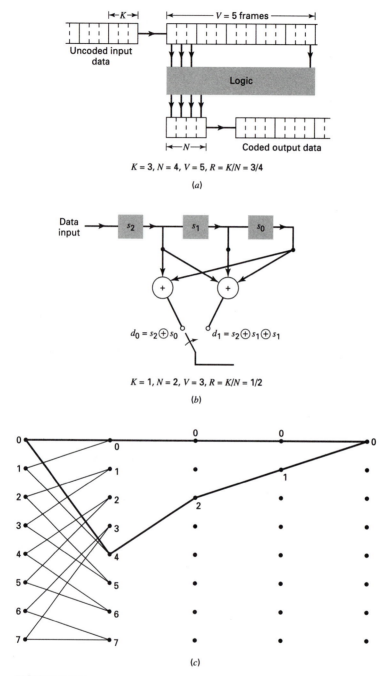

FIGURE 9.1-2

shall extend that approach to convolutional coding. A trellis diagram for the convolutional code represented by Table 9.1 is shown in fig. 9.1-2c. Note that each of the states may be reached from two different states.

Implementation of the Viterbi Algorithm in this instance requires the use of Hamming distance, the number of digits in which two sequences differ, rather than Euclidean distance as the measure of the distance between sequences. Each pair of channel output symbols is compared to output corresponding to the appropriate trellis state in order to compute the incremental distance for the algorithm.

The measure of the error-correcting capability of the convolutional code is measured by the smallest distance between allowable sequences (rather than the smallest distance between codewords in block codes). It is easy to see that the two state sequences outlined in the trellis of fig. 9.1-2c are the two closest possible sequences of states that can be generated by this code. From the point that they diverge until the point that they merge again, the respective outputs are 0000000000 and 0011011100, yielding a distance of 5. Since the error-correcting capability is given by $d = 2e + 1$, the code can correct two errors in a sequence of 10 channel outputs.

Convolutional coding plays an important part in the development of trellis codes.

9.2 INTRODUCING TRELLIS CODES

Trellis coding was first developed by Ungerboeck [Ung82]. Calderbank and Mazo [C&M84] developed a somewhat different way of looking at the issue. This introduction will, in large part, follow Ungerboeck's paper.

According to the plot of channel capacity C vs. SN for two-dimensional signal constellations shown in fig. 8.5-1b, data transmitted at 2 bits/symbol using QPSK as the modulation technique will have a probability of error of 10^{-5}, corresponding to a SN = 12.9 db. According to this graph, if the same 2 bits/symbol were transmitted using 8-PSK it would be theoretically possible to acheive error-free transmission, with suitable coding, with a SN = 5.9 db, a 7 db improvement. It is also seen that efforts to reduce the SN further using arbitrarily complex signaling would yield only an additional 1.2 db. For that reason, efforts have been made only to determine effective ways of transmitting data at M bits/sec using constellations having 2^{M+1} points. The codes that have been used to accomplish this task are called trellis codes.

In order to introduce trellis codes, let us consider the problem of transmitting data at 4800 bps on a channel having a Nyquist bandwidth of 2400 hz. We have already seen that an obvious solution is to use QPSK as the modulation technique since it operates at 2 bits/channel symbol. Our first look at trellis codes involves transmitting these 2 bits/channel symbol using an 8-PSK signal set. These two alternatives are shown in fig. 9.2-1. In fig. 9.2-1b we look inside the 'encoder and modulator' box for the trellis code to see that while two binary channel symbols (x_0, x_1) are required to specify QPSK, three symbols (y_0, y_1, y_2) are required for 8-PSK. The first part of trellis coding is the specification of this transformation.

Although it is theoretically possible to use a higher-order signal set than 8-PSK to transmit the original QPSK signal (for example, we might use 16-QAM and have (y_0, y_1, y_2, y_3) as outputs), we have already seen that the most effective use of trellis coding involves encoding a modulation technique having 2^M points into one having 2^{M+1} points. Consequently the rate of a trellis code is typically $R = (M/M + 1)$.

The transformation from M to $(M + 1)$ binary symbols, (x_0, x_1) to (y_0, y_1, y_2) in our example, is accomplished through the use of a binary convolutional encoder. We have already seen that the convolutional encoder can be arbitrarily complex, having a larger or smaller number of states, while maintaining its rate. Some examples of encoders with different numbers of states will be developed. The second part of the trellis coder is the mapping of the 2^{M+1} encoder outputs onto a 2^{M+1} point signal constellation, in our example (y_0, y_1, y_2) onto the 8-PSK signal set. As examples of trellis codes for successively higher bit-rate systems are developed, involving more complex signal constellations, the mapping of the encoder outputs onto the constellation will also be presented.

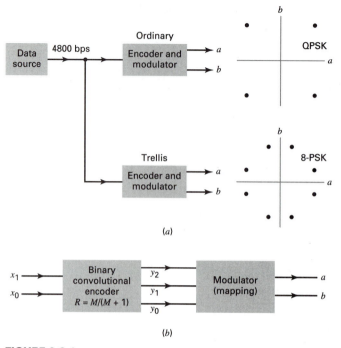

FIGURE 9.2-1

For the current example, the mapping onto the 8-PSK constellation is shown in fig. 9.2-2 together with the partitioning of that constellation into subsets having increasing minimum distance between the points. Since the points in the signal constellation are arranged on a unit circle, it is straightforward to find the minimum distance between points in each of the partitions. In the basic signal set denoted $A0$ this distance is $\Delta_0 = .765$; in the partitions denoted $B0$ and $B1$ the distance is $\Delta_1 = 1.414$; and in the partition denoted $C0$, $C1$, $C2$, $C3$ the distance is $\Delta_2 = 2$.

In order to be able subsequently to determine the logic circuitry for convolutional encoder of fig. 9.2-1b, it is necessary to establish a relationship between its output (y_2, y_1, y_0) and the points in the 8-PSK constellation. This is done in the simplest possible manner:

y_2	y_1	y_0	Signal
0	0	0	0
0	0	1	1
0	1	0	2
0	1	1	3
1	0	0	4
1	0	1	5
1	1	0	6
1	1	1	7

(9.2-1)

where the signals are those of fig. 9.2-2.

With this partitioning it is possible to construct an unlimited number of different codes having increasing memory or numbers of states, allowing us to approach the 7 db improvement indicated above. We shall now continue by showing three examples:

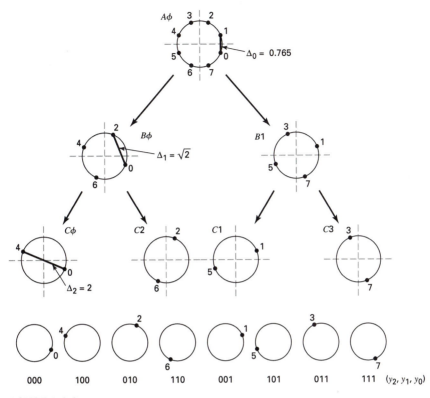

FIGURE 9.2-2

EXAMPLE 1 *Two Trellis States*

A trellis code having two states is shown in fig. 9.2-3. In fig. 9.2-3a we see the basic structure of the convolutional encoder with the logic to be determined subsequently. The dibit inputs to the encoder are denoted as:

x_1	x_0	Input
0	0	ϕ
0	1	1
1	0	2
1	1	3

(9.2-2)

and the states are $q_0 = [s_0 = 0]$ and $q_1 = [s_1 = 1]$. It is clear that an input of either ϕ or 1 will result in the machine going to state q_0 and an input of either 2 or 3 will result in state q_1. The issue is how to choose the corresponding outputs. As shown in fig. 9.2-3b, the outputs are chosen to be pairs of outputs from level C of the partition of fig. 9.2-2. A trellis illustrating this encoder is shown in fig. 9.2-3c. From this trellis we can see that the output partitions $C\phi$ and $C2$ correspond to a 0 and 1 input to the machine while it is in q_0; $C1$ and $C3$ correspond to a 0 and 1 input to the machine while it is in q_1.

To compute the minimum distance d_{free}, the significance of which will be discussed subsequently, we want to first recognize that an input of all 0s leads to transitions from q_0 to itself. We then want to examine all other input sequences that lead from q_0 to q_0 in order to determine the cumulative Euclidean distance between these state trajectories and the trajectory corresponding to the all-0 input sequence. The minimum of these Euclidean distances is d_{free}. First, let us recognize that a transistion from q_0 to q_0 can also occur if

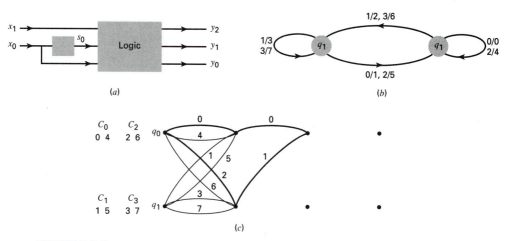

FIGURE 9.2-3

the input is 2 rather than ϕ. The two corresponding outputs are 4 and 0, which are distance $\Delta_2 = 2$ apart. We then consider the path q_0, q_1, q_0 shown in fig. 9.2-3c, and compare it to the all-q_0 trajectory:

$$q_0 \xrightarrow{0} q_0 \xrightarrow{0} q_0$$
$$q_0 \xrightarrow[\Delta_1]{2} q_1 \xrightarrow[\Delta_0]{1} q_0 \qquad (9.2\text{-}3)$$

where Δ_0 and Δ_1 are distances between points in the partitioned signal constellation shown in fig. 9.2-2. Note that in the first pair of transitions, corresponding to the partition $B\phi$, both outputs 0 and 4, associated with a transition to q_0, are distance Δ_1 from outputs 2 and 6, associated with a transition to q_1. The distance between these two sequences is

$$d = \sqrt{\Delta_1^2 + \Delta_0^2} = 1.608 \qquad (9.2\text{-}4)$$

which is actually less than Δ_2, the distance of the apparently simpler q_0-to-q_0 transition. Accordingly, we designate $d_{\text{free}} = 1.608$.

We can also determine the logic for the encoder in fig. 9.2-3a. Recognizing that s_0 is the current state and s_0' is the next state, the truth table for the logic is derived directly from the state diagram.

x_1	x_0	s_0	s_0'	y_2	y_1	y_0	Output
0	0	0	0	0	0	0	ϕ
0	0	1	0	0	1	1	3
0	1	0	1	0	1	0	2
0	1	1	1	0	0	1	1
1	0	0	0	1	0	0	4
1	0	1	0	1	1	1	7
1	1	0	1	1	1	0	6
1	1	1	1	1	0	1	5

(9.2-5a)

so that the logic equations are

$$y_2 = x_1; \qquad y_1 = x_0 \oplus s_0; \qquad y_0 = s_0 \qquad (9.2\text{-}5b)$$

For subsequent discussion of general issues of the structure of trellis codes, we note that two of the three encoder outputs given above are encoded versions of the input data. This is denoted as $N_c = 2$. For a rate $R = 2/3$ code, the maximum value of N_c is 3. ∎

Since the code described above is generated by a sequential machine and is therefore a Markov process, it can be decoded using the Viterbi Algorithm. The significance of d_{free} can be understood in relation to the idea of decoding the received signal in this way.   ecall that the Viterbi Algorithm determines the survivor paths by computing the cumulative Euclidean distance between the received sequence and each of the possible transmitted sequences and then choosing the paths of minimum distance. Consequently d_{free} plays the same role in the Viterbi Algorithm that the minimum distance between signal points plays in symbol-by symbol-decisions. Just as the basic measure for symbol-by-symbol decoding is $Q\,[\sqrt{d_{\text{min}}^2/2\eta}]$, the basic measure for the Viterbi Algorithm is $Q\,[\sqrt{d_{\text{min}}^2/2\eta}]$. The coding gain of the trellis code is defined as

$$\gamma = 10 \log_{10}\left(\frac{d_{\text{free}}^2/E_1}{d_{\text{min}}^2/E_2}\right) \tag{9.2-6}$$

where E_1 and E_2 are the average energies per transmitted symbol in the respective constellations. In our example it is clear that $E_1 = E_2$, but that will not necessarily be the case in the analysis of codes based on more complex signal constellations. For the code of example 1, $d_{\text{free}} = 1.608$ and, for the uncoded QPSK constellation, $d_{\text{min}} = 1.414$. Consequently $\gamma = 1.1$ db.

We shall see in the subsequent examples that it is possible to increase the coding gain significantly by increasing the number of states in the trellis code.

EXAMPLE 2   *Four Trellis States*

The block diagram for a four-state trellis code is shown in fig. 9.2-4a with the states defined as

	s_1	s_0	
q_0	0	0	
q_1	0	1	(9.2-7)
q_2	1	0	
q_3	1	1	

The state diagram for the code is given in fig. 9.2-4b and the trellis diagram is shown in fig. 9.2-4c. Note that each of the states can be reached in two ways, making the states equally likely. We have seen in Chapter 2 that this causes the code to have no effect on the power spectral density of the data. The outputs associated with each state again follow the partitioning of the 8-PSK constellation. An examination of this state diagram easily reveals that in order to find d_{free} we must compare the paths

$$q_0 \xrightarrow{0} q_0 \xrightarrow{0} q_0 \xrightarrow{0} q_0$$

$$q_0 \xrightarrow[\Delta_1]{2} q_1 \xrightarrow[\Delta_0]{1} q_2 \xrightarrow[\Delta_1]{2} q_0 \tag{9.2-8a}$$

having distance

$$d = \sqrt{\Delta_1^2 + \Delta_0^2 + \Delta_1^2} = 2.141 \tag{9.2-8b}$$

with the path q_0 to q_0 under the input 2 which, again, corresponds to distance $\Delta_2 = 2$. In this instance $d_{\text{free}} = 2$ and the coding gain is $\gamma = 3.0$ db, a considerable improvement over the two-state code.

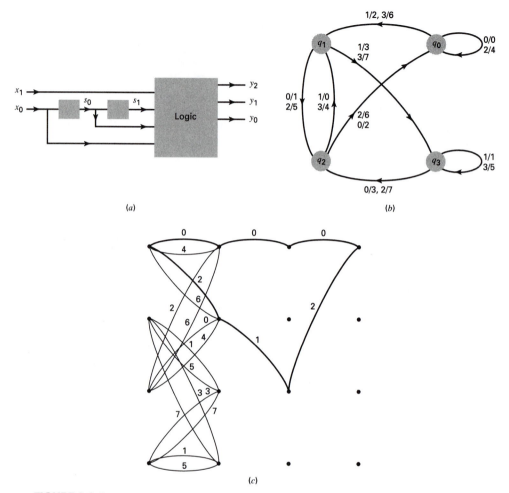

(a)

(b)

(c)

FIGURE 9.2-4

The logic for this code can again be found from the truth table:

x_1	x_0	s_1	s_0	s_1'	s_0'	y_2	y_1	y_0	Output
0	0	0	0	0	0	0	0	0	ϕ
		0	1	1	0	0	0	1	1
		1	0	0	0	0	1	0	2
		1	1	1	0	0	1	1	3
0	1	0	0	0	1	0	1	0	2
		0	1	1	1	0	1	1	3
		1	0	0	1	0	0	0	ϕ
		1	1	1	1	0	0	1	1
1	0	0	0	0	0	1	0	0	4
		0	1	1	0	1	0	1	5
		1	0	0	0	1	1	0	6
		1	1	1	0	1	1	1	7
1	1	0	0	0	1	1	1	0	6
		0	1	1	1	1	1	1	7
		1	0	0	1	1	0	0	4
		1	1	1	1	1	0	1	5

(9.2-9a)

yielding

$$y_2 = x_1; \qquad y_1 = x_0 \oplus s_1; \qquad y_0 = s_0 \qquad (9.2\text{-}9b)$$

We again note that the number of encoded outputs is $N_c = 2$. ∎

EXAMPLE 3 *Eight Trellis States*

We will begin the example of an eight-state trellis coder having the state table

Input/State	ϕ	1	2	3
q_0	$q_0/0$	$q_1/4$	$q_2/2$	$q_3/6$
q_1	$q_4/1$	$q_5/5$	$q_6/3$	$q_7/7$
q_2	$q_0/4$	$q_1/0$	$q_2/6$	$q_3/2$
q_3	$q_4/5$	$q_5/1$	$q_6/7$	$q_7/3$
q_4	$q_0/2$	$q_1/6$	$q_2/0$	$q_3/4$
q_5	$q_4/3$	$q_5/7$	$q_6/1$	$q_7/5$
q_6	$q_0/6$	$q_1/2$	$q_2/4$	$q_3/0$
q_7	$q_4/7$	$q_5/3$	$q_6/5$	$q_7/1$

$$(9.2\text{-}10a)$$

and the trellis diagram shown in fig. 9.2-5a. Since this code does not have two inputs that

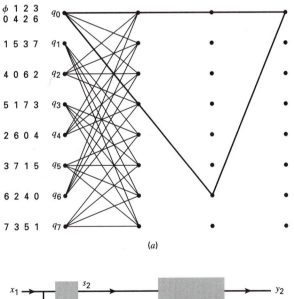

(a)

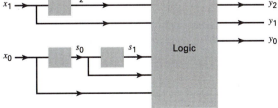

(b)

FIGURE 9.2-5

take q_0 to itself, it is straightforward to see that the two paths that determine d_{free} are

$$q_0 \xrightarrow{0} q_0 \xrightarrow{0} q_0 \xrightarrow{0} q_0$$

$$q_0 \xrightarrow[\Delta_1]{6} q_3 \xrightarrow[\Delta_0]{7} q_6 \xrightarrow[\Delta_1]{6} q_0 \qquad (9.2\text{-}10b)$$

so that

$$d_{\text{free}} = \sqrt{\Delta_1^2 + \Delta_0^2 + \Delta_1^2} = 2.141 \qquad (9.2\text{-}10c)$$

yielding a coding gain of $\gamma = 3.6$ db.

It will be left to the Problems to confirm that this encoder has the form shown in fig. 9.2-5b and the logic equations are

$$y_2 = x_0 \oplus s_2; \qquad y_1 = x_0 \oplus s_1; \qquad y_0 = s_0 \qquad (9.2\text{-}10d)$$

In this example all of the outputs are encoded versions of the input data so that $N_c = 3$. ∎

In the Problems we will examine a 16-trellis state code that yields a coding gain of 4.1 db for the QPSK to 8-PSK transition.

We now want to make some observations and generalizations about the codes shown in the previous examples. These are known as the Ungerboeck Conditions.

1. Under the assumption that the original data stream in each of these codes consists of independent equally likely 1s and 0s, it is clear that in each example each of the states is equally likely. As we have seen in Chapter 2, this is the condition for the power spectral density of the data being unaffected by the coding process. All of the trellis codes with which we deal will have this property.

2. Notice that in Examples 1 and 2 only two of the three output bits (y_2, y_1, y_0) are encoded ($N_c = 2$); in each case $y_2 = x_1$. The consequences of this level of encoding are twofold: first, the fact that inputs ϕ, 2 and the inputs 1, 3 cause the same state transitions leads to the category of *parallel transitions*; second, corresponding to these parallel transitions, the outputs are taken from the second level of partition of the signal set, $C\phi$, $C1$, $C2$, $C3$, each of which contains two points. In Example 3, on the other hand, all of the output bits are encoded, $N_c = 3$, so there are no parallel transitions and the outputs are taken from the lowest, or third, level of the partition, which has only single points.

3. Now let's go back to Example 2 and examine the transitions *from* each of the states. As we can see from the 8-PSK partition, (fig. 9.2-2) the set $B\phi$ contains points 0,2,4,6 with subsets $C\phi$ (0,4) and $C2$ (2,6). Similarly, $B1$ contains points 1,3,5,7 with subsets $C1(1,5)$ and $C3(3,7)$. In fig. 9.2-4b the transitions *from* q_1 have outputs corresponding to $C1$ and $C3$ and the transitions *from* q_2 have outputs from $C\phi$ and $C2$. We do not allow the transitions *from* a particular state to have outputs that come from different supersets. The same condition holds in relation to transitions *to* a particular state. The transitions *to* q_1 come from $B\phi$ and the transitions to q_3 come from $B1$.

Finally, it is important to note that the logic circuitry in all of the example codes given above is linear, that is, consists only of delays and exclusive-OR (modulo-2 addition) gates. Such codes are referred to as Ungerboeck Codes. Their linearity renders them directly

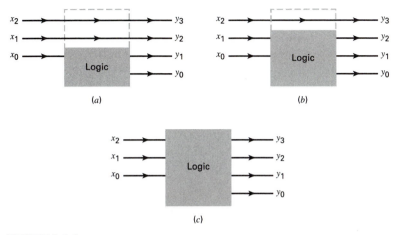

FIGURE 9.2-6

subject to analytic techniques because, consistent with the definition of linearity in the continuous domain, the properties of linearity in the discrete domain include superposition and the separability of the solution of the describing difference equation into a forced and a transient component. Convolutional codes used in communication systems are typically linear. References include Golomb [Gol67] and Viterbi and Omura [V&O79].

It is important to be able to distinguish between the number of states in a code and the number of encoded inputs. It can be seen that Example 2, which has one uncoded input x_1, can be extended to have any number of states simply by increasing the number of delay stages in x_0. Since there is an uncoded input, there will be parallel transitions. On the other hand, both inputs in Example 3 are coded, so there are no parallel transitions and we are operating on the lowest level of the partition.

These conditions may be better understood by extending the discussion to the transmission of 3 bits/symbol (8-PSK) using a 16-QAM constellation. As shown in fig. 9.2-6, the encoder has three data inputs (x_2, x_1, x_0) and four outputs (y_3, y_2, y_1, y_0) that define the point in the 16-QAM constellation to be transmitted. Only one of the inputs is encoded in fig. 9.2-6a, two of the inputs are encoded in fig. 9.2-6b, and all three of the inputs are encoded in fig. 9.2-6c.

The partition of the 16-QAM constellation that results in maximum distance between points in the subsets is shown in fig. 9.2-7. In fig. 9.2-6a there will be four parallel transitions between states, corresponding to the two uncoded input lines. The outputs for these parallel transitions will come from partitions $C\phi$, $C1$, $C2$, $C3$. In fig. 9.2-6b there will be two parallel transitions between states, corresponding to one uncoded input line with the outputs for these transitions coming from $D\phi, \ldots, D7$. Finally, in fig. 9.2-6c, where all of the input lines are encoded and there are no parallel transitions, the outputs come from the lowest level of the partition, which consists of single points in the constellation.

The Problems ask that the encoding for Examples 1, 2, 3 above be extended to develop trellis codes for the 8-PSK to 16-QAM transformation.

Exactly the same methods that have been used for developing trellis codes for two-dimensional signal sets can, as well, be used for one-dimensional signal sets. Examples of some one-dimensional PAM (Pulse Amplitude Modulation) constellations are shown in fig. 9.2-8. These modulation techniques have previously been considered in Chapter 3, Baseband Data Transmission. The partitioning of these signal sets and the design of codes based on the logic of the examples of this section are left to the Problems.

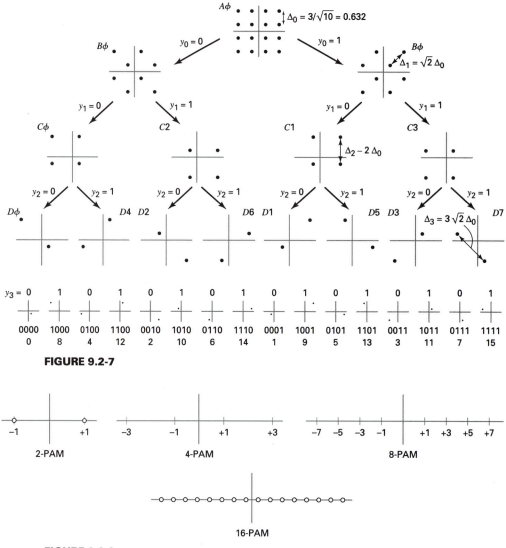

FIGURE 9.2-7

FIGURE 9.2-8

9.3 ROTATIONALLY INVARIANT TRELLIS CODES

As we have seen in Chapter 4, Passband Data Transmission, it is necessary to establish rotational invariance for two-dimensional signal constellations because no carrier-phase recovery system is able to distinguish among quadrants. In order to address this issue, the input data to the modulator in BPSK, QPSK, m-PSK, and 16-QAM were differentially encoded so that phase changes, rather than absolute phases, were transmitted. Although this encoding process somewhat increases the probability of error (single errors propagate to become double errors), it results in an enormous simplification of the receiver.

The difficulty in relation to trellis codes is that the Ungerboeck codes are not rotationally invariant. In order to acheive rotational invariance it is necessary to find nonlinear codes (Wei[Wei84], Ungerboeck [Ung87]).

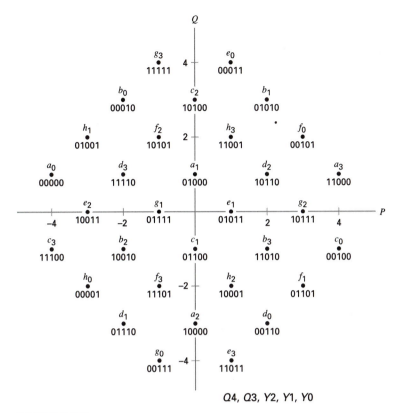

FIGURE 9.3-1

Before discussing the details of some of these codes, we shall first introduce the problem by considering the V.32 modem. This modem, which may be basically characterized by transmission speeds of 9600 bps and 14,400 bps on a channel having a Nyquist bandwidth of 2400 hz, requires, in the absence of any trellis coding, signal sets having 16 points and 64 points repectively. With trellis coding, the requirement on size of the signal set expands to 32 points and 128 points. Further, proper modem design requires that all of these signal sets be compatible and be able to be partitioned in a consistent way for trellis encoding.

We will begin by examining the 32-point constellation, (fig. 9.3-1) that is used for 9600 bps transmission with a rate $R = 4/5$ trellis code. The constellation is shown in relation to solid axes as 32-C , with signal points at ± 1, ± 3, ± 5, and in relation to the dotted axes as 32-QAM, a 45° rotation of 32-QAM. It is left to the Problems to demonstrate that 32-C is more energy-efficient than 32-QAM. The partitioning of this constellation proceeds according to the letters assigned to the respective points:

$$[A, B, C, D, E, F, G, H]$$
$$[A, B, C, D]\,[E, F, G, H]$$
$$[A, B]\,[C, D]\,[E, F]\,[G, H]$$
$$[A]\,[B]\,[C]\,[D]\,[E]\,[F]\,[G]\,[H]$$

(9.3-1)

and there are four points in each of these subsets. Each of the points in the constellation is assigned a five-bit binary number. The logic behind this assignment is key to the operation of the code.

First, all four of the points in each one of the subsets have the same last three bits:

$$A = 000; \quad B = 010; \quad C = 100; \quad D = 110;$$
$$E = 011; \quad F = 101; \quad G = 111; \quad H = 001. \tag{9.3-2}$$

Second, the subsets map into each other under $90°$ counterclockwise rotation according to the pattern:

$$E \rightarrow A \rightarrow G \rightarrow C \rightarrow E$$
$$B \rightarrow H \rightarrow D \rightarrow F \rightarrow B \tag{9.3-3}$$

In this mapping, note that as a point in any of these sets maps, through a $90°$ counterclockwise rotation, into the corresponding point of the appropriate subsequent subset, the first two bits remain the same.

It should be noted that the author has taken the liberty of reversing the order of notation as specified in V.32 in order to make it consistent with the normal ordering of binary numbers.

A block diagram of the encoder is shown in fig. 9.3-2. The two lower-order input data bits (Q_2, Q_1) are differentially encoded in order to produce rotational invariance. This process will be described below. The output of the differential encoder (Y_2, Y_1) is further encoded to generate (Y_2, Y_1, Y_0), which correspond to the bit patterns associated with the subsets $A - H$. The higher-order digits are generated directly from the input data: $Y_4 = Q_4$ and $Y_3 = Q_3$. This accounts for the presence of four different points in each subset.

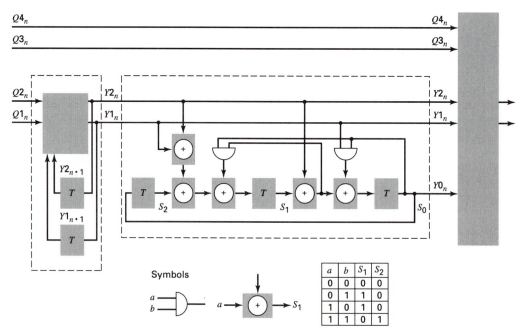

FIGURE 9.3-2

The differential coder is different from the one that was developed for the QPSK and 16-QAM signal sets associated with the V.22 bis modem in Chapter 4. That differential encoder associated an input dibit with a phase change: 00 with 90°, 01 with 0°, 10 with 180°, and 11 with 270°. This differential encoder associates the inputs x_1, x_0 with absolute phases: 00 with 0°, 01 with 90°, 10 with 180°, and 11 with 270° as well as with phase differences. The decoder mirrors the encoder:

Encoder			Decoder		
$Q_2(n)Q_1(n)$	$Y_2(n-1)Y_1(n-1)$	$Y_2(n)Y_1(n)$	$Y_2(n)Y_1(n)$	$Y_2(n-1)Y_1(n-1)$	$Q_2(n)Q_1(n)$
00	00	00	00	00	00
	01	01		01	01
	10	10		10	10
	11	11		11	11
01	00	01	01	00	01
	01	10		01	00
	10	11		10	11
	11	00		11	10
10	00	10	10	00	10
	01	11		01	01
	10	00		10	00
	11	01		11	11
11	00	11	11	00	11
	01	00		01	10
	10	01		10	01
	11	10		11	00

$$(9.3\text{-}4)$$

It is straightforward to find the state transition table for the encoder. The logic equations, determined from fig. 9.3-2, are:

$$S_2(n+1) = S_0(n)$$
$$S_1(n+1) = [S_2(n) \oplus (Y_2 \oplus Y_1)] \oplus [S_0(n) \cdot (S_1(n) \oplus Y_2)] \qquad (9.3\text{-}5)$$
$$S_0(n+1) = [S_1(n) \oplus Y_2] \oplus [S_0(n) \cdot Y_1]$$

The states and inputs in this machine are defined as

State	S_2	S_1	S_0	Input	Y_2	Y_1	
q_0	0	0	0	a	0	0	
q_1	0	0	1	b	0	1	
q_2	0	1	0	c	1	0	
q_3	0	1	1	d	1	1	$(9.3\text{-}6)$
q_4	1	0	0				
q_5	1	0	1				
q_6	1	1	0				
q_7	1	1	1				

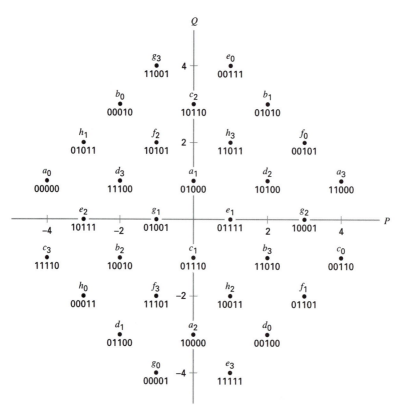

FIGURE 9.3-3

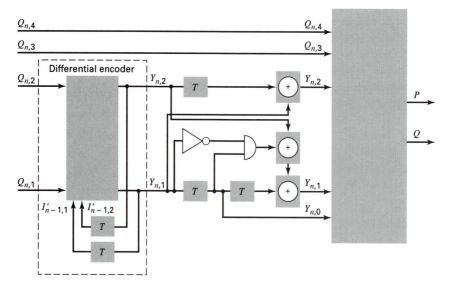

FIGURE 9.3-4

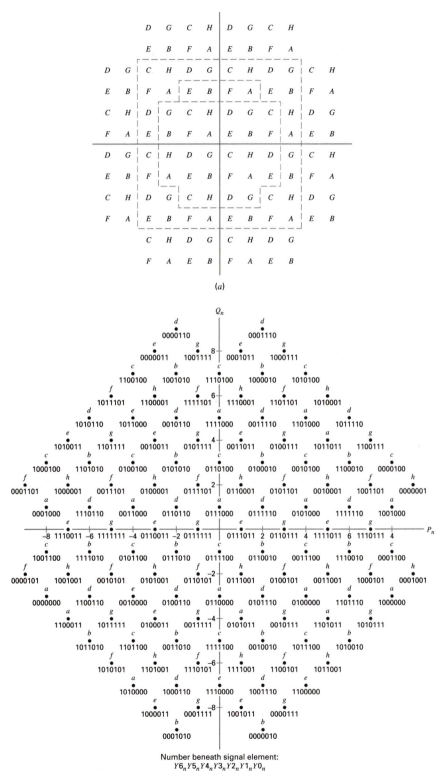

Number beneath signal element:
$Y6_n Y5_n Y4_n Y3_n Y2_n Y1_n Y0_n$

(b)

FIGURE 9.3-5

and the state transition table is

State/Input	a	b	c	d
q_0	q_0/A	q_2/B	q_3/C	q_1/D
q_1	q_4/H	q_7/E	q_5/F	q_6/G
q_2	q_1/A	q_3/B	q_2/C	q_0/D
q_3	q_7/H	q_4/E	q_6/F	q_5/G
q_4	q_2/A	q_0/B	q_1/C	q_3/D
q_5	q_6/H	q_5/E	q_7/F	q_4/G
q_6	q_3/A	q_1/B	q_0/C	q_2/D
q_7	q_5/H	q_6/E	q_4/F	q_7/G

(9.3-7)

The trajectories required to establish the coding gain may be obtained from the state transition table. First we note that there is a sequence of arbitrary length having q_0 returning to itself with output A. Then we have the sequence $q_0 \rightarrow q_2 \rightarrow q_0$ having output BD. Finally there is the sequence $q_0 \rightarrow q_1 \rightarrow q_4 \rightarrow q_0$ having output DHB. The values of d_{free}^2 for the different paths are computed as follows:

A and A	64
AA and BD	$16 + 8 = 24$
AAA and DHB	$8 + 36 + 16 = 60$

(9.3-8)

For the 16-QAM constellation $d_{\text{min}}^2 = 4$ and the ratio of average energy per transmitted symbol in the two constellations plotted to the same scale is 2. It therefore follows that coding gain is $\gamma = 4$ db.

Finally, we want to show that this code is rotationally invariant. A steady input of all zeros into the differential encoder will cause all zeros at its output. This will result in the code only going from $q_0 \rightarrow q_0$ with an output corresponding to point a_0 in partition A. If the signal undergoes a 90° phase shift during transmission, it will be received as point g_0 in partition G. After detection this will result in a repetitive pattern of 0011 from which only the 11 sequence will be applied to the differential decoder. The decoder output under these circumstances is 00, restoring the original data. The reader may make further experiments using this approach in order to confirm the rotational invariance.

An alternative code having the same coding gain is shown in figs. 9.3-3 and 9.3-4. Its disadvantage in relation to the code described above is that Y_2, Y_1, Y_0 are all encoded so that additional effort is required to extract the data from the received signal. Analysis of this code is left to the Problems.

Both of these codes can easily be extended to support data transmission at 12,000 bps and 14,400 bps as provided in modem specification V.33. The extension of the partitioned signal constellation to 64 points and 128 points for codes having rates $R = 5/6$ and $R = 6/7$ respectively is shown in fig. 9.3-5a and the bit assignment is shown in fig. 9.3-5b. The logic of the encoder is trivially extended by adding one or two additional uncoded data lines.

9.4 MULTIDIMENSIONAL SIGNALING AND OPTIMUM SIGNAL CONSTELLATIONS

Among the conclusions of the Channel Capacity Theorem is that communication over a channel having a fixed Nyquist bandwidth is made more efficient as the dimensionality of the transmitted signal is increased. As well, it is possible to optimize the transmitted signal within a given dimensionality.

In order to illustrate the argument about signal dimensionality, consider the comparison between the one-dimensional 4-ASK and the two-dimensional QPSK. Both of these signals carry 2 bits/symbol and we have seen that the expressions for the probabilty of error are:

$$4\text{-ASK} \quad P_e = \frac{3}{2} Q(\sqrt{\rho_m/5})$$

$$\text{QPSK} \quad P_e = 2Q(\sqrt{\rho_m/2})$$

The ratio of the respective values of ρ_m required to acheive the same probabilty of error is approximately 5/2 or 4 db. This means that the QPSK is able to acheive the same error rate as 4-ASK with 4 db less signal energy or, equivalently, with the same energy is able to tolerate 4 db more noise.

The relationship between 16-ASK and 16-QAM makes the same point:

$$16\text{-ASK} \quad P_e = \frac{15}{8} Q(\sqrt{\rho_m/85})$$

$$16\text{-QAM} \quad P_e = 3Q(\sqrt{\rho_m/10})$$

In this instance the ratio is roughly 8.5 or 9.3 db.

This is an indication that increasing the dimensionality of a signal increases its efficiency.

Within two dimensions it has been shown ([FGW73], [FGW74]) that it is possible to choose a 16-point signal set that is 0.5 db better than 16-QAM. This was accomplished through a computer search designed to minimize the average energy in a constellation having a fixed number of points. The resulting constellations are geometrically irregular and have consequently not been adopted in modem standards, but the proof of their existence provides a benchmark for evaluating more regular constellations.

The considerable performance gain that is achieved by going from one to two-dimensional signaling leads to the examination of signaling in higher dimensions in the hope of additional performance increments. There are a number of different ways of implementing multidimensional signals. We have seen in Chapter 5 that FSK can be made to be orthogonal in any number of dimensions. Four dimensions can be achieved in microwave transmission through two bandpass signals having different polarizations. In our applications multiple dimensions will be achieved through successive transmissions of two-dimensional signals, yielding $M = 4, 8, 16$, etc. dimensions. In the next section we shall examine trellis codes that are defined on these expanded signal spaces. In this section we will concentrate on the definition and construction of the spaces themselves.

Whatever the signaling technique that is used to achieve the higher dimesionality, the signal itself may be represented in Euclidean signal space by an M-tuple having appropriate dimension ([W&L74], [G&L84]).

It is simplest to begin the discussion with $M = 4$, so that a point in the signal space looks like (a_1, b_1, a_2, b_2). We could choose each pair (a, b) as coming from the 16-QAM signal set so that we would have a total of 256 possible signals, but this would yield no improvement in minimum distance over simple two-dimensional signaling. The idea is to choose the 256-point 4D constellation in a different way in order to increase $d_{\min}$. The mathematical structure that has been used to accomplish this is a lattice.

An N-dimensional lattice Λ is a set of points (or vectors) in an M-dimensional Euclidean space having the form

$$\mathbf{x} = \sum_{i=1}^{N} d_i \mathbf{x_i} \tag{9.4-1}$$

where the basis vectors have dimension M and are linearly independent so that $N \leq M$. The coefficients d_i are all positive and negative integers. Another way of looking at this process is to form a matrix L having rows that consist of the components of the basis vectors

$$
L = \begin{bmatrix} x_{11} & x_{12} & \dots & x_{1M} \\ x_{21} & x_{22} & \dots & x_{2M} \\ \cdot & & & \\ x_{N1} & x_{N2} & \dots & x_{NM} \end{bmatrix}
\tag{9.4-2}
$$

where these components are not restricted to be integers, and do the multiplication $\mathbf{d}L$ where $\mathbf{d}$ is a vector of integers. Since the components of the basis vectors may be real numbers they can be scaled by a factor α. The lattice may also be translated by an M-dimensional constant $\mathbf{a}$. The lattice contains an infinite number of points in M dimensions.

We are not interested in the entire lattice, only in a subset S of the lattice consisting of a desirable 256-point constellation. (It should be noted that the literature on lattices and their relation to the construction of signal constellations is vast and quite beyond the scope of this book. The interested reader may consult the references given at the end of Chapter 6 of Biglieri et al. [BDMS91].)

We shall briefly examine the 4D constellation developed by Gersho and Lawrence [G&L84]. They start with seven basic points as shown in Table I of fig. 9.4-1a. All permutations of each of the basic points are added to the signal set. Finally, all of the points that are obtained by placing minus signs at some or all of the coordinates are added to the set. This results in 256 points. The assignment of 8-bit data words to these points is shown in Table II of fig. 9.4-1b, where each of the three fields is of variable length: the prefix indicates which of the seven groups shown in Table I has generated the point; the middle field indicates the permutation of the basic point; and the final field shows the distribution of minus signs. Table III of fig. 9.4-1c develops the field assignments of Table II and Table IV of fig. 9.4-1d gives some examples.

It is demonstrated that this constellation is 1.2 db better than 2D 16-QAM. Gersho and Lawrence also develop an 8D constellation that gives an additional 1.2 db of performance. The Shannon limit indicates that an infinite-dimensional signal space could provide 8 db of improvement over 2D 16-QAM.

Implementing this code, which achieves its four dimensions by successively transmitting two 2D QAM signals, requires extra processing at both the transmitter and the receiver. At the transmitter it is necessary to gather together two successive quadbits in order to assemble the eight bits required to determine one of 256 points. The actual transmission requires two successive symbol intervals and the receiver must therefore process these two transmissions before determining the resulting eight bits. The receiver is required to be properly synchronized in order to guarantee that it is processing the correct set of successive transmissions. It is demonstrated that the transmitted spectrum is the same for higher-dimensional constellations as it is for simple 2D QAM signaling.

There has been continuing research in finding and evaluating multidimensional signal constellations. The interested reader may consult de Buda [Bud89], Forney and Wei [F&W89], Forney [For89], and Calderbank and Sloane [C&S87].

These multidimensional contellations based on extensions of QAM signaling present special issues in relation to adaptive equalization and carrier phase recovery.

TABLE I

SEVEN BASIC POINTS

BASIC	POINTS	ENERGY	NO. OF NEIGHBORS
1	1 1 1 1	4	23
2	2 0 0 0	4	22
3	2 2 0 0	8	22
4	2 2 2 0	12	19
5	2 2 2 2	16	15
6	3 1 1 1	12	15
7	3 3 1 1	20	8

(a)

TABLE II

8 BIT WORD

PREFIX	CENTER	SUFFIX
FIELD 1	2	3
(POINT TYPE)	(PERMUTATION)	(SIGN COMBINATION)

(b)

TABLE III

ASSIGNMENT OF BITS

GROUP	PREFIX	CENTER BITS	SIGN BITS	POINT (OMITTING SIGN)
1	0000		4	1111
2	0001		4	2222
3	001	00	3	2220
	001	01	3	2202
	001	10	3	2022
	001	11	3	0222
4	01	00	4	3111
	01	01	4	1311
	01	10	4	1131
	01	11	4	1113
5	10000	00	1	2000
	10000	01	1	0200
	10000	10	1	0020
	10000	11	1	0002
6a	10001	0	2	0022
	10001	1	2	0202
6b	1001	00	2	2200
	1001	01	2	2020
	1001	10	2	2002
	1001	11	2	0220
7a	101	0	4	1133
	101	1	4	1313
7b	11	00	4	3311
	11	01	4	3131
	11	10	4	3113
	11	11	4	1331

(c)

TABLE IV

EXAMPLES OF BIT ASSIGNMENT

00000110	1	−1	−1	1
00001110	−1	−1	−1	1
00011101	−2	−1	2	−2
00010101	2	−2	2	−2
00111111	0	−2	−2	−2
00101101	−2	2	0	−2
01110111	1	−1	1	3
01110000	1	1	1	3
11001110	−3	−3	−1	1
11000001	3	3	1	−1
10000000	2	0	0	0
10000111	0	0	0	−2
10010001	2	−2	0	0
10011111	0	−2	−2	0

(d)

FIGURE 9.4-1

9.5 MULTIDIMENSIONAL TRELLIS CODES

We have now advanced two different approaches for increasing performance within the same bandwidth: trellis coding and multidimensional signaling. The idea now is to combine them. This is, at least conceptually, not an overwhelming idea although the actual implementation gets into a level of mathematical complexity that is beyond the scope of this book.

Recall that in the initial discussion of trellis codes we dealt with the transmission of QPSK using an 8-PSK signal constellation and described a number of different codes having different numbers of states with which we could implement that transmission. In the Problems it is demonstrated that the very same codes could be used in the one-dimensional transmission of 4-ASK using an 8-ASK constellation. It is a crucial point that the code is not directly concerned with the dimensionality of the signal constellation to which it is applied. As long as the total number of signal points remains the same, the code is indifferent to the dimensionality of those points. For example, we have seen that we may transmit 256 points in either two or four dimensions and that better performance will come from the 4D rather than the 2D constellation. Nevertheless, the same trellis code may be used in the expansion of either constellation to 512 points.

The specification of a multidimensional trellis code has three parts:

1. A multidimensional lattice containing the required signal constellation must be defined and partitioned. The partition should have the property that the MSED (mean-squared Euclidean distance) doubles with each partition.
2. A multidimensional signal constellation within that lattice must be chosen.
3. A convolutional code and associated logic for rotational invariance must be determined.

Forney et al. [For84] and Wei [Wei87] have developed some particularly useful and relatively accessible lattices for multidimensional signaling that extend gracefully into trellis codes. Additional work on multidimensional trellis codes has been done by Calderbank and Sloane [C&S85], [C&S86], Pietrobon and Costello [P&C93], and Wang and Costello [W&C96].

Two codes developed by Wei [Wei87] are: a 16-state 4D code having a coding gain of $\gamma = 4.66$ db; a 64-state, 8D code with $\gamma = 5.41$ db. In both of these codes the differential coding and block encoder are designed to guarantee rotational invariance. Readers who wish additional detail may consult Wei's paper, and Chapter 6 of Biglieri et al. [BDMS 91]

These codes are very complex and their deeper analysis is well beyond the scope of this text. The most complex of these codes that is currently in use was developed for the V.34 modem (Forney et al. [For97]), which supports data speeds up to 33.6 kbps, and the V.90, which can support 56 kbps.

9.6 TRELLIS CODES FOR PSK SYSTEMS

We have seen that QAM signals are, in general, more efficient than PSK signals. For example, for a given average energy per transmitted symbol, 16-QAM is approximately 11 db better than 16-PSK. However, for channels having severe nonlinearities there are strong advantages to using a constant-amplitude signal. There is also strong motiviation to recover some of the sacrificed performance though the use of trellis codes.

Trellis codes used in this context clearly must be rotationally invariant, but for higher-order m-PSK signals that requirement becomes more complex than the fourfold 90° invariance that we have examined for QAM. For example, trellis codes for 8-PSK must be invariant under phase ambiguities of multiples of 45° and 16-PSK requires codes that are invariant under multiples of 22.5° ambiguities. Interested readers are referred to Oerder [Oer85], Wilson [Wil86], Wei [Wei89], Divsalar and Simon [D&S88], and Pietrobon et al. [Pie90].

9.7 A FINAL NOTE ON REFERENCES

Only a few years after the appearance of Ungerboeck's seminal paper in 1982, trellis codes were first applied to the V.32 modem. Since that time both the research into trellis codes and their application to real systems has developed rapidly. Although the necessarily cursory treatment of the subject in this text precludes a full listing of the important references, there are a few that should be added, some for their theoretical importance and others for their practical value. These include Calderbank et al. [Cal85], Freeman et al. [Fre88], Benedetto et al. [Ben94], Viterbi et al. [Vit89], and Wolf and Zehavi [W&Z95].

CHAPTER 9 PROBLEMS

9.1 Show that the encoder of Example 3 has the form of fig. 9.2-5b and that the logic equations are
$$y_2 = x_0 \oplus s_2; \qquad y_2 = x_0 \oplus s_1; \qquad y_0 = s_0.$$

9.2 A trellis code that involves a QPSK to 8-PSK transformation has the structure shown in fig. P9-2. Find the state table of this code, d_{free} and the coding gain γ.

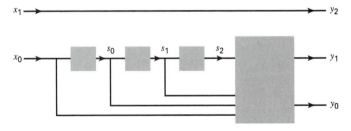

FIGURE P9-2

9.3 A 16-state code that involves a QPSK to 8-PSK transformation is shown in [Ung 82, fig. 7]. Confirm that this code has $d_{\text{free}} = 2.274$ and provides $\gamma = 4.1$ db of coding gain.

9.4 **a.** Use the encoder of Example 1 for an 8-PSK to 16-QAM transformation. Find the coding gain in relation to 8-PSK.

   b. Repeat a. for Example 2.

   c. Repeat a. for Example 3.

9.5 Develop appropriate partitions for the PAM signal sets shown in fig. 9.2-8.

9.6 Use the encoder of Example 2 to construct a code that uses the PAM signal set of fig. 9.2-8 to transmit 3 bits/symbol.

9.7 Show that the 32-CR constellation is more energy-efficient than 32 QAM; i.e., for the average energy per transmitted symbol compare the performance of the two constellations. Find the peak-to-average energy ratio.

9.8   Anayze the code of figs. 9.3-3 and 9.3-4 in the same manner as the code of figs. 9.3-1 and 9.3-2 was analyzed in the text.

REFERENCES

[Ben94]   S. Benedetto et al., "Performance evaluation of trellis-coded modulation schemes," *Proc. IEEE*, **82**, No. 6, June 1994, pp. 833–855.

[BDMS91]   E. Biglieri, D. Divisilar, P. McLane, and M. Simon, *Introduction to Trellis Coded Modulation*, Macmillan, 1991.

[Bud89]   R. de Buda, "Some Optimal Codes Have Structure," *IEEE J. Selected Areas in Communications*, **SAC-7**, No. 6, Aug. 1989.

[Cal85]   A.R. Calderbank et al., "Asymptotic upper bounds on the minimum distance of trellis codes," *IEEE Trans. Communications*, **Com-33**, No. 4, April 1985, pp. 305–309.

[C&M84]   R. Calderbank and J.E. Mazo, "A new description of trellis codes," *IEEE Trans. Information Theory*, **IT-30**, No. 6, Nov. 1984.

[C&S85]   A.R. Calderbank and N.J.A. Sloane, "Four-dimensional modulation with an eight-state trellis code," *AT&T Technical J.*, **64**, No. 5, May-June 1985.

[C&S86]   A.R. Calderbank and N.J.A. Sloane, "An eight-dimensional trellis code," *IEEE Proceedings*, **74**, No. 5, May 1986, p. 757.

[C&S87]   A.R. Calderbank and N.J.A. Sloane, "New trellis codes based on lattices and cosets," *IEEE Trans. Information Theory*, **IT-33**, 1987, p. 177.

[D&S88]   D. Divisilar and M.K. Simon, "Multiple trellis-coded modulation (MTCM)," *IEEE Trans. Communication*, **Com-36**, No. 4, April 1988, pp. 410-419.

[F&W89]   G.D. Forney and L. F. Wei, "Multidimensional constellations—Part I: introduction, figures of merit, and generalized cross constellations," *IEEE J. Selected Areas in Communications*, **SAC-7**, No. 6, Aug. 1989.

[FGW73]   G.J. Foschini, R.D. Gitlin, and S.B. Weinstein, "On the selection of a two-dimensional signal constellation in the presence of phase jitter and Gaussian noise," *Bell System Technical J.*, **52**, July-Aug. 1973, pp. 927–965.

[FGW74]   G.J. Foschini, R.D. Gitlin, and S.B. Weinstein, "Optimization of two-dimensional signal constellations in the presence of Gaussian noise," *IEEE Trans. Communications*, **Com-22**, Jan. 1974, pp. 28–38.

[For84]   G.D. Forney et al., "Efficient modulation for band-limited channels," *IEEE J. Selected Areas in Communications*, **SAC-2**, No. 5, Sept. 1984, pp. 632–647.

[For89]   G.D. Forney, "Multidimensional constellations—Part II: Voronoi constellations," *IEEE J. Selected Areas in Communications*, **SAC-7**, No. 6, Aug. 1989.

[For97]   G.D. Forney, "The V.34 high-speed modem standard," *IEEE Communications Magazine*, Dec. 1997, pp. 28–33.

[Fre88]   G.H. Freeman et al., "Trellis source codes designed by conjugate gradient optimization," *IEEE Trans. Communications*, **Com-36**, No. 1, Jan. 1988, pp. 1–12.

[Gol67]   S. Golomb, *Shift Register Sequences*, Holden-Day, 1967.

[G&L84]   A. Gersho and V.B. Lawrence, "Multidimensional signal constellations for voiceband data transmission," *IEEE J. Selected Areas in Communications*, **SAC-2**, No. 5, Sept. 1984.

[Oer85]   M. Oerder, "Rotationally invariant trellis codes for *m*PSK modulation," *ICC*, June 1985, pp. 18.1.1-5.

[P&C93]   S.S. Pietrobon and D.J. Costello, "Trellis coding with multidimensional QAM signal sets," *IEEE Trans. Information Theory*, **IT-39**, No. 2, March 1993, pp. 325–336.

[Pie90]   S.S. Pietrobon et al., "Trellis-coded multidimensional phase modulation," *IEEE Trans. Inf. Theory.*, **IT-36**, No. 1, Jan. 1990, pp. 63–89.

[Sch97]   C. Schlegel, *Trellis Coding*, IEEE Press, 1997.

[Ung82]   G. Ungerboeck, "Channel coding with multilevel/multiphase signals," *IEEE Trans. Information Theory*, **IT-28**, No. 1, Jan. 1982.

[Ung87]   G. Ungerboeck, "Trellis-coded modulation with redundant signal sets, parts I and II," *IEEE Communications Magazine*, Feb. 1987, **25**, No. 2.

[Vit89]   A.J. Viterbi et al., "A pragmatic approach to trellis-coded modulation," *IEEE Communications Magazine*, July 1989, pp. 11–19.

[V&O79]   A. Viterbi and J. Omura, *Principles of Digital Communications and Coding*, McGraw-Hill, 1979.

[Wei84]   L.F. Wei, "Rotationally invariant channel coding with expanded signal space, parts I and II," *IEEE J. Selected Areas in Communications*, **SAC-2**, Sept. 1984, pp. 659–687.

[Wei87]   L.F. Wei, "Trellis-coded modulation with multidimensional constellations," *IEEE Trans. Information Theory.*, **IT-33**, No. 4, July 1987, pp. 483–501.

[Wei89]   L.F. Wei, "Rotationally invariant trellis-coded modulations with multidimensional M-PSK," *IEEE J. Selected Areas in Communications*, **SAC-7**, No. 9, Dec. 1989, pp. 1281–1295.

[Wil86]   S.G. Wilson, "Rate 5/6 trellis-coded 8-PSK," *IEEE Trans. Communications*, **Com-34**, Oct. 1986, pp. 1045–1049.

[W&C96]   F.Q. Wang and D.J. Costello, "New rotationally invariant four-dimensional trellis codes," *IEEE Trans. Information, Theory*, **IT-42**, No. 1, Jan. 1996, pp. 291–300.

[W&L74]   G.R. Welti and S.L. Lee, "Digital transmission with coherent four-dimensional modulation," *IEEE Trans. Information Theory*, **IT-20**, No. 4, July 1974, pp. 397–502.

[W&Z95]   J.K. Wolf and E. Zehavi, "P^2 Codes: pragmatic trellis codes utilizing punctured convolutional codes," *IEEE Communications Magazine*, Feb. 1995, pp. 94–99.

CHAPTER 10

EQUALIZATION OF DISTORTED CHANNELS

10.0 INTRODUCTION

In the data-transmission systems that we discussed in Chapters 3 and 4 the data, represented as impulse functions, has been applied to a combination of transmit filter, channel, and receiver filter. The output of the receiver filter was sampled at the symbol rate $f_s = 1/T_s$ hz and these samples processed through a carrier-recovery system and a slicer, leading to a decision.

In those chapters we assumed that the channel was perfect, that it had a flat frequency response and a linear phase shift or, equivalently, that the impulse response was itself an impulse function. We also assumed that both the transmitter and receiver filters have the characteristic of the square-root of a raised cosine. The overall end-to-end system was consequently Nyquist. In Chapter 3 we also considered partial-response systems as a variant or extension of Nyquist signaling requiring different kinds of transmitter and receiver filters.

The representation of distorted channels was discussed in Chapter 7 and we saw that such channels could be represented in the frequency domain by their transfer functions $H(\omega)$ and in the time domain by their impulse reponses $h(t)$. If the channel is known in advance, it can be equalized very easily by an analog filter or equalizer, placed after the square-root of raised-cosine receiver filter, having a transfer function $G(\omega) = 1/H(\omega)$. The penalty is that the amplification of the noise in the equalizer filter will degrade the performance of the system.

Before high-speed digital-signal processing was available this was, in fact, the method used for channel equalization for modems on the telephone line and it continues to be used for some magnetic recording channels. Since the analog equalizer is fixed and the telephone channel is unpredictable from call to call, the best that one could do on that system was to equalize some typical or compromise channel. Consequently, equalization was far from perfect. In fact, before DSP for voiceband signals was readily available, high-speed data transmission at 4800 bps and greater was typically confined to leased, dedicated telephone lines where the analog equalizer and the line could be permanently adjusted to each other.

High-speed digital processing of the sampled signal has allowed the equalization process on the telephone line and in wireless systems to be transferred from the analog to the digital domain with the equalization process taking place after the sampling of the signal. This transfer allows for the easy development of adaptive equalizers in which the equalizer 'learns' the particular channel and adjusts its parameters accordingly. Since each channel is then automatically equalized, the performance of the modem increases and with it the ability to transmit data at higher speeds on the ordinary phone line. In the course of a decade this DSP capability has allowed QAM type modems on the ordinary phone line to increase their speed from 2.400 kbps to 28.8 kbps and, more recently, to 56 kbps. In a somewhat different manner, it has also allowed local loop transmission via ISDN to operate at 144 kbs and ADSL at 6 Mbps.

Although the equalization algorithms for digital transmission are the same on the telephone line and on microwave links, the technology involved in implementing these algorithms is completely different. General-purpose voiceband DSP chips, special modem chips, and high speed ASICS (application specific integrated circuits) allow sophisticated algorithms to be relatively easily implemented in even very high-speed modems.

This chapter will analyze a variety of digital equalization strategies and their relative performances. Adaptation techniques for these structures will be discussed in Chapter 11.

10.1 TRANSVERSAL SYMBOL-RATE EQUALIZERS FOR BASEBAND AND BANDPASS CHANNELS

The system of fig. 10.1-1 consists of a square-root of raised-cosine transmitter filter, an unknown distorted channel, and a fixed receiver filter which is also square root of raised cosine. The output of the receiver filter is sampled at the symbol or baud rate. The input to this system is the data sequence

$$x(t) = \sum_n d_n \delta(t - nT_s) \qquad (10.1\text{-}1a)$$

and the receiver filter output is

$$s(t) = \sum_n d_n h(t - nT_s) + n(t) \qquad (10.1\text{-}1b)$$

This is the output that is then sampled and equalized digitally. The equalizer structure that we will examine first is the symbol rate (or baud rate) transversal filter. We will see that this structure is a suboptimal linear receiver, its performance is not as good as the receiver having an analog equalizer.

10.1.1 Baseband Equalization

We will first consider a baseband system. When the sampling takes place at the baud rate (or symbol rate) $f_s = 1/T_s$, the output is written as

$$s(mT_s) = \sum_n d_n h(mT_s - nT_s) + n(mT_s) \qquad (10.1\text{-}2)$$

This suggests that, from the point of view of the sampled output, the overall system may be represented (fig. 10.1-2) by a transversal filter having taps that are the sampled values of the end-to-end impulse response of the system $h(t)$. Our analysis is based on the system shown in fig. 7.1-3, which shows the sampled equivalent of a baseband channel, previously

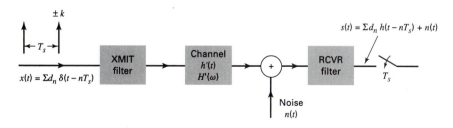

FIGURE 10.1-1

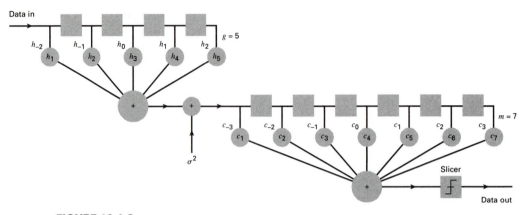

FIGURE 10.1-2

characterized as a transversal FIR filter. Samples of Gaussian noise having variance σ^2 are then added to the data samples.

The sampled impulse response may be represented as the vector

$$H = [h_{-2}, h_{-1}, h_0, h_1, h_2] \qquad (10.1\text{-}3a)$$

For an undistorted Nyquist channel,

$$H = [0, 0, 1, 0, 0]. \qquad (10.1\text{-}3b)$$

and a distorted channel would look like

$$H = [-.05 \quad .15 \quad 1 \quad .1 \quad -.07]. \qquad (10.1\text{-}3c)$$

Note that in the vector H the terms that appear before the main sample, which is normalized to unity, are called 'leading echoes' or precursors, and those that appear afterward are 'trailing echoes' or postcursors. Both sets of echoes constitute the ISI (intersymbol interference) terms. The sampled impulse response vector H may also be represented as a z-transform

$$H(z) = h_{-2}z^2 + h_{-1}z + h_0 + h_1z^{-1} + h_2z^{-2} \qquad (10.1\text{-}4)$$

We can represent the equalization process in the z-domain as well. If the effect of the equalizer is expressed as a polynomial $C(z)$ and the output vector as $E(z)$, then the operation of the equalizer on the channel may be expressed as:

$$H(z)C(z) = E(z) \qquad (10.1\text{-}5a)$$

and for perfect equalization

$$H(z)C(z) = z^{-K} \qquad (10.1\text{-}5b)$$

where $E(z) = z^{-K}$ indicates that the output is an impulse function and K is an integer representing the delay in the equalizer. Perfect equalization then requires that

$$C(z) = z^{-K}/H(z) \qquad (10.1\text{-}5c)$$

indicating the need for an IIR filter structure for the equalizer or, equivalently, an FIR or transversal filter having an infinite number of taps.

Since the ISI terms on even the most distorted channels become too small to consider if they are sufficiently far away from the main sample and because IIR structures are inherently more difficult to work with than FIR structures, the technology for equalizer structures has centered around the transversal filter. Such a transversal equalizer is shown in fig. 10.1-2.

Consider the input to the system of fig. 10.1-2 as being a single, isolated data impulse (as opposed to a stream of data impulses).

We have assumed that the channel has g taps and now we shall assume that the equalizer has m taps, where it is convenient to require that both m and g are odd numbers. The taps of both the channel and the equalizer are shown with two different notations, each of which is appropriate for different purposes. The first shows the taps numbered $1, \ldots, g$ and $1, \ldots, m$. The second shows the center tap with the subscript zero, and the other taps labeled with negative and positive numbers in relation to zero.

Using the first notation, a single isolated data pulse at the input to the system will result in an output sequence

$$E = \lfloor e_1, e_2, \ldots, e_{m+g-1} \rfloor \qquad (10.1\text{-}6a)$$

representing the convolution of the two vectors

$$H = \lfloor h_1, h_2, \ldots h_g \rfloor \quad \text{and} \quad C = [c_1, c_2, \ldots c_m]. \qquad (10.1\text{-}6b)$$

If the channel were perfectly equalized, the output would be

$$E_h = [0\,0 \quad \cdots \quad 0\,1\,0 \quad \cdots \quad 0\,0] \qquad (10.1\text{-}6c)$$

where the length of the vector is $(m + g - 1)$ and the 1 appears in the $(m + g)/2$ or center position.

In matrix terms, the output vector can be computed through the use of the convolution matrix of the channel H_c. An example with $g = 3$, $m = 5$, and $(m + g - 1) = 7$ will demonstrate the idea.

$$[e_{-3} \quad e_{-2} \quad e_{-1} \quad e_0 \quad e_1 \quad e_2 \quad e_3]$$

$$= [c_{-2} \quad c_{-1} \quad c_0 \quad c_1 \quad c_2] \begin{bmatrix} h_{-1} & h_0 & h_1 & 0 & 0 & 0 & 0 \\ 0 & h_{-1} & h_0 & h_1 & 0 & 0 & 0 \\ 0 & 0 & h_{-1} & h_0 & h_1 & 0 & 0 \\ 0 & 0 & 0 & h_{-1} & h_0 & h_1 & 0 \\ 0 & 0 & 0 & 0 & h_{-1} & h_0 & h_1 \end{bmatrix} \qquad (10.1\text{-}7a)$$

or

$$E = CH_c \qquad (10.1\text{-}7b)$$

Note that the convolution matrix H_c has $m = 5$ rows and $(m + g - 1) = 7$ columns. We will subsequently see that in order to acheive effective equalization the length (number of taps) of the equalizer should exceed the length of the channel.

The difference between the actual output E and the ideal output E_h is the error or ISI vector and the magnitude squared of this difference is called the mean-square error (MSE) arising from intersymbol interference

$$\begin{aligned} \text{MSE}_{\text{ISI}} &= \|E - E_h\|^2 = \|\text{ISI}\|^2 = \|CH_c - E_h\|^2 \\ &= (c_{-2}h_{-1})^2 + (c_{-2}h_0 + c_{-1}h_{-1})^2 + (c_{-2}h_1 + c_{-1}h_0 + c_0h_{-1})^2 \\ &\quad + [1 - (c_{-1}h_1 + c_0h_0 + c_1h_{-1})]^2 \\ &\quad + (c_0h_1 + c_1h_0 + c_2h_{-1})^2 + (c_1h_1 + c_2h_0)^2 + (c_2h_1)^2 \end{aligned} \qquad (10.1\text{-}8a)$$

If the ISI terms are small, we may characterize them as independent, more or less identically distributed random variables. Consequently their sum will tend toward being

Gaussian and the sum of the squares, computed in eq. 10.1-8a, becomes the variance of a Gaussian distribution. This variance is the power contained in the ISI terms. This characterization, although not at all mathematically precise, is, nevertheless, essentially valid. It is important because it will allow us to add the MSE or power due to ISI to the power of the output noise to form a single mean-square error that characterizes the system.

The system noise enters the receiver filter, which is the square root of a raised-cosine filter. The power spectral density of the noise is therefore shaped by a full raised-cosine. As a consequence, symbol rate samples of the noise are uncorrelated and the total noise power out of the equalizer is

$$\text{MSE}_{\text{noise}} = \sigma^2 \left(c_1^2 + c_2^2 + \cdots + c_m^2 \right) = \sigma^2 |C|^2 \qquad (10.1\text{-}8\text{b})$$

If the data symbols have values $\pm d$, the total mean square error is the sum of the two

$$\text{MSE} = d^2 |C H_c - E_h|^2 + \sigma^2 |C|^2 \qquad (10.1\text{-}8\text{c})$$

It is this total MSE that determines the SNR at the slicer.

It should be acknowledged that this formulation of the problem based on a single isolated data pulse may seem to be artificial because in an actual system there is a steady stream of data pulses. We shall now demonstrate why it is appropriate to use the expression for mean-square-error due to intersymbol interference given in eq. 10.1-8a.

In the example given above where $g = 3$, $m = 5$, and $(m + g - 1) = 7$, we will imagine a sequence of seven data symbols

$$d_{n-3} \quad d_{n-2} \quad d_{n-1} \quad d_n \quad d_{n+1} \quad d_{n+2} \quad d_{n+3} \qquad (10.1\text{-}9\text{a})$$

applied to the system in succession. When the last one d_{n+3} is applied, the output is

$$\begin{aligned} y &= d_{n+3}[h_{-1}c_{-2}] + d_{n+2}[h_0 c_{-2} + h_{-1}c_{-1}] + d_{n+1}[h_1 c_{-2} + h_0 c_{-1} + h_{-1}c_0] \\ &\quad + d_n[h_{-1}c_1 + h_0 c_0 + h_1 c_{-1}] + d_{n-1}[h_{-1}c_2 + h_0 c_1 + h_1 c_0] \\ &\quad + d_{n-2}[h_0 c_2 + h_1 c_1] + a_{n-3}[h_1 c_2] \end{aligned} \qquad (10.1\text{-}9\text{b})$$

Let us assume that $d_n = 1$ and that there is no decision error so that the decided value is also $\hat{d}_n = 1$. The difference between the decided value and the output is the error

$$\begin{aligned} \text{error} = (1 - y) &= d_{n+3}[h_{-1}c_{-2}] + d_{n+2}[h_0 c_{-2} + h_{-1}c_{-1}] \\ &\quad + d_{n+1}[h_1 c_{-2} + h_0 c_{-1} + h_{-1}c_0] + [1 - h_{-1}c_1 + h_0 c_0 + h_1 c_{-1}] \\ &\quad + d_{n-1}[h_{-1}c_2 + h_0 c_1 + h_1 c_0] + d_{n-2}[h_0 c_2 + h_1 c_1] + d_{n-3}[h_1 c_2] \end{aligned}$$
$$(10.1\text{-}9\text{c})$$

The mean-square error is the expected value of the square of the error

$$\text{MSE}_{\text{ISI}} = E[(\text{error})^2] \qquad (10.1\text{-}9\text{d})$$

where the expectation is taken over a long sequence of outputs. Since the successive data symbols are uncorrelated, the expected value of the cross-product terms that are obtained by squaring the expression for the error in eq. 10.1-9c go to zero. What remains is the sum of the squares of the individual terms and, since the data are assumed to be ± 1, we end up with the same expression as eq. 10.1-8a.

The equivalence of these two ways of finding the MSE is of great importance in the next chapter where we discuss adaptation of the equalizer.

10.1.2 Minimizing MSE, the Mean-Square Error

Returning to eq. 10.1-8c, we want to choose the equalizer taps, the vector C, in order to minimize the value of MSE. In so doing, we will follow the development given by Clark [Cla85].

$$
\begin{aligned}
\text{MSE} &= d^2[CH_c - E_h][CH_c - E_h]^T + \sigma^2 CC^T \\
&= d^2 C H_c H_c^T C^T - 2d^2 E_h H_c^T C^T + d^2 E_h E_h^T + \sigma^2 CC^T \\
&= C[d^2 H_c H_c^T + \sigma^2 I]C^T - 2d^2 E_h H_c^T C^T + d^2
\end{aligned}
\tag{10.10-10a}
$$

Since $H_c H_c^T$ is a real, symmetric, and positive definite, it follows that there is an $(m \times m)$ matrix G such that

$$
GG^T = d^2 H_c H_c^T + \sigma^2 I
\tag{10.10-10b}
$$

so that

$$
\begin{aligned}
\text{MSE} &= C[GG^T]C^T - 2d^2 E_h H_c^T C^T + k^2 \\
&= [CG - d^2 E_h H_c^T G^{T^{-1}}][CG - d^2 E_h H_c^T G^{T^{-1}}]^T \\
&\quad + d^2[1 - d^2 E_h H_c^T G^{T^{-1}} G^{-1} H_c E_h^T]
\end{aligned}
\tag{10.10-10c}
$$

where we have completed the square.

For those who do not remember, completing the square is illustrated by

$$
f(x) = x^2 + 4x - 7 = (x^2 + 4x + 4) - (4 + 7) = (x + 2)^2 - 11
$$

so that the minimum occurs at $x = -2$ and the minimum value is $f(-2) = -11$.

The equalizer tap vector $C_{\min}$ that minimizes the MSE of eq. 10.1-10c is therefore given by:

$$
C_{\min} G = d^2 E_h H_c^T (G^T)^{-1}
\tag{10.1-11a}
$$

or:

$$
\begin{aligned}
C_{\min} &= d^2 E_h H_c^T G^{T^{-1}} G^{-1} = d^2 E_h H_c^T [GG^T]^{-1} \\
&= d^2 E_h H_c^T [d^2 H_c H_c^T + \sigma^2 I]^{-1} \\
&= E_h H_c^T \left[H_c H_c^T + \frac{\sigma^2}{d^2} I \right]^{-1}
\end{aligned}
\tag{10.1-11b}
$$

where $d^2/\sigma^2 = \text{SNR}$, the signal-to-noise ratio of the system. In a multilevel system $E_x = E[d_x^2]$ is the average energy per transmitted symbol and $\text{SNR} = E[d_x^2]/\sigma^2$.

The resulting minimum value of the MSE, which is a combination of ISI and noise, also comes from eq. 10.1-10c as

$$
\begin{aligned}
\text{MSE}_{\min} &= d^2 \lfloor 1 - d^2 E_h H_c^T G^{T^{-1}} G^{-1} H_c E_h^T \rfloor \\
&= d^2 [1 - d^2 E_h H_c^T [d^2 H_c H_c^T + \sigma^2 I]^{-1} H_c E_h^T]
\end{aligned}
\tag{10.1-11c}
$$

or:

$$
\frac{\text{MSE}_{\min}}{d^2} = \left[1 - E_h H_c^T \left[H_c H_c^T + \frac{\sigma^2}{d^2} I \right]^{-1} H_c E_h^T \right] = [1 - C_{\min} H_c E_h^T]
\tag{10.1-11d}
$$

Note that the result is given as the ratio of the energy in the MSE to signal energy. This ratio will be important in computing the probability of error of the equalized system.

Two more important quantities are the Residual ISI and the noise amplification of the equalizer. The Residual ISI is the ISI remaining in output vector after the taps of the equalizer are adjusted to their optimum value.

$$\text{ISI} = \| E_{\min} - E_h \|^2 = \| C_{\min} H_c - E_h \|^2 \tag{10.1-11e}$$

The issue here is that the optimum equalizer in fact operates to minimize the total MSE composed of both ISI and noise. For a given channel, the equalizer will operate differently with large SNR and a small SNR. In the first case the equalizer will tend to operate primarily on the ISI rather than the noise; in the latter case, primarily on the noise rather than the ISI. This will be further explored in the Problems.

We have already established that the equalizer output noise is

$$\sigma_0^2 = \sigma^2 \| C \|^2. \tag{10.1-12a}$$

If a data impulse of magnitude d is applied to the channel then the channel output energy (or equalizer input energy) is $d^2 |H|^2$. The signal-to-noise ratio at the equalizer input is therefore

$$(\text{SNR})_i = d^2 \| H \|^2 / \sigma^2 \tag{10.1-12b}$$

The output energy of the equalized signal is $E_0 = k^2 \| E \|^2$ so that the output signal to noise ratio is

$$(\text{SNR})_o = d^2 \| E \|^2 / \sigma^2 \| C \|^2 \tag{10.1-12c}$$

Therefore the noise amplication in the equalizer is

$$n_{\text{amp}} = (\text{SNR})_i / (\text{SNR})_o = \| C \|^2 \| H \|^2 / \| E \|^2 \tag{10.1-12d}$$

In addition to the MSE there are two other, less frequently used measures of the imperfection of the equalization process. They are:

Peak Distortion

$$D_p = \frac{1}{e_0} \sum_{i \neq 0} |e_i| \tag{10.1-13a}$$

Mean Squared Distortion

$$D_{ms} = \frac{1}{|e_0|^2} \sum_{i \neq 0} |e_i|^2 \tag{10.1-13b}$$

where e_0 is the main sample of the output vector E.

All of this is easily implemented in the MATLAB script EQUAL.M is available on the website cited in the Preface. Exercises that enable the reader to discover some of the salient properties of these equalizers are given in the Problems.

10.1.3   The Bandpass Case

As we have seen in Chapters 4 and 7, the sampled baseband equivalent of a bandpass channel is a vector of complex numbers. For an distortionless Nyquist system (including the transmitter and receiver filters), the sampled complex baseband impulse response vector is

$$H = [0 + j0, \ 0 + j0, \ 1 + j0, \ 0 + j0, \ 0 + j0] \tag{10.1-14a}$$

and for a distorted system this vector would look like

$$H = [-.05 + j.02, .15 + j.08, 1 + j.20, -.10 - j.04, .02 - j.01] \tag{10.1-14b}$$

The basic operational difference between the baseband and the bandpass analysis is that in the bandpass case both the channel and the equalizer tap vectors are composed of complex numbers. In addition, the ideal output vector of the channel plus equalizer is represented as

$$E_h = [0 + j0, \dots, 0 + j0, 1 + j0, 0 + j0, \dots, 0 + j0]. \tag{10.1-14c}$$

The output vector of the channel plus equalizer is $E = C^* H_c$, where $*$ denotes complex conjugate and

$$\text{MSE} = \|E - E_h\|^2 = [E - E_h][E - E_h]^{*T} \tag{10.1-15a}$$

where we have taken the conjugate transpose of $[E - E_h]$. Following this through the same lines as the baseband development, we get the optimum equalizer as

$$C_{\min}^* = E_h H_c^{*T} \left[H_c H_c^{*T} + \frac{\sigma^2}{|d|^2} I \right]^{-1} \tag{10.1-15b}$$

and the minimum mean-square error as

$$\frac{\text{MSE}_{\min}}{|d|^2} = \left[1 - C_{\min}^* H_c E_h^T \right] \tag{10.1-15c}$$

A MATLAB script BPEQUAL.M that implements eqs. 10.1-15b and 10.1-15c is available on the website. This script has been designed to operate with the channel models for the telephone line that were developed in Chapter 7 using the program CHAN768. Recall that the program transforms frequency-response characteristics of the telephone line models (attenuation characteristics A, B, C and delay characteristics 1, 2, 3, 4, 5) into a selection of 15 different (A1, B3, etc.) sampled, baseband equivalent impulse responses. The program is constructed with the transmit and receive filters corresponding to 2400 baud and a rolloff of $\alpha = .15$, the channel model for the V.32 modem.

These impulse response models consist of eight samples per symbol. The program allows choosing the sampling phase (or timing phase) of the impulse response, as shown in fig. 10.1-3. A sampling phase of $n = 0$ corresponds to sampling at the peak of the impulse

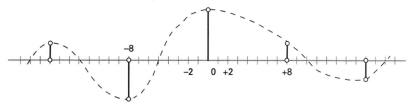

FIGURE 10.1-3

response and sampling phases of $n = \pm 1, \pm 2$ correspond to sampling off the peak by $\pm 45°, \pm 90°$. This allows us to examine the sensitivity of the system to the timing phase, a matter that will be further discussed in the next section. The program also allows for generating a channel model at two samples per symbol. This model will be used subsequently in relation to fractionally spaced equalizers.

10.1.4   The Effect of Channel Distortion on Performance—A Numerical Example

A small example will help in understanding how to evaluate the effect of channel distortion on the performance of the data communication system.

A binary system having signal amplitude equal to $d = \pm 1$ operates with a SNR = 15 db over an ideal channel. Since $\text{SNR}_{db} = 10 \log_{10} \rho$, this translates into $\rho = 31.6$ and since, for this system, $P_e = Q(\sqrt{\rho})$ it follows that $P_e = 10^{-8}$ and the noise power, relative to the signal, is noise $= 1/31.6 = 0.032$.

Now let's say that transmission is taking place on a distorted channel that is equalized according to eq. 10.1-15b and that the equalizer has the following characteristics:

$$\text{noise amplification} = 3 \text{ db}$$

$$\text{residual ISI} = 17 \text{ db below the output equalized signal}$$

It follows that the amplified noise is output noise $= .064$ and the residual ISI is residual ISI $= .02$. These two are added together to give the total output MSE $= .084$ which, in turn, yields a new value of $\rho = 1/.084 = 11.10$. In turn, the new probability of error is $P_e = 2.8 \times 10^{-4}$.

This example has been constructed to demonstrate how dramatically the performance of the system can be affected by ISI and noise amplification. We can see that in order for the residual ISI (after equalization) to have minimal effect on the system performance, it should have been at approximately .003 (compared to the input noise of .032) or 10 db below the input noise. This is, of course, not always achievable. Similarly, there is also a clear necessity to acheive a small noise amplication.

These considerations will be central to our ongoing discussion of the performance characteristics of alternative equalizer structures.

10.1.5   The Autocorrelation Matrix of the Channel

Now let's look at the matrix $R = H_c H_c^T + (\sigma^2/d^2)I$ that appears in eq. 10.1-11b in relation to baseband channels. Consider first the situation where $\sigma^2 = 0$. By way of example, the convolution matrix H_c for $g = 5$ and $m = 7$ is

$$H_c = \begin{bmatrix} h_1 & h_2 & h_3 & h_4 & h_5 & 0 & 0 & 0 & 0 & 0 & 0 \\ 0 & h_1 & h_2 & h_3 & h_4 & h_5 & 0 & 0 & 0 & 0 & 0 \\ 0 & 0 & h_1 & h_2 & h_3 & h_4 & h_5 & 0 & 0 & 0 & 0 \\ 0 & 0 & 0 & h_1 & h_2 & h_3 & h_4 & h_5 & 0 & 0 & 0 \\ 0 & 0 & 0 & 0 & h_1 & h_2 & h_3 & h_4 & h_5 & 0 & 0 \\ 0 & 0 & 0 & 0 & 0 & h_1 & h_2 & h_3 & h_4 & h_5 & 0 \\ 0 & 0 & 0 & 0 & 0 & 0 & h_1 & h_2 & h_3 & h_4 & h_5 \end{bmatrix} \qquad (10.1\text{-}16a)$$

so that

$$R = H_c H_c^T = \begin{bmatrix} R(0) & R(1) & R(2) & R(3) & R(4) & R(5) & R(6) \\ R(1) & R(0) & R(1) & R(2) & R(3) & R(4) & R(5) \\ R(2) & R(1) & R(0) & R(1) & R(2) & R(3) & R(4) \\ R(3) & R(2) & R(1) & R(0) & R(1) & R(2) & R(3) \\ R(4) & R(3) & R(2) & R(1) & R(0) & R(1) & R(2) \\ R(5) & R(4) & R(3) & R(2) & R(1) & R(0) & R(1) \\ R(6) & R(5) & R(4) & R(3) & R(2) & R(1) & R(0) \end{bmatrix} \qquad (10.1\text{-}16b)$$

where

$$\begin{aligned} R(0) &= h_1^2 + h_2^2 + h_3^2 + h_4^2 + h_5^2 \\ R(1) &= h_1 h_2 + h_2 h_3 + h_3 h_4 + h_4 h_5 \\ R(2) &= h_1 h_3 + h_2 h_4 + h_3 h_5 \\ R(3) &= h_1 h_4 + h_2 h_5 \\ R(4) &= h_1 h_5 \\ R(5) &= 0 \\ R(6) &= 0 \end{aligned} \qquad (10.1\text{-}16c)$$

There are two things to be noted: first, these terms comprise the autocorrelation sequence of the discrete version of the channel and, second, that the sequence is symmetric, i.e.

$$R(m) = R(-m). \qquad (10.1\text{-}16d)$$

As a consequence of the symmetry of the autocorrelation function, the matrix R is not only symmetric, $R^T = R$, but the terms on each of the major diagonals are equal so that

$$R_{ij} = R_{i-j} \qquad (10.1\text{-}16e)$$

This matrix is Toeplitz, a name given to matrices of this type in Chapters 2 and 5.

The addition of noise to the system affects only the diagonal elements of the autocorrelation matrix. This makes sense if you consider that a consequence of the noise being white is that successive samples of the noise are uncorrelated.

The autocorrelation matrix of the baseband equivalent of a bandpass channel

$$R = \left[H_c H_c^{*T} + \frac{\sigma^2}{k^2} I \right] \qquad (10.1\text{-}16f)$$

differs somewhat because the autocorrelation function is complex-valued so that

$$(R^*)^T = R \qquad (10.1\text{-}16g)$$

As we have seen, this is the definition of a Hermitian matrix. This matrix is also Toeplitz. We recall from Chapter 5 that Hermitian matrices have real and positive eigenvalues. We will see in Chapter 11 that the eigenvalues of the channel autocorrelation matrix are of considerable importance to our understanding of adaptation process of the equalizer.

There is an additional property that relates the power spectral density to the autocorrelation function [Hay91]. A discrete-time random process has a power spectral density $P(\omega)$ having maximum and minimum values $P_{\max}$ and $P_{\min}$. The eigenvalues of the correlation matrix R of this random process have maximum and minimum values $\lambda_{\max}$ and $\lambda_{\min}$.

Defining $\chi(R)$ as the eigenvalue spread of the matrix, we have

$$\chi(R) = \frac{\lambda_{\max}}{\lambda_{\min}} \leq \frac{P_{\max}}{P_{\min}} \qquad (10.1\text{-}17)$$

Clearly then, the eigenvalue spread is one measure of channel distortion. Note that this measure, because it relates to power spectral density, addresses only amplitude distortion and says nothing about phase distortion.

10.1.6   A Frequency-Domain Approach

We want to relate the results derived above to a frequency-domain approach to the same problem. First we shall rewrite the expression for the optimal transversal tap vector as

$$C_{\min} \left[H_c H_c^{*T} + \frac{\sigma^2}{k^2} I \right]^* = E_h H_c^T \qquad (10.1\text{-}18a)$$

Now we shall represent the channel impulse response vector as a z-transform

$$H(z) = h_1 + h_2 z^{-1} + \cdots + h_g z^{g-1} \qquad (10.1\text{-}18b)$$

so that the autocorrelation function of the channel sequence is

$$R_s(z) = H^*(z) H\left(\frac{1}{z}\right) \qquad (10.1\text{-}18c)$$

where, as previously discussed, the noise term adds only to the $R_s(0)$ term. We use the subscript s in order to identify $R_s(z)$ as the autocorrelation function of a sampled function. It follows that the left-hand side of eq. 10.1-18a can be written as

$$C(z)[R_s(z) + \sigma^2]. \qquad (10.1\text{-}18d)$$

We shall find the right-hand side by example. If $g = 5$ and $m = 7$ as in the previous example, then

$$E_h H_c^T = \begin{bmatrix} 0 & 0 & 0 & 0 & 0 & 1 & 0 & 0 & 0 & 0 & 0 \end{bmatrix}$$

$$\times \begin{bmatrix} h_1 & h_2 & h_3 & h_4 & h_5 & 0 & 0 & 0 & 0 & 0 & 0 \\ 0 & h_1 & h_2 & h_3 & h_4 & h_5 & 0 & 0 & 0 & 0 & 0 \\ 0 & 0 & h_1 & h_2 & h_3 & h_4 & h_5 & 0 & 0 & 0 & 0 \\ 0 & 0 & 0 & h_1 & h_2 & h_3 & h_4 & h_5 & 0 & 0 & 0 \\ 0 & 0 & 0 & 0 & h_1 & h_2 & h_3 & h_4 & h_5 & 0 & 0 \\ 0 & 0 & 0 & 0 & 0 & h_1 & h_2 & h_3 & h_4 & h_5 & 0 \\ 0 & 0 & 0 & 0 & 0 & 0 & h_1 & h_2 & h_3 & h_4 & h_5 \end{bmatrix}^T \qquad (10.1\text{-}19a)$$

so that

$$E_h H_c^T = \begin{bmatrix} h_5 & h_4 & h_3 & h_2 & h_1 \end{bmatrix} \Rightarrow H_c\left(\frac{1}{z}\right). \qquad (10.1\text{-}19b)$$

We therefore have

$$C(z)[R_s(z) + \sigma^2] = H_c\left(\frac{1}{z}\right) \qquad (10.1\text{-}20a)$$

and the output vector of the equalizer is

$$E(z) = C(z)H_c^*(z) = \frac{H_c^*(z)H_c\left(\frac{1}{z}\right)}{[R_s(z) + \sigma^2]} = \frac{R_s(z)}{[R_s(z) + \sigma^2]} \qquad (10.1\text{-}20b)$$

where $E(z)$ is a two-sided z-transform that looks like

$$E(z) = e_{-3}z^{-3} + e_{-2}z^{-2} + e_{-1}z^{-1} + e_0 + e_1z^{-1} + e_2z^{-2} + e_3z^{-3}. \qquad (10.1\text{-}20c)$$

The minimum mean-square error for this system is given by

$$\frac{\text{MSE}_{\min}}{d^2} = \left[1 - C_{\min}^* H_c E_h^T\right] = 1 - \begin{bmatrix} c_1^* & c_2^* & c_3^* & c_4^* & c_5^* \end{bmatrix} \begin{bmatrix} h_5 \\ h_4 \\ h_3 \\ h_2 \\ h_1 \end{bmatrix} = 1 - e_0 \qquad (10.1\text{-}21)$$

where e_0 is the center sample of $E(z)$.

We can find e_0 from eq. 10.1-20b by taking the inverse z-transform

$$e_0 = \frac{1}{2\pi j} \oint \frac{E(z)}{z} dz = \frac{1}{2\pi j} \oint \frac{R_s(z)}{z[R_s(z) + \sigma^2]} dz \qquad (10.1\text{-}22a)$$

and letting $z = e^{j\omega t}$:

$$e_0 = \frac{T_s}{2\pi} \int_{-\pi/T_s}^{\pi/T_s} \frac{R_s\left(e^{j\omega T_s}\right)}{\left[R_s\left(e^{j\omega T_s}\right) + \sigma^2\right]} d\omega \qquad (10.1\text{-}22b)$$

so that

$$\frac{\text{MSE}_{\min}}{d^2} = 1 - e_0 = 1 - \frac{T_s}{2\pi} \int_{-\pi/T_s}^{\pi/T_s} \frac{R_s\left(e^{j\omega T_s}\right)}{\left[R_s\left(e^{j\omega T_s}\right) + \sigma^2\right]} d\omega$$

$$= \frac{T_s}{2\pi} \int_{-\pi/T_s}^{\pi/T_s} \left\{1 - \frac{R_s\left(e^{j\omega T_s}\right)}{\left[R_s\left(e^{j\omega T_s}\right) + \sigma^2\right]}\right\} d\omega$$

$$= \frac{\sigma^2 T_s}{2\pi} \int_{-\pi/T_s}^{\pi/T_s} \frac{d\omega}{\left[R_s\left(e^{j\omega T_s}\right) + \sigma^2\right]} \qquad (10.1\text{-}23)$$

We now want to be able to evaluate $R_s(e^{j\omega T_s})$ in terms of known channel quantities. First let us recall that

$$\left\{R_s(z) = \sum_k r_k z^{-k}\right\} \Rightarrow \left\{R_s\left(e^{j\omega T_s}\right) = \sum_k r_k e^{-j\omega k T_s}\right\} \qquad (10.1\text{-}24a)$$

so that

$$R_s\left(e^{j\omega T_s}\right)e^{j\omega m T_s} = r_m + \sum_{k \neq m} r_k e^{-j\omega(k-m)T_s} \qquad (10.1\text{-}24b)$$

and

$$r_m = \frac{T_s}{2\pi} \int\limits_{-\pi/T_s}^{\pi/T_s} R_s\left(e^{j\omega T_s}\right) e^{j\omega m T_s} \, d\omega.$$ (10.1-24c)

In general we know that the relationship between the spectrum of a signal and its autocorrelation function is

$$R(\tau) = \frac{1}{2\pi} \int\limits_{-\infty}^{\infty} |H(\omega)|^2 \, e^{j\omega\tau} \, d\omega.$$ (10.1-25)

We want to use this expression to evaluate the components of the autocorrelation function of the *sampled version of h(t)*. In order to be able to do this we must first make $\tau = mT_s$ and then make sure that instead of using $|H(\omega)|^2$ we use $|H_s(\omega)|^2$, where $H_s(\omega)$ is the *sampled spectrum*, which includes *the aliasing that comes along with sampling at the baud rate*. Sampling at a higher rate, which we will subsequently do in 'fractionally spaced equalizers' so that the excess-bandwidth tails of the raised cosine do not overlap, will mean that there is no aliasing.

Since the sampled spectrum is periodic with period $\omega_s = 2\pi/T_s$,

$$r_k = \frac{1}{2\pi} \int\limits_{-\pi/T_s}^{\pi/T_s} |H_s(\omega)|^2 e^{j\omega k T_s} \, d\omega$$ (10.1-26a)

It follows then that

$$R\left(e^{j\omega T_s}\right) = \frac{1}{T_s}|H_s(\omega)|^2 = \frac{1}{T_s}\left|\sum_{n=-1}^{1} H\left(\omega + n\frac{2\pi}{T_s}\right)\right|^2$$ (10.1-26b)

and, finally,

$$\frac{\text{MSE}_{\min}}{k^2} = \frac{\sigma^2 T_s}{2\pi} \int\limits_{-\pi/T_s}^{\pi/T_s} \frac{d\omega}{\left[\sigma^2 + \frac{1}{T_s}\left|\sum_{n=-1}^{1} H\left(\omega + n\frac{2\pi}{T_s}\right)\right|^2\right]}$$ (10.1-26c)

We should also note that eq. 10.1-19a allows us to find the baud-rate equalizer that minimizes the MSE as:

$$C(z) = \frac{H_c(z^{-1})}{[R_s(z) + \sigma^2]}$$ (10.1-27)

The equalizer found in eq. 10.1-27 has an infinite number of taps.

One of the important characteristics of baud-rate equalizers is their sensitivity to the timing phase of the samples. Finding the optimum timing phase was addressed in Chapter 6. An examination of the denominators of eqs. 10.1-26c and 10.1-27 reveals why this is the case. Both denominators involve the aliased or 'folded' spectrum of the channel, which is sensitive to timing phase. An example of an undistorted channel having a raised cosine spectrum will make the point. Let

$$H(\omega) = RC(\omega)e^{-j\omega\tau}$$ (10.1-28a)

having an excess bandwidth $\alpha\omega_s/2$ corresponding to a sampling phase of τ, and a shifted version of $H(\omega)$

$$H(\omega - \omega_s) = RC(\omega - \omega_s)e^{-j(\omega-\omega_s)\tau} \qquad (10.1\text{-}28b)$$

where $\omega_s = 2\pi/T_s$ is twice the Nyquist frequency. Setting the origin at the Nyquist frequency, the sum of these spectra in the region of their overlap is

$$H(\omega + \omega_s/2) + H(\omega - \omega_s/2)$$

$$= \left[1 - \mathrm{Sin}\left(\frac{\pi}{\alpha}\frac{\omega}{\omega_s}\right)\right]e^{-j(\omega+\omega_s/2)\tau} + \left[1 + \mathrm{Sin}\left(\frac{\pi}{\alpha}\frac{\omega}{\omega_s}\right)\right]e^{-j(\omega-\omega_s/2)\tau}$$

$$= 2\left\{\mathrm{Cos}\left(\frac{\omega_s}{2}\tau\right) - j\,\mathrm{Sin}\left(\frac{\pi}{\alpha}\frac{\omega}{\omega_s}\right)\mathrm{Sin}\left(\frac{\omega_s}{2}\tau\right)\right\}e^{-j\omega\tau} \qquad (10.1\text{-}28c)$$

showing the behavior of the sampled spectrum around the Nyquist frequency. When the impulse response is sampled at the peak corresponding to $\tau = 0$, the amplitude of the sampled spectrum remains flat. For different values of τ its amplitude will dip because the value of the imaginary part is always equal to zero at $\omega = 0$. When $\tau = T_s/2$ the amplitude goes to zero at $\omega = 0$. This is referred to as a null in the sampled spectrum.

When the channel is distorted, the square-root of raised-cosine receiver filter is not matched to the receiver input. Therefore the receiver filter output, unlike the output of a matched filter, is not a real function of frequency; rather, it is complex. Recall that whereas the magnitude of a transfer function of a real circuit is even around the origin, the phase is odd around the origin. Consequently when a bandlimited received signal is sampled at the symbol rate $f_s = 1/T_s$ the resulting aliasing causes terms of similar amplitudes and opposite phases to add around the Nyquist frequency. We consequently have a situation like

$$[A\,\mathrm{Cos}\,\theta + j A\,\mathrm{Sin}\,\theta] + [A\,\mathrm{Cos}\,\theta - j A\,\mathrm{Sin}\,\theta] = 2A\,\mathrm{Cos}\,\theta \qquad (10.1\text{-}29)$$

tending to decrease the amplitude of the resulting spectrum. In an extreme case this can lead to a spectral null. We will subsequently confirm that sampling the received signal at its peak may very well result in a null and that sampling at some phase off the peak may be a better choice.

Although we must be careful to remember that the amplitude and phase of a bandpass signal are not strictly even and odd around the center frequency, it is also the case that the same kind of circumstances of spectral dips or nulls may occur in demodulated bandpass signals that are sampled at the symbol rate. The equalizer operates to make the equalized sampled spectrum flat, providing gain to compensate for both channel distortion and amplitude dips caused by sampling. In so doing, it amplifies the noise beyond what an analog equalizer would do in equalizing an analog unsampled spectrum. *It is this additional noise amplification that causes the system we have described to be sub-optimal.* This will be illustrated in the Problems.

10.2   THE OPTIMUM LINEAR RECEIVER

In this section we will find the structure of the optimum linear receiver. The proof will follow Gitlin and Weinstein [G&W81]. As shown in fig. 10.2-1, the data stream, assumed to be uncorrelated,

$$d(t) = \sum_n d_n \delta(t - nT_s) \qquad (10.2\text{-}1a)$$

is applied to a transmitter filter having impulse response $p(t)$. This output is then applied to the channel which has an impulse response $q(t)$. The combination of these two impulse

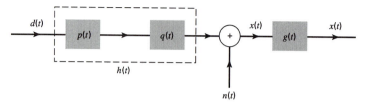

FIGURE 10.2-1

responses is denoted as $h(t)$, which determines the shaping of the signal at the receiver. In addition, there is additive white Gaussian noise $n(t)$ having a variance σ^2 that also appears at the receiver. This means that the overall received signal plus noise is

$$x(t) = \sum_n d_n h(t - nT_s) + n(t) \tag{10.2-1b}$$

where we will assume the data, the impulse response $h(t)$, and the noise $n(t)$ all to be complex-valued. This means that our development will apply as well to baseband and passband systems.

The received signal is then applied to a linear filter having impulse response $g(t)$ so that its output is

$$y(t) = \sum_n d_n f(t - nT_s) + v(t) \tag{10.2-1c}$$

where

$$f(t) = \int_{-\infty}^{\infty} h(\lambda) g^*(t - \lambda) \, d\lambda = \int_{-\infty}^{\infty} g(\lambda) h^*(t - \lambda) \, d\lambda \tag{10.2-1d}$$

and

$$v(t) = \int_{-\infty}^{\infty} n(\lambda) g^*(t - \lambda) \, d\lambda \tag{10.2-1e}$$

The linear filter is followed by a sampler which samples the signal every T_s seconds and makes a decision based on the sample. This sampled output is

$$y(mT_s) = \sum_n d_n f(mT_s - nT_s) + v(mT_s) \tag{10.2-1f}$$

Assuming that the decisions are correct, the mean-square error in this system is the expected value of the magnitude of the error squared. We have already seen that the MSE computed in this manner is statistically equivalent to the MSE computed on the basis of the reponse of the system to a single data impulse.

$$\begin{aligned} \text{MSE} &= E\{[y(mT_s) - d_m][y(mT_s) - d_m]^*\} \\ &= E\{y(mT_s)y^*(mT_s) - d_m y^*(mT_s) - d_m^* y(mT_s) + d_m d_m^*\} \end{aligned} \tag{10.2-2a}$$

We will deal with these terms one at a time:

1.
$$E\{d_m d_m^*\} = E(|d_m|^2). \tag{10.2-2b}$$

2. Using eq. 10.2-2a and recognizing that the data are zero mean and uncorrelated

$$E\{d_m^* y(mT_s)\} = E\left\{d_m^*\left[\sum_n d_n f(mT_s - nT_s) + v(mT_s)\right]\right\}$$

$$= E\left\{\sum_n d_n d_m^* \int_{-\infty}^{\infty} h(\lambda)g^*(mT_s - nT_s - \lambda)\, d\lambda\right\}$$

$$+ E\left\{d_m^* \int_{-\infty}^{\infty} n(\lambda)g^*(mT_s - \lambda)\, d\lambda\right\}$$

$$= E(|d_m|^2) \int_{-\infty}^{\infty} h(\lambda)g^*(-\lambda)\, d\lambda = E(|d_m|^2) \int_{-\infty}^{\infty} h(-\lambda)g^*(\lambda)\, d\lambda$$

$$\text{(10.2-2c)}$$

3. Similarly,

$$E\{d_m y^*(mT_s)\} = E(|d_m|^2) \int_{-\infty}^{\infty} h^*(\lambda)g(-\lambda)\, d\lambda = E(|d_m|^2) \int_{-\infty}^{\infty} h^*(-\lambda)g(\lambda)\, d\lambda$$

$$\text{(10.2-2d)}$$

4. Finally,

$$E\{y(mT_s)y^*(mT_s)\}$$

$$= E\left\{\left[\sum_n d_n f(mT_s - nT_s) + v(mT_s)\right]\left[\sum_j d_j^* f^*(jT_s - nT_s) + v^*(jT_s)\right]\right\}$$

$$= E\left\{\sum_n \sum_j d_n d_j^* f(mT_s - nT_s)f^*(jT_s - nT_s)\right\} + E\{v(mT_s)v^*(jT_s)\}$$

$$= E\{|d_n|^2\} \sum_n f(nT_s)f^*(nT_s)$$

$$+ E\left\{\int_{-\infty}^{\infty} n(\lambda_1)g^*(mT_s - \lambda_1)\, d\lambda_1 \int_{-\infty}^{\infty} n^*(\lambda_2)g(mT_s - \lambda_2)\, d\lambda_2\right\} \quad \text{(10.2-2e)}$$

where the cross products disappear because $E[d_n d_d^*] = 0$ if $n \neq d$. The second term in eq. 10.2-2e can be reduced to

$$E\left\{\int_{-\infty}^{\infty} n(\lambda_1)g^*(mT_s - \lambda_1)\, d\lambda_1 \int_{-\infty}^{\infty} n^*(\lambda_2)g(mT_s - \lambda_2)\, d\lambda_2\right\}$$

$$= \int_{-\infty}^{\infty}\int_{-\infty}^{\infty} E\{n(\lambda_1)n^*(\lambda_2)\}g^*(mT_s - \lambda_1)g(mT_s - \lambda_2)\, d\lambda_1\, d\lambda_2$$

$$= \int_{-\infty}^{\infty}\int_{-\infty}^{\infty} \sigma^2\delta(\lambda_1 - \lambda_2)g^*(\lambda_1)g(\lambda_2)\, d\lambda_1\, d\lambda_2 \quad \text{(10.2-2f)}$$

where σ^2 is the average power of the noise at the receiver input. The first term in eq. 10.2-2e is

$$E\{|d_n|^2\} \sum_n f(nT_s) f^*(nT_s)$$

$$= E\{|d_n|^2\} \sum_n \int_{-\infty}^{\infty} g(\lambda_1) h^*(nT_s - \lambda_1) d\lambda_1 \int_{-\infty}^{\infty} g^*(\lambda_2) h(nT_s - \lambda_2) d\lambda_2$$

$$= E\{|d_n|^2\} \int_{-\infty}^{\infty} \int_{-\infty}^{\infty} g(\lambda_1) g^*(\lambda_2) \left\{ \sum_n h(nT_s - \lambda_2) h^*(nT_s - \lambda_1) \right\} d\lambda_1 \lambda_2$$

$$(10.2\text{-}2g)$$

We can now put all of these terms together to get

$$\text{MSE} = E\{|d_n|^2\} \int_{-\infty}^{\infty} \int_{-\infty}^{\infty} g(\lambda_1) g^*(\lambda_2) \left\{ \sum_n h(nT_s - \lambda_2) h^*(nT_s - \lambda_1) \right.$$

$$\left. + \beta\delta(\lambda_1 - \lambda_2) \right\} d\lambda_1 d\lambda_2 - E(|d_m|^2) \int_{-\infty}^{\infty} h^*(-\lambda) g(\lambda) d\lambda$$

$$- E(|d_m|^2) \int_{-\infty}^{\infty} h(-\lambda) g^*(\lambda) d\lambda + E(|d_m|^2) \qquad (10.2\text{-}3a)$$

where $\beta = \sigma^2 / E\{|d_n|^2\}$.

Now we can find the optimum filter $g(t)$ by taking the gradient or partial derivative of MSE with respect to $g(t)$ and setting the result equal to zero. It should be noted that the operation of partial differentiation with respect to $g(t)$ treats $g^*(t)$ as a constant. As a result, we find that

$$\int_{-\infty}^{\infty} g^*(\lambda_2) \left\{ \sum_n h(nT_s - \lambda_2) h^*(nT_s - t) + \beta\delta(t - \lambda_2) \right\} d\lambda_2 = h^*(-t) \qquad (10.2\text{-}3b)$$

or

$$\sum_n h^*(nT_s - t) \left[\int_{-\infty}^{\infty} g^*(\lambda_2) h(nT_s - \lambda_2) d\lambda_2 \right] + \beta g^*(t) = h^*(-t) \qquad (10.2\text{-}3c)$$

and letting

$$z_n = \left[\int_{-\infty}^{\infty} g^*(\lambda_2) h(nT_s - \lambda_2) d\lambda_2 \right] \qquad (10.2\text{-}3d)$$

so that

$$\sum_n z_n h^*(nT_s - t) + \beta g^*(t) = h^*(-t) \qquad (10.2\text{-}3e)$$

so that the optimum receiver filter has the form

$$g(t) = \sum_m c_m^* h(mT_s - t) \qquad (10.2\text{-}3f)$$

where the c_n coefficients are as yet unspecified. The form of this filter is a filter matched to $h(t)$, the received signal, followed by a baud-rate transversal equalizer.

Let us note at this point that the combination of $h(t)$ and a filter matched to $h(t)$ will have an output that is the autocorrelation function of $h(t)$. We shall call this output $R_D(\tau)$ where

$$R_D(\tau) = \int_{-\infty}^{\infty} h^*(t)h(t - \tau)\,dt \tag{10.2-3g}$$

and $P_D(\omega)$ is the Fourier Transform of $R_D(\tau)$ and the power spectral density of $h(t)$.

In order to find the equalizer tap coefficents we will substitute eq. 10.2-3f into eq. 10.2-3c, yielding:

$$h^*(-t) = \sum_n h^*(nT_s - t)\left[\int_{-\infty}^{\infty} \sum_m c_m h^*(mT_s - \lambda_2)h(nT_s - \lambda_2)\,d\lambda_2\right]$$

$$+ \beta \sum_n c_n h^*(nT_s - t) \tag{10.2-3h}$$

where the sampled autocorrelation function of the channel impulse response $h(t)$ is

$$R_D(m - n) = \int_{-\infty}^{\infty} h^*(mT_s - \lambda_2)h(nT_s - \lambda_2)\,d\lambda_2 \tag{10.2-3i}$$

so that eq. 10.2-3h becomes

$$h^*(-t) = \sum_n \sum_m c_m R_D(n - m)h^*(nT_s - t) + \beta \sum_n c_n h^*(nT_s - t) \tag{10.2-3j}$$

The Fourier Transform of eq. 10.2-3j is:

$$H(\omega) = \sum_n \sum_m c_m R_D(n - m)H(\omega)e^{jn\omega T_s} + \beta \sum_n c_n^* H(\omega)e^{jn\omega T_s} \tag{10.2-3k}$$

Dividing both sides by $H(\omega)$ and rearranging the summations, we get:

$$1 = \sum_k R_D(k)e^{jk\omega T_s} \sum_n c_n e^{jn\omega T_s} + \beta \sum_n c_n e^{jn\omega T_s}$$

$$= \left[\beta + \sum_k R_D(k)e^{jk\omega T_s}\right]\sum_n c_n e^{jn\omega T_s}$$

$$= \left[\beta + \frac{1}{T_s}\sum_k P_D(\omega + k\,2\pi/T_s)\right]C(\omega) \tag{10.2-3m}$$

so that the baud-rate equalizer transfer function is

$$C(\omega) = \frac{1}{\left[\beta + \frac{1}{T_s}\sum_k P_D(\omega + k\,2\pi/T_s)\right]} \tag{10.2-3n}$$

and, following eq. 10.1-27c, we can find the MSE of the optimum system as

$$\text{MSE} = \frac{\beta T_s}{2\pi} \int_{-\pi/T_s}^{\pi/T_s} \frac{d\omega}{\left[\beta + \frac{1}{T_s}\sum_k P_D\left(\omega + k\,2\pi/T_s\right)\right]}$$

$$= \frac{\beta T_s}{2\pi} \int_{-\pi/T_s}^{\pi/T_s} \frac{d\omega}{\left[\beta + \frac{1}{T_s}\sum_k |H\left(\omega + k\,2\pi/T_s\right)|^2\right]} \tag{10.2-3p}$$

It is very important to notice the difference between eq. 10.2-3p for the optimum linear receiver and eq. 10.1-27c for the suboptimum baud-rate equalizer

$$\frac{\text{MSE}_{\min}}{d^2} = \frac{\sigma^2 T_s}{2\pi} \int_{-\pi/T_s}^{\pi/T_s} \frac{d\omega}{\left[\sigma^2 + \frac{1}{T_s}\left|\sum_{n=-1}^{1} H\left(\omega + n\,2\pi/T_s\right)\right|^2\right]} \tag{10.1-27c}$$

The placement of the magnitude sign outside the summation in the denominator of eq. 10.1-26c allows cancellation due to phase distortion in the suboptimum system. The placement of the magnitude sign inside the summation in the denominator of eq. 10.2-3p excludes the possibility of phase interaction in the summation.

As an example, consider the system of fig. 10.2-2. The received signal $h(t)$, which combines the transmit filter and the channel, is shown as the response of a lowpass RC circuit to a rectangular data pulse. The Laplace Transform of this signal is

$$H(s) = A\frac{1}{s(s+a)}\left(1 - e^{-sT_s}\right) \tag{10.2-4a}$$

and the Fourier Transform is

$$H(\omega) = A\frac{1}{j\omega(j\omega + a)}\left(1 - e^{-j\omega T_s}\right). \tag{10.2-4b}$$

If the the receiver filter $g(t)$ is matched to the actual received signal then

$$G(\omega) = H^*(\omega)e^{j\omega T_s} \tag{10.2-5a}$$

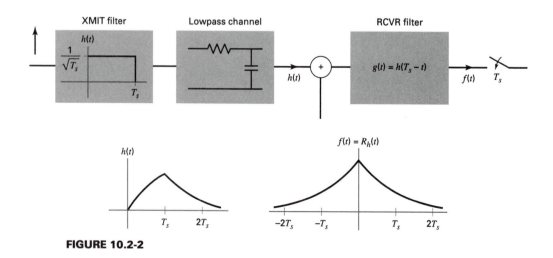

FIGURE 10.2-2

and the Laplace Transform of the output is

$$F(\omega) = H(\omega)H^*(\omega)e^{-j\omega T_s} = \frac{4A^2}{\omega^2(\omega^2 + a^2)}\mathrm{Sin}^2\frac{\omega T_s}{2}e^{-j\omega T_s} \qquad (10.2\text{-}5b)$$

which, except for a time delay, is *completely real*. When this output is sampled at the symbol rate, the resulting sampled spectrum will not suffer from the spectral nulls generated by the phase characteristic in the region of spectral overlap or aliasing. On the other hand, as shown in fig. 10.2-2, the output $f(t)$ contains intersymbol interference terms and must be followed by a baud-rate transversal equalizer.

It is very useful to illustrate this issue from the point of view of the discrete channel models that we have generated using the program CHAN768. This exercise will provide not only a numerical example but also the basis for a theoretical understanding of the mathematical techniques that we will use to develop further the ideas of optimum data transmission.

Consider the sampled impulse response of the B3 channel, corresponding to a timing phase of 0 (sampling at the peak), as generated by CHAN768, where the sampling is done every T_s sec. (fig. 10.2-3a) and every $T_s/2$ sec. (fig. 10.2-3b).

In comparing these two impulse responses, note that *b3ch3b*, corresponding to baud-rate sampling, already includes aliasing because we, in fact, have baud-rate samples of the impulse response. The Fourier Transform of these samples is the aliased, sampled spectrum of the channel including the raised cosine associated with the transmitter and receiver filters. The impulse response *b3ch3f*, corresponding to samples taken at twice the baud rate, contains the samples of *b3ch3b* as a subset. Since there is no aliasing at this sampling rate, the Fourier Transform will be equivalent to the analog spectrum.

```
b3ch3b                          load  b3ch3f.m
                                b3ch3f

b3ch3b  =
   -0.0006      -0.0278         b3ch3f  =
    0.0035       0.0382            -0.0006      -0.0278
    0.0002      -0.0506             0.0309      -0.0031
   -0.0436       0.0600             0.0035       0.0382
    0.0997      -0.4854            -0.0401       0.0048
    1.0000      -0.0486             0.0002      -0.0506
    0.0589       0.1222             0.0515       0.0022
   -0.1166      -0.2791            -0.0436       0.0600
    0.0131       0.2296            -0.1488      -0.1357
    0.0207      -0.1448             0.0997      -0.4854
   -0.0280       0.0908             0.6767      -0.5020
  in-phase      quadrature          1.0000      -0.0486
                                    0.6728       0.3168
     (a) Ts samples                 0.0589       0.1222
                                   -0.2280      -0.2845
                                   -0.1166      -0.2791
                                    0.0218       0.0811
                                    0.0131       0.2296
                                   -0.0089       0.0192
                                    0.0207      -0.1448
                                    0.0151      -0.0392
                                   -0.0280       0.0908

                                    (b) Ts/2 samples
```

FIGURE 10.2-3

The best baud-rate bandpass equalizer for a bandpass channel was found in eq. 10.1-12 (the Matlab script BPEQUAL.m), where we were, in fact, equalizing an aliased channel such as *b3ch3b*. This was not a situation in which the receiver filter was matched to the received signal; rather, both the transmitter and receiver filter are square-root of raised cosine.

We have seen that the optimum linear receiver requires a matched filter followed by a baud-rate equalizer. The matched-filter output is the autocorrelation function of the input. This autocorrelation function must be computed from impulse response samples representing the inverse Fourier Transform of the unaliased channel frequency response. The autocorrelation function must then be sampled at the baud rate. We can accomplish this using the $T_s/2$ samples in *b3ch3f* through a Matlab script MATFILT.m (available on the website), which computes the baud-rate samples of the matched filter output.

For example, consider the vector H representing the $T_s/2$ samples of the channel impulse response $h(t)$:

$$H = [h_{-2,0} \quad h_{-2,1} \quad h_{-1,0} \quad h_{-1,1} \quad h_{0,0} \quad h_{0,1} \quad h_{1,0} \quad h_{1,1} \quad h_{2,0} \quad h_{2,1}] \quad (10.2\text{-}6a)$$

where the second subscript in each of the samples is 0 for an on-beat sample, 1 for an off-beat sample.

The correlation of this vector with itself will give us $T_s/2$ samples of the autocorrelation function of $h(t)$ which we have labeled $R_D(t)$. The baud rate or T_s samples that make up the vector R_D correspond to those correlation products that involve products of on-beat samples of H with themselves and off-beat samples with themselves. For example,

```
load   b3ch3f . m
fschan  =  b3ch3f;
matfilt


RD  =
  −0.0025 − 0.0008i
   0.0080 + 0.0009i
  −0.0176 − 0.0004i
   0.0322 − 0.0043i
  −0.0865 + 0.0023i
   0.1303 + 0.0997i
  −0.1234 − 0.1942i
   0.1309 + 0.2846i
  −0.1674 − 0.4979i
  −0.0351 + 1.1657i
   2.9024
  −0.0351 − 1.1657i
  −0.1674 + 0.4979i
   0.1309 − 0.2846i
  −0.1234 + 0.1942i
   0.1303 − 0.0997i
  −0.0865 − 0.0023i
   0.0322 + 0.0043i
  −0.0176 + 0.0004i
   0.0080 − 0.0009i
  −0.0025 + 0.0008i
```

Matched Filter Output Sampled at the Baud Rate

FIGURE 10.2-4

this shifting of the vectors does not give a baud-rate sample:

$$
\begin{array}{cccccccccc}
h_{-2,0} & h_{-2,1} & h_{-1,0} & h_{-1,1} & h_{0,0} & h_{0,1} & h_{1,0} & h_{1,1} & h_{2,0} & h_{2,1} \\
 & h_{-2,0} & h_{-2,1} & h_{-1,0} & h_{-1,1} & h_{0,0} & h_{0,1} & h_{1,0} & h_{1,1} & h_{2,0} & h_{2,1}
\end{array} \quad \text{(10.2-6b)}
$$

but this one does:

$$
\begin{array}{cccccccccc}
h_{-2,0} & h_{-2,1} & h_{-1,0} & h_{-1,1} & h_{0,0} & h_{0,1} & h_{1,0} & h_{1,1} & h_{2,0} & h_{2,1} \\
 & & h_{-2,0} & h_{-2,1} & h_{-1,0} & h_{-1,1} & h_{0,0} & h_{0,1} & h_{1,0} & h_{1,1} & h_{2,0} & h_{2,1}
\end{array}
$$
$$\text{(10.2-6c)}$$

The sampled matched filter output in the example of *b3ch3f* is shown in fig. 10.2-4. Since the impulse response is complex, the autocorrelation function satisfies the condition that $R_D(-nT_s) = R_D^*(nT_s)$. A crucial property of this autocorrelation function is that it is independent of the sampling phase of the $T_s/2$ samples because the power spectral density of the of an unaliased signal is independent of the signal phase. We will confirm this property in the Problems. The task is to equalize this output, the autocorrelation function R_D, with a baud-rate equalizer.

It is these observations that lead to the idea of the *fractionally spaced equalizer*.

10.3   THE FRACTIONALLY SPACED EQUALIZER

In addition to [G&W81], fractionally spaced equalizers are discussed in [Ung76], [Q&F77], [GMW82]. The structure of a receiver that uses a fractionally spaced equalizer is as follows: the receiver filter is typically the square-root of raised cosine but the output of this filter is sampled at a rate higher than the symbol rate, every $(m/n)T_s$ seconds where $m < n$ are integers, rather than every T_s seconds. The sampling rate is chosen to avoid aliasing in the sampled spectrum and consequently the interference between adjacent spectra that leads to nulls and excessive noise amplification.

In theory, the sampling rate could be chosen according to the rolloff of the raised cosine so that the adjacent patterns in the sampled spectrum just touch each other. For example, if the rollof $\alpha = .25$ then $m/n = 2/5$. These samples are entered into a transversal filter at the sample rate but the output is taken at the symbol rate $f_s = 1/T_s$.

A small bit of reflection will indicate that if the ratio m/n is anything but very simple, the digital signal-processing problem will escalate in a major way. For this reason, the simplest possible structure, sampling at $T_s/2$, is most commonly used, as shown in fig. 10.3-1. This is not only computationally easy (the equalizer output is sampled once for

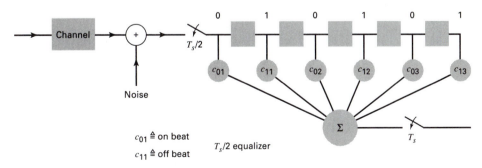

FIGURE 10.3-1

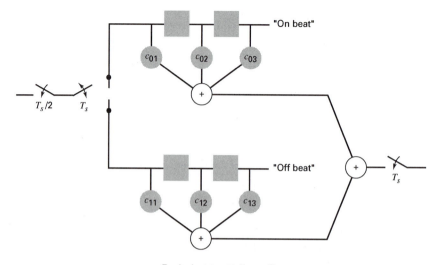

Equivalent to a $T_s/2$ equalizer

FIGURE 10.3-2

every two inputs), but it also accommodates all rolloffs up to $\alpha = 1.0$. Because it is so easy, the balance of the universe requires that there be a penalty and there is. As α decreases from unity toward zero, the amount of "dead space" between repetitions of the spectrum increases and, as we shall later see, the eigenvalues of the autocorrelation matrix have a greater spread. This will slow down the adaptation process of the equalizer.

The idea is that the fractionally spaced equalizer, by making the sampled received signal Nyquist, synthesizes both the matched filter and the baud-rate equalizer required to make an optimum linear receiver. Again, the key issue here is that the fractionally spaced equalizer is able to do this because the sampled spectrum on which it is operating does not have any aliasing and therefore does not have any spectral nulls.

We shall again proceed, Clark [Cla85], with a matrix analysis of the $T_s/2$ structure. The channel impulse response vector H, composed of samples taken at a $T_s/2$ rate, is divided into two subvectors composed of alternating samples H_0, the "on-beat" portion and H_1, the "off-beat" portion. According to fig. 10.3-2, these two subvectors are applied to two 'separate' equalizers in parallel. In fact, these two separate equalizers are a single transversal filter into which data is entered every $T_s/2$ seconds. But since data is taken from the equalizer every T_s seconds, it is as if the on-beat and off-beat portions of the equalizer were in parallel. These two equalizer sections are denoted C_0 and C_1. We should note that in each of the two parallel sections of this equalizer the noise is sampled every T_s seconds so that these noise samples are uncorrelated. In order to demonstrate how to formulate the problem, we will do an example.

Let the channel impulse response vector be

$$H = [h_{01} \quad h_{11} \quad h_{02} \quad h_{12} \quad h_{03} \quad h_{13}] \qquad \text{where } g = 6 \qquad (10.3\text{-}1a)$$

and where h_{01} is the first on-beat term of the impulse response and h_{11} is the first off-beat term. In general, the length g of the impulse response vector is even and for h_{in}

$$i = \begin{cases} 0 & \text{on beat} \\ 1 & \text{off beat} \end{cases} \quad n = 1, 2, \ldots, \left(\frac{g}{2}\right); \quad g \text{ even} \qquad (10.3\text{-}1b)$$

and for bandpass channels h_{in} is complex.

Similarly, we have two equalizer tap vectors

$$C_0 = [c_{01} \quad c_{02} \quad c_{03} \quad c_{04} \quad c_{05}] \qquad \text{the on beat} \tag{10.3-1c}$$

and

$$C_1 = [c_{11} \quad c_{12} \quad c_{13} \quad c_{14} \quad c_{15}] \qquad \text{the off beat} \tag{10.3-1d}$$

where the length of each of these vectors is $m/2$ where m is even (in this example $m = 10$).

The equalizer output is computed as $E = E_0 + E_1$ where E_0 is the convolution of H_0 and C_0, and E_1 is the convolution of H_1 and C_1. In matrix terms, $E = CH_c$:

$$E = [c_{01} \quad c_{11} \quad c_{02} \quad c_{12} \quad c_{03} \quad c_{13} \quad c_{04} \quad c_{14} \quad c_{05} \quad c_{15}]$$

$$\times \begin{bmatrix} h_{01} & h_{02} & h_{03} & 0 & 0 & 0 & 0 \\ h_{11} & h_{12} & h_{13} & 0 & 0 & 0 & 0 \\ 0 & h_{01} & h_{02} & h_{03} & 0 & 0 & 0 \\ 0 & h_{11} & h_{12} & h_{13} & 0 & 0 & 0 \\ 0 & 0 & h_{01} & h_{02} & h_{03} & 0 & 0 \\ 0 & 0 & h_{11} & h_{12} & h_{13} & 0 & 0 \\ 0 & 0 & 0 & h_{01} & h_{02} & h_{03} & 0 \\ 0 & 0 & 0 & h_{11} & h_{12} & h_{13} & 0 \\ 0 & 0 & 0 & 0 & h_{01} & h_{02} & h_{03} \\ 0 & 0 & 0 & 0 & h_{11} & h_{12} & h_{13} \end{bmatrix} \tag{10.3-1e}$$

where the channel length is $g = 6$, the equalizer has $m = 10$ taps and the matrix H_c has m rows and $(m + g - 2)/2$ columns. The ideal output vector is

$$E_h = [0 \quad 0 \quad 0 \quad 1 \quad 0 \quad 0 \quad 0] \tag{10.3-1f}$$

The analysis proceeds in much the same way as for the symbol rate system. The autocorrelation matrix is again defined as

$$R = H_c H_c^{*T} + \frac{\sigma^2}{k^2} I \tag{10.3-1g}$$

and the optimum equalizer tap vector is

$$C_{\min} = E_h H_c^{T*} \left[H_c H_c^{*T} + \frac{\sigma^2}{k^2} I \right]^{-1} \tag{10.3-1h}$$

In the z-transform domain eq. 10.3-1h becomes

$$C_{\min}(z) = \frac{H_c \left(\frac{1}{z} \right)}{H_c(z) H_c \left(\frac{1}{z} \right) + \frac{\sigma^2}{k^2}} \tag{10.3-1i}$$

It is of interest to note that if the ISI is small with respect to the noise then the optimum tap vector is simply a scaled reversal of the channel impulse response vector

$$C_{\min}(z) = \frac{k^2}{\sigma^2} H_c \left(\frac{1}{z} \right) \tag{10.3-1j}$$

This makes the equalizer into a matched filter.

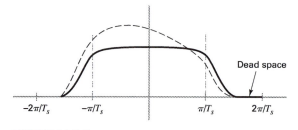

FIGURE 10.3-3

On the other hand, if the ISI is much larger than the noise then

$$C_{\min}(z) = \frac{1}{H_c(z)} \tag{10.3-1k}$$

which has the equalizer simply forcing the ISI terms to be zero.

All this is summarized in the Matlab script FSEQUAL.M which is available on the website. Theoretically, the performance of the fractionally spaced equalizer should not vary with the timing phase because it is supposed to synthesize the matched filter and the following infinite-length symbol-rate equalizer corresponding to the optimum linear receiver. In fact, there is a slight variation of performance with sampling phase in a finite-length $T/2$ equalizer. This as well as other performance questions will be examined in a subsequent example.

The finite length of the fractionally spaced equalizer also figures prominently into whether the fractionally spaced equalizer has a unique tap vector that minimizes the mean-square error. The problem is that the channel that is to be equalized, shown in fig. 10.3-3, has 'dead space' where the channel response is zero and the only energy present is that of noise. An analog filter that tries to equalize this channel to be Nyquist would have a lot of trouble because of the dead space. In that region the equalizer would be trying to minimize the noise while not trying to increase the noise amplification. As the noise level goes down, the problem becomes more difficult. In the limit, as the noise goes to zero, we run into uniqueness problems.

Gitlin and Weinstein [G&W81] have shown that as long as the equalizer has a finite number of taps, even as the noise gets arbitrarily small, there will be a unique set of equalizer tap coefficents. Correspondingly, the correlation matrix of eq. 10.3-1g will always be invertible.

Nevertheless, there is a penalty to be paid for the dead space. We have seen that the eigenvalue spread

$$\chi(R) = \frac{\lambda_{\max}}{\lambda_{\min}} \le \frac{P_{\max}}{P_{\min}} \tag{10.1-18}$$

increases with the ratio of maximum to minimum in the power spectral density of the channel and therefore this ratio will be higher for a fractionally spaced than for a baud-rate system. We will see in Chapter 11 that the greater the eigenvalue spread, the more difficult it is to get an adaptive equalizer to converge. Therefore, along with the better performance of the fractionally-spaced equalizer comes a greater difficulty of convergence.

In some applications there may also be a hardware obstacle. Until the appearance of high-speed voiceband DSP chips in the mid-1980s, the advantage was to baud-rate equalizers because they require only half of the computations of a $T/2$ equalizer having the same span of the impulse response. This is no longer a constraint in voiceband modems but may be a constraint in digital microwave systems.

Another implementation of fractionally spaced equalizers, operating directly on the received real passband input rather than on its complex baseband version, has been proposed by Mueller and Werner [M&W82] and further analyzed by Ling and Qureshi [L&Q90].

10.4 WIENER FILTERING AND THE WIENER–HOPF EQUATION

In finding the tap vectors that minimize the mean-square error in both baud rate and fractionally spaced transversal equalizers, we were, without using the name, using the approach of Wiener filtering. Since we will be using these ideas again, and because some of the formalisms of the theory are worth noting, we will take a small side trip in which we will develop the theory in a somewhat broader framework. This section may be skipped on the first reading.

Let $x(t)$, a WSS (wide-sense stationary random process), be the input to a linear filter having impulse response $h(t)$ and output

$$\hat{y}(t) = \int_{-\infty}^{\infty} h(\lambda)x(t-\lambda)\,d\lambda \tag{10.4-1a}$$

which, according to our discussion in Chapter 2, is also WSS.

We would like to choose the impulse response $h(t)$ so that the filter output is a WSS random process $y(t)$ but we cannot achieve that exactly. Instead, we will choose $h(t)$ to minimize the mean-square error between the actual and the desired outputs. In order to do this let us first define the error

$$e(t) = y(t) - \hat{y}(t) \tag{10.4-1b}$$

so that the idea is to find the optimum $h(t)$ that minimizes the expectation $E[e^2(t)]$.

THEOREM 1

The impulse response $h(t)$ will minimize $E[e^2(t)]$ if and only if $x(t)$ and $e(t)$ are uncorrelated, i.e.

$$\bar{R}_{xe}(\tau) = E[x(t)e(t+\tau)] = E[x(t-\tau)e(t)] = 0 \tag{10.4-1c}$$

If $x(t)$ and $e(t)$ are both zero-mean processes, this means that they are orthogonal. ∎

Proof:

Let $\bar{R}_{xe}(\tau) = 0$. Now let $g(t)$ be the impulse response of another filter having $x(t)$ as input and $z(t)$ as an optimal output. The errors in these two systems are

$$e_h(t) = y(t) - \hat{y}(t)$$
$$e_g(t) = y(t) - z(t) \tag{10.4-2a}$$

so that

$$\begin{aligned}
E\left[e_g^2(t)\right] &= E\{[y(t) - z(t)]^2\} = E\{[y(t) - \hat{y}(t) + \hat{y}(t) - z(t)]^2\} \\
&= E\{[e_h(t) + \hat{y}(t) - z(t)]^2\} \\
&= E\left\{e_h^2(t) + 2e_h(t)[\hat{y}(t) - z(t)] + [\hat{y}(t) - z(t)]^2\right\}
\end{aligned} \tag{10.4-2b}$$

The cross-product term in eq. 10.4-2b can be written as

$$
E\{2e_h(t)[\hat{y}(t) - z(t)]\} = E\left\{2e_h(t)\int_{-\infty}^{\infty} x(t-\lambda)[h(\lambda) - g(\lambda)]\,d\lambda\right\}
$$

$$
= 2\int_{-\infty}^{\infty} E\{e_h(t)x(t-\lambda)\}\,[h(\lambda) - g(\lambda)]\,d\lambda
$$

$$
= 2\int_{-\infty}^{\infty} R_{xe_h}(\lambda)[h(\lambda) - g(\lambda)]\,d\lambda \qquad (10.4\text{-}2c)
$$

If $R_{xe_h}(\tau) = 0$ then from eq. 10.4-2b,

$$
E\left[e_g^2(t)\right] = E\left\{e_h^2(t)\right\} + E\{[\hat{y}(t) - z(t)]^2\} \qquad (10.4\text{-}2d)
$$

Since all of these terms are non-negative, it follows that $E[e_g^2(t)] \geq E\{e_h^2(t)\}$ with equality only if $g(t) = h(t)$. This proves the 'if' part.

Now let $g(t) = h(t)$. It then follows from eq. 10.4-2b that

$$
E\left[e_g^2(t)\right] = E\left\{e_h^2(t)\right\} + R_{xe_h}(\tau) \qquad (10.4\text{-}2e)
$$

an equality that can only be satisfied if $R_{xe_h}(\tau) = 0$.   *end of proof*

The summary of this Theorem is that in a Wiener filter the input and the error are orthogonal. From eq. 10.4-2e, we have seen that if the filter is optimal, i.e. $R_{xe_h}(\tau) = 0$, then

$$
R_{xe}(\tau) = E\{e(t)x(t-\lambda)\} = E\{[y(t) - \hat{y}(t)]x(t-\lambda)\}
$$
$$
= R_{xy}(\tau) - R_{x\hat{y}}(\tau) = 0
$$

or

$$
R_{xy}(\tau) = R_{x\hat{y}}(\tau) \qquad (10.4\text{-}3a)
$$

where $y(t)$ is the desired output process and $\hat{y}(t)$ is the actual output process. Continuing:

$$
R_{xy}(\tau) = E\{\hat{y}(t)x(t-\tau)\} = E\left\{x(t-\tau)\int_{-\infty}^{\infty} x(t-\lambda)h(\lambda)\,d\lambda\right\}
$$

$$
= \int_{-\infty}^{\infty} E\{x(t-\tau)x(t-\lambda)\}h(\lambda)\,d\lambda = \int_{-\infty}^{\infty} R_x(t-\lambda)h(\lambda)\,d\lambda \quad (10.4\text{-}3b)
$$

epeating eq. 10.4-3b and taking its Fourier Transform, we get the two forms of the *Wiener–Hopf Condition*:

$$
R_{xy}(\tau) = \int_{-\infty}^{\infty} R_x(t-\lambda)h(\lambda)\,d\lambda
$$

$$
P_{xy}(\omega) = P_x(\omega)H(\omega) \qquad (10.4\text{-}3c)
$$

which defines the transfer function $H(\omega)$ of the optimal filter in terms of the power spectral density of the input random process and the cross power spectral density of the input process and the desired output process.

For $R_{xe}(\tau) = 0$, the mean-square error in the optimal system is:

$$E\{e^2(t)\} = E\{e(t)[y(t) - \hat{y}(t)]\} = E\{e(t)y(t)\} - E\{e(t)\hat{y}(t)\}$$

$$= E\{[y(t) - \hat{y}(t)]y(t)\} - E\left\{ e(t) \int_{-\infty}^{\infty} x(t - \lambda)h(\lambda)\, d\lambda \right\}$$

$$= E\{y(t)y(t)\} - E\left\{ y(t) \int_{-\infty}^{\infty} x(t - \lambda)h(\lambda)\, d\lambda \right\} - \int_{-\infty}^{\infty} E\{e(t)x(t - \lambda)\}h(\lambda)\, d\lambda$$

$$= \overline{R_y}(0) - \int_{-\infty}^{\infty} \overline{R_{xy}}(\lambda)h(\lambda)\, d\lambda - \int_{-\infty}^{\infty} \overline{R_{xe}}(\lambda)h(\lambda)\, d\lambda \qquad (10.4\text{-}3\text{d})$$

so that

$$E\{e^2(t)\} = \overline{R_y}(0) - \int_{-\infty}^{\infty} \overline{R_{xy}}(\lambda)h(\lambda)\, d\lambda \qquad (10.4\text{-}3\text{e})$$

The mean-square error can also be expressed in frequency-domain terms. Thus:

$$\overline{R_y}(0) = \frac{1}{2\pi} \int_{-\infty}^{\infty} P_y(\omega)\, d\omega \qquad (10.4\text{-}3\text{f})$$

and from Parseval's Theorem

$$\int_{-\infty}^{\infty} \overline{R_{xy}}(\lambda)h(\lambda)\, d\lambda = \frac{1}{2\pi} \int_{-\infty}^{\infty} P_{xy}^*(\omega)H(\omega)\, d\omega \qquad (10.4\text{-}3\text{g})$$

so that

$$E\{e^2(t)\} = \frac{1}{2\pi} \int_{-\infty}^{\infty} [P_y(\omega) - P_{xy}^*(\omega)H(\omega)]\, d\omega \qquad (10.4\text{-}3\text{h})$$

and substituting eq. 10.4-38c yields

$$E\{e^2(t)\} = \frac{1}{2\pi} \int_{-\infty}^{\infty} \left[P_y(\omega) - \frac{|P_{xy}(\omega)|^2}{P_x(\omega)} \right] d\omega \qquad (10.4\text{-}3\text{i})$$

These results form the basis for the theory of optimum linear filtering and prediction.

EXAMPLE 1   *Extracting an Analog Signal from Noise*

Let the received signal $x(t)$ be the sum of the transmitted signal $s(t)$ and noise $n(t)$:

$$x(t) = s(t) + n(t) \qquad (10.4\text{-}4\text{a})$$

where $s(t)$ and $n(t)$ are uncorrelated, and the desired output $y(t) = s(t)$.

In order to find the optimum filter we must find the appropriate correlation functions for the Wiener–Hopf equation.

$$\overline{R_x}(\tau) = E\{[s(t)+n(t)][s(t+\tau)+n(t+\tau)]\}$$

$$= E\{s(t)s(t+\tau)\} + E\{n(t)n(t+\tau)\}$$

$$= \overline{R_s}(\tau) + \overline{R_n}(\tau) \tag{10.4-4b}$$

so that

$$P_x(\omega) = P_s(\omega) + P_n(\omega) \tag{10.4-4c}$$

Similarly,

$$\overline{R_{xy}}(\tau) = E\{[s(t)+n(t)][s(t+\tau)]\}$$

$$= E\{s(t)s(t+\tau)\} + E\{n(t)s(t+\tau)\} = \overline{R_s}(\tau) \tag{10.4-4d}$$

so that

$$P_{xy}(\omega) = P_s(\omega) \tag{10.4-4e}$$

It follows from eq. 10.4-38c that the transfer function of the optimum linear filter is

$$H(\omega) = \frac{P_{xy}(\omega)}{P_x(\omega)} = \frac{P_s(\omega)}{P_s(\omega) + P_n(\omega)} \tag{10.4-4f}$$

and from eq. 10.4-3i that the mean square error is

$$E\{e^2(t)\} = \frac{1}{2\pi} \int_{-\infty}^{\infty} \left[P_s(\omega) - \frac{P_s^2(\omega)}{P_s(\omega) + P_n(\omega)} \right] d\omega$$

$$= \frac{1}{2\pi} \int_{-\infty}^{\infty} \left[\frac{P_s(\omega)P_n(\omega)}{P_s(\omega) + P_n(\omega)} \right] d\omega \tag{10.4-4g}$$

Note that, if the signal and noise do not overlap in frequency, the optimal $H(\omega)$ is an ideal filter surrounding $S(\omega)$ and having infinite attenuation in the region of the noise. Under these circumstances the mean-square error is zero. ∎

EXAMPLE 2 *A Discrete-Time Transversal Filter*

In this example we will demonstrate that the approach that we have used to optimize transversal equalizers is equivalent to Wiener filtering. Let the data in the transversal equalizer be the random variables $x_1, x_2, \ldots, x_N$ and the equalizer taps be $c_1, c_2, \ldots, c_N$ so that the output is

$$x = \sum_{n=1}^{N} x_n c_n \tag{10.4-5a}$$

and the desired output is y. The mean square error is

$$\text{MSE} = E\{[y-x]^2\} = E\left\{ \left[y - \sum_{n=1}^{N} x_n c_n \right]^2 \right\} \tag{10.4-5b}$$

which we want to minimize with respect to the tap coefficients. This requires us to set the partial derivative of MSE with respect to each one of the tap coefficients equal to zero. For $n = m$,

$$
\frac{\partial[\text{MSE}]}{\partial c_m} = \frac{\partial}{\partial c_m} E\left\{\left[y - \sum_{n=1}^{N} x_n c_n\right]^2\right\}
$$

$$
= E\left\{2\left[y - \sum_{n=1}^{N} x_n c_n\right]\frac{\partial}{\partial c_m}\left[y - \sum_{n=1}^{N} x_n c_n\right]\right\}
$$

$$
= E\left\{-2x_m\left[y - \sum_{n=1}^{N} x_n c_n\right]\right\} = 0 \qquad \text{(10.4-5c)}
$$

or:

$$
\overline{R_{ym}} = \sum_{n=1}^{N} c_n \overline{R_{nm}} \qquad \text{for } m = 1, \dots, N \qquad \text{(10.4-5d)}
$$

This can be written as the matrix equation

$$
\begin{bmatrix} \overline{R_{y1}} \\ \overline{R_{y2}} \\ \cdot \\ \overline{R_{yN}} \end{bmatrix} = \begin{bmatrix} \overline{R_{11}} & \overline{R_{12}} & \cdots & \overline{R_{1N}} \\ \overline{R_{21}} & \overline{R_{22}} & \cdots & \overline{R_{2N}} \\ \cdot & & \cdot & \\ \overline{R_{N1}} & \overline{R_{N2}} & \cdots & \overline{R_{NN}} \end{bmatrix} \begin{bmatrix} c_1 \\ c_2 \\ \cdot \\ c_N \end{bmatrix} \qquad \text{(10.4-5e)}
$$

or

$$
\mathbf{R}_{xy} = R_{xx}\mathbf{C}_{\text{opt}} \qquad \text{(10.4-5f)}
$$

so that

$$
\mathbf{C}_{\text{opt}} = \mathbf{R}_{xx}^{-1}R_{xy} \qquad \text{(10.4-5g)}
$$

This is a restatement, in vector-matrix terms, of the Wiener–Hopf condition of eq. 10.4-3c. It is also a restatement of the basic results that we obtained for transversal equalizers earlier in this chapter. Although perhaps somewhat repetitious, recognizing that our results are part of a larger framework is useful. ■

10.5 DECISION–FEEDBACK EQUALIZATION

Both the baud-rate and the fractionally spaced equalizer approaches to overcoming channel distortion involve linear equalization of the sampled signal using transversal feed-forward filters. We have seen that fractionally spaced equalizers realize the structure for optimum linear equalization of the signal, a receiver filter matched to the channel output followed by a baud-rate equalizer.

We have also seen that optimum linear equalization still results in noise amplification and the resulting system therefore does not attain the performance bound of the matched filter receiver on a noisy distortionless channel. In order to come closer to that bound we will now consider decision-feedback equalization.

10.5.1 The Basic Structure of a Decision-Feedback Equalizer

The idea of decision-feedback equalization has a long history. The interested reader may consult Austin [Aus67], Monsen [Mon71], George et al. [GBS71], Price [Pri71] and [Pri72], and Belfiore and Park [B&P79]. In order to understand the structure of the decision-feedback

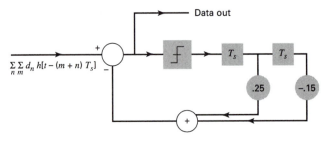

FIGURE 10.5-1

equalizer (DFE), let us first consider a channel having a sampled baud-rate impulse response that consists only of trailing echoes or postcursors such as

$$H = [h_0, h_1, h_2] = [1, .25, -.15]$$

The basic DFE structure is shown in fig. 10.5-1. The idea is that the incoming noisy data is sliced and the decisions that constitute the output are also stored in a symbol rate transversal filter. The taps of this transversal are made equal to the respective postcursors so that the ISI generated by these previous data symbols is subtracted from the current data before the slicing operation. The nonlinearity of the DFE comes from the use of sliced data in the feedback path.

The integrity of this process depends on the slicer decisions being correct, because feedback may cause propogation of an incorrect decision or error. This potential for error propagation is the most serious drawback of decision feedback equalization and will be discussed subsequently. The great advantage of the DFE is that the equalization is accomplished through the feedback of sliced, and therefore noiseless, data. Consequently this equalization process does not amplify the noise.

Another advantage to the DFE is computational. Since the stored data is sliced, it is possible to replace the operation of multiplying data with taps to a simpler shift-and-add operation. This is particularly useful in very high-speed systems such as digital microwave where it is often not realistic to have an authentic multiply. Decision feedback is also used in ISDN modem chips both because the impulse response of the local loop does not contain significant leading echoes and because of the simple structure of the DFE.

In order to use a DFE, we must have either a channel that has no leading echoes *or* a way of getting rid of leading echoes before the impulse response arrives at the DFE. In order to equalize the leading echoes before the DFE we must have a prefilter. There are a variety of approaches to the structure of this prefilter and we shall examine some of these. In order to do this, we shall first examine some theoretical foundations of decision-feedback equalization.

10.5.2 Optimum Decision-Feedback Equalization

In order to find the optimum prefilter for equalizing the leading echoes in a DFE structure (fig. 10.5-2), we will follow Salz [Sal73]. The analysis will proceed on the assumption that all of the decisions are correct. This assumption, in effect, linearizes the problem and allows a solution of the problem of finding the receiver structure that minimizes mean-square error. The uncorrelated data stream

$$d(t) = \sum_n d_n \delta(t - nT_s) \tag{10.5-1a}$$

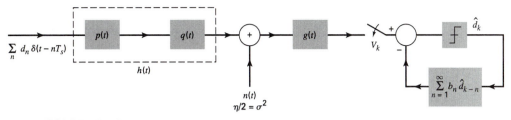

FIGURE 10.5-2

is applied to a transmitter filter having impulse response $p(t)$. This output is then applied to the channel, which has an impulse response $q(t)$. The combination of these two impulse responses is denoted as $h(t)$, which determines the shaping of the signal at the receiver. In addition, there is additive white Gaussian noise $n(t)$ having a variance σ^2 that also appears at the receiver. This means that the overall received signal plus noise is

$$x(t) = \sum_n d_n h(t - nT_s) + n(t) \tag{10.5-1b}$$

where we will assume the data, the impulse response $h(t)$, and the noise $n(t)$ all to be complex-valued. This means that our development will apply as well to baseband and passband systems. The received signal is then applied to a linear filter having impulse response $g(t)$ so that its output is

$$y(t) = \sum_{n=-\infty}^{\infty} d_n f(t - nT_s) + v(t) \tag{10.5-1c}$$

where

$$f(t) = \int_{-\infty}^{\infty} h(\lambda) g^*(t - \lambda) \, d\lambda = \int_{-\infty}^{\infty} g(\lambda) h^*(t - \lambda) \, d\lambda \tag{10.5-1d}$$

and

$$v(t) = \int_{-\infty}^{\infty} n(\lambda) g^*(t - \lambda) \, d\lambda \tag{10.5-1e}$$

The output of the DFE section is subtracted from the symbol-rate-sampled output of the linear filter. The decision is based on this sample, which can be expressed as

$$z_m = \sum_{n=-\infty}^{\infty} f_n^* d_{m-n} - \sum_{n=1}^{\infty} b_n^* d_{m-n} + v_m \tag{10.5-1f}$$

so that the mean-square error is

$$\text{MSE} = E\{[d_m - z_m][d_m - z_m]^*\} \tag{10.5-2a}$$

or

$$\text{MSE} = E\left\{ \left[(f_0^* - 1)d_m + \sum_{n=-\infty}^{-1} f_n^* d_{m-n} + \sum_{n=1}^{\infty} (f_n^* - b_n^*) d_{m-n} + v_m \right] \right.$$
$$\left. \times \left[(f_0 - 1)d_m^* + \sum_{n=-\infty}^{-1} f_n d_{m-n}^* + \sum_{n=1}^{\infty} (f_n - b_n) d_{m-n}^* + v_m^* \right] \right\} \tag{10.5-2b}$$

Since the data terms in each portion of eq. 10.5-2a are different and therefore uncorrelated, the expectation of the cross-product terms goes to zero so that only the squares of the individual terms remain. Therefore:

$$\text{MSE} = E\{d_m d_m^*\}(f_0^* - 1)(f_0 - 1) + E\left\{\left[\sum_{n=-\infty}^{-1} f_n^* d_{m-n}\right]\left[\sum_{n=-\infty}^{-1} f_n d_{m-n}^*\right]\right\}$$

$$+ E\left\{\left[\sum_{n=1}^{\infty}(f_n^* - b_n^*)d_{m-n}\right]\left[\sum_{n=1}^{\infty}(f_n - b_n)d_{m-n}^*\right]\right\} + E\{v_m v_m^*\} \quad (10.5\text{-}2c)$$

and continuing with the same idea

$$\text{MSE} = E\{|d_m|^2\}[(f_0 - 1)(f_0^* - 1)] + E\{|d_m|^2\}\left[\sum_{n=-\infty}^{-1} f_n f_n^*\right]$$

$$= + E\{|d_m|^2\}\left[\sum_{n=1}^{\infty}(f_n - b_n)(f_n^* - b_n^*)\right] + E\{v_m v_m^*\} \quad (10.5\text{-}2d)$$

Examination of eq. 10.5-2d makes it clear that the first step in minimizing the MSE is to set the feedback coefficients b_n equal to the postcursors (or causal components) f_n of the baud-rate samples of the output of $g(t)$, the linear filter. The next step is to choose the linear filter so that the remaining terms

$$\text{MSE} = E\{|d_m|^2\}[(f_0 - 1)(f_0^* - 1)] + E\{|d_m|^2\}\left[\sum_{n=-\infty}^{-1} f_n f_n^*\right] + E\{v_m v_m^*\}$$

$$= E\{|d_m|^2\}\left\{1 - f_0 - f_0^* + \left[\sum_{n=-\infty}^{0} f_n f_n^*\right] + E\{v_m v_m^*\}\right\} \quad (10.5\text{-}2e)$$

are minimized. According to eqs. 10.5-1e and 10.5-2e, this can be rewritten as

$$\text{MSE} = E\{|d_m|^2\}\left\{1 - \int_{-\infty}^{\infty} g(\lambda)h^*(-\lambda)\,d\lambda - \int_{-\infty}^{\infty} g^*(\lambda)h(-\lambda)\,d\lambda\right.$$

$$+ \sum_{n=-\infty}^{0}\left[\int_{-\infty}^{\infty} g(\lambda)h^*(nT_s - \lambda)\,d\lambda\right]\left[\int_{-\infty}^{\infty} g^*(\lambda)h(nT_s - \lambda)\,d\lambda\right]$$

$$+ \left.\frac{\sigma^2}{E\{|d_m|^2\}}\int_{-\infty}^{\infty} g(\lambda)g^*(\lambda)\,d\lambda\right\} \quad (10.5\text{-}3a)$$

In order to choose the filter $g(t)$ that minimizes the MSE, we will take the partial derivative of the MSE with respect to $g(t)$ and set it equal to zero:

$$\frac{\partial[\text{MSE}]}{\partial g(\lambda)} = E\{|d_m|^2\}\left\{-h^*(-t) + \int_{-\infty}^{\infty} g^*(\lambda)\sum_{n=-\infty}^{0}[h^*(nT_s - \lambda)h(nT_s - t)]\,d\lambda\right.$$

$$+ \left.\frac{\sigma^2}{E\{|d_m|^2\}}g^*(t)\right\} = 0 \quad (10.5\text{-}3b)$$

so that

$$h^*(-t) = \int_{-\infty}^{\infty} g^*(\lambda) \sum_{n=-\infty}^{0} [h(nT_s - \lambda)h^*(nT_s - t)] \, d\lambda + \frac{\sigma^2}{2E\{|d_m|^2\}} g^*(t) \qquad (10.5\text{-}3c)$$

Now let

$$U_n = \int_{-\infty}^{\infty} g^*(\lambda) h(nT_s - \lambda) \, d\lambda \qquad (10.5\text{-}3d)$$

so that

$$h^*(-t) = \sum_{n=-\infty}^{0} U_n h^*(nT_s - t) + \frac{\sigma^2}{2E\{|d_m|^2\}} g^*(t) \qquad (10.5\text{-}3e)$$

and

$$g^*(t) = \frac{2E\{|d_m|^2\}}{\sigma^2} \left\{ (1 - U_0)h^*(-t) + \sum_{n=-\infty}^{-1} U_n h^*(nT_s - t) \right\} \qquad (10.5\text{-}3f)$$

or

$$g(t) = \sum_{n=-\infty}^{0} g_n^* h(nT_s - t) \qquad (10.5\text{-}3g)$$

This expression indicates that $g(t)$ is comprised of a filter matched to $h(t)$ followed by a transversal filter sampled at the symbol rate that operates only on the precursors or anti-causal components of the impulse response $h(t)$. The resulting postcursors are cancelled by the nonlinear feedback structure. This configuration is an extension of the optimum linear receiver, which has a matched filter followed by a baud-rate transversal that equalizes its entire output.

We now want to take a series of steps that will enable us to compute the tap values of both the one-sided transversal and the DFE section.

10.5.3   The Whitened Matched Filter

We now want to demonstrate how to find the tap coefficents of both the forward and the feedback portions of the decision feedback equalizer in order to achieve optimum performance. We will then compute the performance and compare it to that of the optimum linear receiver. This development will follow Cioffi et al. [CDEF 95].

Let us start by recalling that the combined transmitter and channel have an impulse response $h(t) \Leftrightarrow H(\omega)$ so that the matched filter has an impulse response $h(-t)$ and the combined impulse response is the convolution

$$\int_{-\infty}^{\infty} h(\lambda) h^*(-t + \lambda) \, d\lambda = R_h(t)$$

where $R_h(\tau)$ is the autocorrelation function of $h(t)$ and $R_h(\tau) = R_h^*(-\tau)$. Since $h(t)$ is a finite energy signal, the Fourier Transform of $R_h(\tau)$ is the energy spectral density $E_h(\omega)$.

The matched filter output is sampled at the baud rate, resulting in the sampled signal $R_D(z)$ where $z = e^{j\theta} = e^{j\omega T_s}$,

$$R_D(z) = \sum_k R_{h,k} z^{-k} \qquad (10.5\text{-}4a)$$

and where

$$E_D(\omega) = R_D\left(e^{j\omega T_s}\right) = \sum_m \left| H\left(\omega + m\frac{2\pi}{T_s}\right)\right|^2 \qquad -\frac{\pi}{T_s} \le \omega \le \frac{\pi}{T_s} \qquad \text{(10.5-4b)}$$

The input data sequence to the system is

$$x(t) = \sum_m d_m \delta(t - mT_s) \Leftrightarrow X(z) = \sum_m d_m z^{-m} \qquad \text{(10.5-4c)}$$

where $E_x = E(|d_m|^2)$ is the average energy per transmitted symbol. The sampled output of the matched filter is

$$Y(z) = X(z)R_D(z) + n'(z) \qquad \text{(10.5-4d)}$$

where $n'(z)$ is a sequence of samples of the noise process that results from passing the white Gaussian noise $n(t)$, having a two-sided power spectral density $\eta/2$, through the matched filter $h^*(-t)$. The time-average autocorrelation of $n'(z)$ is

$$\overline{R_{n'}(z)} = \frac{\eta}{2}R_D(z) \qquad \text{(10.5-5a)}$$

and its average power is

$$P_{n'} = \frac{\eta}{2}R_{h,0}. \qquad \text{(10.5-5b)}$$

Recall from Chapter 1 that if the unit impulse response of the transmitted signal $h(t)$ is constrained to have energy $E = 1$, then the peak value of its autocorrelation function is $R_h(0) = 1$ and the noise output power of the matched filter is simply $\eta/2$. The average energy per output symbol is $E_x R_{h,0}^2$.

As a reference point, the average energy per transmitted symbol divided by the average power of the output noise in an *ideal matched filter* system having a single transmitted data point is

$$\text{SNR} = \frac{E_x R_{h,0}^2}{\frac{\eta}{2}R_{h,0}} = \left[\frac{2E_x}{\eta}\right] R_{h,0} \qquad \text{(10.5-6a)}$$

which we have seen in Chapter 4 is the same as

$$\text{SNR} = \left[\frac{P_x}{\sigma^2}\right] R_{h,0} \qquad \text{(10.5-6b)}$$

The constraint of a single data point avoids the issue of ISI. We may therefore collapse that part of fig. 10.5-2 to the left of the sampler into fig. 10.5-3.

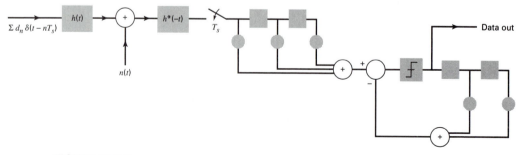

FIGURE 10.5-3

To continue the discussion, let us now address the issue of *spectral factorization of autocorrelation functions*. We shall proceed by example.

EXAMPLE

A discrete signal $x(n)$ is defined as

$$x(n) = \begin{cases} (.25)^{-n} & n > 0 \\ 1 & n = 0 \\ (2)^n & n < 0 \end{cases}$$

so that its z-transform is

$$X(z) = \frac{1}{1 - \frac{z^{-1}}{4}} + \frac{1}{1 - \frac{z}{2}} - 1 = \frac{\frac{7}{8}}{\left(1 - \frac{z^{-1}}{4}\right)\left(1 - \frac{z}{2}\right)}$$

Note that $X(z)$ has poles both inside and outside the unit circle. Those outside indicate the noncausal character of $x(n)$, which can be made causal by adding a delay (in this case an infinite delay) to the system. The autocorrelation function is

$$R_x(z) = X(z)X(z^{-1}) = \frac{\frac{7}{8}}{\left(1 - \frac{z^{-1}}{4}\right)\left(1 - \frac{z}{2}\right)} \frac{\frac{7}{8}}{\left(1 - \frac{z^1}{4}\right)\left(1 - \frac{z^{-1}}{2}\right)}$$

$$= \left\{ \frac{\left(\frac{7}{8}\right)}{\left(1 - \frac{z^{-1}}{4}\right)\left(1 - \frac{z^{-1}}{2}\right)} \right\} \left\{ \frac{\left(\frac{7}{8}\right)}{\left(1 - \frac{z}{4}\right)\left(1 - \frac{z}{2}\right)} \right\} = E_0 X'(z)X'(z^{-1})$$

∎

We assert that within broad conditions, every autocorrelation function that we will encounter may be factored in this way:

$$R_x(z) = X(z)X(z^{-1}) = E_0 X'(z)X'(z^{-1}) \tag{10.5-7a}$$

where $X'(z)$ is *canonical* (is causal, with $x_0 = 1$, and has its poles and zeros inside the unit circle, making it minimum-phase) and $X'(z^{-1})$ is *anticanonical* (is anticausal, with $x_0 = 1$, and has its poles and zeros outside the unit circle, making it non-minimum phase).

We state without proof that

$$\log E_0 = \frac{T_s}{2\pi} \int_{-\pi/T_s}^{\pi/T_s} \log E_x(\omega)\, d\omega = \frac{T_s}{2\pi} \int_{-\pi/T_s}^{\pi/T_s} \log R_x\left(e^{j\omega T_s}\right) d\omega \tag{10.5-7b}$$

where the logarithm can be to any base.

We are now ready to proceed with finding the equalizer tap coefficients and computing the performance. The general form of the optimum DFE is shown in fig. 10.5-4. The received sequence at the slicer is given by

$$U(z) = a(z)Y(z) - [b(z) - 1]X'(z) \tag{10.5-8a}$$

where $X'(z)$ is the estimate of the original transmitted data sequence. If there are no decision errors, our usual assumption in analyzing the performance of decision-feedback equalizers, then the error sequence at the slicer is

$$e(z) = U(z) - X(z) = a(z)Y(z) - [b(z) - 1]X(z) - X(z)$$
$$= a(z)Y(z) - b(z)X(z) \tag{10.5-8b}$$

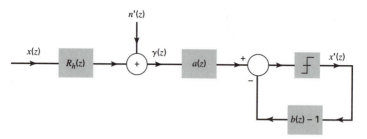

FIGURE 10.5-4

For a given $b(z)$ we can optimize $a(z)$ by invoking the orthogonality principle of the Wiener–Hopf condition, namely that the cross-correlation of the input sequence $Y(z)$ and the error sequence $e(z)$ must be zero. Therefore:

$$R_{ey}(z) = a(z)R_{yy}(z) - b(z)R_{xy}(z) = 0 \qquad (10.5\text{-}9a)$$

and

$$a(z) = b(z)\frac{R_{xy}(z)}{R_{yy}(z)} = b(z)\frac{\overline{R_{xy}(z)}}{\overline{R_{yy}(z)}} \qquad (10.5\text{-}9b)$$

where we have implicitly divided numerator and denominator by T_s in order to make time-average autocorrelation functions.

Since we obtain a sample of $y(t)$ by passing a sample of $x(t)$ through $h(t)$ and its matched filter $h^*(-t)$ and sampling the result, it follows by our previous arguments that

$$R_{xy}(z) = E_x R_D(z) \Leftrightarrow \overline{R_{xy}}(z) = P_x R_D(z) \qquad (10.5\text{-}10)$$

where P_x is the average power of the data stream. Now note that we obtain the autocorrelation function of $Y(z)$ by convolving the autocorrelation sequence R_h with itself, and add to that the channel noise passed through the matched filter, as shown in fig. 10.5-5:

$$\begin{aligned}\overline{R_{yy}}(z) &= P_x[R_D(z)]^2 + \sigma^2 R_D(z) \\ &= R_D(z)[P_x R_D(z) + \sigma^2]\end{aligned} \qquad (10.5\text{-}11a)$$

Using the spectral factorization process described above

$$\overline{R_{yy}}(z) = P_0 R_D(z) f(z) f^*\left(z^{-1}\right) = P_0 R_D(z) R_f(z) \qquad (10.5\text{-}11b)$$

where:

$$\log P_0 = \frac{1}{2\pi} \int\limits_{-\pi/T_s}^{\pi/T_s} \log\left[P_x R_h\left(e^{j\omega T_s}\right) + \sigma^2\right] d\omega \qquad (10.5\text{-}11c)$$

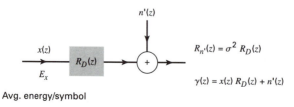

Avg. energy/symbol

FIGURE 10.5-5

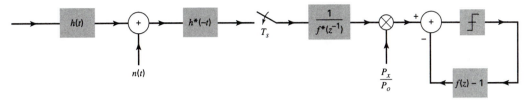

FIGURE 10.5-6

It follows from eq. 10.5-9b that

$$a(z) = b(z)\frac{\bar{R}_{xy}(z)}{\bar{R}_{yy}(z)} = b(z)\frac{P_x R_h(z)}{P_0 R_h(z)R_f(z)}$$

$$= b(z)\left(\frac{P_x}{P_0}\right)\frac{1}{R_f(z)} = b(z)\left(\frac{P_x}{P_0}\right)\frac{1}{f(z)f^*(z^{-1})} \qquad (10.5\text{-}12)$$

We have already established that the feedback taps $b(z)$ should be causal and the forward taps $a(z)$ should be anticausal. The spectral factorization into causal and anticausal portions indicates that these should be assigned to $a(z)$ and $b(z)$ accordingly:

$$a(z) = \left(\frac{P_x}{P_0}\right)\frac{1}{f^*\left(z^{-1}\right)} \qquad (10.5\text{-}13)$$

$$b(z) = f(z)$$

The resulting configuration is shown in fig. 10.5-6.

In order to find the MSE, we will start with eq. 10.5-8b. Using eq. 10.5-13:

$$e(z) = b(z)\left\{\left(\frac{P_x}{P_0}\right)\frac{1}{R_f(z)}Y(z) - X(z)\right\} = b(z)e'(z) \qquad (10.5\text{-}14\text{a})$$

and

$$\bar{R}_e(z) = b(z)b^*\left(z^{-1}\right)\bar{R}_{e'}(z) = R_f(z)\bar{R}_{e'}(z) \qquad (10.5\text{-}14\text{b})$$

The autocorrelation function of $e'(z)$ can then be found by squaring $e'(z)$ and taking the expectation

$$R_{e'}(z) = \left(\frac{P_x}{P_0}\right)^2\frac{R_{yy}(z)}{[R_f(z)]^2} - 2\left(\frac{P_x}{P_0}\right)\frac{R_{xy}(z)}{R_f(z)} + S_x \qquad (10.5\text{-}14\text{c})$$

where the autocorrelation function $R_x(z) = P_x$ because the data are uncorrelated. It follows that

$$R_{e'}(z) = \left(\frac{P_x}{P_0}\right)^2\frac{P_0 R_h(z)R_g(z)}{[R_f(z)]^2} - 2\left(\frac{P_x}{P_0}\right)\frac{P_x R_h(z)}{R_f(z)} + P_x$$

$$= P_x\left\{-\left(\frac{P_x}{P_0}\right)\frac{R_h(z)}{R_f(z)} + 1\right\} = \left(\frac{P_x}{P_0}\right)\left\{\frac{-P_x R_h(z) + P_0 R_f(z)}{R_f(z)}\right\}$$

$$= \left(\frac{P_x}{P_0}\right)\left\{\frac{\sigma^2}{R_f(z)}\right\} \qquad (10.5\text{-}14\text{d})$$

It therefore follows that

$$R_e(z) = \left(\frac{P_x}{P_0}\right)\sigma^2 \qquad (10.5\text{-}15\text{a})$$

This is actually quite an important result. First, it demonstrates that the autocorrelation sequence is a single term corresponding to $n = 0$, so that the power spectral density of the mean-square error is flat. Second, it shows that the mean-square error for the optimum DFE is:

$$\text{MSE} = \left(\frac{P_x}{P_0}\right)\sigma^2 \tag{10.5-15b}$$

and, finally, we can compute the signal-to-noise ratio for the optimum DFE as:

$$\text{SNR} = \frac{P_x}{\text{MSE}} = \left(\frac{P_0}{\sigma^2}\right) \tag{10.5-15c}$$

The structure of a matched filter followed by a one sided transversal that equals $a(z)$ in eq. 10.5-13 is known as *the whitened matched filter*. It turns out to be a quite important structure that we will come back to in relation to maximum likelihood sequence estimation (MLSE), another detection approach that will be addressed later in this chapter.

We can now compare this result to the performance of the *optimum linear equalizer*. If we set the feedback taps $b(z) = 1$, we get the structure of the optimum linear receiver. From eq. 10.5-14 we see that

$$\bar{R}_e(z) = \bar{R}_{e'}(z) = \left(\frac{P_x}{P_0}\right)\left\{\frac{\sigma^2}{R_f(z)}\right\} = \frac{P_x}{\left[\dfrac{P_x}{\sigma^2}R_h(z) + 1\right]} \tag{10.5-16a}$$

and consequently

$$\text{MSE} = \frac{P_x}{2\pi}\int_{-\pi/T_s}^{\pi/T_s}\frac{d\omega}{\left[\dfrac{P_x}{\sigma^2}R_h(z) + 1\right]} \tag{10.5-16b}$$

and

$$\text{SNR} = \left[\frac{1}{2\pi}\int_{-\pi/T_s}^{\pi/T_s}\frac{d\omega}{\left[\dfrac{P_x}{\sigma^2}R_h(z) + 1\right]}\right]^{-1} \tag{10.5-16c}$$

In comparing eqs. 10.5-15a and 10.5-16a, it is apparent that the MSE for the DFE is less than that for the optimum linear receiver.

10.5.4   Optimum Decision Feedback Equalizers—A Matlab Script

We have already developed a Matlab script FSEQUAL that computes the performance of the $T_s/2$ fractionally spaced equalizer, realizing the optimum linear receiver. This script was developed before we discussed the similarity in structure between the optimum linear receiver and the optimum decision-feedback receiver. We shall now develop another script that directly compares the optimum linear receiver and the optimum decision-feedback receiver based on the identity of the front end structures, a filter matched to the overall channel impulse response $h(t)$ followed by a baud-rate sampler. Therefore the impulse response of the channel plus matched filter is the autocorrelation function $R_h(t)$ and the task is to equalize, one way or another, $R_D(z)$, the baud-rate samples of this output. These baud-rate samples have already been generated by the script MATFILT. That script will be our starting point. We will first equalize these samples with a transversal filter, as previously done in the script BPEQUAL. This operation will give us the optimum linear receiver. Then we will equalize only the leading echoes of $R_h(t)$ using a one-sided transversal. The trailing echoes will, we assume, be cancelled by the decision feedback portion.

The process of equalizing only the leading echoes is illustrated by an example. Consider eq. 10.5-17a, which shows a channel impulse response having five taps and a one-sided transversal equalizer with three taps. The impulse response, which is identified as the samples of $R_h(t)$, is shown sliding past the transversal taps.

$$
\begin{array}{ccccccccc}
 & & & & c_{-2} & c_{-1} & c_0 & & \\
r_2 & r_1 & r_0 & r_{-1} & r_{-2} & & & & \\
 & r_2 & r_1 & r_0 & r_{-1} & r_{-2} & & & \\
 & & r_2 & r_1 & r_0 & r_{-1} & r_{-2} & & \\
 & & & r_2 & r_1 & r_0 & r_{-1} & r_{-2} & \\
 & & & & r_2 & r_1 & r_0 & r_{-1} & r_{-2} \\
 & & & & & r_2 & r_1 & r_0 & r_{-1} & r_{-2} \\
 & & & & & & r_2 & r_1 & r_0 & r_{-1} & r_{-2}
\end{array}
\tag{10.5-17a}
$$

Notice that it is only on the fifth output that the main sample h_0 of the channel impulse response lines up with c_0 on the transversal, corresponding to the main sample of the output. In matrix terms the output vector is:

$$
E_1 = [e_{-4} \quad e_{-3} \quad e_{-2} \quad e_{-1} \quad e_0 \quad e_1 \quad e_2]
$$

$$
= [c_{-2} \quad c_{-1} \quad c_0]
\begin{bmatrix}
r_{-2} & r_{-1} & r_0 & r_1 & r_2 & | & 0 & 0 \\
0 & r_{-2} & r_{-1} & r_0 & r_1 & | & r_2 & 0 \\
0 & 0 & r_{-2} & r_{-1} & r_0 & | & r_1 & r_2
\end{bmatrix}
= C R_1 \tag{10.5-17b}
$$

Ideally, the equalized output vector of the one-sided transversal would be

$$
E_h = [0 \quad 0 \quad 0 \quad 0 \quad 1 \quad e_1 \quad e_2] \tag{10.5-17c}
$$

where the DFE has the job of canceling e_1 and e_2. We assume that this cancellation will be complete so in describing only the transversal part of the equalizer we truncate eq. 10.5-17b to

$$
E_2 = [e_{-4} \quad e_{-3} \quad e_{-2} \quad e_{-1} \quad e_0]
$$

$$
= [c_{-2} \quad c_{-1} \quad c_0]
\begin{bmatrix}
r_{-2} & r_{-1} & r_0 & r_1 & r_2 \\
0 & r_{-2} & r_{-1} & r_0 & r_1 \\
0 & 0 & r_{-2} & r_{-1} & r_0
\end{bmatrix}
= C R_2 \tag{10.5-17d}
$$

where the ideal output vector from the one-sided transversal is $E_{h_2} = [0 \ 0 \ 0 \ 0 \ 1]$. It is important to note that the equalization of the leading echoes plus noise by the one-sided transversal affects the magnitude of the trailing echoes. This can be viewed as the working out, by the equalizer, of the spectral factorization of the system autocorrelation function.

From our previous work we know that the form of the solution to the equalizer taps is

$$
C_{\text{opt}} = E_{h_2} R_2 A_f^{-1} \tag{10.5-17e}
$$

where A_f is the correlation matrix of the system.

As in our previous examples of this sort, the system autocorrelation function has two parts, one relating to the ISI and one to the noise. When the noise input to the equalizer was simply the output noise of a square-root of raised-cosine filter, then it was white, as in the systems described by BPEQUAL and FSEQUAL, so that the noise autocorrelation matrix was $R_n = \sigma^2 I$, where I is the identity matrix. In our current model, the noise input to the equalizer is not white, having passed through the matched filter $h^*(-t)$. The noise therefore has the autocorrelation function $R_h(\tau)$. It therefore follows that the system correlation matrix for the one-sided transversal is written as

$$
R_f =
\begin{bmatrix}
r_{-2} & r_{-1} & r_0 & r_1 & r_2 \\
0 & r_{-2} & r_{-1} & r_0 & r_1 \\
0 & 0 & r_{-2} & r_{-1} & r_0
\end{bmatrix}
\begin{bmatrix}
r_{-2} & 0 & 0 \\
r_{-1} & r_{-2} & 0 \\
r_0 & r_{-1} & r_{-2} \\
r_1 & r_0 & r_{-1} \\
r_2 & r_1 & r_0
\end{bmatrix}
+ \sigma^2
\begin{bmatrix}
r_0 & r_1 & r_2 \\
r_{-1} & r_0 & r_1 \\
r_{-2} & r_{-1} & r_0
\end{bmatrix}
$$

$$(10.5\text{-}17f)$$

In computing R_f for the optimum linear filter, the noise is included in a similar manner.

With this, we can compute the transversal tap vector C_{opt} as shown in eq. 10.5-17e and we can compute the transversal output vector from eq. 10.5-17b.

The Matlab script is called OPTIMUM.m and is available on the website.

The extent to which the DFE provides improved performance compared to the fractionally spaced equalizer is strongly dependent on the channel. One can argue that since the channel phase distortion does not play a part in the performance of a matched filter receiver, the issue comes down to which of the two structures is better able to handle amplitude distortion. This will be addressed in the example in Section 10.7 and in the Problems.

If the channel is known in advance, the optimum DFE can be readily implemented. The matched filter can be implemented with a $T_s/2$ equalizer structure, taking care to use this structure only for the matched filter and not for any equalization. The baud-rate output of the matched filter can then be applied to the one-sided transversal and DFE as indicated above.

If the channel is not known in advance, it is not apparent how to obtain the optimum DFE structure through the adaptive process. Under these circumstances there may be some advantage in considering some suboptimal DFE structures.

10.6 SUBOPTIMAL DECISION-FEEDBACK STRUCTURES

The problem of not acheiving the optimum DFE structure through adaptation opens the door to looking for suboptimal decision-feedback structures that are adaptive. We will briefly discuss three alternatives having increasing computational complexity. In all of these alternatives the receiver filter is square root of raised cosine or perhaps some compromise equalizer, but certainly not a matched filter, followed by a baud-rate sampler.

ALTERNATIVE 1

The equalizer is a one-sided baud-rate transversal followed by a DFE. This is the same structure used in the optimum receiver but differs because it works on baud-rate samples of the channel rather than on samples of the matched filter output. We would expect the performance of this structure to depend heavily on the timing phase. The implementation of this structure has been discussed above. ■

ALTERNATIVE 2

Let us recall that in both the analog and digital worlds a phase equalizer is realized by an all-pass circuit that has flat amplitude response across the band of interest and therefore

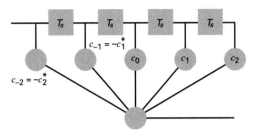

Alternative #2 DFE prefilter

FIGURE 10.6-1

does not amplify any noise. Consider trying to use such an equalizer to attack *only the leading echoes*, treating them *as if* they arose only from phase distortion. The idea is to have very little noise amplification in the transversal part, perhaps at the expense of making the DFE work a little harder.

As a first attempt to accomplish this, consider a transversal filter with taps having negative conjugate symmetry as shown in fig. 10.6-1. In this transversal the transfer function would be

$$C(\omega) = c_{-2} + c_{-1}e^{-j\omega T_s} + c_0 e^{-j\omega 2T_s} + c_1 e^{-j\omega 3T_s} + c_2 e^{-j\omega 4T_s}$$

$$c_{-1} = -c_1^* = a_1 + jb_1,$$

$$c_{-2} = -c_2^* = a_2 + jb_2$$

$$(10.6\text{-}1a)$$

so that

$$C(\omega) = \left\{ c_0 + |c_{-1}| \left[e^{j(\omega T_s + \theta_1)} - e^{-j(\omega T_s + \theta_1)} \right] \right.$$
$$\left. + |c_{-2}| \left[e^{j(\omega 2T_s + \theta_2)} - e^{-j(\omega 2T_s + \theta_2)} \right] \right\} e^{j\omega 2T_s} \qquad (10.6\text{-}1b)$$

or

$$C(\omega) = \{ c_0 + 2j[|c_{-1}|\text{Sin}(\omega T_s + \theta_1) + c_{-2}\text{Sin}(2\omega T_s + \theta_2)]\} e^{j\omega 2T_s} \qquad (10.6\text{-}2a)$$

Eliminating the delay term $e^{j2\omega T_s}$, we see that for modest values of c_{-1} and c_{-2} the value of the magnitude $|C(\omega)|$ does not vary appeciably with ω but the phase is

$$\theta(\omega) = \tan^{-1} \left[\left| \frac{c_{-1}}{c_0} \right| \text{Sin}(\omega T_s + \theta_1) + \left| \frac{c_{-2}}{c_{-1}} \right| \text{Sin}(2\omega T_s + \theta_2) \right] \qquad (10.6\text{-}2b)$$

indicating that the phase is quite variable so that it is plausible to think of this structure as an approximation to a phase equalizer.

An example of how this strategy is implemented is given below with a channel having $g = 5$ baud rate samples and a transversal equalizer having $m = 5$ complex taps.

$$[c_{-2} \quad c_{-1} \quad c_0 \quad -c_{-1}^* \quad -c_{-2}^*] \begin{bmatrix} h_1 & h_2 & h_3 & h_4 & h_5 & 0 & 0 & 0 & 0 \\ 0 & h_1 & h_2 & h_3 & h_4 & h_5 & 0 & 0 & 0 \\ 0 & 0 & h_1 & h_2 & h_3 & h_4 & h_5 & 0 & 0 \\ 0 & 0 & 0 & h_1 & h_2 & h_3 & h_4 & h_5 & 0 \\ 0 & 0 & 0 & 0 & h_1 & h_2 & h_3 & h_4 & h_5 \end{bmatrix}$$

$$= [e_1 \quad e_2 \quad e_3 \quad e_4 \quad e_5 \quad e_6 \quad e_7 \quad e_7 \quad e_9] = E \qquad (10.6\text{-}3a)$$

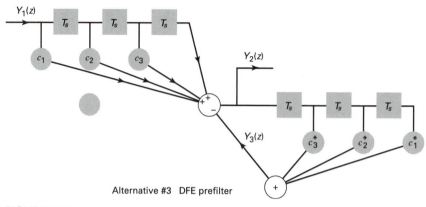

FIGURE 10.6-2

or, taking into account the relations between the taps,

$$[c_{-2} \quad c_{-1} \quad c_0] \begin{bmatrix} h_1 & h_2 & h_3 & h_4 & (h_5 - h_1^*) & | & -h_2^* & -h_3^* & -h_4^* & -h_5^* \\ 0 & h_1 & h_2 & (h_3 - h_1^*) & (h_4 - h_2^*) & | & (h_5 - h_3^*) & -h_4^* & -h_5^* & 0 \\ 0 & 0 & h_1 & h_2 & h_3 & | & h_4 & h_5 & 0 & 0 \end{bmatrix} = E$$

(10.6-3b)

In this example, only the first five columns of the matrix are used to form the correlation matrix R_h and the ideal output vector is $E_h = [0 \ 0 \ 0 \ 0 \ 1]$. ∎

ALTERNATIVE 3

This approach continues the idea of attempting to equalize the leading echoes using a phase equalizer, but this time we use an authentic all-pass filter structure as shown in fig. 10.6-2. This structure is discussed in detail in Regalia et al [Reg88]. Notice that the leading and the trailing taps are complex conjugate pairs. The transfer function of the filter is determined in the usual manner:

$$(c_1 + c_2 z^{-1} + c_3 z^{-2} + z^{-3})Y_1(z) - Y_3(z) = Y_2(z)$$

(10.6-4a)

and

$$Y_3(z) = (c_3^* z^{-1} + c_2^* z^{-2} + c_1^* z^{-3})Y_2(z)$$

(10.6-4b)

so that

$$H(z) = \frac{Y_2(z)}{Y_1(z)} = \frac{(c_1 + c_2 z^{-1} + c_3 z^{-2} + z^{-3})}{(1 + c_3^* z^{-1} + c_2^* z^{-2} + c_1^* z^{-3})}$$

(10.6-4c)

Since this filter is recursive or IIR, it is not possible to find its characteristics as an equalizer by matrix inversion. We must instead actually do a step-by-step covergence. This will be undertaken in Chapter 11. ∎

10.7 COMPARING THE PERFORMANCE OF EQUALIZER STRUCTURES

It is useful to have a comparative performance of all of the equalizer structures that have been developed above. This comparison is made using the telephone line B3 channel for five different timing phases.

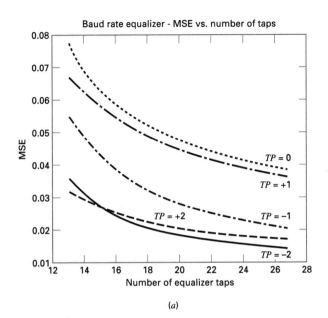

(a)

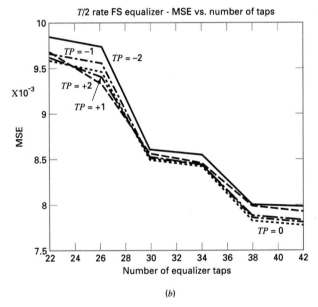

(b)

FIGURE 10.7-1

The MSE vs. number of equalizer taps is shown for SNR = 20 db and five timing phases for a baud-rate transversal in fig. 10.7-1a and for a T_s/s transversal in fig. 10.7-1b. The performance is clearly strongly dependent on timing phase for the baud-rate equalizer and almost independent of timing phase for the fractionally spaced equalizer. The fractionally spaced equalizer requires 50% more taps but performs around 3 db better, reflecting its smaller noise amplification. The relationship of timing phase to performance is shown in another way in fig. 10.7-2a and fig. 10.7-2b, where it is clear that it is not optimum to sample at the peak in a baud-rate system and that the sampling phase is not important in a fractionally spaced equalizer. Recall that we used band-edge timing recovery [God78] in

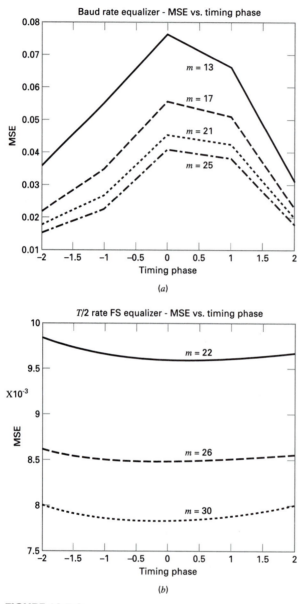

FIGURE 10.7-2

Section 6.10 to generate the optimum timing phase for a baud-rate equalizer. By way of contrast, the issue in a fractionally spaced equalizer is simply the stability of the timing phase. Ungerboeck [Ung76] proposes that the timing phase be chosen so that the difference of the sum of the squares of the leading taps and the sum of the squares of the training taps is driven to zero.

Although the relationship of the eigenvalues of the autocorrelation matrix to the rate of equalizer convergence will not be addressed until the next chapter, it is useful to examine the relative magnitudes of these eigenvalues. The eigenvalue spread $\chi = \lambda_{max}/\lambda_{min}$ is plotted in fig. 10.7-3 as a function of the number of equalizer taps and for different

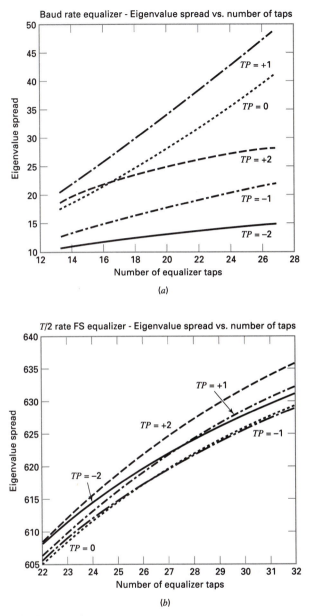

FIGURE 10.7-3

values of timing phase. As expected, the value of the spread is much smaller but much more dependent on timing phase for the baud-rate than the fractionally spaced equalizer. Finally, in fig. 10.7-4, we see the dependence of the eigenvalue spread on the SNR. This spread varies only modestly for the baud-rate equalizer but increases dramatically with the SNR for the fractionally spaced equalizer. This phenomenon was discussed in relation to the difficulty in equalizing the dead space in a fractionally spaced system in a low-noise environment. Reducing this dead space, and therefore the eigenvalue spread, is the primary reason for using more complex fractionally spaced equalizers such as $3T_s/4$ structures.

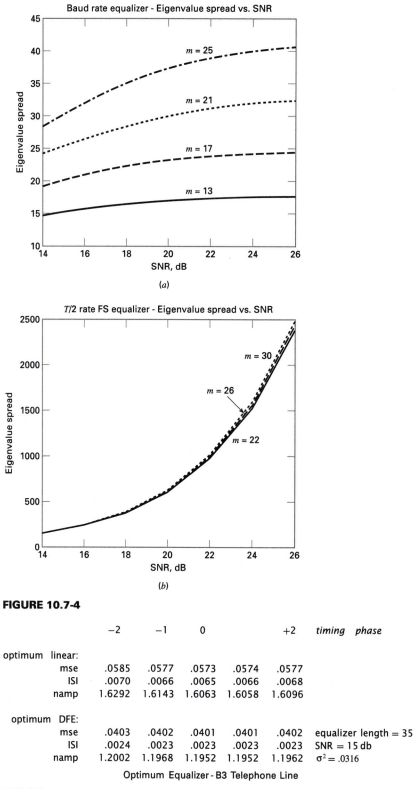

FIGURE 10.7-4

	−2	−1	0		+2	*timing phase*
optimum linear:						
mse	.0585	.0577	.0573	.0574	.0577	
ISI	.0070	.0066	.0065	.0066	.0068	
namp	1.6292	1.6143	1.6063	1.6058	1.6096	
optimum DFE:						
mse	.0403	.0402	.0401	.0401	.0402	equalizer length = 35
ISI	.0024	.0023	.0023	.0023	.0023	SNR = 15 db
namp	1.2002	1.1968	1.1952	1.1952	1.1962	$\sigma^2 = .0316$

Optimum Equalizer - B3 Telephone Line

FIGURE 10.7-5

	−2	−1	0	+1	+2	*timing phase*
pure transversal:						
mse	.0985	.1076	.1234	.1268	.1135	
ISI	.0196	.0294	.0482	.0485	.0284	
namp	2.495	2.472	2.378	2.475	2.69	
One sided trans + dfe:						
mse	.0827	.0909	.1031	.1160	.1275	
ISI	.0112	.0153	.0203	.0232	.0245	
namp	2.26	2.39	2.62	2.935	3.25	
constrained trans + dfe:						
mse	.0749	.0772	.0757	.0726	.0697	equalizer length = 21
ISI	.0078	.0080	.0077	.0075	.0073	SNR = 15 db
namp	2.1223	2.188	2.152	2.059	1.97	$\sigma^2 = .0316$

BPCLASS   Baud Rate Equalization - B3 Telephone Line

FIGURE 10.7-6

A comparison of the performance of the optimum linear equalizer and the optimum DFE is shown in fig. 10.7-5. Recall that the optimum linear equalizer, which is equivalent to the fractionally spaced equalizer, consists of a matched filter followed by a baud-rate equalizer that operates on the entire ISI. The optimum DFE follows the matched filter with a baud-rate equalizer that operates only on the leading echoes and cancels the trailing echoes using a DFE structure. The performance of both structures is essentially independent of timing phase, but the DFE is, as expected, clearly superior.

Finally, in fig. 10.7-6, we compare the baud-rate transversal to the suboptimal DFE structures. These suboptimal structures have better performance than the pure transversal and are less dependent on timing phase.

The results that have been presented are consistent with those developed by Salazar [Sal74] and Salz [Sal77]. The use of decision feedback in QAM systems was developed by Falconer and Foschini [F&F73] and Falconer [Fal76]. Lin [Lin91] has considered a DFE having a fractionally spaced forward filter. Decision feedback has found favor in mobile radio receivers (Leclert and Vandamme [L&V85] and Chen et al. [CMMG95]) because of its ability to equalize deep nulls without noise amplification.

10.8   ERROR PROPAGATION IN DECISION-FEEDBACK EQUALIZERS

Since decision-feedback equalizers work by using past decisions to cancel postcursor ISI terms, an incorrect decision (i.e. − a slicer error) can make it easier for subsequent errors to occur by feeding back wrong data. In order to illustrate this phenomenon, we shall follow the work of Beaulieu [Bea94]. As our model, we will use fig. 10.8-1a the channel $H(z)$ incorporates the actual channel followed by the whitened matched filter discussed above. Consequently the impulse response seen by the feedback section is entirely causal, consisting only of postcursor ISI terms. We will also assume that $H(z)$ is finite in duration, consisting of K terms so that

$$H(z) = \sum_{i=0}^{K} h_i z^{-i} \qquad (10.8\text{-}1)$$

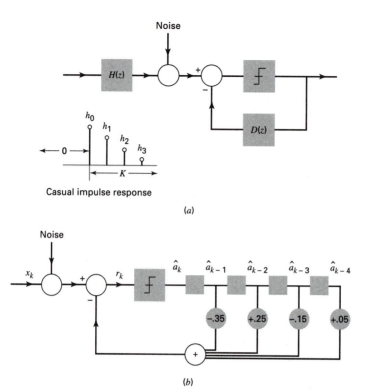

FIGURE 10.8-1

where, for the sake of simplicity, we will allow $h_0 = 1$. It is assumed that the noise is white.

By way of example, let us represent $H(z)$ as the finite vector

$$H = [1, -0.35, 0.25, -0.15, 0.05]$$

so that the appropriate DFE structure is shown in fig. 10.8-1b with the values of the feedback taps equal to the appropriate postcursor.

In our example, let the input data be binary, $a_k = \pm 1$ (life gets more complicated when the input data is multilevel) and let the current channel output correspond to an input datum having value $a_k = +1$ with the previous data having values $a_{k-1} = +1$, $a_{k-2} = -1$, $a_{k-3} = +1$, $a_{k-4} = -1$:

a_k	a_{k-1}	a_{k-2}	a_{k-3}	a_{k-4}
1	1	−1	1	−1

The resulting channel output x_k would then involve all of the ISI terms subtracting from the main sample so that

$$x_k = 1 - 0.35 - 0.25 - 0.15 - 0.05 = 0.20.$$

If all the previous slicer decisions had been correct

r_k	$\hat{a}_{k-1}$	$\hat{a}_{k-2}$	$\hat{a}_{k-3}$	$\hat{a}_{k-4}$
	1	−1	1	−1

then all of the ISI terms, which are embodied in the DFE taps, would be added back to the distorted output so that

$$r_k = x_k + 0.35 + 0.25 + 0.15 + 0.05 + \text{noise} = 1 + \text{noise}$$

The slicer would then simply be making a normal decision against a normal threshold. On the other hand, if the previous decision had been incorrect but all the others correct,

r_k	$\hat{a}_{k-1}$	$\hat{a}_{k-2}$	$\hat{a}_{k-3}$	$\hat{a}_{k-4}$
	-1	-1	1	-1

the input to the slicer would be

$$r_k = x_k - 0.35 + 0.25 + 0.15 + 0.05 + \text{noise} = 0.30 + \text{noise}$$

Under these circumstances, the probability of making a decision error would be substantial because the threshold against the noise had been lowered from the expected value of $r_k = 1$ to the distorted value of $r_k = 0.30$.

If an error were made in this second decision, the data pattern stored in the feedback filter would become

r_k	$\hat{a}_{k-1}$	$\hat{a}_{k-2}$	$\hat{a}_{k-3}$	$\hat{a}_{k-4}$
	-1	-1	-1	1

If the next input datum is $+1$, then the recent transmitted data pattern would be

a_k	a_{k-1}	a_{k-2}	a_{k-3}	a_{k-4}
1	1	1	-1	1

The channel output would be

$$x_k = 1 - 0.35 + 0.25 + 0.15 + 0.05 = 1.10$$

and the subsequent value of the slicer input would be

$$r_k = x_k - 0.35 + 0.25 - 0.15 - 0.05 + \text{noise} = 0.80 + \text{noise}$$

Here we see that the value of the signal portion of the slicer input has again been reduced, making the likelihood of a subsequent error high.

In sum, a decision error has a tendency to propagate. Although this is a problem, it should not be overstated because the issue of error propogation is very dependent on the data pattern. To see this, let's back up and examine what would have happened if the previous piece of data had been -1. We would have $x_k = -0.90$ and $r_k = -1.20$, so that the probability of an error on that piece of data would have decreased. So we can also say that even though there is a tendency for a decision error to propagate, there is also a tendency for the DFE to clear itself of errors.

An analysis of this phenomenon is quite complex and the best that has been done is to establish some upper bounds on the average number of transmissions required to clear the equalizer of errors once an error has been made.

We shall state Beaulieu's major result for binary transmission. Readers interested in more general results may consult the reference. The channel impulse response is rewritten as

$$H(z) = \sum_{i=0}^{K-r} h_i z^{-i} + \sum_{i=K-r+1}^{K} h_i z^{-i} \qquad (10.8\text{-}2a)$$

where r is the largest value such that

$$2 \sum_{i=K-r+1}^{K} |h_i| < h_0. \tag{10.8-2b}$$

The average value of the recovery time (in symbols) from a decision error in a channel with a relatively large SNR (greater than 10 db) is upper-bounded by

$$T_{K,r}^U = \frac{4}{3}(2^{K-r} - 1) + 1 \tag{10.8-2c}$$

so if $K = 10$ and $r = 5$, the upper bound is 41 symbols. In more complex systems, with multilevel signalling, the recovery time can be substantially greater. The interested reader may also consult [R&B94].

10.9 TOMLINSON-HARASHIMA PRECODING

If we consider that the purpose of an equalizer is to correct for a distorted channel, there is no conceptual problem in placing the equalizer in either the receiver or the transmitter. Tomlinson [Tom71] and Harashima and Miyakawa [H&M72] separately and more or less simultaneously devised the same method of designing a simple, stable digital equalizer to be placed in the transmitter of a data modem. The theory has been further developed by Messerschmitt [Mes75] and Mazo and Salz [M&S75]. We shall refer to this approach as THP (Tomlinson–Harashima Precoding).

Let's start with the system of fig. 10.9-1a, which shows a baseband equivalent channel consisting of a main sample and postcursor ISI

$$H(z) = 1 + h_1 z^{-1} + h_2 z^{-2} + \cdots = 1 + \sum_{n=1}^{N} h_n z^{-n} \tag{10.9-1}$$

followed by a decision feedback equalizer that cancels that ISI. If the equalizer is to be moved to the transmitter it must satisfy

$$P(z) = 1/H(z) \tag{10.9-2}$$

which may be realized as a transversal filter having infinite length or as an IIR filter, as shown in fig. 10.9-1b. The problem with the IIR approach is that the zeros of $H(z)$ may very well result in poles of $P(z)$ that result in instability. For example,

$$H(z) = \left(1 - \frac{10}{9}z^{-1}\right)\left(1 + \frac{3}{9}z^{-1}\right) = \left(1 - \frac{7}{9}z^{-1} + \frac{30}{81}z^{-2}\right) \tag{10.9-3a}$$

which is a perfectly reasonable channel impulse response. Yet its inverse

$$P(z) = \frac{z^2}{\left(z - \frac{10}{9}\right)\left(z + \frac{3}{9}\right)} \tag{10.9-3b}$$

is clearly unstable.

The solution to the problem of instabilty is the key to the precoding idea. Let the data be $a_k = \pm 1, \pm 3, \ldots, \pm(M-1)$ where M is an even number; $M = 2$ corresponds to binary transmission, $M = 4$ corresponds to four-level transmission, etc. Now follow the adder in

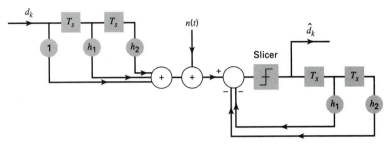

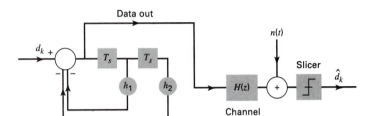

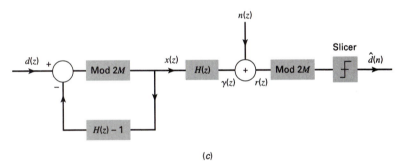

FIGURE 10.9-1

the transmitter filter of fig. 10.9-1b with a modulo-$2M$ device and follow the output with a modulo-$2M$ device. This new system is shown in fig. 10.9-1c.

The modulo-$2M$ operation operates on a number x to produce a number $x_M \in (-M, M]$ such that

$$x = x_M + b(2M) \qquad (10.9\text{-}4)$$

where k is a positive or negative integer. For example, let $M = 2$ and $x = 2.5$ so that $b = -1$ and $x_M = -1.5$. As another example, let $M = 2$ and $x = -6.5$ so that $k = b$ and $x_M = 1.5$. Note that for any M and x, b is a unique integer. Clearly this mod-$2M$ operation limits the transmitter filter output and prevents instability.

The operation of this system can be described in terms of z-transforms as follows:

$$X(z) = \{d(z) - X(z)[F(z) - 1]\} + 2Mb(z)$$

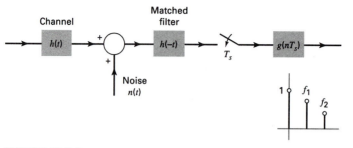

FIGURE 10.9-2

or

$$X(z)F(z) = a(z) + 2Mb(z) = Y(z) \tag{10.9-5a}$$

where $Y(z)$ is the channel output before the noise is added and $b(z)$ is simply the sequence of integers b that generate the modulo operation.

The received sequence $r(z)$ is $Y(z)$ corrupted by noise $n(z)$ so that

$$r(z) = X(z)F(z) + n(z) = a(z) + 2Mb(z) + n(z) \tag{10.9-5b}$$

Now let us note that the data sequence a_k lies in the range $(-M, M]$, as does the transmitted sequence x_k. The received sequence clearly can have magnitude greater than M but if we subject $r(z)$ to a second modulo-$2M$ operation we get

$$r(z)_M = a(z) + n(z) \xrightarrow{\text{slicer}} \hat{a}(z) \tag{10.9-5c}$$

as an estimate of $a(z)$.

The fact that the second modulo-$2M$ operation restores the data to its original level may not be intuitively obvious. It will therefore be demonstrated in the Problems by a writing a simple program.

Based on our previous discussions, the system in which this precoding would be useful consists of a channel $h(t)$ and a matched filter $h(-t)$, followed by a symbol-rate sampler $g(nT_s)$ that equalizes the leading echoes of the matched filter output (fig. 10.9-2). This is our definition of the *whitened matched filter*. The remaining ISI terms, the trailing echoes or postcursors, can be cancelled using decision feedback equalization or the precoding described above. Since the receiver front end is a whitened matched filter operating on white Gaussian noise, the noise sequence $n(z)$ that accompanies the signal to the slicer will also be white.

This technique can easily be expanded to include complex data transmitted on complex channels. The modulo-$2M$ operation is simply expanded to operate independently on the real and imaginary parts of the data.

There is clearly a relationship between this precoding and partial reponse signaling. We pointed out in Chapter 3 that there is a choice in partial response signaling between precoding the data according to the mirror image of the partial response polynomial, mod 2, or cancelling the controlled ISI using decision feedback in the receiver. Clearly, THP is a generalization of this approach. Consequently it is of considerable interest to examine how two important characteristics of partial response transmission extend to THP.

We have seen that partial response coding causes binary data to arrive at the receiver with multiple integer levels. For example, a binary input to both Class 1 and Class 4 results in receiver levels of 0, ±2. A bit of trial calculation together with the computer simulation mentioned above leads to the conclusion that with non-integer ISI, a random input data stream $d(z)$ results in $x(z)$ (fig. 10.9-1c) being approximately uniformly distributed over the interval $(-M, M]$. Consequently there is no discernible pattern in the transmitter output. The channel $H(z)$ causes the signal points to move beyond the range $(-M, M]$. The second modulo-$2M$ operation brings the data back into line for the slicer. The confinement of $x(z)$ to $(-M, M]$ also addresses the second issue, how the use of THP modifies the power spectrum of the transmitted signal. Recall that partial response precoding was demonstrated not to affect this power spectral density. Mazo and Salz [M&S76] and Forney and Eyuboglu [F&E91] make the point that the uniform distribution of $x(z)$ means that the transmitted sequence is the sum of independent, identically distributed random variables that tend toward being in distributed in a Gaussian manner and have a flat power spectral density. The average power in this uniform distribution is $M^2/3$. If the signal were transmitted by one-dimensional PAM where the data is $a_k = \pm 1, \pm 3, \ldots, \pm(M-1)$ then the power would be $(M-1)^2/3$. Consequently the ratio of the transmitter power required for THP vs. that for standard PAM transmission is $(M^2/M - 1)$. This is considered to be a minimal effect on the transmitter power for large M.

THP can be used only when the channel characteristics are known to the transmitter. This turns out to be the case in many high-speed modems since part of the connect sequence involves 'sounding the channel'. As a result of the sounding the transmitter may make the computations for establishing the precoder and the receiver may make the computations for approximating the whitened matched filter, although the latter may also be established through adaptive equalization, to be discussed in Chapter 11.

The great advantage of THP is realized when it is used in conjunction with trellis coding. We have seen that the appropriate receiver structure for TCM (trellis coded modulation) on an ideal channel is MLSE (maximum likelihood sequence estimation) using the Viterbi Algorithm. Since it has been shown to be superior to linear equalization on a distorted channel, it is desirable to be able to use decision feedback equalization in conjunction with TCM. The difficulty is that decision feedback operates on current decisions on the data whereas the Viterbi Algorithm delays those decisions by the number of symbols in the buffer. Although there have been efforts to reconcile this contradiction through the use of complicated equalizer structures ([D-H&H89], [C&E89], [E&Q89]), the use of THP to place the decision feedback into the transmitter is the most elegant solution.

Advances in precoding and its use for DSL modems are described by Fisher and Huber [F&H97].

Precoded implementation of single carrier decision-feedback QAM and multitone have emerged as competing technologies for transmission on the ADSL. Reference has already been made to articles by Bingham [Bin90], Kalet [Kal89], and Cioffi [Cio97] on multicarrier. The argument for the single-carrier implementation has been advanced by Kalet [Kal87] Zervos and Kalet [Z&K89] and Zervos [Zer90]. Saltzberg [Sal98] has written a very accessible summary of this debate.

10.10 OPTIMUM SIGNAL DETECTION USING MLSE

We have seen that both the optimum linear receiver and the optimum decision feedback receiver require a matched filter followed by a baud-rate equalizer in order to minimize the mean-square error caused by the combination of ISI and noise. In the case of the optimum linear receiver, the baud-rate equalizer is a *two-sided* transversal equalizer that operates on both the anti-causal precursor ISI terms and the causal postcursor ISI terms. In the case of the optimum decision feedback equalizer, the anti-causal terms are equalized by a one-sided transversal filter and the causal terms are cancelled by the nonlinear decision feedback structure. These two sets of ISI terms are obtained through spectral factorization of the autocorrelation function of the channel plus noise, which is the same as the noisy output of the matched filter.

Both the theoretical development and the experimental results have demonstrated that the optimum decision feedback receiver will perform better in the presence of amplitude distortion that will the optimum linear receiver. Purely phase distortion will be corrected by the matched filter with no need for further equalization.

The development of the optimum decision feedback equalizer gave rise to the idea of the *whitened matched filter*, which consists of the receiver matched filter followed by the one-sided baud-rate transversal that equalizes the anti-causal portion of the ISI. It was demonstrated that the output noise of the whitened matched filter will have the same power spectral density as the input noise. It follows from the previous discussion that the impulse response of the channel plus noise plus whitened matched filter, as shown in fig. 10.9-2, is entirely causal.

The basic argument of this section is that the MLSE (maximum likelihood sequence estimation), which was discussed in Chapter 5, is a more effective way of 'equalizing' the postcursors than decision feedback. In order to grasp this issue, consider that the effect of channel distortion is to disperse the energy of the original data impulse (a distorted channel is often called a dispersive channel) and that the purpose of equalization is to gather that dispersed energy back into an impulse. An isolated data pulse in a matched filter receiver accomplishes that purpose. (The reason for considering an isolated data pulse is so that we need not worry about ISI.) The matched filter bound is optimum. Recall that for binary transmission, the matched filter bound is

$$P_e = Q(\sqrt{\rho_M}) \qquad \text{where} \qquad \rho_M = \frac{2E_s}{\eta} = \frac{d_{\min}^2}{2\sigma^2} \qquad (10.10\text{-}1)$$

The problem with decision feedback is that by cancelling the postcursors rather than using their energy in order to help reconstruct the signal, that approach undermines the goal of attaining the matched filter bound in the presence of continuous data.

We have previously demonstrated that a distorted Nyquist channel having ISI and a partial response channel can be represented as finite-state machines representing a Markov process—that is, as channels having memory—and we have seen that it is possible, using MLSE, to reconstruct the transmitted data from the noisy, distorted received signal provided that the channel model is known.

MLSE, implemented with the Viterbi Algorithm, operates on the postcursors nonlinearly in such a way that the matched filter bound comes close to being attained. In effect it does this by reconstructing the original signal from the dispersed postcursors. The classical references on this topic are both by Forney [For72], [For73], who has demonstrated that using this approach we can attain a probability of error that is exponentially equal to that of single pulse matched filter transmission:

$$K_0 Q(\sqrt{\rho_M}) \le P_e \le K_1 Q(\sqrt{\rho_M}) \qquad (10.10\text{-}2)$$

Falconer and Magee [F&M76] considered the two alternatives for voiceband data transmission. They concluded that although the Viterbi algorithm performs better in relation to linear distortion, decision feedback was more effective in the presence of carrier phase jitter.

The issue of suboptimality arises in two different circumstances: 1) when the Viterbi Algorithm is asked to deal with an impulse response having a large number of terms; and 2) when, as in most realistic situations, the receiver front end is not a whitened matched filter. In both instances the problem is how to design a prefilter to the VA that produces an impulse response having a length that is computationally manageable and, simultaneously, does not contribute significantly to amplifying the noise.

One variation of this problem is the use of the Viterbi Algorithm to simultaneously deal with ISI and the decoding of trellis-coded modulation. In this case the idea is to generate a shortened impulse response (SIR) which, together with the trellis code, has a manageable number of states. This issue was initially addressed by Falconer and Magee [F&M73] and is discussed in detail in [C&E89], [E&Q88], [G&L-N96], [Bin88], art 9.3.1, [A-D&C96] and the interested reader is directed to these references.

Not surprisingly, a particular use of MLSE detection is in digital magnetic recording. We have already seen that advanced recording techniques use partial response signaling to achieve higher bit density and that partial response can be represented by a state machine. This makes MLSE an alternative to precoding. The particular advantage of using MLSE with partial response lies in the fact, established in Chapter 3, that simple binary transmission performs 2.1 db better than Class 1 and Class 4 partial response. This is a consequence of the signal energy devoted to establishing the controlled ISI. The effect of MLSE is effectively to restore the dispersed energy of the signal to a single impulse. Some useful references on this issue are [Kob71-1], [Kob71-2], [W&P86], [M&C90].

CHAPTER 10 PROBLEMS

10.1 This problem is an exercise in using the MATLAB script EQUAL.M in order to develop an understanding of some basic properties of baseband transversal equalizers.

Using the channel model base $= [.02, -.15, 1.0, .21, .03]$, find the baseband transversal equalizer under the following conditions:

a. For a very low-noise system (SNR $= 100$ db), plot a graph of the residual MSE due to ISI for transversal equalizers having a number of taps: $g = 3, 5, 7, 9, 11, \ldots$ until increasing 'g' no longer decreases the residual MSE.

b. Repeat (a) for SNR $= 30$ db, 20 db, 15 db.

c. For an optimum equalizer length, as defined in (a), find the noise amplification in db for SNR $= 30$ db, 20 db, 15 db.

d. Repeat (a) for the channel model base $= [.02, .15, 1.0, .21, .03]$ and compare the performance of equalizers on the two channels. Note that the difference between the two channels is a change of the sign of one of the terms. This indicates no change in the amplitude characteristic but a lot of change in the phase characteristic.

e. For SNR $= 100$ db and an optimum equalizer tap length as defined above, compare the frequency responses of the two channels, the two equalizers, and the two equalized channels. You may use the FFT function of MATLAB to easily make these computations.

10.2 Consider the four impulse responses

$$H_1(z) = (z - 1/2)\left(z^{-1} - \frac{5}{4}\right)$$

$$H_2(z) = (z - 1/2)\left(z^{-1} - \frac{4}{5}\right)$$

$$H_3(z) = (z + 1/2)\left(z^{-1} - \frac{5}{4}\right)$$

$$H_4(z) = (z + 1/2)\left(z^{-1} - \frac{4}{5}\right)$$

Using the MATLAB script EQUAL.M, compare the performance of a baud-rate transversal equalizer with various forms of decision feedback equalizer for each of these channels.

Compute the frequency response of each of the channels and make some connections between equalizer performance and frequency response.

10.3   This problem will use the complex impulse response corresponding to a C3 telephone line as an example. The files for these channel models at various timing phases, both for baud-rate and fractionally spaced $T/2$ transversal equalizers, are attached to the MATLAB files shown in Appendices B and C and available on the website.

   a.   Through experimentation, find the number of equalizer taps N appropriate for each of the systems when the SNR = 30 db and the timing phase is centered at 0. Present your results as a graph of residual MSE vs. number of taps for the two systems. Comment on the results and, in particular, indicate why you chose the number of taps that you did.

   b.   For the number of taps chosen in (a), find the residual MSE for each of the timing phases, 0, ± 1, ± 2, for the two systems. Plot the results together. Comment on the sensitivity of the two systems to timing phase.

10.4   Using the FFT function of MATLAB, find the frequency response of the sampled channel, the equalizer, and the equalized channel for the cases of Problem 10-3a. Present your results in a manner that indicates the differences between baud-rate and $T/2$ equalization. Comment on these differences. Can you identify dips or nulls in the sampled spectrum in the baud-rate system?

10.5   We want to investigate the eigenvalue spread $\chi(R)$ in the two systems of Problem 10-2 as a function of equalizer length and of timing phase. Compute and present the values of $\chi(R)$ for the following conditions:

$$\text{equalizer length} = N(\text{from Problem 10-2})$$
$$\text{SNR} = 20 \text{ db}$$
$$\text{Timing phase} = 0, \pm 1$$

Make some observations about the variation of $\chi(R)$. Discuss the variations that you have computed with what is anticipated by the theory.

10.6   For the C3 channel, compare the performance of the optimum DFE and the one-sided transversal plus DFE. Choose an appropriate number of taps and SNR. Consider the performance at different timing phases.

10.7   This problem uses the three channel models generated in Problem 7-2. These channels, simulating Rayleigh fading, have 20 db gain dips near the lower cutoff frequency, the center frequency, and the upper cutoff frequency of the band.

   For each of these channels, compare the performance of a baud-rate transversal, a fractionally spaced transversal, and a one-sided baud-rate transversal plus decision feedback.

10.8   A data communication system has a channel plus whitened matched filter response $F(z) = 1 - 0.3z^{-1} + 0.2z^{-2}$. Design a Tomlinson precoder to equalize the ISI in the transmitter. Choose a random sequence of binary data symbols $a_k = \pm 1$ of appropriate length and apply that sequence to a combination of precoder, channel, and mod(2) in the receiver. Trace the system data and find the output sequence. Repeat using the four-level alphabet $a_k = \pm 1, \pm 3$.

REFERENCES

[A-D&C96]   N. Al-Dhahir and J. Cioffi, "Efficiently computed reduced-parameter input-aided MMSE equalizers for ML detection: a unified approach," *IEEE Trans. Information Theory*, **IT-42**, No. 3, May 1996, pp. 903–915.

[Aus67]   M. Austin, "Decision-feedback equalization for digital communication over dispersive channels," M.I.T. Res. Lab. Electron., Tech. Rep. 461, Aug. 1967.

[B&P79]   C. Belfiore and J. Park, "Decision feedback equalization," *Proc. IEEE*, **67**, No. 8, Aug. 1979, pp. 1143–1156.

[Bea94]   N.C. Beaulieu, "Bounds on recovery times of decision feedback equalizers," *IEEE Trans. Communications*, **Com-42**, Oct. 1994, pp. 2786–2794.

[Bin88]   J.A.C. Bingham, "The theory and practice of modem design," Wiley Interscience, 1988.

[Bin90]   J.A.C. Bingham, "Multicarrier modulation for data transmission: an idea whose time has come," *IEEE communications Magazine*, May 1990.

[C&E89]   P. Chevillat and E. Eleftheriou, "Decoding of trellis-encoded signals in the presence of intersymbol interference and noise," *IEEE Trans. Communications*, **Com-37**, No. 7, July 1989, pp. 669–676.

[CDEF95] J. Cioffi, G. Dudevoir, M.V. Eyuboglu, and G.D. Forney, "MMSE decision feedback equalizers and coding: Part I: equalization results, Part II: coding results," *IEEE Trans. Communication*, **Com-43**, No. 10, Oct. 1995, pp. 2582–2604.

[Cio97]   J. Cioffi, "Asymmetric digital subscriber lines," in J.D. Gibson ed., *The Communications Handbooks*, CRC Press, 1997, pp. 450–479.

[CMMG95] S. Chen et al., "Adaptive Bayesian decision feedback equalizer for dispersive mobile radio channels," *IEEE Trans. Communications*, **Com-43**, No. 5, May 1995, pp. 1937–1945.

[Cla85]   A.P. Clark, *Equalizers for Digital Modems*, Halsted Press, 1985.

[D-H&H89] A. Duel-Hallen and C. Heegard, "Delayed decision-feedback equalization," *IEEE Trans. Communications*, **Com-37**, No. 5, May 1989, pp. 428–436.

[E&Q88]   M.V. Eyuboglu and S.U. Qureshi, "Reduced-state sequence estimation with set partitioning and decision feedback," *IEEE Trans. Communications*, **Com-36**, No. 1, Jan. 1988, pp. 13–20.

[E&Q89]   M.V. Eyuboglu and S.U. Qureshi, "Reduced-state sequence estimation for coded modulation on intersymbol interference channels," *IEEE J. Selected Areas of Communications*, **SAC-7**, Aug. 1989, pp. 989–995.

[F&E91]   G.D. Forney Jr. and M. Vedat Eyuboglu, "Combined equalization and coding using precoding," *IEEE Communications Magazine*, **30**, No. 12, Dec. 1991, pp. 25–35.

[F&F73]   D. Falconer and G. Fosschini, "Theory of minimum mean square error QAM systems employing decision feedback equalization," *Bell System Technical J.*, **52**, Dec. 1973, pp. 1821–1849.

[F&H97]   R. Fisher and J. Huber, "Comparison of precoding schemes for digital subscriber lines," *IEEE Trans. Communications*, **COM-45**, No. 3, March 1997, pp. 334–343.

[F&M76]   D. Falconer and F. Magee, "Evaluation of decision feedback equalization and viterbi algorithm detection for voiceband data transmission—Part I," *IEEE Trans. Communications*, **Com-24**, No. 10, Oct. 1976, pp. 1130–1139; Part II, Nov. 1976, pp. 1238–1245.

[F&M73]   D. Falconer and F. Magee, "Adaptive channel memory truncation for maximum-likelihood sequence estimation," *Bell System Technical J.*, **52**, Nov. 1973, pp. 1541–1562.

[Fal76]   D. Falconer, "Application of passband decision feedback equalization in two dimensional data communication systems," *IEEE Trans. Communications*, **Com-24**, No. 10, Oct. 1976, pp. 1159–1166.

[For72]   G. David Forney, "Maximum-likelihood sequence estimation of digital sequences in the presence of intersymbol interference," *IEEE Trans. Information Theory*, **IT-18**, May 1972, pp. 363–378.

[For73]   G. David Forney, "The viterbi algorithm," *Proceed. IEEE*, March 1973, pp. 268–278.

[G&L-N96] Y. Gu and T. Le-Ngoc, "Adaptive combined DFE/MLSE techniques for ISI channels," *IEEE Trans. Communications*, **COM-44**, No. 7, July 1996, pp. 847–857.

[G&W81] R. Gitlin and S. Weinstein, "Fractionally-spaced equalization: an improved digital transversal equalizer," *Bell System Technical J.*, Feb. 1981.

[GBS71] D. George, R. Bowen, and J. Storey, "An adaptive decision feedback equalizer," *IEEE Trans. Commun. Technol.*, **COM-19**, June 1971, pp. 281–292.

[GMW82] R. Gitlin, H. Meadors, and S. Weinstein, "The tap-leakage algorithm: an algorithm for the stable operation of a digitally implemented, fractionally spaced adaptive equalizer," *Bell System Technical J.*, Oct. 1982, pp. 1817–1831.

[H&M72] H. Harashima and H. Miyakawa, "Matched-transmission technique for channels with intersymbol interference," *IEEE Trans. Communications*, **Com-20**, No. 4, Aug. 1972, pp. 774–780.

[Hay91] Simon Haykin, *Adaptive Filter Theory*, 2nd ed., Prentice Hall, 1991.

[Kal87] I. Kalet, "Optimization of linearly equalized QAM," *IEEE Trans. Communications*, **Com-35**, No. 11, Nov. 1987, pp. 1234–1236.

[Kal89] I. Kalet, "The multitone channel," *IEEE Trans. Communications*, **Com-37**, No. 2, Feb. 1989, pp. 119–124.

[Kob71-1] H. Kobayashi, "Application of probabilistic decoding to digital magnetic recording systems," *IBM J. Research and Development*, Jan. 1971, pp. 64–74.

[Kob71-2] H. Kobayashi, "Correlative level coding and maximum-likelihood decoding," *IEEE Trans. Information Theory*, **IT-17**, No. 5, Sept. 1971, pp. 586–594.

[L&Q90] F. Ling and S. Qureshi, "Convergence and steady-state behavior of a phase-splitting fractionally spaced equalizer," *IEEE Trans. Communications*, **Com-38**, April 1990, pp. 418–425.

[L&V85] A. Leclert and P. Vandamme, "Decision feedback equalization of dispersive radio channels," *IEEE Trans. Communications*, **Com-33**, No. 7, July 1985, pp. 676–684.

[Lin91] D. Lin, "Minimum mean-squared error decision-feedback equalization for digital subscriber line transmission with possibly correlated line codes," *IEEE Trans. Communications*, **39**, No. 8, Aug. 1991, pp. 1197–1206.

[M&C90] J. Moon and L.R. Carley, "Performance comparison of detection methods in magnetic recording," *IEEE Trans. Magnetics*, **Mag-26**, No. 6, Nov. 1990, pp. 3155–3172.

[M&S75] J. Mazo and J. Salz, "On the transmitted power in generalized partial response," *IEEE Trans. Communications*, **24**, No. 3, March 1976, pp. 348–351.

[M&S76] J.E. Mazo and J. Salz, "On the transmitted power in generalized partial response," *IEEE Trans. Communications*, **Com-24**, No. 3, March 1976, pp. 348–352.

[M&W82] K.H. Mueller and J.J. Werner, "A hardware efficient passband equalizer structure for data transmission," *IEEE Trans. Communications*, **Com-30**, March 1982, pp. 538–541.

[Mes75] D. Messerschmitt, "Generalized partial response for equalized channels with rational spectra," *IEEE Trans. Communications*, **Com-23**, No. 11, Nov. 1975, pp. 1251–1258.

[Mon71] P. Monsen, "Feedback equalization for fading dispersive channels," *IEEE Trans. Information Theory*, **IT-17**, Jan. 1971, pp. 56–64.

[Pri71] R. Price, "Optimized PAM vs. capacity for low-noise filter channels," *Record IEEE Int. Conf. Comm.* (1971), Montreal, June 14–16, 1971, pp. 5–7–5–13.

[Pri72] R. Price , "Nonlinearly feedback-equalized PAM vs. capacity for noisy filter channels," *Record IEEE Int. Conf. Comm.* (1972), Philadelphia, PA, pp. 22–12 to 22–17.

[Q&F77] S. Qureshi and G.D Forney, "Performance and properties of a $t/2$ equalizer," NTC 77 pp. 11:1-1–11:1-7.

[R&B94] M. Russell and J. Bergmans, "A technique to reduce error propagation in Mary decision feedback equalization," *IEEE Trans. Communications*. **Com-42**, No. 10, Oct. 1994, pp. 2582–2604.

[Reg88]   Regalia et al, "The digital all-pass filter," *Proc. IEEE*, **76**, No. 1, Jan. 1988, pp. 19–39.

[Sal73]   J. Salz, "Optimum mean-square decision feedback equalization," *Bell System Technical J.*, **52**, No. 8, Oct. 1973, pp. 1341–1373.

[Sal74]   A.C. Salazar, "Design of transmitter and receiver filters for decision feedback equalization," *Bell System Technical J.*, **53**, No. 3, March 1974, pp. 503–523.

[Sal77]   J. Salz, "On mean-square decision feedback and timing phase," *IEEE Trans. Communications*, **Com-25**, No. 12, Dec. 1977, pp. 1471–1476.

[Sal98]   B. Saltzberg, "Comparison of single-carrier and multitone digital modulation for ADSL applications," *IEEE Communications Magazine*, **37**, No. 11, Dec. 1998, pp. 114–121.

[Tom71]   M. Tomlinson, "New automatic equalizer using modulo arithmetic," *Electronics Letters*, **7**, Nos. 5/6, March 25, 1971, pp. 138–139.

[Ung76]   G. Ungerboeck, "Fractionally tap-spacing equalizer and consequences for clock recovery in data modems," *IEEE Trans. Communications*, **Com-24**, Aug. 1976, pp. 856–864.

[W&P86]   R. Wood and D. Petersen, "Viterbi detection of Class IV partial response on a magnetic recording channel," *IEEE Trans. Communications*, **Com-34**, No. 5, May 1986, pp. 454–461.

[Z&K89]   N. Zervos and I. Kalet, "Optimized decision feedback equalization versus optimized orthogonal frequency division multiplexing for high speed data transmission over the local cable network," *Proc. Intl. Comm. Conf.*, 1989, pp. 10.80–10.85.

[Zer90]   N. Zervos, "High-speed carrierless passband transmission over the local cable network," *Globecom* 1990, pp. 1188–1195.

CHAPTER 11

ADAPTIVE EQUALIZATION AND ECHO CANCELLATION

11.0 INTRODUCTION

Until now we have looked at the perfomance of digital equalizers from an analytic point of view. The unstated, but underlying, assumption has been that the channel is known and we are trying to find an appropriate equalizer by computational means. This assumption is sometimes valid. One example is a channel connecting a disc to a CPU. Another is a private, fixed telephone connection between two particular points. Most of the time, however, the channel is unknown in advance. In an ordinary telephone system, for example, there is simply no way of knowing in advance just which connection is going to be made and, in fact, there is a virtually limitless number of connections. In a wireless link the channel is not only unknown but will also be changing significantly with time. For this reason the equalizer must be able to adjust itself to the network automatically, or 'learn' the network. One way to do this is to have the multiplying taps of the various equalizer structures follow an adjustment algorithm that leads the taps to the optimum values that we have previously computed. Another way is to begin the transmission by 'sounding' the channel in order to ascertain its amplitude and delay distortion and then use DSP algorithms to establish the equalizer coefficients.

We will first discuss the gradient algorithm, which, although not a practical solution in most circumstances, provides the analytic core for understanding the process of equalizer convergence. We will then move to the computationally simple LMS algorithm, the workhorse of equalizer convergence, extending the modem simulation program to include adaptive equalization. The LMS algorithm will be applied to both Nyquist and partial response systems.

Next we will examine the RLS (recursive least squares) or Kalman algorithm, which is computationally complex but converges very rapidly, and finally the technique of cyclic equalization, which allows rapid convergence of the equalizer at startup. Finally, we will present the Fast Kalman and lattice approaches to adaptive equalization. General references for these techniques, which provide an extensive set of references, are survey articles by Qureshi [Qur85] and Cioffi and Byun [C&B93].

As an extension of adaptive equalization, we will look at adaptive echo cancellation. As we shall see, echo cancellation is required for full-duplex operation of high-speed modems.

11.1 THE GRADIENT ALGORITHM FOR EQUALIZER CONVERGENCE

Our first look at the convergence of the equalizer will be from the point of view of the gradient algorithm. Examination of this algorithm will yield considerable insight into the dynamics of equalizer convergence, even though the algorithm itself turns out not to be of any practical value since it assumes that the channel is known in advance. The trick is to be able to adapt to the channel without knowing it in advance. This is accomplished by the LMS algorithm, a statistical sibling of the gradient algorithm, that will be discussed in the next section.

We saw in Chapter 10 that from an matrix point of view, the output vector of the equalizer can be computed through the use of the convolution matrix of the channel H_c

$$E = C^* H_c \tag{11.1-1a}$$

where the channel impulse response vector H has length g, the transversal equalizer has length m, the output vector has length $(m + g - 1)$, and the convolution matrix H_c has g rows and $(m + g - 1)$ columns. An example with $g = 3$, $m = 5$, and $(m + g - 1) = 7$ is shown in eq. 10.1-8a. The ideal output vector of a perfectly equalized system is

$$E_h = [0 + j0, \ldots, 0 + j0, 1 + j0, 0 + j0, \ldots, 0 + j0] \tag{11.1-1b}$$

where the length of the vector is $(m + g - 1)$ and the 1 appears in the $(m + g)/2$ or center position.

The difference between the actual output E and the ideal output E_h is the error or ISI vector and the magnitude-squared of this difference is called the mean-square error arising from intersymbol interference:

$$
\begin{aligned}
S_{\text{ISI}} = \|E - E_h\|^2 &= \|C^* H_c - E_h\|^2 = [C^* H_c - E_h][C^* H_c - E_h]^{*T} \\
&= C^* H_c H_c^{*T} C^T - C^{**} H_c E_h^T - E_h H_c^{*T} C^T + 1 \\
&= C^* R C^T - C^{**} H_c E_h^T - E_h H_c^{*T} C^T + 1
\end{aligned}
\tag{11.1-2a}
$$

In the example of eq. 10.1-8a, we can also see that the mean-square error due to ISI is also given by

$$
\begin{aligned}
S_{\text{ISI}} = (c_{-2}^* h_{-1})^2 &+ (c_{-2}^* h_0 + c_{-1}^* h_{-1})^2 + (c_{-2}^* h_1 + c_{-1}^* h_0 + c_0^* h_{-1})^2 \\
&+ [1 - (c_{-1}^* h_1 + c_0^* h_0 + c_1^* h_{-1})]^2 \\
&+ (c_0^* h_1 + c_1 h_0 + c_2 h_{-1})^2 + (c_1 h_1 + c_2 h_0)^2 + (c_2 h_1)^2
\end{aligned}
\tag{11.1-2b}
$$

The expressions of eqs. 11.1-2a and 11.1-2b for the MSE can be considered as arising from a single data pulse making its way through the combination of channel and equalizer. In the next section, in the context of the LMS algorithm, we will show that when the data stream is continuous the computed MSE is the same as that of eq. 11.1-2b.

The mean-square error can be seen from eq. 11.1-2a to be a quadratic form function of the equalizer tap vector C. As shown in fig. 11.1-1 (with a two-dimensional tap vector), this quadratic form is a smooth surface having no local minima and one global minimum. The

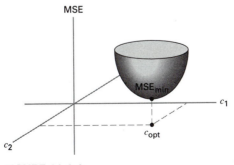

FIGURE 11.1-1

global minimum is the minimum mean-square error $S_{min} = (1 - \mathbf{h}^* R^{-1} \mathbf{h}^T)$ corresponding to the optimum tap vector $C_{opt} = E_h H_c^T [H_c^* H_c^T]^{-1}$.

Our object is to be able to start from any initial tap setting, corresponding to any point on the surface, and, following the gradient of the surface at each step, walk down the surface to the optimum tap setting corresponding to the minimum mean-square error. This means that the taps will be adjusted according to a *gradient algorithm* (or steepest-descent algorithm) so that the kth tap adjustment is made according to

$$C_k = C_{k-1} - \frac{1}{2}\mu\nabla[S_k] \tag{11.1-3}$$

In order to proceed with the algorithm, it is necessary to find the gradient of the mean square error (eq. 11.1-2a) at any equalizer tap setting.

THEOREM

The gradient vector of the complex quadratic form

$$S_{ISI} = C^* RC^T - C^* H_c E_h^T - E_h H_c^{*T} C^T + 1 \tag{11.1-4a}$$

with respect to the equalizer tap coefficient vector C is

$$\nabla(S_{ISI}) = 2[CH_c^* H_c^T - E_h H_c^T]$$
$$= 2[CH_c^* - E_h]H_c^T = 2[CR - \mathbf{h}] \tag{11.1-4b}$$

where $\mathbf{h} = E_h H_c^T$ and $R = H_c^* H_c^T$. ∎

Proof:

The proof will be by example. Consider a system having two complex channel taps and two complex equalizer taps. According to eq. 11.1-4a,

$$S_{ISI} = (c_1 - j\hat{c}_1 \quad c_2 - j\hat{c}_2)\begin{bmatrix} a_{11} & a_{12} \\ a_{12}^* & a_{11} \end{bmatrix}\begin{pmatrix} c_1 + j\hat{c}_1 \\ c_2 + j\hat{c}_2 \end{pmatrix}$$

$$- (c_1 - j\hat{c}_1 \quad c_2 - j\hat{c}_2)\begin{bmatrix} h_1 & h_2 & 0 \\ 0 & h_1 & h_2 \end{bmatrix}\begin{pmatrix} 0 \\ 1 \\ 0 \end{pmatrix} - (0 \quad 1 \quad 0)\begin{bmatrix} h_1 & 0 \\ h_2 & h_1 \\ 0 & h_2 \end{bmatrix}^*\begin{pmatrix} c_1 + j\hat{c}_1 \\ c_2 + j\hat{c}_2 \end{pmatrix} + 1$$

The gradient vector is composed of a real part plus an imaginary part

$$\nabla(S_{ISI}) = \nabla(S_{ISI})_{\begin{pmatrix} c_1 \\ c_2 \end{pmatrix}} + j\nabla(S_{ISI})_{\begin{pmatrix} \hat{c}_1 \\ \hat{c}_2 \end{pmatrix}}$$

1. $C^* RC^T$.

$$C^* RC^T = (c_1 - j\hat{c}_1 \quad c_2 - j\hat{c}_2)\begin{bmatrix} a_{11} & a_{12} \\ a_{12}^* & a_{11} \end{bmatrix}\begin{pmatrix} c_1 + j\hat{c}_1 \\ c_2 + j\hat{c}_2 \end{pmatrix}$$

$$= a_{11}(c_1^2 + \hat{c}_1^2) + a_{12}^*(c_2 - j\hat{c}_2)(c_1 + j\hat{c}_1) + a_{12}(c_2 + j\hat{c}_2)(c_1 - j\hat{c}_1)$$
$$+ a_{11}(c_2^2 + \hat{c}_2^2)$$

$$\frac{\partial(C^* RC^T)}{\partial c_1} = 2a_{11}c_1 + a_{12}^*(c_2 - j\hat{c}_2) + a_{12}(c_2 + j\hat{c}_2)$$

$$\frac{\partial(C^* RC^T)}{\partial \hat{c}_1} = 2a_{11}\hat{c}_1 + ja_{12}^*(c_2 - j\hat{c}_2) - ja_{12}(c_2 + j\hat{c}_2)$$

so that

$$\frac{\partial(C^* R C^T)}{\partial c_1} + j\frac{\partial(C^* R C^T)}{\partial \hat{c}_1} = 2a_{11}(c_1 + j\hat{c}_1) + 2a_{12}(c_2 + j\hat{c}_2)$$

Similarly,

$$\frac{\partial(C^* R C^T)}{\partial c_2} = 2a_{11}c_2 + a_{12}^*(c_1 + j\hat{c}_1) + a_{12}(c_1 - j\hat{c}_1)$$

$$\frac{\partial(C^* R C^T)}{\partial \hat{c}_2} = 2a_{11}\hat{c}_2 - ja_{12}^*(c_1 + j\hat{c}_1) + ja_{12}(c_1 - j\hat{c}_1)$$

and therefore

$$\frac{\partial(C^* R C^T)}{\partial c_2} + j\frac{\partial(C^* R C^T)}{\partial \hat{c}_2} = 2a_{11}(c_2 + j\hat{c}_2) + 2a_{12}^*(c_1 + j\hat{c}_1)$$

It follows that this portion of the gradient vector is

$$2\lfloor[a_{11}(c_1 + j\hat{c}_1) + a_{12}(c_2 + j\hat{c}_2)]\lfloor a_{11}(c_2 + j\hat{c}_2) + a_{12}^*(c_1 + j\hat{c}_1)\rfloor\rfloor$$
$$= 2C H_c^* H_c^T = 2CR$$

2. $C^* H_c E_h^T + E_h H_c^{*T} C^T$

$$(c_1 - j\hat{c}_1 \quad c_2 - j\hat{c}_2)\begin{bmatrix} h_1 & h_2 & 0 \\ 0 & h_1 & h_2 \end{bmatrix}\begin{pmatrix} 0 \\ 1 \\ 0 \end{pmatrix} + (0 \quad 1 \quad 0)\begin{bmatrix} h_1^* & 0 \\ h_2^* & h_1^* \\ 0 & h_2^* \end{bmatrix}\begin{pmatrix} c_1 + j\hat{c}_1 \\ c_2 + j\hat{c}_2 \end{pmatrix}$$

$$= h_2(c_1 - j\hat{c}_1) + h_1(c_2 - j\hat{c}_2) + h_2^*(c_1 + j\hat{c}_1)h_1^*(c_2 + j\hat{c}_2) = A$$

The gradient of this portion of the MSE is the row vector

$$\left[\left(\frac{\partial A}{\partial c_1} + j\frac{\partial A}{\partial \hat{c}_1}\right)\left(\frac{\partial A}{\partial c_2} + j\frac{\partial A}{\partial \hat{c}_2}\right)\right] = 2[h_2 \quad h_1] = 2\mathbf{h} = 2E_h H_c^T$$

end of proof

Note that the gradient is equal to zero when

$$C = E_h H_c^T \left[H_c^* H_c^T\right]^{-1} = C_{\text{opt}} \tag{11.1-5}$$

corresponding to the optimum equalizer taps.

Now we shall discuss the process of convergence, [Cha71], [Ung72] of the equalizer based on the gradient algorithm. Define the difference between the optimum tap vector C_{opt} and the actual tap vector C_k, at time $t = k$, as

$$B_k = C_k - C_{\text{opt}} = C_k - \mathbf{h}R^{-1}. \tag{11.1-6a}$$

Since

$$S_{\text{ISI}} = C^* R C^T - C^* H_c E_h^T - E_h H_c^{*T} C^T + 1 \tag{11.1-4a}$$

and $\mathbf{h} = E_h H_c^T$, it follows that corresponding to the tap vector C_k we have the mean-square error S_k:

$$S_k = C_k^* R C_k^T - C_k^* \mathbf{h}^T - \mathbf{h}^* C_k^T + 1$$
$$= (B_k + \mathbf{h}R^{-1})^* R(B_k + \mathbf{h}R^{-1})^T - (B_k + \mathbf{h}R^{-1})^* \mathbf{h}^T - \mathbf{h}^*(B_k + \mathbf{h}R^{-1})^T + 1$$
$$= B_k^* R B_k^T + (1 - \mathbf{h}^* R^{-1}\mathbf{h}^T) = B_k^* R B_k^T + S_{\text{min}} \tag{11.1-6b}$$

The mean-square error at the kth step is seen from eq. 11.1-4b to be

$$\nabla[S_k] = 2[C_{k-1}R - \mathbf{h}] \tag{11.1-6c}$$

It follows from eq. 11.1-3 that

$$C_k = C_{k-1} - \mu[C_{k-1}R - \mathbf{h}] \tag{11.1-6d}$$

or

$$C_k = C_{k-1}[I - \mu R] + \mu\mathbf{h} \tag{11.1-6e}$$

Substituting eq. 11.1-6a into eq. 11.1-6e, we get

$$[B_k + \mathbf{h}R^{-1}] = [B_{k-1} + \mathbf{h}R^{-1}][I - \mu R] + \mu\mathbf{h} \tag{11.1-6f}$$

or

$$B_k = B_{k-1}[I - \mu R] \tag{11.1-6g}$$

so that recursively, if B_0 is the initial equalizer tap error vector, then

$$B_k = B_0[I - \mu R]^k \tag{11.1-6h}$$

Substituting eq. 11.1-6h into eq. 11.1-6b, the mean-square error at the kth iteration is

$$S_k = B_k^* R B_k^T + S_{\min} = B_0[I - \mu R]^k R[[I - \mu R]^k]^{*T} B_0^{*T} + S_{\min} \tag{11.1-7a}$$

In order to be able to discuss the convergence properties of the equalizer, it will be necessary to diagonalize the channel correlation matrix R in order to find its eigenvalues.

Let us briefly review a few relevant issues in how R is characterized.

1. A matrix H is *Hermitian* if it equals its conjugate transpose, $H = H^{*T}$. An example of a Hermitian matrix is

$$H = \begin{bmatrix} 4 & 1+j_2 & 2+j & 1-j \\ 1-j_2 & 3 & 2-j & 1+j \\ 2-j & 2+j & 3 & 1+2j \\ 1+j & 1-j & 1-2j & 1 \end{bmatrix}$$

2. A matrix T is *Toeplitz* if the elements on each of the diagonals are equal, $t_{ij} = t_{j-i}$. An example of a Toeplitz matrix is

$$T = \begin{bmatrix} 4 & 2+j & 1+j_3 & 2-j_2 \\ 2-j_5 & 4 & 2+j & 1+j_3 \\ 2-j_3 & 2-j_5 & 4 & 2+j \\ 3+j_2 & 2-j_3 & 2-j_5 & 4 \end{bmatrix}$$

It is important to note that a Toeplitz matrix is not necessarily Hermitian and a Hermitian matrix is not necessarily Toeplitz.

3. Autocorrelation functions of complex random processes satisfy $R(\tau) = R^*(-\tau)$. In the discrete domain, where $R(z) = \sum_{n=-\infty}^{\infty} r_n z^{-n}$, then $r_n = r_{-n}^*$. This makes the correlation matrix R of a channel output, *sampled at the baud-rate T_s, both Hermitian*

and Toeplitz. An example of such a correlation matrix R is:

$$R = \begin{bmatrix} 4 & 2+j & 1+j_3 & 2-j_2 \\ 2-j & 4 & 2+j & 1+j_3 \\ 1-j_3 & 2-j & 4 & 2+j \\ 2+j_2 & 1-j_3 & 2-j & 4 \end{bmatrix}$$

We shall see in a subsequent section that the correlation matrix of a channel whose output is sampled faster than the baud rate, say every $T_s/2$ sec, is *Hermitian but not Toeplitz.*

For the moment, we will concentrate on those properties of the mean-square error that derive from the correlation matrix R being Hermitian.

A Hermitian matrix has positive and real eigenvalues $\{\lambda_i; i = 1, \ldots, M\}$, where M is the order of the matrix, and corresponding eigenvectors $\mathbf{u}_i$, which are assumed to be row vectors. According to well-known results on Hermitian matrices, the matrix R can be written as $R = QDQ^{*T}$, where D is the diagonal matrix of the eigenvalues, Q is a matrix having rows of corresponding eigenvectors, and Q has the property (for Hermitian matrices) that $Q^{-1} = Q^{*T}$. As a consequence, eq. 11.1-7a can be rewritten as

$$S_k = B_0 Q[I - \mu D]^k D[[I - \mu D]^k]^{*T} Q^{*T} B_0^{*T} + S_{\min} \tag{11.1-7b}$$

or

$$S_k = \sum_{i=1}^{M} \left(B_0 u_i^T \right)^2 \lambda_i [1 - \mu\lambda_i]^{2k} + S_{\min} \tag{11.1-7c}$$

In order for the mean-square error to approach its minimum as the number of iterations k increases, it is necessary for each one of the terms in the summation to go to zero. This happens if

$$|1 - \mu\lambda_i| < 1 \qquad \text{for each } \lambda_i \tag{11.1-7d}$$

corresponding to

$$0 < \mu < \frac{2}{\lambda_{\max}} \tag{11.1-7e}$$

where $\lambda_{\max}$ is the largest eigenvalue. The parameter μ is called the *step size.*

Gersho [Ger69] has noted that if

$$\mu = \frac{2}{\lambda_{\max} + \lambda_{\min}} \tag{11.1-8a}$$

then the terms in eq. 11.1-7c corresponding to the maximum and minimum eigenvalues will converge at the same rate:

$$[1 - \mu\lambda_{\max}]^{2k} = [1 - \mu\lambda_{\min}]^{2k} = \left[\frac{\rho - 1}{\rho + 1} \right]^{2k} \tag{11.1-8b}$$

where $\rho = \lambda_{\max}/\lambda_{\min}$. All the other terms in the summation will converge faster.

Gitlin et al. [GMT73] have more closely investigated the optimum rate of convergence of this equalizer. Recall that B_k is the difference between the actual equalizer tap vector and the optimum equalizer vector. From eq. 11.1-6h, X_{n+1} is defined as the norm of the tap error vector B_k:

$$\begin{aligned} X_{k+1} = B_{k+1} B_{k+1}^{*T} &= B_k[I - \mu R]\lfloor I - \mu R^{*T} \rfloor B_k^{*T} \\ &= B_k[I - 2\mu\mathrm{Re}(R) + \mu^2 R R^{*T}] B_k^{*T} \end{aligned} \tag{11.1-9a}$$

According to the bound

$$\lambda_{\min} B_k B_k^{*T} \leq B_k R B_k^{*T} \leq \lambda_{\max} B_k B_k^{*T} \tag{11.1-9b}$$

which is well established for a Hermitian matrix R, it follows that

$$X_{k+1} \leq \left(1 - 2\mu\lambda_{\min} + \mu^2\lambda_{\max}^2\right)X_k = \gamma X_k \tag{11.1-9c}$$

and convergence of the tap vector is guaranteed if $\gamma < 1$. This corresponds to

$$\mu < \frac{2}{\lambda_{\max}}\left(\frac{\lambda_{\min}}{\lambda_{\max}}\right). \tag{11.1-9d}$$

The smallest value of γ, for the most rapid convergence, is obtained by setting the derivative

$$\frac{d\gamma}{d\mu} = (-2\lambda_{\min} + 2\mu\lambda_{\max}^2) = 0 \Rightarrow \mu_{\text{opt}} = \frac{1}{\lambda_{\max}}\left(\frac{\lambda_{\min}}{\lambda_{\max}}\right) \tag{11.1-9e}$$

Note that this value of μ is half the bound of eq. 11.1-9d , which is, in turn, smaller than the bound of eq. 11.1-8e. For this optimum value

$$X_n = \gamma X_{n-1} = \left[1 - \left(\frac{\lambda_{\min}}{\lambda_{\max}}\right)^2\right]X_{n-1} = \gamma^n X_0 = \left[1 - \left(\frac{\lambda_{\min}}{\lambda_{\max}}\right)^2\right]^k X_0 \tag{11.1-10a}$$

The bound of eq. 11.1-10a may be easily reconciled with the Gersho bound of eq. 11.1-8a. From eq 11.1-10a,

$$\left[1 - \left(\frac{\lambda_{\min}}{\lambda_{\max}}\right)^2\right]^k = \left[1 - \frac{1}{\rho^2}\right]^2 = \left[\frac{(\rho+1)(\rho-1)}{\rho^2}\right]^k \geq \left[\frac{(\rho-1)}{(\rho+1)}\right]^k \tag{11.1-10b}$$

and the two bounds become equal for large ρ. Note that the larger the value of ρ, the slower the convergence. Since the Gitlin bound is inferior to the Gersho bound, it is not at all obvious in this context why it should be considered. The reason is that the methodology used in finding the Gitlin bound will be extended to find convergence bounds for the LMS algorithm.

Now let's go back to an example to see why a pure gradient algorithm presents difficulties. Recall that

$$\nabla(S_{\text{ISI}}) = 2[CH_c^* - E_h]H_c^T \tag{11.1-11a}$$

where $CH_c^* = E = [e_1 \ e_2 \ e_3 \ e_4 \ e_5 \ e_6 \ e_7]$ is the output vector and, using the example of eq. 10.1-8a,

$$\nabla(S_{\text{ISI}}) = 2[e_1 \ e_2 \ e_3 \ e_4 \ -1 \ e_5 \ e_6 \ e_7]\begin{bmatrix} h_1 & 0 & 0 & 0 & 0 \\ h_2 & h_1 & 0 & 0 & 0 \\ h_3 & h_2 & h_1 & 0 & 0 \\ 0 & h_3 & h_2 & h_1 & 0 \\ 0 & 0 & h_3 & h_2 & h_1 \\ 0 & 0 & 0 & h_3 & h_2 \\ 0 & 0 & 0 & 0 & h_3 \end{bmatrix} \tag{11.1-11b}$$

The gradient is a row vector, in this case having five elements, equal to the number of taps in the transversal equalizer, which tells us how to update each equalizer tap according to

$$C_{\text{new}} = C_{\text{old}} - \frac{1}{2}\nabla(S_{\text{ISI}}) \tag{11.1-11c}$$

Unless we know the channel in advance, computing the gradient in this way is not possible. In order to use this algorithm it is therefore necessary to have a way of obtaining or estimating the gradient without knowing the channel. This is the basis for turning to the LMS algorithm.

Finally, before moving on, we want to examine the distribution of eigenvalues and their relation to the power spectral density in a baud-rate sampled system. Recall from Section 8.7 that the DFT of the sampled autocorrelation function becomes the Fourier Transform of the continuous autocorrelation function and consequently the eigenvalues of the correlation matrix become the magnitude of $P(\omega)$. This may be expressed mathematically first with the finite-order eigenvalue relationship

$$R\mathbf{u} = \lambda\mathbf{u} \tag{11.1-12a}$$

of the Mth-order matrix correlation matrix R. As $M \to \infty$, the Fourier Transform of both sides yields

$$P(f)\mathbf{U}(f) = \lambda\mathbf{U}(f); \qquad |f| \le \frac{1}{2T_s} \tag{11.1-12b}$$

where $P(f)$ is the magnitude-squared of the folded spectrum of signal plus noise. The only way that eq. 11.1-12b can be satisfied is if $\mathbf{U}(f) = \delta(f - f_c)$, the Fourier Transform of a sinusoid, and λ_c is the magnitude of $P(f_c)$. This leads to the classic result, proved by Haykin [Hay91]:

$$\rho = \frac{\lambda_{\max}}{\lambda_{\min}} \le \frac{P_{\max}(f)}{P_{\min}(f)} \tag{11.1-12c}$$

which indicates that the more distorted the channel, the longer it will take to converge.

11.2 THE LMS ALGORITHM FOR BAUD-RATE EQUALIZER CONVERGENCE

The LMS algorithm [Wid70] is the key to computationally easy adaptive equalization. Instead of trying to compute the gradient, we instead compute an estimate of the gradient. When the data is continuous, the input data stream can be represented as

$$d(t) = \sum_n d_n\delta(t - nT_s) \tag{11.2-1a}$$

If the channel taps are denoted as h_i, $i = -M/2, \ldots, M/2$, then the channel output is

$$y(t) = \sum_n y_n\delta(t - nT_s) = \sum_n \sum_{i=-M/2}^{M/2} d_{n-i}h_i\delta(t - nT_s) \tag{11.2-1b}$$

and the output of a baud-rate transversal equalizer having taps c_j, $j = -N/2, \ldots, N/2$, is:

$$z(t) = \sum_n z_n\delta(t - nT_s) = \sum_n \sum_{j=-N/2}^{N/2} y_{n-j}c_j^*\delta(t - nT_s) \tag{11.2-1c}$$

where

$$z_n = \sum_{j=-N/2}^{N/2} y_{n-j}c_j^* = \sum_{i=-M/2}^{M/2} \sum_{j=-N/2}^{N/2} d_{n-j-i}h_i c_j^* \tag{11.2-1d}$$

and y_n is the data entering the equalizer.

At the output of the equalizer we have a slicer that makes a decision on each of the outputs, deciding that the output z_n is really $\hat{d}_n$, an estimate of the data. The slicer error is $e_n = z_n - \hat{d}_n$ and we shall define the mean-square error as

$$\text{MSE}_{\text{ISI}} = E[|z_n - \hat{d}_n|^2] = E[|e_n|^2] = E[e_n e_n^*] \tag{11.2-2}$$

In the example of eq. 10.1-8a where $g = 3$, $m = 5$, and $(m + g - 1) = 7$, we will imagine a sequence of seven data symbols

$$d_{n-3} \quad d_{n-2} \quad d_{n-1} \quad d_n \quad d_{n+1} \quad d_{n+2} \quad d_{n+3} \tag{11.2-3a}$$

applied to the system in succession. When the last one d_{n+3} is applied, the output is

$$
\begin{aligned}
z_n = {} & d_{n+3}[h_{-1}c_{-2}] + d_{n+2}[h_0c_{-2} + h_{-1}c_{-1}] + d_{n+1}[h_1c_{-2} + h_0c_{-1} + h_{-1}c_0] \\
& + d_n[h_{-1}c_1 + h_0c_0 + h_1c_{-1}] + d_{n-1}[h_{-1}c_2 + h_0c_1 + h_1c_0] \\
& + d_{n-2}[h_0c_2 + h_1c_1] + d_{n-3}[h_1c_2]
\end{aligned} \tag{11.2-3b}
$$

Let us assume that $d_n = 1$ and that there is no error in the slicer, so that the decided value is also $\hat{d}_n = 1$. Then the difference between the decided value and the output is:

$$
\begin{aligned}
e_n = {} & (z_n - \hat{d}_n) = (z_n - 1) \\
= {} & d_{n+3}(h_{-1}c_{-2}) + d_{n+2}(h_0c_{-2} + h_{-1}c_{-1}) + d_{n+1}(h_1c_{-2} + h_0c_{-1} + h_{-1}c_0) \\
& + [(h_{-1}c_1 + h_0c_0 + h_1c_{-1}) - 1] + d_{n-1}(h_{-1}c_2 + h_0c_1 + h_1c_0) \\
& + d_{n-2}(h_0c_2 + h_1c_1) + d_{n-3}(h_1c_2)
\end{aligned} \tag{11.2-3c}
$$

In performing the computation of eq. 11.2-2 we note that, since the successive data symbols are uncorrelated, the expected value of the cross-product terms that are obtained by squaring eq. 11.2-3c go to zero. What remains is the sum of the squares of the individual terms and since the data are assumed to be ± 1 we end up with the same expression as eq. 11.1-2b.

In order to get to the LMS algorithm, we take the gradient of eq. 11.2-2 with respect to the equalizer tap coefficients:

$$\nabla_c(\text{MSE}_{\text{ISI}}) = E[e_n^* \nabla_c(e_n)]. \tag{11.2-4a}$$

From eq. 11.2-1d we get

$$e_n = (z_n - \hat{d}_n) = \left[\sum_{j=-N/2}^{N/2} y_{n-j}c_j^* \right] - \hat{d}_n \tag{11.2-4b}$$

where y_{n-j} is the data at the jth stage of the equalizer. It follows that

$$\nabla_c(e_n) = \left[\frac{\partial e_n}{\partial c_{-N/2}} \quad \cdots \quad \frac{\partial e_n}{\partial c_{N/2}} \right] = \left[y_{-N/2} \quad \cdots \quad y_{N/2} \right] = Y \tag{11.2-4c}$$

where Y is the vector of data in the equalizer. It follows then that

$$\nabla_c(\text{MSE}_{\text{ISI}}) = E[e_n^* \nabla_c(e_n)] = E[e_n^* Y] \tag{11.2-4d}$$

so that the LMS algorithm tap update algorithm for a transversal equalizer is:

$$C_{n+1} = C_n - \mu E[e_n^* Y] \tag{11.2-4e}$$

The meaning of this algorithm is as follows. At each symbol interval the output of the equalizer is sent to the slicer, which both makes a decision and computes an error $e_n = (z_n - \hat{d}_n)$. In order to update a particular equalizer tap, we multiply the complex conjugate of the error with the data in that location, multiply by μ, and subtract the result from the current tap value. This is done separately for each one of the taps at each symbol interval.

It can be easily shown (see the Problems) that the LMS algorithm for updating the taps of a decision feedback equalizer is

$$F_{n+1} = F_n + \mu E[e_n^* D] \qquad (11.2\text{-}4f)$$

where $D = [d_{n-1} \ d_{n-2} \ \dots \ d_{n-L}]$ is the vector of past decisions that are the contents of the L stage DFE. It should be noted that in the equalizer configuration that includes both a transversal and decision feedback, the error that drives the two sections of the equalizer is the same. The values of μ for the transversal and the decision feedback need not be the same.

The code for implementing this algorithm is very simple. In a modem, which requires real-time processing, the computation must be completed in less than one symbol interval. The standard way of measuring the computation required in a DSP algorithm is by the number of multiply and accumulate operations (MAC) involved. Computing the output of an N-stage complex transversal filter requires N complex MACs or $4N$ real MACs. An additional $4N$ MACs are required in order to update the equalizer taps at each piece of data, making a total of $8N$ MACs. Some overhead, typically 10%–20%, will have to be added to this total to take into account the other operations required.

The convergence behavior of the LMS algorithm can be analyzed by methods that are similar to those used in relation to the gradient algorithm. The mathematics are quite involved and the interested reader is referred to Gitlin et al. [GMT73], [G&M73], [Maz79], Mazo [Maz80], Gitlin, Hayes, and Weinstein [GH&W92], and Gardner [Gar84]. We shall only summarize the results of their analysis, which, in many ways, reveals that the convergence properties of the LMS algorithm differ considerably from those of the gradient algorithm.

Define the average value of the eigenvalues of the channel autocorrelation matrix as

$$\bar{\lambda} = \frac{\sum_{i=1}^{M} \lambda_i}{M} = \frac{1}{M} \text{trace}(R) \qquad (11.2\text{-}5a)$$

and the ratio

$$\hat{\rho} = \lambda_{\max}/\bar{\lambda} \qquad (11.2\text{-}5b)$$

The upper bound on the value of μ to guarantee the convergence of the equalizer is

$$\mu < \mu_{\max} = \frac{1}{M} \frac{2}{\lambda_{\max}} = \frac{1}{M} \frac{2}{\hat{\rho}\bar{\lambda}} \qquad (11.2\text{-}5c)$$

which reduces by a factor of M, the equalizer length, the maximum value of μ for the gradient algorithm, eq. 11.1-9e.

The mean-square error in the system is given by

$$S_k = S_{\mu,k} + S_{\min} \qquad (11.2\text{-}6a)$$

where

$$S_{\min} = 1 - \mathbf{c}_{\text{opt}}^* H_c E_h^T \qquad (11.2\text{-}6b)$$

and where $S_{\mu,k}$, the excess mean-square error, follows the recursive relationship

$$S_{\mu,k+1} = [1 - 2\mu\bar{\lambda} + \mu^2 M \hat{\rho}\bar{\lambda}^2]S_{\mu,k} + \mu^2 M \hat{\rho}\bar{\lambda}^2 S_{\min} \qquad (11.2\text{-}6c)$$

It has been shown if we allow the value of μ to be changed at each iteration then, at each iteration $S_{\mu,k}$ is minimized if

$$\mu_k = \frac{S_{\mu,k}}{S_{\mu,k} + S_{\min}} \frac{1}{M\hat{\rho}\bar{\lambda}} \tag{11.2-6d}$$

Recognizing that initially $S_{\mu,k} \gg S_{\min}$, it follows that the initial step size should be

$$\mu_0 = \frac{1}{M\hat{\rho}\bar{\lambda}} \tag{11.2-6e}$$

half the value of $\mu_{\max}$ in eq. 11.2-5b. Under these circumstances, the steady-state excess mean-square error is shown to be

$$S_\mu = \frac{\mu M\hat{\rho}\bar{\lambda}}{2 - \mu M\hat{\rho}\bar{\lambda}} S_{\min} \xrightarrow{\mu = \mu_0} S_\mu = S_{\min} \tag{11.2-7a}$$

resulting in a 3 db increase in the mean square error over the minimum. In order to decrease that excess, it is necessary to decrease the value of μ. For the excess MSE to be reduced to a fraction γ of the minimum MSE, the final value of μ should be

$$\mu = \frac{2\gamma}{1 + \gamma} \mu_0 \Rightarrow S_\mu = \gamma S_{\min} \tag{11.2-7b}$$

The convergence process, which also implements the expectation operation that is fundamental to the LMS approach, is consequently very noisy. This will be examined through simulation in the Problems.

This additional component has a complicated relationship to the step size. According to Gitlin and Weinstein [G&W79], as long as the digital arithmetic implementing the equalizer is of sufficiently high precision (at least 14 bits), then decreasing the step size will decrease this additional MSE. Since contemporary modems for the telephone line are implemented with 16- or 32-bit DSP chips, there is no problem with exceeding this precision.

These considerations produce the convergence strategy of beginning the LMS algorithm with the largest possible step size in order to achieve as rapid convergence as possible and ending up with a step size that is small enough to minimize the residual MSE but large enough to track a time-varying channel. The issue of varying the step size during the process of convergence is addressed by Schonfeld and Schwartz [S&S71-1], [S&S71-2].

When it is not possible to achieve the required precision (in very high-speed modems for digital microwave, for example) because of limited DSP capability, it is necessary to take into account some problems arising from making the step size too small. The issue is that if the tap correction factor $\mu e_n y_j$ for tap j is less than half the least significant bit of the tap, then the convergence ceases. This may actually result in a 'converged' state that has a larger MSE than what is possible with the limited-precision arithmetic. The solution is to increase the value of the final step size μ.

According to Gitlin and Weinstein [G&W79], if the representation of numbers in the processing is limited to B bits lying between ± 1, then convergence will continue to take place as long as

$$\mu\sqrt{S_k \frac{\bar{\lambda}}{2}} \geq 2^{-B} \tag{11.2-8a}$$

Referring to eq. 11.2-18b, in order to achieve a given value of γ, the requirement on B

is that

$$2^B \geq M_\rho \frac{1+\gamma}{\gamma} \sqrt{\frac{\bar{\lambda}}{2S_k}}. \tag{11.2-8b}$$

In an extreme circumstance where DSP capabilities are so limited that no real-time multiplications are possible, the LMS tap update algorithm may be modified to be

$$c_{\text{new}} = c_{\text{old}} - \delta \cdot \text{sgn}(e_n y) \tag{11.2-9a}$$

for each equalizer tap. In this approach, δ is a small fixed step that may be changed at different stages in the process of convergence and is either added to or subtracted from the current (or old) tap according to the sign of the product of the data at that tap and the slicer error. The generation of the sgn function can be accomplished with the exclusive-OR operation. In a passband equalizer where the taps, the error, and the data are complex, a suitable extension of this algorithm might be

$$\delta \cdot \text{sgn}(e_n y) = \delta \cdot \text{sgn}[(e_1 + je_2)(y_1 + jy_2)]$$
$$= \delta \cdot \text{sgn}(e_1 y_1 - e_2 y_2) + j\delta \cdot \text{sgn}(e_1 y_2 + e_2 y_1) \tag{11.2-9b}$$

where, in order to avoid multiplications, we define

$$\delta \cdot \text{sgn}(e_1 y_1 - e_2 y_2) = \begin{cases} 0; & \text{if } \text{sgn}(e_1 y_1) = \text{sgn}(e_2 y_2) \\ +\delta; & \text{if } \text{sgn}(e_1 y_1) = +1 \text{ and } \text{sgn}(e_2 y_2) = -1 \\ -\delta; & \text{if } \text{sgn}(e_1 y_1) = -1 \text{ and } \text{sgn}(e_2 y_2) = +1 \end{cases} \tag{11.2-9c}$$

and

$$\delta \cdot \text{sgn}(e_1 y_2 + e_2 y_1) = \begin{cases} 0; & \text{if } \text{sgn}(e_1 y_2) = -\text{sgn}(e_2 y_1) \\ +\delta; & \text{if } \text{sgn}(e_1 y_2) = +1 \text{ and } \text{sgn}(e_2 y_1) = +1 \\ -\delta; & \text{if } \text{sgn}(e_1 y_2) = -1 \text{ and } \text{sgn}(e_2 y_1) = -1 \end{cases} \tag{11.2-9d}$$

As might be imagined, with this algorithm the rate of convergence is slow and the residual MSE is high but convergence does take place without the use of any multiplies.

In sum, a fundamental difference between the gradient algorithm and the LMS algorithm is the issue of step size. In the gradient algorithm the choice of step size depends only on the eigenvalues of the correlation matrix and the step size remains constant during convergence. In the LMS algorithm the optimum step size depends on the eigenvalues, the number of equalizer taps (which also affect the eigenvalues), and the digital precision of number representation, as well as the point in the convergence process and the desired level of excess MSE.

11.3 THE LMS ALGORITHM FOR FRACTIONALLY SPACED EQUALIZERS

The major theoretical issue to be addressed in relation to the convergence of fractionally spaced equalizers is their eigenvalue spread. We will begin by reviewing the properties of the autocorrelation matrices for baud-rate systems and for $T_s/2$ systems.

The convolution matrix for a baud-rate sampled channel having three taps and a baud-rate transversal equalizer having five taps would look like

$$
H_c = \begin{bmatrix}
.1+.05j & .9-.2j & -.1+.2j & 0 & 0 & 0 & 0 \\
0 & .1+.05j & .9-.2j & -.1+.2j & 0 & 0 & 0 \\
0 & 0 & .1+.05j & .9-.2j & -.1+.2j & 0 & 0 \\
0 & 0 & 0 & .1+.05j & .9-.2j & -.1+.2j & 0 \\
0 & 0 & 0 & 0 & -1+.05j & .9-.2j & -.1+.2j
\end{bmatrix}.
$$

$$(11.3\text{-}1a)$$

and the autocorrelation matrix $R_1 = H_c^* H_c'$ is

$$
R_1 = \begin{bmatrix}
.9125 & -.05+.095j & .025j & 0 & 0 \\
-.05-.095j & .9125 & -.05+.095j & .025j & 0 \\
-.025j & -.05-.095j & .9125 & -.05+.095j & .025j \\
0 & .025j & -.05-.095j & .9125 & -.05+.095j \\
0 & 0 & -.025j & -.05-.095j & .9125
\end{bmatrix}
$$

$$(11.3\text{-}1b)$$

This matrix is both Toeplitz and Hermitian.

Now let's look at the convolution matrix of the same channel, sampled at $T_s/2$, having a 10-tap equalizer. Referring to eq. 10.3-1e, we find

$$
H_c = \begin{bmatrix}
.1+.05j & .9-.2j & -.1+.2j & 0 & 0 & 0 & 0 \\
-.1+.02j & .8+.1j & .1-.1j & 0 & 0 & 0 & 0 \\
0 & .1+.05j & .9-.2j & -.1+.2j & 0 & 0 & 0 \\
0 & -.1+.02j & .8+.1j & .1-.1j & 0 & 0 & 0 \\
0 & 0 & .1+.05j & .9-.2j & -.1+.2j & 0 & 0 \\
0 & 0 & -.1+.02j & .8+.1j & .1-.1j & 0 & 0 \\
0 & 0 & 0 & .1+.05j & .9-.2j & -.1+.2j & 0 \\
0 & 0 & 0 & -.1+.02j & .8+.1j & .1-.1j & 0 \\
0 & 0 & 0 & 0 & .1+.05j & .9-.2j & -.1+.2j \\
0 & 0 & 0 & 0 & -.1+.02j & .8+.1j & .1-.1j
\end{bmatrix}
$$

$$(11.3\text{-}2a)$$

and the correlation matrix $R_2 = H_c^* H_c'$ is

$$
R_2 = \begin{bmatrix}
.9125 & .661-247j & -.05+.095j & -.154+.172j & .025j & .014-.018j & 0 & 0 & 0 & 0 \\
.661+.247j & .6804 & .195-.10j & -.008-.116j & .005-.015j & -.012+.008j & 0 & 0 & 0 & 0 \\
-.05-.095j & .195+.10j & .9125 & .661-.247j & -.05+.095j & -.154+.172j & .025j & .014-.018j & 0 & 0 \\
-.154-.172j & -.008+.116j & .661+.247j & .6804 & .195-.10j & -.008-.116j & .005-.015j & -.012+.008j & 0 & 0 \\
-.025j & .005+.015j & -.05-.095j & .195+.10j & .9125 & .661-.247j & -.05+.095j & -.154+.172j & .025j & .014-.018j \\
.014+.018j & -.012-.008j & -.154-.172j & -.008+.116j & .661+247j & .6804 & .195-.10j & -.008-.116j & .005-.015j \\
0 & 0 & -.025j & .005+.015j & -.05-.095j & .195+.10j & .9125 & .661-247j & -.05+.095j & -.154+.172j \\
0 & 0 & .014+.018j & -.012-.008j & -.154-.172j & -.008+.116j & .661+247j & .6804 & .195-10j & -.008-.116j \\
0 & 0 & 0 & 0 & -.025j & .005+.015j & -.05-.095j & .195+.10j & .9125 & .661-.247j \\
0 & 0 & 0 & 0 & .014+.018j & -.012-.008j & -.154-.172j & -.008+.116j & .661+247j & .6804
\end{bmatrix}
$$

$$(11.3\text{-}2b)$$

where N, the order of R_2, is always even.

R_2 is clearly Hermitian but it is also clearly not Toeplitz. The eigenvalue equation for this matrix is

$$R_2 \mathbf{u} = \lambda \mathbf{u}$$

$$(11.3\text{-}3a)$$

where, for this example, $\mathbf{u}$ has 10 rows and one column. Following Gitlin and Weinstein [G&W81], eq. 11.3-3a may be partitioned into two equations

$$\sum_{n=0,2,4,\ldots} R_2(m,n)u(n) + \sum_{n=1,3,5,\ldots} R_2(m,n)u(n) = \lambda u(m); \quad m \text{ even} \qquad (11.3\text{-}3b)$$

or

$$R_2^{ee}\mathbf{u}^e + R_2^{eo}\mathbf{u}^o = 2\mathbf{u}^e \qquad (11.3\text{-}3c)$$

and

$$\sum_{n=0,2,4,\ldots} R_2(m,n)u(n) + \sum_{n=1,3,5,\ldots} R_2(m,n)u(n) = \lambda u(m); \quad m \text{ odd} \qquad (11.3\text{-}3d)$$

or

$$R_2^{oe}\mathbf{u}^e + R_2^{oo}\mathbf{u}^o = \lambda \mathbf{u}^o \qquad (11.3\text{-}3e)$$

where, for example, R_2^{eo} in eq. 11.3-3c refers to the submatrix of R_2 consisting of the even-numbered rows and odd-numbered columns and $\mathbf{u}^e$ is the subvector of the vector $\mathbf{u}$ consisting of the even-numbered rows.

In order to illustrate the point, we will use the matrix of eq. 11.3-2b as an example. From eq. 11.3-3c, we get

$$\begin{bmatrix} .9125 & -.05+.095j & .025j & 0 & 0 \\ -.05-.095j & .9125 & -.05+.095j & .025j & 0 \\ -.025j & -.05-.095j & .9125 & -.05+.095j & .025j \\ 0 & -.025j & -.05-.095j & .9125 & -.05+.095j \\ 0 & 0 & -.025j & -.05-.095j & .9125 \end{bmatrix} \begin{bmatrix} U(0) \\ U(2) \\ U(4) \\ U(6) \\ U(8) \end{bmatrix}$$

$$+ \begin{bmatrix} .661-.247j & -.154+.172j & .014-.018j & 0 & 0 \\ .195+.10j & .661-.247j & -.154+.172j & .014-.018j & 0 \\ .005+.015j & .195+.10j & .661-.247j & -.154+.172j & .014-.018j \\ 0 & .005+.015j & .195+.10j & .661-.247j & -.154+.172j \\ 0 & 0 & .005+.015j & .195+.10j & .661-.247j \end{bmatrix}$$

$$\times \begin{bmatrix} U(1) \\ U(3) \\ U(5) \\ U(7) \\ U(9) \end{bmatrix} = \lambda \begin{bmatrix} U(0) \\ U(2) \\ U(4) \\ U(6) \\ U(8) \end{bmatrix} \qquad (11.3\text{-}4a)$$

and from eq. 11.3-3e we get

$$\begin{bmatrix} .661+.247j & .095-.10j & .005-.015j & 0 & 0 \\ -.154-.172j & .661+.247j & .095-.10j & .005-.015j & 0 \\ .014+.018j & -.154-.172j & .661+.247j & .095-.10j & .005-.015j \\ 0 & .014+.018j & -.154-.172j & .661+.247j & .095-.10j \\ 0 & 0 & .014+.018j & -.154-.172j & .661+.247j \end{bmatrix} \begin{bmatrix} U(0) \\ U(2) \\ U(4) \\ U(6) \\ U(8) \end{bmatrix}$$

$$
+ \begin{bmatrix} .6804 & -.008 - .116j & -.012 + .008j & 0 & 0 \\ -.008 + .116j & .6804 & -.008 - .116j & -.012 + .008j & 0 \\ -.012 - .008j & -.008 + .116j & .6804 & -.008 - .116j & -.012 + .008j \\ 0 & -.012 - .008j & -.008 + .116j & .6804 & -.008 - .116j \\ 0 & 0 & -.012 - .008j & -.008 + .116j & .6804 \end{bmatrix}
$$

$$
\times \begin{bmatrix} U(1) \\ U(3) \\ U(5) \\ U(7) \\ U(9) \end{bmatrix} = \lambda \begin{bmatrix} U(1) \\ U(3) \\ U(5) \\ U(7) \\ U(9) \end{bmatrix}
\tag{11.3-4b}
$$

Note that although only two of these four matrices are Hermitian, all four of them are Toeplitz.

The decompositions of eqs. 11.3-3b and 11.3-3d may be written more directly in relation to the channel autocorrelation function $R(\tau = t_2 - t_1) = R(t_1, t_2)$ as

$$
\sum_{n=0}^{N/2} R(mT_s, nT_s)u(nT_s) + \sum_{n=0}^{N/2} R\left(mT_s, \left(nT_s + \frac{T_s}{2}\right)\right)u\left(nT_s + \frac{T_s}{2}\right) = \lambda u(mT_s)
\tag{11.3-5a}
$$

$$
\sum_{n=0}^{N/2} R\left(\left(mT_s + \frac{T_s}{2}\right), nT_s\right)u(nT_s) + \sum_{n=0}^{N/2} R\left(\left(mT_s + \frac{T_s}{2}\right), \left(nT_s + \frac{T_s}{2}\right)\right)u\left(nT_s + \frac{T_s}{2}\right)
$$
$$
= \lambda u\left(mT_s + \frac{T_s}{2}\right)
\tag{11.3-5b}
$$

where $m = 0, 1, \ldots, N/2$.

All of the terms in eqs. 11.3-3a and 11.3-3b are sampled every T_s sec (some of the samples are displaced from others by $T_s/2$ sec) so that we can now use the limiting property of Toeplitz matrices stated in Chapter 8 to find their Fourier Transforms. The unsampled channel transfer function $X(f)$ has less than 100% excess bandwidth. For $|f| \le 1/2T_s$, define:

$$
X_{eq}(f) = X(f) + X\left(f - \frac{1}{T_s}\right) + X\left(f + \frac{1}{T_s}\right)
\tag{11.3-6a}
$$

$$
\tilde{X}_{eq}(f) = X(f) - X\left(f - \frac{1}{T_s}\right) - X\left(f + \frac{1}{T_s}\right)
\tag{11.3-6b}
$$

The Discrete Fourier Transform of the eigenvector $\mathbf{u}$ in eq. 11.3-3a is, with $f_n = n/T_s$,

$$
U(f_n) = \sum_{n=0}^{N-1} u\left(n\frac{T_s}{2}\right) e^{-jf_n T_s/2 \, 2\pi/N} = \sum_{n=0}^{N-1} u\left(n\frac{T_s}{2}\right) e^{-jn\pi/N}
$$
$$
= U_{eq}(f_n) + e^{-jn\pi/N}\tilde{U}_{eq}(f_n)
\tag{11.3-6c}
$$

where

$$
U_{eq}(f) = U(f) + U\left(f - \frac{1}{T_s}\right) + U\left(f + \frac{1}{T_s}\right)
\tag{11.3-6d}
$$

$$
\tilde{U}_{eq}(f) = U(f) - U\left(f - \frac{1}{T_s}\right) - U\left(f + \frac{1}{T_s}\right)
\tag{11.3-6e}
$$

for $|f| \le 1/2T_s$.

With this, the Fourier Transforms of eqs. 11.3-3c and 11.3-3e are

$$|X_{eq}(f_n)|^2 U_{eq}(f_n) + X_{eq}(f_n)\tilde{X}_{eq}^*(f_n)\tilde{U}_{eq}(f_n) = \lambda U_{eq}(f_n) \qquad (11.3\text{-}7a)$$

$$\tilde{X}_{eq}(f_n)X_{eq}^*(f_n)U_{eq}(f_n) + |\tilde{X}_{eq}(f_n)|^2 \tilde{U}_{eq}(f_n) = \lambda \tilde{U}_{eq}(f_n) \qquad (11.3\text{-}7b)$$

where solutions exist only when the determinant is zero, or

$$[|X_{eq}(f_n)|^2 - \lambda][|\tilde{X}_{eq}(f_n)|^2 - \lambda] - X_{eq}(f_n)\tilde{X}_{eq}^*(f_n)\tilde{X}_{eq}(f_n)X_{eq}^*(f_n) = 0 \qquad (11.3\text{-}7c)$$

It follows that the eigenvalues must satisfy

$$\lambda_n^2 - \lambda_n[|X_{eq}(f_n)|^2 + |\tilde{X}_{eq}(f_n)|^2] = 0 \qquad (11.3\text{-}7d)$$

and in turn

$$\lambda_n = 0;$$

$$\lambda_n = [|X_{eq}(f_n)|^2 + |\tilde{X}_{eq}(f_n)|^2] = \sum_k \left| X\left(f_n + \frac{k}{T_s}\right) \right| \qquad (11.3\text{-}7e)$$

The conclusion to be drawn is that half of the eigenvalues of the correlation matrix R are zero and half are samples of the aliased magnitude-square of the channel transfer function.

Since finding the optimum tap vector of the equalizer requires finding the inverse of the correlation matrix R (we found this inverse in Chapter 10 and the LMS algorithm effectively leads to the optimum tap vector in a recursive manner) and depends on having non-zero eigenvalues, there is the temptation to conclude that there is *not a unique solution* solution for the optimum tap vector in a fractionally spaced equalizer.

This would be a mistaken conclusion for a number of reasons:

1. The argument that led to eq. 11.3-7d assumes that the equalizer has an infinite number of taps and no noise. If there were noise, the resulting eigenvalues would all be greater than zero and there would be a unique inversion of R.

2. It has been shown [G&W81] that if the equalizer has a finite number of taps, then there is a unique solution even if there is no noise. This is computationally confirmed by the Problems in Chapter 10.

Nevertheless, the results of eq. 11.3-7d are an indicator of a problem in the convergence of fractionally spaced equalizers. Even though the eigenvalues never become zero in a finite-length $T_s/2$ equalizer, it is the case that the eigenvalue spread $\rho = \lambda_{max}/\lambda_{min}$ becomes very large. This too has been confirmed computationally in Chapter 10 and will be reconfirmed in the Problems in this chapter. The large value of ρ leads to two problems. The first is that the larger the value of ρ, the longer the convergence time. This problem will be examined through simulation. The simulation program will be discussed Section 11.5.

The second problem that arises from a large eigenvalue spread combined with finite-precision arithmetic is the 'tap-wandering' phenomenon described by Gitlin, Meadors, and Weinstein [GM&W82]. The consequence of this phenomenon, if uncorrected, is the saturation of one or more tap coefficients and the consequent catastrophic loss of equalization. It is also the case that this phenomenon takes place in a deterministic manner.

In order to understand the character of the problem, recall that the quadratic form for the mean-square error is a generalization to higher dimensions of the paraboloid of revolution shown in fig. 10.1-1. Horizontal cross-sections of this paraboloid, corresponding to different values of the MSE, are shown in fig. 11.3-1. The ellipses of constant MSE

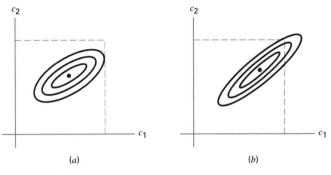

(a)   (b)

FIGURE 11.3-1

that constitute these cross-sections have eccentricity that depends on the value of $\rho = \lambda_{max}/\lambda_{min}$. In fig. 11.3-1a, we show these ellipses for moderate values of ρ, corresponding to typical baud-rate equalizers, and in fig. 11.3-1b we show them for large values of ρ typical of fractionally spaced equalizers. In both instances we show the ellipses in relation to the region of tap saturation. Clearly, the fractionally spaced equalizer, by virtue of its large value of ρ, intersects the saturation region, whereas the baud rate equalizer does not.

We have also seen that one of the properties of the LMS algorithm is that the minimum value of the MSE is never acheived; there is always an excess MSE. While the minimum value of MSE corresponds to a unique tap setting, any value above the minimum corresponds to a continuum of tap settings along one of the contours. This creates the possibility of the taps drifting with the MSE remaining the same. Note that with a small value of ρ that drift will not drive the tap values into saturation but with a large value of ρ, it will.

The final issue is the finite-precision character of the arithmetic, which is represented by an equalizer tap bias vector B, with same length as the equalizer tap vector B, having equal components of half a quantization. It is shown that after the equalizer has fully converged this vector has the effect of deterministically driving the tap coefficents around the contour and therefore eventually into saturation.

This is clearly a very important issue for those who are designing fractionally spaced equalizers. The 'tap-leakage' algorithm that has been developed is a modification of the LMS tap adjustment algorithm and comes in two slightly different versions:

$$C_{n+1} = (1 - \mu\delta)C_n - \mu e_n^* Y_n \tag{11.3-8a}$$
$$C_{n+1} = C_n - \mu\delta \, \text{sgn}[C_n] - \mu e_n^* Y_n \tag{11.3-8b}$$

where μ is the LMS step size, δ is a small constant of the tap leakage algorithm, and Y_n is the data vector in the equalizer at the nth iteration. Interested readers are referred to [GM&W82].

11.4 THE RELATIONSHIP BETWEEN THE ADAPTIVE ROTATOR AND THE ADAPTIVE EQUALIZER

At this point we wish to return to a topic that was left unfinished in Chapter 6.4, the relationship of the adaptive equalizer and the adaptive rotator in the Hilbert Transform receiver. The placement of the adaptive equalizer before the adaptive rotator in fig. 11.4-1

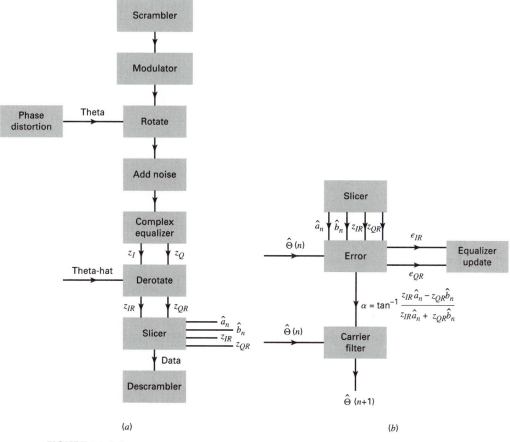

FIGURE 11.4-1

was first proposed by Falconer [Fal82], who analyzed the perfomance of this configuration in the presence of carrier-phase jitter and noise. We can see that the data in the equalizer is rotated with respect to the decision regions defined by the slicer. This means that the errors computed by the slicer to drive the equalizer adaptation are not properly oriented to the data. They must therefore be rotated back to the data orientation by PLL output $\hat{\theta}$, theta-hat.

In fig. 11.4-1 the output of the Complex Equalizer is the complex data pair (z_I, z_Q), the inphase and quadrature outputs. This data pair is then applied to Derotate in order to compensate for the phase rotation that took place previously. The rotate operation was $e^{j\theta}$ where θ is the total phase distortion of all kinds at that symbol time. Consequently the data pair (z_I, z_Q) should be derotated by $e^{j\hat{\theta}}$ where $\hat{\theta}$ is an estimate of θ that was computed in the last symbol interval. This derotation yields the new data pair (z_{IR}, z_{QR}) where:

$$z_{IR} = z_I \, \mathrm{Cos}(\hat{\theta}) + z_Q \, \mathrm{Sin}(\hat{\theta})$$
$$z_{QR} = -z_I \, \mathrm{Sin}(\hat{\theta}) + z_Q \, \mathrm{Cos}(\hat{\theta})$$
(11.4-1)

The Slicer operates on (z_{IR}, z_{QR}) to produce the decisions $(\hat{a}_n, \hat{b}_n)$ and it outputs both pairs to the block Error, fig. 11.4-1b, that computes the errors that drive the equalizer adaptation and the carrier recovery. For the equalizer, Error first computes

$$
\begin{aligned}
e_I &= z_{IR} - \hat{a}_n \\
e_Q &= z_{QR} - \hat{b}_n
\end{aligned}
\tag{11.4-2a}
$$

and then rotates these error terms back to the phase reference before the Derotate operation:

$$
\begin{aligned}
e_{IR} &= e_I \cos(\hat{\theta}) - e_Q \sin(\hat{\theta}) \\
e_{QR} &= e_I \sin(\hat{\theta}) + e_Q \cos(\hat{\theta})
\end{aligned}
\tag{11.4-2b}
$$

As indicated in Chapter 6.4, the block Error then computes

$$
\alpha = \tan^{-1} \left[\frac{z_{QR}\hat{a}_n - z_{IR}\hat{b}_n}{z_{IR}\hat{a}_n + z_{QR}\hat{b}_n} \right]
\tag{11.4-3}
$$

which goes to the Carrier Filter. The output of Carrier Filter will be the next value of $\hat{\theta}$. Note once again that the one-symbol delay that is required for the PLL has been established by using a value of $\hat{\theta}$ computed in a previous cycle in order to compute the value of $\hat{\theta}$ to be used in the next cycle.

It has been shown by Gitlin et al. [GH&M73] that in the absence of carrier recovery the taps of a complex equalizer will attempt to rotate to meet the carrier-phase distortion. Since the carrier-phase jitter has zero average value and varies faster than the equalizer can respond, the equalizer rotation will typically not reflect phase jitter. The degree to which the equalizer precesses in order to compensate for frequency offset depends on the characteristics of the adaptive rotator loop. If, as shown in Chapter 6.5, the loop filter has a pole at $z = +1$ then the rotator will follow the frequency offset perfectly. Otherwise there will be some rotation by the equalizer. This will mitigate against optimum equalizer convergence. The equalizer will not have the problem of being able to adjust itself to a constant-phase offset. Consequently, as Falconer points out, the combination of adaptive equalizer and adaptive rotator will not lead to a unique equalizer tap setting. There is a clear advantage to preventing the equalizer from rotating at all in response to frequency offset. Such rotation would keep the equalizer in a constant state of adaptation, thereby preventing it from fully minimizing the mean-square error.

11.5 SOFTWARE IMPLEMENTATION OF THE LMS ALGORITHM

The program Equalizer which accompanies the text is designed to test and evaluate adaptive equalization algorithms. It is a variation of the simulation program SNRTEST developed in Chapter 4 for measuring the SNR performance of different modulation techniques.

Although Equalizer contains more than the LMS algorithm, it will be discussed at this point in relation to that algorithm. In order to simplify the code to address only the problem of equalization, the choice of modulation techniques has been limited to QPSK. Since it is the channel rather than the modulation technique that is being equalized, this does not limit the equalization process. In fact, modems that employ adaptive equalization typically use less complex signal sets, having larger distances between signal points at startup in order to facilitiate equalization before switching to the data mode.

The equalizer structures that may be used are:

1. Two-sided baud-rate transversal equalizer
2. One-sided baud-rate transversal equalizer plus decision feedback

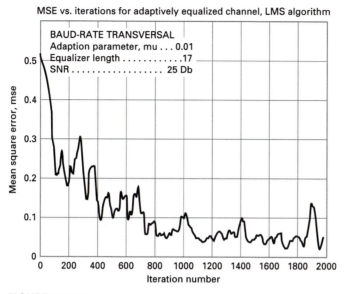

MSE vs. iterations for adaptively equalized channel, LMS algorithm

BAUD-RATE TRANSVERSAL
Adaption parameter, mu . . . 0.01
Equalizer length 17
SNR . . . . 25 Db

FIGURE 11.5-1

3. Constrained tap, two-sided baud-rate transversal plus decision feedback

4. IIR baud rate all-pass plus decision feedback

5. Fractionally spaced $T_s/2$ transversal equalizer

The reader is urged to examine the structure of the code, which is available on the website, in order to see how the various algorithms are implemented.

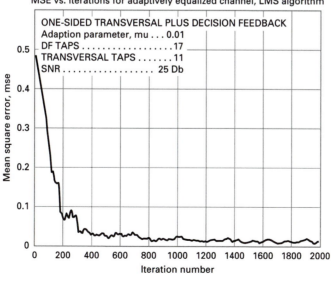

MSE vs. iterations for adaptively equalized channel, LMS algorithm

ONE-SIDED TRANSVERSAL PLUS DECISION FEEDBACK
Adaption parameter, mu . . . 0.01
DF TAPS 17
TRANSVERSAL TAPS 11
SNR . . . . 25 Db

FIGURE 11.5-2

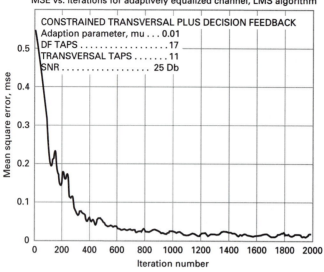

FIGURE 11.5-3

A number of experiments using this simulation software are given in the Problems. These experiments will aid the reader to understand the dynamics of equalizer convergence and performance.

Typical results for these equalizer structures are shown in figs. 11.5-1–11.5-5 for the convergence of the different types of equalizer structure using the LMS algorithm on the B3 channel. This channel was used as the basis for extensive examples of steady-

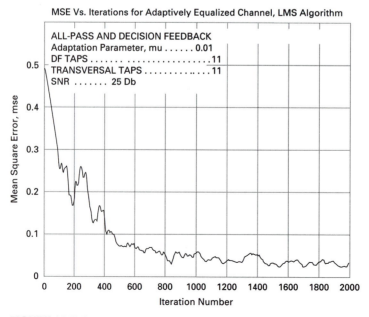

FIGURE 11.5-4

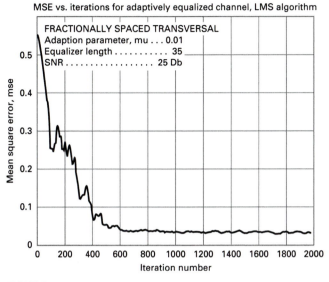

FIGURE 11.5-5

state performance in Chapter 10. It is worth noting that the fractionally spaced transversal equalizer is considerably slower than the baud-rate transversal.

It will be left to the Problems to provide a full exploration of the convergence characteristics of these equalizers.

11.6 THE RAM-DFE DECISION FEEDBACK EQUALIZER

An interesting variation of the decision feedback equalizer, which has found application in both the ISDN and in magnetic recording, uses a AM instead of a transversal filter in the feedback path. The forward path remains a transversal equalizer having one of the configurations that has been discussed. This structure, shown in fig. 11.6-1, is particularly useful in equalizing nonlinearly generated ISI. As will be demonstrated, it converges more slowly than a standard DFE structure.

The basic idea of the AM-DFE is that the ISI that is actually generated at a given sampling instant depends on the data pattern surrounding that instant. If the forward

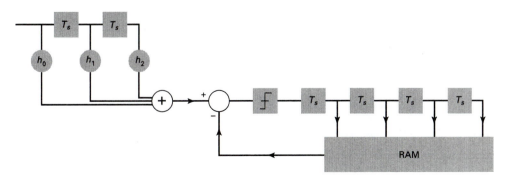

FIGURE 11.6-1

equalizer removes the leading echoes, then the remaining ISI depends only on the trailing echoes and, consequently, on the past data. If one were to consider the shift register that stores past decisions in the feedback path as an address register into a lookup table, then the contents at each address would be the total ISI value corresponding to that data pattern. The longer the past data pattern stored in the shift register, the larger the number of memory locations, and the more precise the ISI term will be.

Since the entries in the lookup table are simply associated with the addresses rather than being generated by the data, there is no constraint on these entries being in any way linearly related to the data pattern. This structure is then uniquely suited to cancel nonlinearly generated ISI.

The use of the AM-DFE for the ISDN is described in Van Gerwen et al. [VGV&C84] and Lin and Tzeng [L&T88], who propose this structure for both equalization and echo cancellation (to be discussed subsequently). The application to magnetic recording is discussed by Cioffi et al. [Cio90-1], and Fisher et al. [Fis91].

In order to make the structure adaptive, the lookup table is implemented by a AM with the shift register containing past decisions operating as the memory address register. A typical configuration for magnetic channels is an address length of $M = 8$ and a corresponding memory size of $2^M = 256$ locations. The updating of the AM-DFE is accomplished through an LMS algorithm. If $\langle M \rangle$ is the contents of memory location M, and e_k is the slicer error at the kth iteration, then

$$\langle M \rangle_{k+1} = \langle M \rangle_k + \mu_R e_k \qquad (11.6\text{-}1)$$

where e_k is the slicer error and μ_R is a step size for the LMS algorithm. The algorithm can also be implemented with a fixed increment following if that is required by processing limitiations.

The adaptation is slow because the basic algorithm allows only one memory location to be updated at a single time. This requires a long training time to have a rich mixture of of all of the possible data sequences in order to make sure that every memory location is properly adjusted before regular data transmission takes place.

Adaptation can be considerably speeded up by a method termed *broadcasting*. If the memory address is $M = 8$, then the initial adjustments are made according to the $M_1 = 4$ most significant bits of the address, presumably corresponding to most significant ISI terms. This means that each 4-bit address will update 16 locations simultaneously to the same values. The next set of adjustments are made with the $M_2 = 6$ most significant bits of the address so that each more refined update changes four locations. Finally the update algorithm is switched to the full $M = 8$ bits of address.

Fisher et al. [Fis91] demonstrate that the AM-DFE can give considerable improvement in performance on a representative nonlinear magnetic channel over that of a conventional DFE. They also demonstrate that using broadcasting, the convergence time of the AM-DFE comes very close to that of the conventional DFE.

The nonlinearities of the subscriber loop make the AM-DFE an attractive alternative in the ISDN as well.

11.7 EQUALIZER CONVERGENCE FOR PARTIAL RESPONSE SYSTEMS

Perhaps the most salient distinction that can be made between Nyquist signaling and the varieties of partial response signaling is that baud-rate samples of a Nyquist data stream are designed to be uncorrelated but baud-rate samples of a partial response data stream are designed to be correlated in some fashion. As a consequence, the correlation matrix of Nyquist data on channel having no distortion will be diagonal with all of the eigenvalues equal to unity. On the other hand, the correlation matrix of a partial response signal, even on an undistorted channel, will have a large spread of eigenvalues. This would indicate difficulty of equalizer convergence even on a mildly distorted channel. For this reason, the early efforts to investigate how to speed up the convergence of equalizers began with partial respose systems. The insights and techniques of the resulting approaches were subsequently applied to the more general problem of rapid convergence. We shall follow the approach of Chang [Cha71].

It is first useful to get an idea of how the eigenvalues behave in a typical partial response system. We will use the Class 1 (Duobinary) as an example. It was established in Chapter 3 that the spectrum of a Duobinary partial response pulse is

$$G(\omega) = 1 + e^{-j\omega T_s} = 2\cos\left(\frac{\omega T_s}{2}\right)e^{-j\omega T_s/2} \qquad (11.7\text{-}1\text{a})$$

so that the energy spectral density is

$$|G(\omega)|^2 = 4\cos^2\left(\frac{\omega T_s}{2}\right) = 2[1 + \cos(\omega T_s)] \qquad (11.7\text{-}1\text{b})$$

within the band $-\pi/T_s \le \omega \le \pi/T_s$. It follows that the autocorrelation function is

$$R(\tau) = \frac{1}{2\pi}\int_{-\infty}^{\infty} |G(\omega)|^2 e^{j\omega\tau}\,d\omega = \frac{1}{\pi}\int_{-\pi/T_s}^{\pi/T_s} [1 + \cos(\omega T_s)]\cos(\omega\tau)\,d\omega \qquad (11.7\text{-}1\text{c})$$

and at the sampling instants

$$R(nT_s) = \begin{cases} 1; & n = 0 \\ 1/2; & n = \pm 1 \\ 0; & \text{otherwise} \end{cases} \qquad (11.7\text{-}1\text{d})$$

It follows that the correlation matrix will be of the form

$$R = \begin{bmatrix} 1 & 1/2 & 0 & 0 & 0 \\ 1/2 & 1 & 1/2 & 0 & 0 \\ 0 & 1/2 & 1 & \cdot & 0 \\ 0 & 0 & \cdot & \cdot & 1/2 \\ 0 & 0 & 0 & 1/2 & 1 \end{bmatrix} \qquad (11.7\text{-}1\text{e})$$

Finding the correlation matrix for Class 4 Partial Response will be left to the Problems.

We now want to find the eigenvalues of the Duobinary system for a general Nth-order correlation matrix. These are the roots of the characteristic equation which is the determinant

$$\Delta_N = \det \begin{bmatrix} 1-\lambda & 1/2 & 0 & 0 & 0 \\ 1/2 & 1-\lambda & 1/2 & 0 & 0 \\ 0 & 1/2 & 1-\lambda & \cdot & 0 \\ 0 & 0 & \cdot & \cdot & 1/2 \\ 0 & 0 & 0 & 1/2 & 1-\lambda \end{bmatrix} \qquad (11.7\text{-}2\text{a})$$

where

$$\Delta_N = (1 - \lambda) \det \begin{bmatrix} 1 - \lambda & 1/2 & 0 & 0 & 0 \\ 1/2 & 1 - \lambda & 1/2 & 0 & 0 \\ 0 & 1/2 & 1 - \lambda & \cdot & 0 \\ 0 & 0 & \cdot & \cdot & 1/2 \\ 0 & 0 & 0 & 1/2 & 1 - \lambda \end{bmatrix}$$

$$- \frac{1}{2} \det \begin{bmatrix} 1/2 & 1/2 & 0 & 0 & 0 \\ 0 & 1 - \lambda & 1/2 & 0 & 0 \\ 0 & 1/2 & 1 - \lambda & \cdot & 0 \\ 0 & 0 & \cdot & \cdot & 1/2 \\ 0 & 0 & 0 & 1/2 & 1 - \lambda \end{bmatrix} \qquad (11.7\text{-}2b)$$

or

$$\Delta_N = (1 - \lambda)\Delta_{N-1} - \frac{1}{4}\Delta_{N-2} \qquad (11.7\text{-}2c)$$

which is a difference equation having initial conditions $\Delta_0 = 1$ and $\Delta_1 = (1 - \lambda)$ that recursively generates the determinant. Allowing $\text{Cos}(x) = (1 - \lambda)$, eq. 11.7-2c can be rewritten as

$$\Delta_{N+2} - \text{Cos}(x)\Delta_{N+1} + \frac{1}{4}\Delta_N = 0 \qquad (11.7\text{-}2d)$$

and taking the z-transform, we find

$$\Delta(z)\left[z^2 - z\,\text{Cos}(x) + \frac{1}{4}\right] = z^2 \qquad (11.7\text{-}2e)$$

The mechanics of finding the inverse z-transform will be left to the Problems. The result is

$$\Delta_N = \frac{\text{Sin}[(N+1)x]}{2^N\,\text{Sin}(x)} \qquad (11.7\text{-}2f)$$

which must be set equal to zero to find the N roots. It is easy to see that these roots correspond to

$$x_k = \frac{k}{N+1}\pi; \qquad k = 1, 2, \ldots, N \qquad (11.7\text{-}2g)$$

so that the eigenvalues are

$$\lambda_k = 1 - x_k = 1 - \text{Cos}\left(\frac{k}{N+1}\pi\right); \qquad k = 1, 2, \ldots, N \qquad (11.7\text{-}2h)$$

It is clear that as N increases, so does the ratio $\rho = \lambda_{\max}/\lambda_{\min}$. In short, the controlled ISI in partial response signaling operates to increase ρ just as does the ISI resulting from channel distortion.

Chang's approach is use the knowledge that we have about the ISI to filter the received signal in such a way as to reduce the value of ρ. With perfect knowledge of the ISI, it would be possible to cause $\rho = 1$ so that equalization could take place very rapidly. Instead of a conventional transversal equalizer, consider the structure of fig. 11.7-1 where the input to each multiplying tap c_n is the linear combination of data in the entire transversal rather

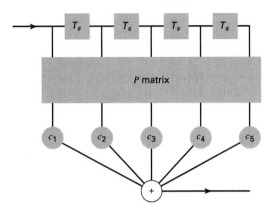

FIGURE 11.7-1

than simply the data in the nth stage. Following the argument starting with eq. 11.1-6a, the expression for the mean-square error becomes

$$S_{\text{ISI}} = C^* P^* R P^T C^T - C^{**} P^* H_c E_h^T - E_h H_c^{*T} C^T P^T + 1 \tag{11.7-3a}$$

and the gradient vector becomes

$$\delta(S_{\text{ISI}}) = 2[C P^* R P^T - P^* \mathbf{h}] \tag{11.7-3b}$$

and the tap vector error becomes

$$S_k = B_0 [I - \mu P^* R P^T]^k P^* R P^T [[I - \mu P^* R P^T]^k]^{*T} B_0^{*T} + S_{\text{min}} \tag{11.7-3c}$$

If $P^* R P^T = I$, then it should be clear that all of the terms in eq. 11.7-3c above are equal and the convergence can be completed in a single step. The consequence of the matrix P has been to orthogonalize the system by making all of the eigenvalues equal.

This idea of system orthogonalization turns out to be the key to understanding the algorithms for rapid convergence of equalizers. It will be pursued further in the next section.

At this point we will simply find the required matrix P on the assumption that we know the eigenvalues of R. We have already seen that $R = Q D Q^{*T}$, where D is a diagonal matrix of the eigenvalues of R. Now let

$$H = \begin{bmatrix} \sqrt{\lambda_1} & 0 & 0 & 0 \\ 0 & \sqrt{\lambda_2} & 0 & 0 \\ 0 & 0 & \cdot & 0 \\ 0 & 0 & 0 & \sqrt{\lambda_N} \end{bmatrix} \tag{11.7-3d}$$

so that $D = H^* H^T$ and

$$P R P^{*T} = (P Q H)(P Q H)^{*T} = I \tag{11.7-3e}$$

where $P Q H = G$, an orthogonal matrix satisfying $G^{-1} = G^{*T}$. Since there are many such orthogonal matrices, the solution for P is not unique but for any such matrix G,

$$P = G H^{-1} Q^{*T} \tag{11.7-3f}$$

The operation of the LMS algorithms with this structure is quite direct. Instead of updating the tap vector according to

$$C_{n+1} = C_n - \mu e_n^* Y \tag{11.7-4a}$$

where Y is the data vector in the equalizer, we would instead update according to

$$C_{n+1} = C_n - \mu e_n^* P Y \qquad (11.7\text{-}4b)$$

correlating the slicer error with the output vector of the orthogonalizing matrix P. Variations on this basic approach are given by Mueller [Mue75] and Bingham [Bin87].

The point that is to be made with this approach is that the effect of the orthogonalizing matrix P is to equalize the incoming signal to have a Nyquist shape. In that respect, undesirable channel distortion is treated in the same way as partial response shaping: something has spread out the eigenvalues and the purpose of the P matrix is to make the eigenvalues equal. This means that the equalizer causes an incoming partial response signal to lose its partial response coding by the time that it gets to the output.

11.8 CYCLIC EQUALIZATION

We have seen that the process of transversal equalizer convergence using random data and the LMS algorithm can be lengthy for both baud-rate and fractionally spaced equalizers. There are applications in which such a long convergence time is not acceptable. One example is where a computer in a central location is periodically polling a large number of remote modems over the telephone line for relatively short blocks of data. Even if the phone lines are dedicated, each new connection will require a new convergence of the receiver equalizer. We have seen that convergence on random data may take 1000–2000 symbols and if the data block is of the same order of magnitude then there is advantage to having a computationally simple startup procedure that will guarantee more rapid convergence.

The basic idea that will be developed below is to send a binary training sequence

$$\mathbf{x} = (x_0 x_1 \cdots x_{N-1}) \qquad (11.8\text{-}1a)$$

which is N symbol intervals long, where N is the number of symbols spanned by the transversal equalizer in the receiver, and where the sequence is one of the maximal length shift-register sequences that was discussed in Section 3.18. These sequences, also known as pseudo-random sequences, have an autocorrelation function

$$R(n) = \begin{cases} 1; & n = 0 \\ -1/N; & n \neq 0, \ -\frac{N-1}{2} < n < \frac{N-1}{2} \end{cases} \qquad (3.18\text{-}2b)$$

which can be approximated as

$$\frac{1}{N}\sum_{n=0}^{N-1} x_n x_k = \delta_{nk} = \begin{cases} 1, & n = k \\ 0, & n \neq k \end{cases} \qquad (11.8\text{-}1b)$$

The transmitted sequence is assumed to be known at the receiver and the autocorrelation properties of this sequence provides for rapid convergence of the equalizer as shown below.

The first use of these training sequences was by Chang and Ho [C&H72] but their approach was considerably improved by Mueller and Spalding [M&S75], who originated the term *cyclic equalization*.

The equalizer tap vector at time kT_s is the column vector

$$\mathbf{c}(k) = [c_0(k), c_1(k), \dots, c_{N-1}(k)]^T$$
$$= \{c_n(k)\}; \quad n = 0, 1, \dots (N-1) \qquad (11.8\text{-}2a)$$

and the received data in the equalizer at time kT_s is the column vector

$$\mathbf{y}(k) = [y_0(k), y_1(k), \ldots, y_{N-1}(k)]^T$$
$$= \{y_n(k)\}; \quad n = 0, 1, \ldots (N-1) \tag{11.8-2b}$$

Note that if the input data sequence $\mathbf{x}$ is periodic, the noiseless received data will also be periodic:

$$y_0(k+1) = y_{N-1}(k)$$
$$y_n(k+1) = y_{n-1}(k) \tag{11.8-2c}$$

By way of example, let the channel impulse reponse $\mathbf{h}$, the equalizer tap vector $\mathbf{c}$, the periodic input sequence $\mathbf{x}$, and the channel output sequence $\mathbf{y}$ all have length $N = 5$. Then at time k the vector $\mathbf{y}$ is

$$\mathbf{y}(k) = \begin{bmatrix} y_0 \\ y_1 \\ y_2 \\ y_3 \\ y_4 \end{bmatrix} = \begin{bmatrix} x_0 & x_4 & x_3 & x_2 & x_1 \\ x_1 & x_0 & x_4 & x_3 & x_2 \\ x_2 & x_1 & x_0 & x_4 & x_3 \\ x_3 & x_2 & x_1 & x_0 & x_4 \\ x_4 & x_3 & x_2 & x_1 & x_0 \end{bmatrix} \begin{bmatrix} h_0 \\ h_1 \\ h_2 \\ h_3 \\ h_4 \end{bmatrix} = X\mathbf{h} \tag{11.8-3a}$$

Let U be the cyclic shift matrix

$$U = \begin{bmatrix} 0 & 1 & 0 & 0 & 0 \\ 0 & 0 & 1 & 0 & 0 \\ 0 & 0 & 0 & 1 & 0 \\ 0 & 0 & 0 & 0 & 1 \\ 1 & 0 & 0 & 0 & 0 \end{bmatrix} \tag{11.8-3b}$$

where $U^T U = U^5 = U^0 = I$, so that

$$\mathbf{y}(k+1) = \begin{bmatrix} y_4 \\ y_0 \\ y_1 \\ y_2 \\ y_3 \end{bmatrix} = [UX]\mathbf{h} \tag{11.8-3c}$$

and, in general,

$$\mathbf{y}(k+n) = \lfloor U^n X \rfloor \mathbf{h}; \quad n = 0, 1, \ldots, 4 \tag{11.8-3d}$$

Define the matrix Y as consisting of the column vectors

$$Y = \lfloor \mathbf{y}(k) \quad \mathbf{y}(k+1) \quad \cdots \quad \mathbf{y}(k+4) \rfloor \tag{11.8-3e}$$

so that the $N = 5$ successive equalizer outputs are $Y^{*T} \mathbf{c}_p$ where the subscript p has been introduced in order to make clear that these equalizer taps correspond to a periodic input. If the channel is noiseless and the equalizer is properly adjusted, these N outputs will be equal to the input vector $\mathbf{x}$ so that

$$\mathbf{x} = Y^{*T} \mathbf{c}_p \tag{11.8-3f}$$

It will be left to the Problems to demonstrate that if the input sequence $\mathbf{x}$ has the properties of eq. 11.8-1b, then eq. 11.8-3f will be reduced to

$$\mathbf{x} = \begin{bmatrix} x_0 \\ x_1 \\ x_2 \\ x_3 \\ x_4 \end{bmatrix} = \begin{bmatrix} h_0 & h_1 & h_2 & h_3 & h_4 \\ h_4 & h_0 & h_1 & h_2 & h_3 \\ h_3 & h_4 & h_0 & h_1 & h_2 \\ h_2 & h_3 & h_4 & h_0 & h_1 \\ h_1 & h_2 & h_3 & h_4 & h_0 \end{bmatrix} \begin{bmatrix} c_{p0} \\ c_{p1} \\ c_{p2} \\ c_{p3} \\ c_{p4} \end{bmatrix} = H_p \mathbf{c}_p \qquad (11.8\text{-}3\text{g})$$

so that there is enough information after N inputs to find the desired tap vector $\mathbf{c}_p$. This equation does not guarantee that the taps will be in the 'correct' order but they can be subsequently shifted to have the largest tap be at the center. We shall subsequently show how to achieve the matrix inversion of eq. 11.8-3g sequentially in N steps.

The periodic autocorrelation matrix of the system is

$$R_p = H_p^{*T} H_p = \begin{bmatrix} r_{p0} & r_{p1} & r_{p2} & r_{p3} & r_{p4} \\ r_{p4} & r_{p0} & r_{p1} & r_{p2} & r_{p3} \\ r_{p3} & r_{p4} & r_{p0} & r_{p1} & r_{p2} \\ r_{p2} & r_{p3} & r_{p4} & r_{p0} & r_{p1} \\ r_{p1} & r_{p2} & r_{p3} & r_{p4} & r_{p0} \end{bmatrix} \qquad (11.8\text{-}4\text{a})$$

which is circulant as well as Toeplitz. Consequently it can be diagonalized with the cyclic shift matrix U to look like

$$R_p = U D_p U^{-1} \qquad (11.8\text{-}4\text{b})$$

where the diagonal elements are the eigenvalues of R_p: $\lambda_{p,0}, \ldots, \lambda_{p,N-1}$.

$$\sum_{n=0}^{N-1} r_{pi} e^{-j2\pi ni/N}$$

First it is important to understand the significance of the tap vector $\mathbf{c}_p$ and its relationship to the optimum tap vector $\mathbf{c}_{\text{opt}}$ for the equalizer under the conditions of random rather than periodic data. The eigenvalues of R_p are the coefficients of the DFT (Discrete Fourier Transform) of the periodic autocorrelation function:

$$\lambda_{pi} = \sum_{n=0}^{N-1} r_{pi} e^{-j2\pi ni/N} = \left| \sum_{k=0}^{N-1} H\left(\frac{i-k}{N}\right) \right|^2, \qquad i = 0, 1, \ldots, N-1 \qquad (11.8\text{-}5\text{a})$$

and it can be shown that the frequency response of the equalizer having taps $\mathbf{c}_p$ is the inverse of the channel spectrum at N uniformly spaced discrete frequencies

$$C_p(n/N) = \frac{1}{\sum_{k=0}^{N-1} H\left(\frac{n-k}{N}\right)}, \qquad n = 0, 1, \ldots, N-1 \qquad (11.8\text{-}5\text{b})$$

This means that the tap vector $\mathbf{c}_p$ does not provide full equalization but equalization only at a discrete set of frequencies. The argument is that this measure of equalization is sufficient to open the eye and allow essentially error-free decision-directed equalization to fully converge the equalizer.

Let us start the convergence argument with the idea of adjusting the equalizer taps once every N symbols intervals according to the algorithm

$$c_{pn}(k+1) = c_{pn}(k) - \mu \sum_{j=0}^{N-1} e(k+j)y_n^*(K+j), \qquad n = 0, 1, \ldots, (N-1) \quad (11.8\text{-}6a)$$

where the product or correlation of the slicer error and the data in each of the equalizer stages is being accumulated over N time intervals. The N values of the error in this interval constitute the column vector

$$\mathbf{e}(k) = \{e(k+j)\}, \qquad j = 0, 1, \ldots, (N-1) \tag{11.8-3b}$$

and each of the components of this vector is

$$e(k+j) = \mathbf{y}^{*T}(k+j)\mathbf{c}(k) - x(k+j) \tag{11.8-3c}$$

where $x(k+j)$ is a member of the periodic data sequence $\mathbf{x}$. It is central to the cyclic equalization technique that there is no necessity for any kind of synchronization between the received input vector $\mathbf{y}(k)$ and the data vector $\mathbf{x}$ reproduced at the receiver. To be more specific, there is no slicer or decision implied in eq. 11.8-3c; rather, the error is generated by subtracting the current component of the data vector $\mathbf{x}$ from the equalizer output.

Using an $N \times N$ matrix of the N data input vectors as $Y(k)$ of eq. 11.8-3e, the error vector can be written as

$$\mathbf{e}(k) = Y^{*T}(k)\mathbf{c}_p(k) - \mathbf{x}(k) \tag{11.8-7a}$$

and the tap update algorithm as

$$\mathbf{c}_p(k+1) = \mathbf{c}_p(k) - \mu \lfloor Y^{*T}(k)Y(k)\mathbf{c}_p(k) - Y^{*T}(k)\mathbf{x}(k) \rfloor \tag{11.8-7b}$$

or

$$\mathbf{c}_p(k+1) = [I - \mu N R_p]\mathbf{c}_p(k) + \mu N \mathbf{h}_N^*(-i) \tag{11.8-7c}$$

where

$$Y^{*T}(k)\mathbf{x}(k) = N\mathbf{h}_N^*(-i) \tag{11.8-7d}$$

Since the form of this recursion equation is similar to that of the gradient algorithm, we can conclude that the optimum value of the tap vector is

$$\mathbf{c}_p(\text{opt}) = R_p^{-1}\mathbf{h}_N^*(-i) \tag{11.8-6a}$$

so that

$$\lfloor \mathbf{c}_p(k+1) - \mathbf{c}_p(\text{opt}) \rfloor = \lfloor I - \mu N R_p \rfloor \lfloor \mathbf{c}_p(k) - \mathbf{c}_p(\text{opt}) \rfloor \tag{11.8-6b}$$

We can now get some useful results if instead of updating every N symbols, we update the taps every symbol. Under these circumstances eq. 11.8-6b becomes

$$\lfloor \mathbf{c}_p(k+1) - \mathbf{c}_p(\text{opt}) \rfloor = \lfloor I - \mu R_p \rfloor \lfloor \mathbf{c}_p(k) - \mathbf{c}_p(\text{opt}) \rfloor \tag{11.8-7a}$$

and if we define

$$B_k = I - \mu R_p \tag{11.8-7b}$$

then from eq. 11.8-3 we find

$$B_{k+1} = U^{-1} B_k U \tag{11.8-7c}$$

It follows from eq. 11.8-7a that the result of the first N updates can be written as

$$\lfloor \mathbf{c}_p(N) - \mathbf{c}_p(\text{opt}) \rfloor = B_{N-1} B_{N-2} \dots B_0 \lfloor \mathbf{c}_p(0) - \mathbf{c}_p(\text{opt}) \rfloor \qquad (11.8\text{-}7d)$$

or

$$[\mathbf{c}_p(N) - \mathbf{c}_p(\text{opt})] = (U B_0)^N [\mathbf{c}_p(0) - \mathbf{c}_p(\text{opt})] \qquad (11.8\text{-}7e)$$

Using the approach of eq. 11.7-2, the characteristic equation can be reduced to

$$\det(\lambda I - U B_0) = \lambda^N + \mu N \sum_{n=0}^{N-1} \lambda^n r_{pn} - 1 = 0 \qquad (11.8\text{-}8)$$

Qureshi [Qur85] shows that if the largest solution of eq. 11.8-8, $\lambda_{\max} < 1$, then convergence in N samples will be sufficient to open the eye.

11.9 THE RLS (RECURSIVE LEAST SQUARES) ALGORITHM FOR EQUALIZER CONVERGENCE

In addition to the problem of the initial convergence of adaptive equalizers, there is the problem of the tracking of time-varying channels. This problem is particularly acute in fading channels, as discussed in Chapter 7. Whether in microwave applications or in   ayleigh fading situations, the problem can be characterized as the sudden change from one baseband-equivalent channel model to another. The object is to equalize (or re-equalize) the channel very rapidly is order to minimize the interruption to the data flow.

One of the approaches that has been developed is to transmit data in blocks of a given number of symbols, each block having a short preamble. The idea is to be able to re-equalize the channel during the preamble of each of the blocks.

  apid equalization of this kind typically requires considerably more computation for updating the equalizer taps per symbol than the LMS algorithm which, as we have seen, requires $4N$ multiplies where N is the equalizer length. The first algorithm that we will examine, developed by Godard [God74] and refined by Gitlin and Magee [G&M77], is based on the techniques of Kalman filtering. A general discussion of this approach is outside the scope of this book and the reader is referred to Haykin [Hay91]. As we shall see, this approach, which has been termed the   LS (recursive least squares) algorithm, requires of the order of N^2 multiplies per symbol for updating the equalizer taps.

The formal statement of the least-squares criterion for equalizer adaptation is, at each time nT_s, to find the equalizer tap vector $\mathbf{c}(n)$ that minimizes the quantity

$$\sum_{k=0}^{n} \lambda^{n-k} |\mathbf{y}^{*T}(k)\mathbf{c}(n) - x(k)|^2 \qquad (11.9\text{-}1)$$

where $\mathbf{y}(k)$ is the data vector in the equalizer at time $k \le n$ and $x(k)$ is the decision at time k. The idea is that the errors are computed as if $\mathbf{c}(n)$ were the tap vector for all time up to nT_s. The parameter $\lambda \le 1$ allows the early terms in the quantity to be minimized to decay exponentially. This is called *exponential windowing* and allows easier response to time variations in the channel.

We have already seen from eq. 10.1-16b that the optimum noiseless transversal equalizer taps are given by

$$C_{\text{opt}} = \left[H_c H_c^{*T} \right]^{-1} E_h H_c^T = R^{-1} H_c E_h^T \qquad (11.9\text{-}2)$$

where R is the correlation matrix of the channel and $H_c E_h^T = \mathbf{h}$ is a column vector of the sampled channel impulse response. epeating an example given earlier, if the channel has $m = 3$ samples h_1, h_2, h_3, and the transversal equalizer has $g = 5$ taps $c_1, \ldots, c_5$, then

$$\mathbf{h} = H_c E_h^T = \begin{bmatrix} h_1 & h_2 & h_3 & 0 & 0 & 0 & 0 \\ 0 & h_1 & h_2 & h_3 & 0 & 0 & 0 \\ 0 & 0 & h_1 & h_2 & h_3 & 0 & 0 \\ 0 & 0 & 0 & h_1 & h_2 & h_3 & 0 \\ 0 & 0 & 0 & 0 & h_1 & h_2 & h_3 \end{bmatrix} \begin{bmatrix} 0 \\ 0 \\ 0 \\ 1 \\ 0 \\ 0 \\ 0 \end{bmatrix} = \begin{bmatrix} 0 \\ h_3 \\ h_2 \\ h_1 \\ 0 \end{bmatrix} \qquad (11.9\text{-}3)$$

and the correlation matrix is given by

$$R = \mathbf{h}\mathbf{h}^T \qquad (11.9\text{-}4)$$

where R is $N \times N$. Clearly, if we can find the correlation matrix R and its inverse, we will have overcome the major obstacle to finding the optimum equalizer tap vector.

The LMS algorithm does not attempt to find that inverse. As we have seen, it instead computes an estimate to the gradient and follows the gradient down to the optimum setting. In so doing, its speed is limited, having only a single convergence variable, the step size μ, to deal with a system having multiple eigenvalues. Consequently, convergence can take a long time.

The LS algorithm operates to estimate R^{-1} on a recursive basis. As we shall see, this estimation process requires on the order of N^2 operations in each symbol interval as compared to the N operations per symbol interval for the LMS algorithm. A deep analysis of the LS algorithm reveals that the effect of the resulting computational complexity is to cause the trajectories corresponding to each of the eigenvalues to be excited simultaneously. One can understand the LS algorithm as recursively performing the system orthogonalization dicussed in the previous section. The result is very rapid covergence, typically within a number of symbols equal to twice the length of the equalizer.

We have already seen that the data contained in the transversal equalizer at some time k can be represented as a column vector $\mathbf{Y}$ where, in our five-tap example,

$$\begin{aligned} y_1 &= h_1 d_{k-1} + h_2 d_{k-2} + h_3 d_{k-3} \\ y_2 &= h_1 d_k + h_2 d_{k-1} + h_3 d_{k-2} \\ y_3 &= h_1 d_{k+1} + h_2 d_k + h_3 d_{k-1} \\ y_4 &= h_1 d_{k+2} + h_2 d_{k+1} + h_3 d_k \\ y_5 &= h_1 d_{k+3} + h_2 d_{k+2} + h_3 d_{k+1} \end{aligned} \qquad (11.9\text{-}5)$$

Now let's form the matrix $\hat{R} = \mathbf{Y}\mathbf{Y}^T$, which is clearly equal to the correlation matrix R (corresponding to the terms that are multiplied by $d_k d_k$) plus additional terms (corresponding to the terms $d_i d_j$, where $i \neq j$). Since $E[d_i d_j] = 0$, if we accumulate these matrices as either an expectation or a time average (since the data random process is assumed to be ergodic, it makes no difference), these latter terms will average out to zero leaving an authentic correlation matrix.

Define a time-average correlation matrix

$$\bar{R}(m) = \sum_{i=1}^{m} \lambda^{m-i} \mathbf{Y}(i)\mathbf{Y}(i)^{*T} \qquad (11.9\text{-}6a)$$

where λ is a positive constant having value somewhat less than unity. Accordingly, computations that were made at the beginning of the process of accumulation tend to be 'forgotten'. This makes possible tracking of time-varying channels. Eq. 11.9-6a can be rewritten in the form of a recursion as:

$$\bar{R}(m) = \lambda \left[\sum_{i=1}^{m-1} \lambda^{m-1-i} \mathbf{Y}(i)\mathbf{Y}(i)^{*T} \right] + \mathbf{Y}(m)\mathbf{Y}(m)^{*T}$$

$$= \lambda \bar{R}(m-1) + \mathbf{Y}(m)\mathbf{Y}(m)^{*T} \tag{11.9-6b}$$

We have seen from the discussion of the LMS algorithm that the term $H_c E_h^T$ in eq. 11.9-2 can be approximated by the cross-correlation of the equalizer data vector U and the slicer output d. Using the same approach as for eqs. 11.9-6a and 11.9-6b, define

$$\mathbf{V}(m) = \sum_{i=1}^{m} \lambda^{m-i} \mathbf{Y}(i)d^*(i)$$

$$= \lambda \left[\sum_{i=1}^{m-1} \lambda^{m-1-i} \mathbf{Y}(i)d(i)^* \right] + \mathbf{Y}(m)d^*(m)$$

$$= \lambda \mathbf{V}(m-1) + \mathbf{Y}(m)d^*(m) \tag{11.9-7}$$

The basic idea is that we want to update the tap coefficient vector at each symbol interval according to

$$\mathbf{C}(m) = \bar{R}^{-1}(m)\mathbf{V}(m) \tag{11.9-8}$$

but we want to do it in a way that minimizes the computation involved in finding the matrix inverse. This is accomplished through a combined use of the recursion developed in eq. 11.9-6b and a theorem that will now be introduced, the Matrix Inversion Lemma.

THE MATRIX INVERSION LEMMA

Let A and B be positive definite $M \times M$ matrices, C an $M \times N$ matrix, and D a positive definite $N \times N$ matrix that are related by

$$A = B^{-1} + CDC^{*T} \tag{11.9-9a}$$

Then, according to the Matrix Inversion Lemma,

$$A^{-1} = B - BC(D + C^{*T}BC)^{-1} + C^{*T}B \tag{11.9-9b}$$

∎

We now observe that eq. 11.9-6b is in form of eq. 11.9-9a where

$$A = \bar{R}(m), \qquad B^{-1} = \lambda \bar{R}(m-1), \qquad C = \mathbf{Y}(m), \qquad D = 1 \tag{11.9-10a}$$

It then follows that

$$\bar{R}^{-1}(m) = \lambda^{-1}\bar{R}^{-1}(m-1) - \frac{\lambda^{-2}\bar{R}^{-1}(m-1)\mathbf{Y}(m)\mathbf{Y}^{*T}(m)\bar{R}^{-1}(m-1)}{[1 + \lambda^{-1}\mathbf{Y}^{*T}(m)\bar{R}^{-1}(m-1)\mathbf{Y}(m)]} \tag{11.9-10b}$$

where the matrix inverse term on the right-hand side of eq. 11.9-8b turns out to be a scalar that can more easily be placed in the denominator.

We now want to simplify eq. 11.9-10b in order to reduce the recursion process to an easily implemented algorithm. Define the tap gain vector

$$\mathbf{k}(m) = \frac{\lambda^{-1}\bar{R}^{-1}(m-1)\mathbf{Y}(m)}{[1 + \lambda^{-1}\mathbf{Y}^{*T}(m)\bar{R}^{-1}(m-1)\mathbf{Y}(m)]} \qquad (11.9\text{-}11a)$$

so that eq. 11.9-9b becomes

$$\bar{R}^{-1}(m) = \lambda^{-1}\bar{R}^{-1}(m-1) - \mathbf{k}(m)\lambda^{-1}\mathbf{Y}^{*T}(m)\bar{R}^{-1}(m-1) \qquad (11.9\text{-}11b)$$

Rearranging eq. 11.9-10a yields

$$\mathbf{k}(m)\lfloor 1 + \lambda^{-1}\mathbf{Y}^{*T}(m)\bar{R}^{-1}(m-1)\mathbf{Y}(m)\rfloor = \lambda^{-1}\bar{R}^{-1}(m-1)\mathbf{Y}(m) \qquad (11.9\text{-}12a)$$

or

$$\mathbf{k}(m) = \lfloor \lambda^{-1}\bar{R}^{-1}(m-1) - k(m)\lambda^{-1}\mathbf{Y}^{*T}(m)\bar{R}^{-1}(m-1)\rfloor\mathbf{Y}(m) = \bar{R}^{-1}(m)\mathbf{Y}(m) \qquad (11.9\text{-}12b)$$

We can now find the tap update algorithm. From eq. 11.9-7 and eq. 11.9-9b, we find

$$\mathbf{C}(m) = \bar{R}^{-1}(m)\mathbf{V}(m) = \bar{R}^{-1}(m)[\lambda\mathbf{V}(m-1) + \mathbf{Y}(m)d^*(m)] \qquad (11.9\text{-}13a)$$

or

$$
\begin{aligned}
\mathbf{C}(m) &= \lfloor \lambda^{-1}\bar{R}^{-1}(m-1) - \mathbf{k}(m)\lambda^{-1}\mathbf{Y}^{*T}(m)\bar{R}^{-1}(m-1)\rfloor\lambda\mathbf{V}(m-1) + \bar{R}^{-1}(m)\mathbf{Y}(m)d^*(m) \\
&= [\bar{R}^{-1}(m-1) - \mathbf{k}(m)\mathbf{Y}^{*T}(m)\bar{R}^{-1}(m-1)]\bar{V}(m-1) + \bar{R}^{-1}(m)\mathbf{Y}(m)d^*(m) \\
&= \bar{R}^{-1}(m-1)\mathbf{V}(m-1) - k(m)\mathbf{Y}^{*T}(m)\bar{R}^{-1}(m-1)\mathbf{V}(m-1) + \bar{R}^{-1}(m)\mathbf{Y}(m)d^*(m) \\
&= \mathbf{C}(m-1) - k(m)\mathbf{Y}^{*T}(m)\mathbf{C}(m-1) + \bar{R}^{-1}(m)\mathbf{Y}(m)d^*(m) \qquad (11.9\text{-}13b)
\end{aligned}
$$

and using eq. 11.9-12b,

$$
\begin{aligned}
\mathbf{C}(m) &= \mathbf{C}(m-1) - \mathbf{k}(m)\mathbf{Y}^{*T}(m)\mathbf{C}(m-1) + \mathbf{k}(m)d^*(m) \\
&= \mathbf{C}(m-1) + \mathbf{k}(m)[d^*(m) - \mathbf{Y}^{*T}(m)\mathbf{C}(m-1)] \\
&= \mathbf{C}(m-1) + \mathbf{k}(m)\alpha^*(m) \qquad (11.9\text{-}13c)
\end{aligned}
$$

where

$$\alpha(m) = d(m) - \mathbf{Y}^T(m)\mathbf{C}^*(m-1) \qquad (11.9\text{-}13d)$$

These equations are summarized below in the order in which they constitute the RLS algorithm.

From eq. 11.9-11a, we define

$$\mathbf{P}(m) = \mathbf{Y}^{*T}(m)\bar{R}^{-1}(m-1) \qquad (11.9\text{-}14a)$$

which is a row vector. It follows that $P(m)U(m)$ is a scalar, so that we can then compute

$$\mathbf{k}(m) = \frac{\bar{R}^{-1}(m-1)\mathbf{Y}(m)}{[\lambda + \mathbf{P}(m)\mathbf{Y}(m)]} \qquad (11.9\text{-}14b)$$

which is a column vector. We can also compute eq. 11.9-12d:

$$\alpha(m) = d(m) - \mathbf{Y}^T(m)\mathbf{C}^*(m-1) \tag{11.9-14c}$$

where $d(m)$, the slicer output, and $\alpha(m)$ are complex scalars. The transversal taps are then updated according to

$$\mathbf{C}(m) = \mathbf{C}(m-1) + \mathbf{k}(m)\alpha^*(m) \tag{11.9-14d}$$

and, finally, from eq. 11.9-10b,

$$\bar{R}^{-1}(m) = \lambda^{-1}\bar{R}^{-1}(m-1) - \mathbf{k}(m)\lambda^{-1}\mathbf{P}(m) \tag{11.9-14e}$$

which is a matrix. Recall that if the channel impulse response has length N then its correlation function has length $2N$, so that the order of the computation required in eq. 11.9-14e is $4N^2$.

There must be initial conditions established on $\bar{R}^{-1}(m-1)$, $\mathbf{Y}(m)$, and $\mathbf{C}(m)$ in order to begin the recursion process. These have been established to be

$$\bar{R}^{-1}(0) = \delta I \qquad \text{where } \delta \approx .01$$
$$\mathbf{C}(0) = \mathbf{0}$$
$$\mathbf{Y}(0) = \begin{bmatrix} 0 \\ 0 \\ \lambda^{(-M+1)/2}\delta^{1/2} \end{bmatrix} \tag{11.9-15}$$

There is a considerable literature included in the references about the RLS algorithm in which a variety of convergence properties of the algorithm are explored as well as the performance of the algorithm in tracking time-varying channels, [E&F86],[F&M82].

11.10   THE FAST KALMAN ALGORITHM FOR CHANNEL EQUALIZATION

The major obstacle to the use of the RLS algorithm is the computational complexity of the order of N^2 multiplies per symbol in order to update the N equalizer taps. In the Fast Kalman Algorithm, Falconer, Morf, and Ljung [F&L78], [LMF78] have reduced this to the order of $20N$ for complex taps and Cioffi and Kailath [C&K84] have further reduced this to the order of $14N$.

The key to the reduction of computational complexity in the Fast Kalman Algorithm is the recognition that the data vector $\mathbf{Y}$ in the equalizer (eq. 11.9-4) does not change completely from symbol to symbol. Rather, there is a shift in the data with old data dropping off and new data being entered. This allows the recursive process of computing the updates to the tap coefficients to proceed without the multiplication of an $N \times N$ matrix by a vector, rather simply by the multiplication of vectors. Consequently the order of the multiplications required is reduced from N^2 to N. The proof of this algorithm requires an understanding of Kalman filtering and is therefore beyond the scope of this book. Readers interested in such a proof may consult [Hay91], [Mue81-1], [Mue81-2], [Lin84], [C&K84], and [L&K71].

Following Falconer and Ljung [F&L78] and Lin [Lin84], we shall present the algorithm via applications to a $T_s/2$ fractionally spaced equalizer, a baud-rate transversal equalizer, and a baud rate transversal followed by a decision feedback equalizer.

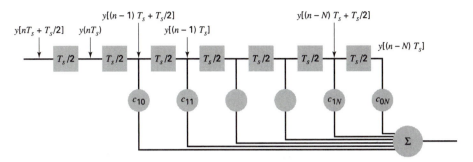

FIGURE 11.10-1

11.10.1   A Fractionally Spaced Transversal Equalizer

At time nT_s the data in the $T_s/2$ equalizer of fig. 11.10-1 can be represented as a column vector $\mathbf{Y}_N(n)$ having $2N$ rows

$$\mathbf{Y}_N(n) = \begin{bmatrix} y[(n-1)T_s + T_s/2] \\ y[(n-1)T_s] \\ \cdot \\ \cdot \\ y[(n-N)T_s + T_s/2] \\ y[(n-N)T_s] \end{bmatrix} \tag{11.10-1a}$$

and the next two data inputs in the next symbol interval as a corresponding column vector $a_2(n)$

$$\mathbf{a}_2(n) = \begin{bmatrix} y[nT_s + T_s/2] \\ y[nT_s] \end{bmatrix} \tag{11.10-1b}$$

The expanded data vector is defined as

$$\hat{\mathbf{Y}}_N(n) = \begin{bmatrix} \mathbf{a}_2(n) \\ \mathbf{Y}_N(n) \end{bmatrix} = \begin{bmatrix} \mathbf{Y}_N(n+1) \\ \mathbf{b}_2(n) \end{bmatrix} \tag{11.10-1c}$$

where

$$\mathbf{b}_2(n) = \begin{bmatrix} y[(n-N)T_s + T_s/2] \\ y[(n-N)T_s] \end{bmatrix} \tag{11.10-1d}$$

The $2N$ equalizer tap vector is defined consistently with the data matrix of eq. 11.10-1a

$$\mathbf{C}_N(n) = \begin{bmatrix} c_{11}(n) \\ c_{10}(n) \\ \\ c_{N1}(n) \\ c_{N0}(n) \end{bmatrix} \tag{11.10-2}$$

where, in the second element of the subscript, 0 refers to the on-beat tap and 1 to the off-beat tap. Note that both data and taps are assumed to be complex numbers.

The slicer error is first computed in the standard manner:

$$e(n) = \mathbf{C}_N^{*T}(n)\mathbf{Y}_N(n) - d(n) \tag{11.10-3}$$

where $d(n)$ is the decision made on the output data.

The algorithm requires the definition of some intermediate variables, which will be done as the algorithm is presented.

Step 1
Define the matrix $A(n)$, the *forward predictor*, as having $2N$ rows and 2 columns. The initial value $A(0)$ is all zeros. Then the Nth-order *forward prediction error* is the column vector $\mathbf{e}_2(n)$ having two rows:

$$\mathbf{e}_2(n, n-1) = \mathbf{a}_2(n) + A^{*T}(n-1)\mathbf{Y}_N(n) \tag{11.10-4a}$$

Step 2
Define the column vector $\mathbf{k}(n)$, the *tap update vector*, as having $2N$ rows and an initial value $\mathbf{k}(0)$ of all zeros. We then update the *forward predictor*

$$A(n) = A(n-1) - \mathbf{k}^*(n)\mathbf{e}_2^T(n) \tag{11.10-4b}$$

Step 3
We now update the *forward prediction error*

$$\mathbf{e}_2(n, n) = \mathbf{a}_2(n) + A^{*T}(n)\mathbf{Y}_N(n) \tag{11.10-4c}$$

Step 4
The *autocorrelation matrix of the forward prediction error* $E(n)$, a 2×2 matrix, has an initial value $E(0)$ of δI_2 where δ is small. The value of $\lambda < 1$, the 'forgetting factor', is an experimentally determined constant. Find

$$E(n) = \lambda E(n-1) + \mathbf{e}_2^*(n, n)\mathbf{e}_2^T(n, n) \tag{11.10-4d}$$

and then compute the column vector

$$\mathbf{d}_2(n) = [E^{-1}(n)]^*\mathbf{e}_2(n) \tag{11.10-4e}$$

Step 5
The *extended tap-update vector* $\mathbf{K}(n)$ is defined as having $2N + 2$ rows and one column

$$\mathbf{K}(n) = \left[\frac{\mathbf{d}_2(n)}{\mathbf{k}(n) + A^*(n)\mathbf{d}_2^T(n)} \right] = \left[\begin{matrix} \mathbf{m}(n) \\ \mathbf{u}(n) \end{matrix} \right] \tag{11.10-4f}$$

and is repartitioned with the vector $\mathbf{m}(n)$ having $2N$ rows and the vector $\mathbf{u}(n)$ having two rows.

Step 6
Define the matrix $B(n)$, the *backward predictor*, as having $2N$ rows and two columns. The initial value $B(0)$ is all zeros. Then the Nth-order *backward prediction error* is the

column vector $\mathbf{g}_2(n)$ having two rows.

$$\mathbf{g}_2(n, n - 1) = \mathbf{b}_2(n) + B^{*T}(n - 1)\mathbf{Y}_N(n + 1) \tag{11.10-4g}$$

where $\mathbf{Y}_N(n + 1)$ is defined in eq. 11.10-1c.

Step 7
Update the *backward predictor*

$$B(n) = \lfloor B(n - 1) - \mathbf{m}^*(n)\mathbf{g}_2^T(n, n - 1)\rfloor\lfloor I_2 - \mathbf{u}^*(n)\mathbf{g}_2^T(n, n - 1)\rfloor \tag{11.10-4h}$$

Step 8
Increment $\mathbf{k}(n)$, the *tap update vector*

$$\mathbf{k}(n + 1) = \mathbf{m}(n) + B^*(n)\mathbf{u}(n) \tag{11.10-4j}$$

Step 9
Update the equalizer tap coefficients

$$\mathbf{C}_N(n + 1) = \mathbf{C}_N(n) - \mathbf{k}(n + 1)e(n) \tag{11.10-4k}$$

where $e(n)$ is the slicer error computed in eq. 11.10-3.

Although the Fast Kalman Algorithm is more involved than the RLS algorithm, there is no multiplication of a matrix by a vector. The Fast Algorithm only multiplies vectors together, reducing the order of the complexity from N^2 to N. From the point of view of the simulation software, the Fast Algorithm simply takes the place of the LMS tap update algorithm.

There is a problem of the stability of the algorithm. As noted by Lin [Lin84], this instability can be rectified by a periodic reinitialization of the algorithm.

11.10.2 A Symbol Rate Transversal Equalizer

Having developed the Fast Kalman Algorithm for a $T_s/2$ fractionally spaced transversal equalizer, it is a simple matter to specialize it to a baud-rate transversal. The data vector and the tap vector each have N rows:

$$\mathbf{Y}_N(n) = \begin{bmatrix} y[(n - 1)T_s] \\ \cdot \\ \cdot \\ \cdot \\ y[(n - N)T_s] \end{bmatrix} \tag{11.10-5a}$$

$$\mathbf{C}_N(n) = \begin{bmatrix} c_1(n) \\ \cdot \\ \cdot \\ c_N(n) \end{bmatrix} \tag{11.10-5b}$$

and the new data input in the next symbol interval is a scalar $\mathbf{a}_1(n) = y[nT_s]$ so that the expanded data vector is defined as

$$\hat{\mathbf{Y}}_N(n) = \begin{bmatrix} \mathbf{a}_1(n) \\ \mathbf{Y}_N(n) \end{bmatrix} = \begin{bmatrix} \mathbf{Y}_N(n + 1) \\ \mathbf{b}_1(n) \end{bmatrix} \tag{11.10-5c}$$

where $\mathbf{b}_1(n) = y[(n - N)T_s]$.

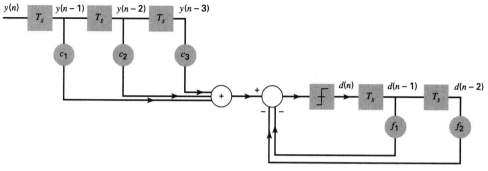

FIGURE 11.10-2

Consequently, the forward and backward predictors $A(n)$ and $B(n)$ may be redefined as having N rows and one column and the forward and backward prediction errors become the scalars $e_1(n)$ and $g_1(n)$. The rest of the algorithm is adjusted accordingly.

11.10.3 A Decision Feedback Equalizer

With a small bit of adjustment, the Fast Kalman Algorithm may also be used for the decision feedback equalizer of fig. 11.10-2. This stucture is similar to the $T_s/2$ transversal equalizer in the sense that the input to the baud-rate transversal prefilter and the input to the decision feedback portion constitute two inputs that occur in the same symbol interval. We shall proceed by example with $N_1 =$ three forward taps and $N_2 =$ two feedback taps and $N_1 + N_2 = N$. The data vector is then

$$\mathbf{Y}_N(n) = \begin{bmatrix} y[(n-1)T_s] \\ y[(n-2)T_s] \\ y[(n-3)T_s] \\ d[(n-1)T_s] \\ d[(n-2)T_s] \end{bmatrix} \tag{11.10-6a}$$

where the contents of the feedback portion are the previous decisions. The corresponding tap vector is

$$\mathbf{C}_N(n) = \begin{bmatrix} c_1(n) \\ c_2(n) \\ c_3(n) \\ f_1(n) \\ f_2(n) \end{bmatrix} \tag{11.10-6b}$$

The new data input is defined as

$$\mathbf{a}_2(n) = \begin{bmatrix} y[nT_s] \\ d[nT_s] \end{bmatrix} \tag{11.10-6c}$$

and the old data out is

$$\mathbf{b}_2(n) = \begin{bmatrix} y[(n-3)T_s] \\ d[(n-2)T_s] \end{bmatrix} \tag{11.10-6d}$$

The expanded data vector is defined as

$$\hat{\mathbf{Y}}_N(n) = \begin{bmatrix} y[nT_s] \\ y[(n-1)T_s] \\ y[(n-2)T_s] \\ y[(n-3)T_s] \\ d[nT_s] \\ d[(n-1)T_s] \\ d[(n-2)T_s] \end{bmatrix} \tag{11.10-6e}$$

We may now define the permutation matrices

$$S = \begin{bmatrix} 1 & 0 & 0 & 0 & 0 & 0 & 0 \\ 0 & 0 & 0 & 0 & 1 & 0 & 0 \\ 0 & 1 & 0 & 0 & 0 & 0 & 0 \\ 0 & 0 & 1 & 0 & 0 & 0 & 0 \\ 0 & 0 & 0 & 1 & 0 & 0 & 0 \\ 0 & 0 & 0 & 0 & 0 & 1 & 0 \\ 0 & 0 & 0 & 0 & 0 & 0 & 1 \end{bmatrix}, \quad Q = \begin{bmatrix} 1 & 0 & 0 & 0 & 0 & 0 & 0 \\ 0 & 1 & 0 & 0 & 0 & 0 & 0 \\ 0 & 0 & 1 & 0 & 0 & 0 & 0 \\ 0 & 0 & 0 & 0 & 1 & 0 & 0 \\ 0 & 0 & 0 & 0 & 0 & 1 & 0 \\ 0 & 0 & 0 & 1 & 0 & 0 & 0 \\ 0 & 0 & 0 & 0 & 0 & 0 & 1 \end{bmatrix} \tag{11.10-6f}$$

where $S^{-1} = S^T$ and $Q^{-1} = Q^T$. It is easily seen that

$$S\hat{\mathbf{Y}}_N(n) = \begin{bmatrix} \mathbf{a}_2(n) \\ \mathbf{Y}_N(n) \end{bmatrix} \quad \text{and} \quad Q\hat{\mathbf{Y}}_N(n) = \begin{bmatrix} \mathbf{Y}_N(n+1) \\ \mathbf{b}_2(n) \end{bmatrix} \tag{11.10-6g}$$

With this, Step 5 of the Fast Kalman Algorithm (eq. 11.10-4f) is broken into two parts:

Step 5a

$$\mathbf{K}(n) = S\left[\frac{\mathbf{d}_2(n)}{\mathbf{k}(n) + A^*(n)\mathbf{d}_2^T(n)} \right] \tag{11.10-7a}$$

Step 5b

$$Q\mathbf{K}(n) = \begin{bmatrix} \mathbf{m}(n) \\ \mathbf{u}(n) \end{bmatrix} \tag{11.10-7b}$$

with the vector $\mathbf{m}(n)$ having $2N$ rows and the vector $\mathbf{u}(n)$ having two rows.

The rest of the algorithm remains the same.

The Fast Kalman Algorithm is implemented in the program fastkal.c, available on the website, for all of the equalizer structures discussed above. Exploring the behaviour of the algorithm will be left to the Problems.

11.11 LATTICE EQUALIZERS

Lattice filters have been imported to adaptive equalization from speech prediction by Makhoul [Mak77], [Mak78] and Satorius [S&A79], [S&P81] and have been summarized by

Qureshi [Qur85]. This equalizer structure shares with the RLS and Fast Kalman equalizers the property of very rapid convergence, although it acheives that property in a considerably different way. As we have seen, starting with Chang [Cha71], the achievement of rapid convergence of a transversal filter was based on the development of algorithms that had the purpose of orthogonalizing the equalizer taps, as indicated in eq. 11.7-4b. The same idea was implemented in the RLS algorithm, eq. 11.9-13d, and in the Fast Kalman Algorithm, eq. 11.10-4k.

Adaptive lattice equalizers instead take the approach of orthogonalizing the signals into the equalizer tap coefficients. Lattice equalizers come with two different update algorithms, the gradient algorithm and the least-squares algorithm. We shall begin by discussing the gradient algorithm.

11.11.1 The Gradient Lattice Equalizer

The structure of a gradient transversal lattice equalizer [S&A79] (fig. 11.11-1a) differs from that of a standard transversal equalizer in a number of ways. First, it substitutes the lattice structure of fig. 11.11-1b for the simple delay element. Second, the tap update does not

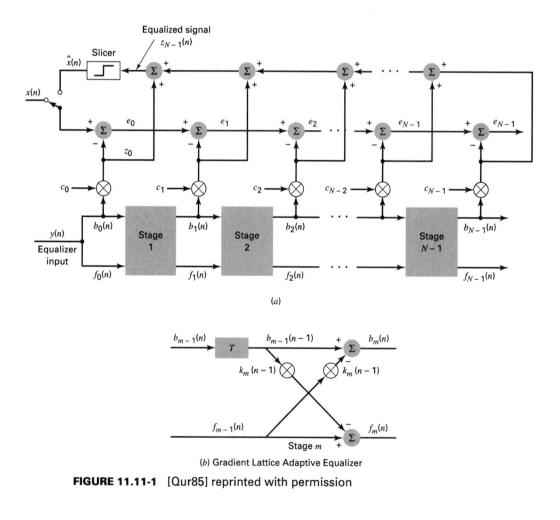

(a)

(b) Gradient Lattice Adaptive Equalizer

FIGURE 11.11-1 [Qur85] reprinted with permission

depend on a single error computed at the slicer. Rather, a different error is computed at each tap according to

$$e_m(n) = e_{m-1}(n) - c_m(n-1)b_m(n-1) \tag{11.11-1a}$$

where

$$e_0(n) = \hat{x}(n-1) - c_0(n-1)b_0(n-1) \tag{11.11-1b}$$

and where $\hat{x}(n)$ is the slicer output and the current equalizer input $y(n) = b_0(n)$. The tap update algorithm is

$$c_m(n) = c_m(n-1) - (2e_m(n)b_m^*(n)/v_m(n)) \tag{11.11-1c}$$

where the computation of $v_m(n)$ will be discussed below.

The orthogonalization of the signal at the various stages of the equalizer can be understood structurally in terms of the Gram-Schmidt procedure discussed in Section 5.1. In that procedure, just as in the lattice structure, the vector is decomposed by successively subtracting orthogonal components, leaving a remainder (or errors) that are reduced in dimension by one. Since successive lattice stages operate on orthogonal components of the signal, the number of such stages can be increased without in any way affecting the coefficients of the previous stages.

The lattice stages of fig. 11.11-1b are updated according to the algorithm

$$\begin{aligned} f_m(n) &= f_{m-1}(n) - k_m(n-1)b_{m-1}(n-1) \\ b_m(n) &= b_{m-1}(n-1) - k_m(n-1)f_{m-1}(n) \end{aligned} \tag{11.11-2a}$$

for $m = 1, \ldots, N - 1$. The signals f_m and b_m are, respectively, the forward and backward prediction errors. The reflection coefficient k_m is updated according to

$$k_m(n) = k_m(n-1) + \lfloor f_{m-1}^*(n)b_m(n) + b_{m-1}^*(n)f_m(n) \rfloor/v_m(n) \tag{11.11-2b}$$

where

$$v_m(n) = \lambda v_m(n-1) + \{|f_{m-1}^*(n)|^2 + |b_{m-1}(n-1)|^2\} \tag{11.11-2c}$$

and $f_0(n) = b_0(n) = y(n)$ and $k_m(0) = 0$. As in the Fast Kalman Algorithm, the value of $\lambda \leq 1$.

The order of the algorithm is

1. eq. 11.11-2c
2. eq. 11.11-2b
3. eq. 11.11-2a
4. eq. 11.11-1a
5. eq. 11.11-1b

This algorithm has been implemented in the program gradlat.c and appropriate simulations will be left to the Problems.

11.11.2 The Least-Squares Lattice Equalizer

The more widely used version of the lattice equalizer is based on the least squares criterion of eq. 11.9-1, which converges even faster than the gradient lattice algorithm but

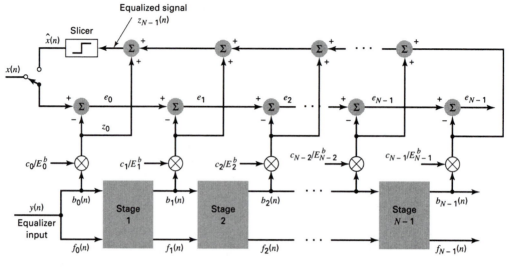

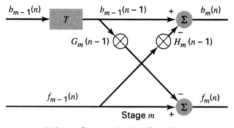

(b) Least Squares Lattice Equalizer

FIGURE 11.11-2

with a somewhat heavier computational load. The reader interested in the mathematical development of the algorithm should consult [Fri82], [Hay91], and [Pro95].

The distinctions between the least-squares lattice equalizer (fig. 11.11-2) and the gradient lattice equalizer first can be seen by comparing comparing both the two different lattices and the computation of the output. The forward and backward reflection coefficients of the least-squares lattice are unequal and are separately updated. The computation of the output also involves weighting the taps with the backward prediction errors E_m^b. The ordering of the algorithm follows [Qur85], [S&P81].

For each successive value of the time or iteration variable n and for an equalizer having $N - 1$ lattices numbered $m = 1, \ldots, N - 1$, the following steps are to be taken. Initial conditions will be discussed subsequently.

1. As shown in fig. 11.11-2a, the new input data $y(n)$ is set equal to the inputs to the first lattice stage $f_0(n)$ and $b_0(n)$ and then the forward and backward prediction errors $E_0^f(n)$ and $E_0^b(n)$ are computed as

$$E_0^f(n) = E_0^b(n) = \lambda E_0^f(n-1) + |y(n)|^2 \qquad (11.11\text{-}3a)$$

2. At each stage, from $m = 1, \ldots, N - 1$, compute

$$K_m(n) = \lambda K_m(n-1) + t_m(n-1)f_{m-1}(n) \qquad (11.11\text{-}3b)$$

3. We now compute the forward and backward prediction errors for $m = 1, \ldots, N-1$:

$$f_m(n) = f_{m-1}(n) - G_m(n-1)b_{m-1}(n-1) \qquad (11.11\text{-}4a)$$
$$b_m(n) = b_{m-1}(n-1) - H_m(n-1)f_{m-1}(n) \qquad (11.11\text{-}4b)$$

In making this computation it is useful to examine the lattice of fig 11.11-2b closely to see that, for a given value of n, there is an undelayed propogation along the forward prediction error rail.

4. The next set of computations must be done in the following **for** loop (note that $E_0^f(n)$-$E_0^b(n)$ was computed in eq. 11.11-3a):

$$\text{for } m = 1 : (N-1)$$
$$\{ \qquad G_m(n) = K_m(n)/E_{m-1}^b(n-1) \qquad (11.11\text{-}5a)$$
$$H_m(n) = K_m^*(n)/E_{m-1}^f(n) \qquad (11.11\text{-}5b)$$
$$E_m^f(n) = E_{m-1}^f(n) - G_m(n)K_m^*(n) \qquad (11.11\text{-}6a)$$
$$E_m^b(n) = E_{m-1}^b(n) - H_m(n)K_m(n) \qquad (11.11\text{-}6b)$$
$$t_m(n) = [1 - \gamma_{m-1}(n)]b_{m-1}^*(n) \qquad (11.11\text{-}7)$$
$$\gamma_m(n) = \gamma_{m-1}(n) + |t_m(n)|^2 / E_{m-1}^b(n) \qquad (11.11\text{-}8)$$
$$\}$$

5. We may then compute the equalizer output, again noting that the output $z_{N-1}(n)$ is the sum of the outputs of the individual stages. For $m = 1, \ldots, N-1$,

$$z_m(n) = z_{m-1}(n) + \left[c_m(n-1)/E_{m-1}^b(n-1) \right] b_{m-1} \qquad (11.11\text{-}9)$$

6. The tap-update error computation in the LS lattice differs significantly from that in the gradient lattice (eq. 11.11-1). Here each error is the difference between the decided data $\hat{x}(n)$ and the output of each stage:

$$e_m(n) = x(n) - z_m(n) \qquad (11.11\text{-}10)$$

7. The taps are then updated according to

$$c_m(n) = \lambda c_m(n-1) + t_m(n)e_{m-1}(n) \qquad (11.11\text{-}11)$$

All the variables have an intial value of zero except for $E_m^f(0) = E_m^b(0) = \delta$. In order to maintain the numerical stability, the update algorithm should not begin until the lattice chain has filled with data and the update algorithm should periodically be reinitialized. The algorithm has been implemented in the program lslat.c.

Sartorius and Pack [S&P81] have done some computer simulations comparing the LMS algorithm, the gradient lattice algorithm, and the least-squares lattice algorithm for two channels, one having an eigenvalue ratio $\rho = 11$ and the other with $\rho = 21$. These results are shown in fig. 11.11-3. Using the LMS algorithm, a larger value of ρ has the expected result of a longer convergence time. The convergence time of both of the lattice algorithms is essentially independent of ρ and each one is much faster than the LMS algorithm. The least-squares algorithm converges in approximately one third of the time of the gradient algorithm.

It should be reemphasized that this LS lattice structure is, in fact, simply a tap update algorithm for a transversal equalizer. Although the use of this algorithm makes equalizer convergence much faster, the steady-state transversal equalizer performance remains unchanged. It therefore continues to be advantageous to use both decision feedback and

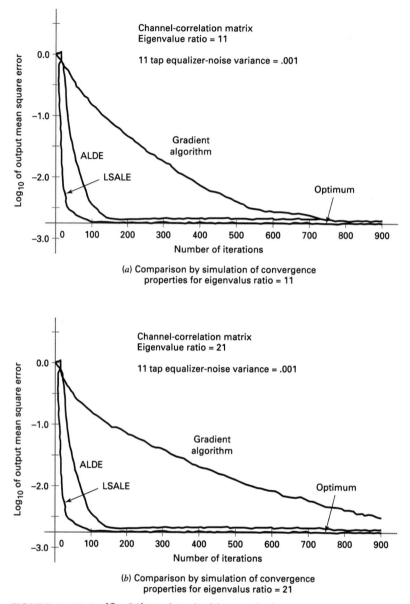

(a) Comparison by simulation of convergence
properties for eigenvalus ratio = 11

(b) Comparison by simulation of convergence
properties for eigenvalus ratio = 21

FIGURE 11.11-3   [Stp81] reprinted with permission

fractionally spaced equalizers and, of course, to be able to apply the LS lattice algorithm to
those structures.

Fractionally spaced equalizers using the LS lattice structure for tap update are described by Mueller [L&M80], [Mue81-1], [Mue81-2]. The modifications to the algorithm
involve changing some of the scalar parameters to vectors in a manner similar to Sec. 11.10a,
the Fast Kalman Algorithm for fractionally spaced equalizers. Decision-feedback equalizers using the LS lattice algorithm are addressed by Shensa [She80], Ling and Proakis
[L&P85](who allow unequal numbers of transversal and feedback taps), [E&F87] and
[Pro95]. Researching these algorithms and writing programs for their implementation is
referred to the Problems.

Finally, many of the references compare the numerical complexity of the different convergence algorithms that have been discussed in terms of the number of multiplies required for each update of an N-tap equalizer. These can be summarized for a complex baud-rate equalizer as

LMS	$4N$
RLS (Kalman)	$4N^2$
Fast Kalman	$14N$
Gradient Lattice	$26N$
LS Lattice	$40N$

This presentation has only briefly alluded to the issue of the numerical stability of these algorithms. Again, the interested reader is directed to the references. However it should be noted that the LS lattice algorithm, although the most complex of the group, also is the most numerically stable.

11.12 ADAPTIVE ECHO CANCELLATION

The issue of the generation of echoes and the need to cancel them in order to achieve full duplex transmission on the subscriber loop has been introduced in Section 4.3. This section will develop those ideas only somewhat, with the overriding concern of showing the similarity of behavior between adaptive echo cancellers and adaptive equalizers. Much deeper general treatments of echo cancelling can be found in [GH&W92], [S&B80], and [Mes84]. Some introductory survey articles on the subject include [G&L84] [Wein87], [Mof87], and [MUA90]. and some specific applications include [G&P84], [Lin90], [H&S81] and [IUL89].

Recall that echos are generated in the telephone line (fig. 11.12-1a) in the first place as a result of the imperfection of the hybrids in separating the transmitted and received signals at the joining of two-wire and four-wire lines at the home (Station A) and at the central office (Station B). The echo consists, in part, of leakage of the transmitted signal through hybrid A into the receiver and, in part, the reflection of the transmitted signal from hybrid B back to the receiver at A, as shown in fig. 11.12-1b. These two portions are, together, called the *near echo*. The *far (distant) echo* results when the signal goes from the central office onto some kind of carrier system and reflects back from a distant hybrid or antenna through the same carrier system. Each of these pieces of the echo (fig. 11.12-1c) is typically approximately 5 msec long. The far echo will suffer from frequency offset and phase jitter and will be separated in time from the near echo by a fixed (bulk) delay which may be in the range of 5 msec to 600 msec. The larger delay will come from the round-trip time in satellite communications.

Let us begin with the simpler problem of cancelling only the near echo. It was originally proposed [K&W73] (fig. 11.12-2a) that the replica of the echo be generated in an analog transversal filter from the analog transmitted signal. It was subsequently recognized [FMW76], [Mue76] that the canceller can be realized as an digital adaptive tranversal filter (fig 11.12-2b) having the transmitted data as input. When the transversal has converged its taps will be samples of the received echo. The idea is that is that the output of the canceller is subtracted from the total received signal leaving the desired received signal. The subtraction can be carried out in a number of different ways:

1. The received signal is sampled and converted to digital form with an A/D converter (fig. 11.12-3). The output of the echo canceller remains in digital form and is subtracted

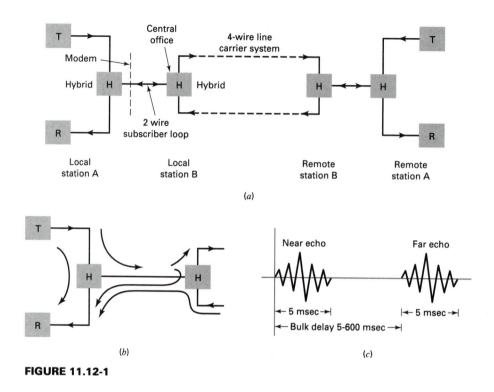

FIGURE 11.12-1

from the samples of the received signal leaving samples of the desired data signal. This is the structure of the ISDN modem [VESG79], [Mes86], [L&T88]. In this system the sampling of the signal is at the symbol rate and the canceller contains symbol rate samples of the echo [Mue76]. This creates a problem of choosing a timing phase that is correct for both the desired received signal and the echo. The problem is avoided in the ISDN by slaving the transmitter clock of station A in fig. 11.11-1a to that of station B. The timing recovery is based on these baud rate samples of the remaining desired signal as described in Chapter 6 and further in [THM86] and [B&F86]. Recall that the ISDN is a baseband system requiring a Nyquist bandwidth of 80 khz. Recall from Chapter 4 that the transmitted and received data have been made statistically independent of each other by passing them through different scramblers. This means that when the slicer error in fig. 11.12-3 is correlated with the transmitted data in order to adapt the echo canceller, the received data appears as noise. This guarantees that the echo canceller will not adapt to cancel the received data as well. We shall return to the implementation of echo cancellation and its relation to adaptive equalization in the ISDN subsequently.

  2. As we have seen, data transmission on the ordinary telephone line requires much less bandwidth than the ISDN but is a bandpass system with a nominal carrier of 1800 hz and a typical symbol rate of 2400 baud. Since the received data signal has its origin at a remote location rather than at the local central office as in the case of the ISDN, it is not possible to have an arrangement in which the clocks of the local transmitted data signal and the remote transmitted data signal are locked to each other. Consequently there is the need to have separate timing references for the transmitted data signal and the desired received data signal. Under these circumstances baud rate sampling of the received signal, as implemented in fig. 11.12-3, is inappropriate. The approach proposed by Weinstein [Wein77] (fig. 11.12-4a) is to sample the total received signal, desired data signal plus echo, at four times the symbol

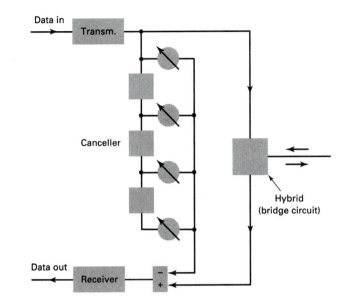

(a) Analog Echo Canceller

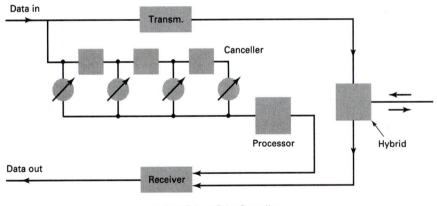

(b) Data Driven Echo Canceller

FIGURE 11.12-2   [FMW76] reprinted with permission

rate or 9600 hz. This is greater than twice the highest frequency in the signal (recall that ordinary phone transmission is confined to 300 hz–3300 hz), so that we may consider this baseband rather than passband sampling. The output of the fractionally spaced $T_s/4$ echo canceller is applied to a D/A converter at the transmitter clock rate. The input signal is also sampled at the transmitter clock rate and is held between samples. The subtraction takes place in the analog domain and the result, presumably the received signal with the echo absent, is the input to a Hilbert Transform receiver. Since the original signal was sampled at 4 fs hz, the input to the H.T. receiver remains a bandpass signal centered at 1800 hz. This filtered result is processed to derive a receiver timing signal. Note that if the adaptive equalizer in the receiver is fractionally spaced then the receiver timing signal need only be stable and is not required to have a particular timing phase for optimum performance. The box labeled Passband Transversal (fig. 11.12-4b) consists of two separate $T_s/4$ real transversal filters

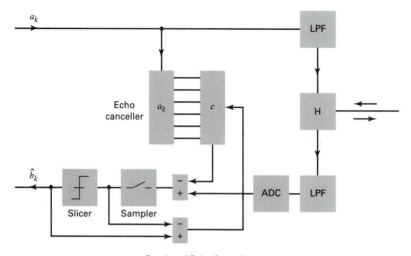

Baseband Echo Canceller

FIGURE 11.12-3 [FMW76] reprinted with permission

that generate sampled versions of the in-phase and quadrature portions of the replica of the transmitted signal. The outputs of these two transversals are added to generate the sampled replica of the total transmitted signal. By way of contrast, a complex baseband transversal (fig. 11.12-4c) represents samples of a signal that has already been demodulated.

3. In the echo canceller structures that have been discussed so far, it has assumed that the indicated error signals, the difference between the actual and replicated echo, would be used to converge the echo canceller in the manner described earlier in this chapter, using one or another of the adaptation algorithms. Although this assumption is valid, we now want to take a more careful look at the error signal and, in particular, note that this error includes not only the difference between the actual and replicated echo but also ordinary noise as well as the desired data signal masquerading as noise by virtue of its scrambled character. It is this latter term that now draws our attention.

The idea is that if we can somehow subtract the desired data signal from the error that goes to the echo canceller, that error will be less 'noisy' and the echo canceller will exhibit better convergence. An implementation of this idea was first advanced by Mueller [Mue79], as shown in (fig. 11.12-5). This is a baseband system in which it is assumed that the transmitted signal and the desired received data signal are locked to the same clock. Starting at the sampler and quantizer (or slicer), the error between the decided data and the actual signal is used to drive the convergence of both the echo canceller and the decision feedback equalizer. The error contains components appropriate to both of these structures and it is assumed that each one of these will converge separately. With both converged, both the echo and the trailing ISI components of the desired signal will be removed at the combiner. Mueller discusses the conditions for this joint convergence as well as the performance of the algorithm. Note that this combination does not address leading ISI terms at all.

This issue was next addressed by Falconer [Fal82] with an approach called Adaptive Reference Echo Cancellation (fig. 11.12-6). The received signal is shown as being sampled every $T_s/2$ sec, although it could just as well be sampled every $T_s/4$ sec as discussed above. The subtraction of the echo takes place as in fig. 11.12-4a. Again, the received signal with the echo removed is filtered and the result $y(t)$ is applied to a standard Hilbert Transform receiver which contains a fractionally spaced adaptive equalizer and a carrier-phase recovery

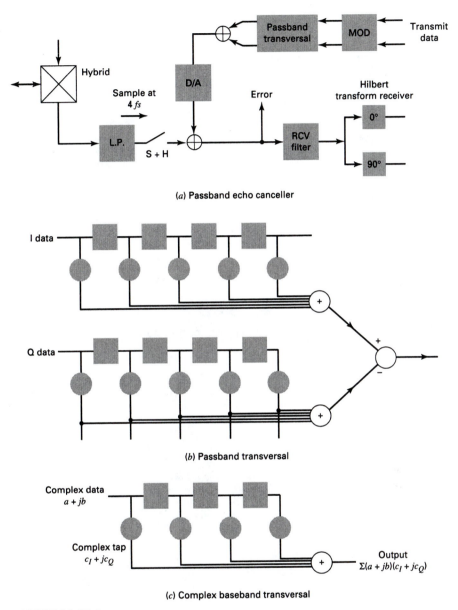

(a) Passband echo canceller

(b) Passband transversal

(c) Complex baseband transversal

FIGURE 11.12-4

system. The delay in this receiver, including the Hilbert Transform filter and the equalizer, is known to be DT_s. This delay is inserted into the error path. As shown in fig. 11.12-6, the receiver decisions are fed into a $T_s/2$ (or $T_s/4$) adaptive transversal filter that has an output that should be samples of the data portion of the received signal. The error $e(n)$ that drives the adaptation of both the echo canceller and the adaptive reference former is the difference between $y(t)$ and the output of the reference former. Falconer demonstrates that this coupled convergence in fact operates to improve the performance of the system.

4. We now will turn to the cancellation of the far echo which is understood to include frequency offset and phase jitter. The first step is to separate the cancellation of the near echo and the far echo (fig. 11.12-7a) into two different echo cancellors. The near echo

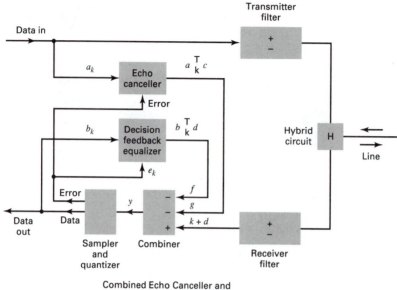

Combined Echo Canceller and
Decision Feedback Equalizer

FIGURE 11.12-5 [Mue78] reprinted with permission

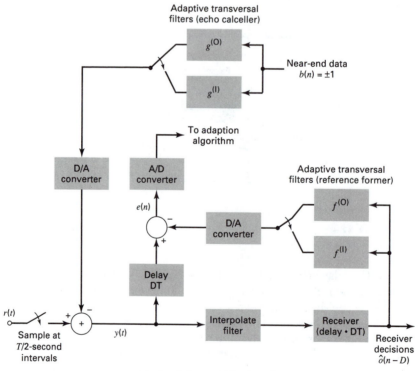

Adaptive Reference Echo Cancellation

FIGURE 11.12-6 [Fal82] reprinted with permission

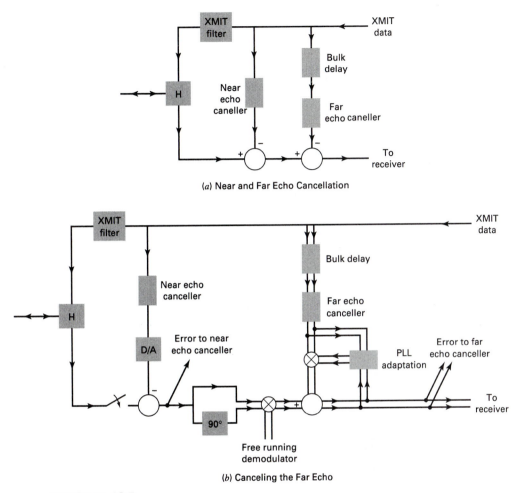

(a) Near and Far Echo Cancellation

(b) Canceling the Far Echo

FIGURE 11.12-7

canceller will not be able to follow the phase variations in the far echo. The far-echo canceller is preceded by a bulk delay that may be determined during the startup procedure of the modem by sounding the channel. The idea is that if the near echo is first cancelled, it will then be possible to determine the appropriate instantaneous phase orientation of the far echo in order to enable its cancellation. This issue has been addressed by [Wein77], [M&P86], [W&W88], and [Q&L89].

An approach to this problem is shown in fig. 11.12-7b. The near echo canceller operates as in fig. 11.12-4a. After the near echo is cancelled, the signal is applied to the front end of a Hilbert Transform receiver and changed to a baseband-equivalent complex analytic signal by passing it through a free-running demodulator. The far echo is replicated by a complex baseband transversal as shown in fig. 11.12-4c. This replica of the far-end echo must be rotated to match the phase of the signal after the free-running demodulator. The phase of this rotator is established by comparing the phase of the output of the far-echo canceller with the phase of the output after subtraction of the far echo.

With the structure outlined above, both operations of echo cancelling suffer from the presence of extraneous error signals in the adaptation process. The philosophy of adaptive reference echo cancellation may be used to create a more complex receiver structure that will remove these portions of the error from the actual error signals that drive the two portions of the echo canceller.

Until this point, the echo canceller structure has been assumed to be a tranvsversal filter in both bandpass and baseband transmission. We have seen in Section 11.6 that a decision feedback equalizer for use on the ISDN may be implemented with a RAM. Since there is no far echo on the ISDN, the echo canceller for that system and other baseband systems [VESG79] may also be implemented with a RAM [L&T88] or other nonlinear structures [AMH82] [Mes84], [Mes86]. The advantage of these nonlinear structures is their ease of implementation on a chip as well as their ability to compensate for nonlinearities in the system. The disadvantage is the same as the RAM-DFE, the slow rate of convergence.

On the issue of convergence, since the echo cancellers are adaptive transversal filters, their convergence behaviour mirrors that of transversal filters used as adaptive equalizers. If the LMS algorithm is used for convergence it has been shown [Mes84] that the best step size $\mu = 1/NR_0$ where N is the number of canceller taps and R_0 is the average power in the received signal. Again, since the structure of the echo canceller is the same as the structure of the equalizer, it is also possible to use the various techniques associated with the RLS algorithm [Hon85], [C&K85], [D&S90], to have more rapid convergence. Additional techniques for rapid convergence of echo cancellers have been proposed by Salz [Sal83], who uses pn sequences such as those discussed in cyclic equalization, and Cioffi [Cio90-2] who proposes a frequency-domain approach.

Finally, high-speed data transmission on the subscriber loop, HDSL, ADSL, and VDSL, has generated research on echo cancellation in those systems [YRL94], [C&B94].

CHAPTER 11 PROBLEMS

11.1 Show that if $\mu = 2/(\lambda_{max} + \lambda_{min})$ then the terms in eq 11.1-7c corresponding to the largest and smallest eigenvalues converge at the same rate.

11.2 Using the gradient algorithm, how many iterations will be required for the MSE to drop by 20 db for $\rho = 1.5, 2, 4$ (ρ is defined in eq. 11.1-8b)?

11.3 Explain why the LMS tap update algorithm for the forward taps of a transversal equalizer is $C_{n+1} = C_n - \mu E[e_n^* Y]$ but the algorithm for updating the decision feedback taps is $F_{n+1} = F_n + \mu E[e_n^* D]$.

11.4 **a.** Show that eq. 11.3-7d follows from eq. 11.3-7c.

      b. In eq. 11.3-7e show that

$$[|X_{eq}(f_n)|^2 - |\tilde{X}_{eq}(f_n)|^2] = \sum_k \left| X\left(f_n + \frac{k}{Ts}\right)\right|$$

11.5 Use the program MODEM and QPSK modulation to examine the convergence properties of the C3 channel for a 0 timing phase. Consider the following equalizer configurations:

    1. Two-sided baud-rate transversal equalizer
    2. One-sided baud-rate transversal equalizer plus decision feedback
    3. Constrained tap, two-sided baud-rate transversal plus decision feedback
    4. IIR baud-rate all-pass plus decision feedback
    5. Fractionally spaced $T_s/2$ transversal equalizer.

Consider the appropriate step size for convergence in the different equalizer configurations and comment on the relative speeds of convergence. Use the results of Problem 10-2 to guide your choice of the appropriate number of equalizer taps.

11.6 Repeat Problem 11-5 for the Rayleigh fading channels of Problem 10-6 using

1. Two-sided baud-rate transversal equalizer
2. One-sided baud-rate transversal equalizer plus decision feedback
3. Fractionally spaced $T_s/2$ transversal equalizer.

11.7 Confirm eqs. 11.7-2e and 11.7-2f.

11.8 Using the approach of eq. 11.7-2, find the eigenvlaues for Class-4 partial response.

11.9 Using MATLAB, find the P matrix of eq 11-7 for Duobinary, $N = 8$.

11.10 Referring to Problem 6-1, rewrite the the Matlab script of Sec. 6.5, Example 1, to include the FIR-ALE structure of Sec. 6.6. Examine the response of this structure to 60 hz carrier-phase jitter as a function of the number of taps in the ALE. What is the best number of taps? Add 120 hz jitter and see how the system responds in the presence of both. Use the FFT facility of Matlab to examine the frequency response of the FIR-ALE.

11.11 Using the program MODEM as a platform, write code that implements the Fast Kalman Algorithm for:

a. A fractionally spaced $T_s/2$ transversal equalizer.

b. A baud-rate transversal equalizer

c. A one-sided baud-rate plus decision feedback equalizer

Through computer simulation, compare the convergence characteristics of the Fast Kalman with that of the LMS algorithm for the Rayleigh fading channels of Problem 11-6.

11.12 Repeat Problem 11-12 for the gradient Lattice Algorithm.

11.13 Repeat Problem 11-12 for the Least Squares Lattice Algorithm.

11.14 Verify eq. 11.8-3g.

REFERENCES

[AMH82] O. Agazzi, D. Messerschmitt, and D. Hodges, "Nonlinear echo cancellation of data signals," *IEEE Trans. Communications*, **Com-30**, No. 11, Nov. 1982, pp. 2421–2433.

[B&F86] S. Brophy and D. Falconer, "Investigation of synchronization parameters in a digital subscriber loop transmission system," *IEEE J. Selected Areas in Communications*, **SAC-4**, No. 8, Nov. 1986, pp. 1312–1316.

[Bin87] J.A.C. Bingham, "Improved methods of accelerating the convergence of adaptive equalizers for partial-response signals," *IEEE Trans. Communications*, **Com-35**, No. 3, March 1987, pp. 257–260.

[C&B93] J. Cioffi and Y.-S. Byun, "Adaptive filtering," in *Handbook for Digital Signal Processing*, S.K. Mitra and J.F. Kaiser eds., Wiley 1993, pp. 1085–1142.

[C&B94] J. Cioffi and J.A.C. Bingham, "A data-driven multitone echo canceller," *IEEE Trans. Communications*, **Com-42**, No. 10, Oct. 1994, pp. 2853–2869.

[C&H72] R. Chang and E. Ho, "On fast start-up data communication systems using pseudo-random training sequences," *Bell System Technical J.*, Nov. 1972, pp. 2013–2027.

[C&K84] J. Cioffi and T. Kailath, "Fast recursive-least-squares transversal filters for adaptive filtering," *IEEE Trans. Acoustics, Speech and Signal Processing*, **ASSP-32**, No. 2, April 1984, pp. 304–337.

[C&K85] J. Cioffi and T. Kailath, "An efficient RLS data-driven echo canceller for fast initialization of full-duplex data transmission," *IEEE Trans. Communication*, **COM-33**, No. 7, July 1985, pp. 601–611.

[Cha71]   R.W. Chang, "A new equalizer structure for fast start-up digital communication," *Bell System Technical J.*, July-Aug. 1971, pp. 1969–2014.

[Cio90-1]   J.M. Cioffi, W.L. Abbott, H.K. Thapar, C.M. Melas, and K.D. Fisher, "Adaptive equalization in magnetic-disc storage channels," *IEEE Communications Magazine*, Feb. 1990, **28**, No. 2, pp. 14–29.

[Cio90-2]   J. Cioffi, "A fast echo canceller initialization method for the CCITT V.32 modem," *IEEE Trans. Communications*, **COM-38**, No. 5, May 1990, pp. 629–638.

[Cla85]   A.P. Clark, *Equalizers for Digitial Modems*, Halsted, 1985.

[D&S90]   A. Dembo and J. Salz, "On the least squares tap adjustment algorithm in adaptive digital echo cancellers," *IEEE Trans. Commuications*, **COM-38**, No. 5, May 1990, pp. 622–628.

[E&F86]   E. Eleftheriou and D. Falconer, "Tracking properties and steady state performance of RLS adaptive filter algorithms," *IEEE Trans. Acoustics, Speech and Signal Processing*, **ASSP-34**, No. 5, Oct. 1986, pp. 1097–1110.

[E&F87]   E. Eleftheriou and D. Falconer, "Adaptive equalization techniques for HF channels," *IEEE J. Selected Areas in Communications*, **SACV-5**, No. 2, Feb. 1987, pp. 238–247.

[F&L78]   D.D. Falconer and L. Ljung, "Application of fast Kalman estimation to adaptive equalization," *IEEE Trans. Communications*, **Com-26**, No. 10, Oct. 1978, pp. 1439–1446.

[F&M82]   B. Friedlander and M. Morf, "Least squares algorithms for adaptive linear-phase filtering," *IEEE Trans. Acoustics, Speech and Signal Processing*, **ASSP-30**, No. 3, June 1982, pp. 381–389.

[FMW76]   D. Falconer, K. Mueller, S. Weinstein, "Echo cancellation techniques for full-duplex data transmission on two wire lines," NTC 1976, paper No. 8.3.

[Fal82]   D. D. Falconer, "Adaptive reference echo cancellation," *IEEE Trans. Communications*, **COM-30**, No. 9, Sept. 1982, pp. 2083–2094.

[Fis91]   K.D. Fisher, J.M. Cioffi, W.L. Abbott, P.S. Bednarz, and C.M. Melas, "An adaptive RAM-DFE for storage channels," *IEEE Trans. Communications*, **39**, No. 11, Nov. 1991, pp. 1559–1561.

[Fri82]   B. Friedlander, "Lattice filters for adaptive processing," *Proc. IEEE*, **70**, No. 8, Aug. 1982, pp. 829–867.

[G&L84]   C. W. K. Gritton and D. W. Lin, "Echo cancellation algorithms," *IEEE ASSP Mag.*, pp. 30–38, April 1984.

[G&M73]   R.D. Gitlin and J.E. Mazo, "Comparison of some cost functions for automatic equalization," *IEEE Trans. Communications*, March 1973; pp. 233–237.

[G&M77]   R.D. Gitlin and F.R. Magee, "Self-orthogonalizing adaptive equalization algorithms," *IEEE Trans. Communications*, **Com-25**, No. 7, July 1977, pp. 666–672.

[G&P84]   L. Guidoux and B. Peuch, "Binary passband echo canceller in a 4800 Bit/s two-wire duplex modem," *IEEE Journal on Selected Areas in Communications*, **SAC-2**, No. 5, Sept. 1984, pp. 711–721.

[G&W78]   R.D. Gitlin and S.B. Weinstein, "The effects of large interference on the tracking capability of digitally echo cancelers," *IEEE Trans. Communications*, **Com-26**, June 1978, pp. 833–839.

[G&W79]   R.D. Gitlin and S.B. Weinstein, "On the required tap-weight precision for digitally implemented, adaptive, mean-squared equalizers," *Bell System Technical J.*, Feb. 1979, pp. 301–321.

[G&W81]   R.D. Gitlin and S.B. Weinstein, "Fractionally-spaced equalization: an improved digital transversal equalizer," *Bell System Technical J.*, Feb. 1981, pp. 275–296.

[GH&M73]   R.D. Gitlin, E.Y. Ho, and J.E. Mazo, "Passband equalization of differentially phase-modulated data signals," *Bell System Technical J.*, **52**, No. 2, Feb. 1973, pp. 219–238.

[GM&W82] R.D. Gitlin, H.C. Meadors Jr. and S.B. Weinstein, "The tap-leakage algorithm: an algorithm for the stable operation of a digitally implemented, fractionally spaced adaptive equalizer," *Bell System Technical J.*, Oct. 1982, pp. 1817–1839.

[GH&W92] R.D. Gitlin, J.F. Hayes, and S.B. Weinstein, *Data Communications Principles*, Plenum, 1992.

[GMT73] R.D. Gitlin, J.E. Mazo, and M.G. Taylor, "On the design of gradient algorithms for digitally implemented adaptive filters," *IEEE Trans. Circuit Theory*, March 1973, pp. 125–136.

[Gar84] W.A. Gardner, "Learning characteristics of stochastic-gradient-descent algorithms," *Signal Processing*, **6**, 1984, pp. 113–133.

[Ger69] A. Gersho, "Adaptive equalization of highly dispersive channels for data transmission," *Bell System Technical J.*, Jan. 1969, pp. 55–70.

[God74] D. Godard, "Channel equalization using a Kalman filter for fast data transmission," *IBM J. Res. Develop.*, May 1974, pp. 267–273.

[Gra72] R.M. Gray, "On the asymptotic eigenvalue distribution of Toeplitz matrices," *IEEE Trans. Information Theory*, **IT-18**, No. 6, Nov. 1972, pp. 725–730.

[H&S81] N. Holte and S. Stueflotten, "A new digital echo canceler for two-wire subscriber lines," *IEEE Trans. Communications*, **COM-29**, No. 11, Nov. 1981, pp. 1573–1581.

[Hay91] S. Haykin, *Adaptive Filters*, 2nd edition, Prentice Hall, 1991.

[Hon85] M. Honig, "Echo cancellation of voiceband data signals using recursive least squares and stochastic gradient algorithms," *IEEE Trans. Communications*, **COM-33**, No. 1, Jan. 1985, pp. 65–73.

[IUL89] G. H. Im, L.K. Un, and J.C. Lee, "Performance of a class of adaptive data-driven echo cancelers," *IEEE Trans. Communications*, **COM-37**, No. 12, Dec. 1989, pp. 1254–1263.

[K&W73] V.G. Koll and S.B. Weinstein, "Simultaneous two-way data transmission over a two-wire circuit," *IEEE Trans. Communications*, **COM-21**, No. 2, Feb. 1973, pp. 143–147.

[L&K71] R. Lawrence and H. Kaufman, "The Kalman filter for the equalization of a digital communications channel," *IEEE Trans. Communications*, **Com-19**, No. 6, Dec. 1971, pp. 1137–1141.

[L&M80] T. Lim and M. Mueller, "Rapid equalizer startup using least squares algorithms," *IEEE International Conf. on Communications*, Seattle, 1980, Paper 57.7.

[L&P85] F. Ling and J. Proakis, "Lattice decision-feedback equalizers and their application to fading dispersive channels," *IEEE Trans. Communications*, **Com-33**, No. 4, April 1985.

[L&T88] N.S. Lin and C.P.J. Tzeng, "Full-duplex data transmission over local loops," *IEEE Communications Magazine*, **26**, No. 2, Feb. 1988, pp. 31–47.

[LMF78] L. Ljung, M. Morf, and D. Falconer, "Fast calculation of gain matrices for recursive estimation schemes," *Int. J. Control*, **27**, No. 1, 1978, pp. 1–19.

[Lin84] D. Lin, "On digital implementation of the fast Kalman algorithms," *IEEE Trans. Acoustics, Speech and Signal Processing*, **ASSP-32**, No. 5, Oct. 1984, pp. 998–1005.

[Lin90] D. Lin, "Minimum mean-squared error echo cancellation and equalization for digital subscriber line transmission: part I—theory and computation; part II—a simulation study," *IEEE Trans. Communications*, **COM-38**, No. 1, Jan. 1990, pp. 31–45.

[M&P86] O. Macchi and K.H. Park, "An echo canceller with controlled power for frequency offset correction," *IEEE Trans. Communications*, **COM-34**, No. 4, April 1986, pp. 408–411.

[M&S75] K. Mueller and D. Spaulding, "Cyclic equalization—a new rapidly converging equalization technique for synchronous data communication," *Bell System Technical J.*, Feb. 1975, pp. 369–406.

[MKL76]   M. Morf, T. Kailath, and L. Ljung, "Fast algorithms for recursive identification," *IEEE Conf. on Decision and Control*, Dec. 1976, pp. 916–921.

[MUA90]   K. Murano, S. Unagami, F. Amano, "Echo cancellation and applications," *IEEE Communications Magazine*, **28**, No. 1, Jan. 1990, pp. 49–55.

[Mak77]   J. Makhoul, "Stable and efficient lattice methods for linear prediction," *IEEE Trans. Acoustics, Speech and Signal Processing*, **ASSP-25**, No. 7, Oct. 1977, pp. 423–428.

[Mak78]   J. Makhoul, "A Class of all-zero lattice digital filters: properties and applications," *IEEE Trans. Acoustics, Speech and Signal Processing*, **ASSP-26**, No. 4, Aug. 1978, pp. 304–314.

[Mak81]   J. Makhoul, "On the eigenvectors of symmetric Toeplitz matrices," *IEEE Trans. Acoustics, Speech, and Signal Processing*, **ASSP-29**, No. 4, Aug. 1981, pp. 868–872.

[Maz79]   J.E. Mazo, "On the independence theory of equalizer convergence," *Bell System Technical J.*, May-June 1979, pp. 963–993.

[Maz80]   J.E. Mazo, "Analysis of decision-directed equalizer convergence," *Bell System Technical J.*, Dec. 1980, pp. 1857–1876.

[Mes84]   D.G. Messerschmitt, "Echo cancellation in speech and data transmission," *IEEE J. Selected Areas in Commonunications*, **SAC-2**, No. 2, March 1984, pp. 283–297.

[Mes86]   D.G. Messerschmitt, "Design issues in the ISDN U-interface transceiver," *IEEE J. Selected Areas in Communications*, **SAC-4**, Nov. 1986, pp. 1281–1293.

[Mof87]   R.H. Moffett, "Echo and delay problems in some digital communication systems," *IEEE Communications Magazine*, **25**, No. 1, Jan. 1987, pp. 41–47.

[Mue75]   K.H. Mueller, "A new, fast-converging mean-square algorithm for adaptive equalizers with partial-response signaling," *Bell System Technical J.*, Jan. 1975, pp. 143–153.

[Mue76]   K. H. Mueller, "A new digital echo canceler for two-wire full-duplex data transmission," *IEEE Trans. Communications*, **COM-24**, No. 9, 1976, pp. 956–962.

[Mue79]   K. H. Mueller, "Combining echo cancellation and decision feedback equalization," *Bell System Technical J.*, **58**, Feb. 1979, pp. 491–500.

[Mue81-1]   M.S. Mueller, "Least-squares algorithms for adaptive equalizers," *Bell System Technical J.*, **60**, No. 8, Oct. 1981, pp. 1905–1925.

[Mue81-2]   M.S. Mueller, "On the rapid initial convergence of least-squares equalizer adjustment algorithms," *Bell System Technical J.*, **60**, No. 10, Dec. 1981, pp. 2345–2358.

[Pro95]   J.G. Proakis, "Digital Communications," 3rd ed., McGraw Hill, 1995.

[Q&L89]   T. Quatieri and G. O'Leary, "Far-echo cancellation in the presence of frequency offset," *IEEE Trans. Communications*, **COM-37**, No. 6, June 1989, pp. 635–644.

[Qur85]   S. Qureshi, "Adaptive equalization," *IEEE Proceedings*, **73**, No. 9, Sept. 1985, pp. 1349–1387.

[S&A79]   E. H. Satorius and S.T. Alexander, "Channel equalization using adaptive lattice algorithms," *IEEE Trans. Communications*, **Com-27**, No. 6, June 1979, pp. 899–905.

[S&B80]   M. Sondhi and D. Berkley, "Silencing echoes on the telephone network," *Proc. IEEE*, **68**, No. 8, Aug. 1980, pp. 948–963.

[S&P81]   E. H. Satorius and J.D Pack, "Application of least-squares lattice algorithms to adaptive equalization," *IEEE Trans. Communications*, **Com-29**, No. 2, Feb. 1981, pp. 136–142.

[S&S71-1]   T. Schonfeld and M. Schwartz, "A rapidly converging first-order training algorithm for an adaptive equalizer," *IEEE Trans. Information Theory*, **IT-17**, No. 4, July 1971, pp. 431–439.

[S&S71-2]   T. Schonfeld and M. Schwartz, "A rapidly converging second-order tracking algorithm for adaptive equalization," *IEEE Trans. Information Theory*, **IT-17**, No. 5, Sept. 1971, pp. 572–579.

[S&T72]   R.T. Sha and D.T. Tang, "A new class of automatic equalizers," *IBM J. Research and Development*, Nov. 1972, pp. 556–566.

[Sal83]   J. Salz, "On the start-up problem in digital echo cancelers," *Bell System Technical J.*, **62**, No. 6, Part 1, July-Aug. 1983, pp. 1353–1364.

[She80]   M. Shensa, "A least squares lattice decision feedback equalizer," *IEEE International Conference on Communications*, Seattle, 1980, Paper 57.6.

[THM86]   C.-P. Tzeng, D. Hodges, and D. Messerschmitt, "Timing recovery in digital subscriber loops using baud-rate sampling," *IEEE J. Selected Areas in Communications*, **SAC-4**, No. 8, Nov. 1986, pp. 1302–1311.

[Ung72]   G. Ungerboeck, "Theory on the speed of convergence in adaptive equalizers for digital communication," *IBM J. Research and Development*, Nov. 1972, pp. 546–555.

[VESG79]   N.A.M. Verhoeckx, H.C. Van den Elzen, F.A.M. Snijders, and P.J. Van Gerwen, "Digital echo cancellation for baseband data transmission," *IEEE Trans. Acoustics, Speech, and Signal Processing*, **ASSP-27**, No. 6, Dec. 1979.

[VGV&C84] P.J. Van Gerwen, N. Verhoeckx, T. Classen, "Design considerations for a 144 kbit/sec digital transmission unit for the local telephone network," *IEEE J. Selected Areas in Communications*, **SAC-2**, No. 2, March 1984, pp. 314–323.

[W&W88]   J. Wang and J. Werner, "Performance analysis of an echo-cancellation arrangement that compensates for frequency offset in the far echo," *IEEE Trans. Communications*, **COM-36**, No. 3, March 1988, pp. 364–372.

[Wein77]   S.B. Weinstein, "A passband data-driven echo canceler for full duplex transmission on two-wire circuits," *IEEE Trans. Communications*, **COM-25**, No. 7, July 1977, pp. 654–666.

[Wein87]   S.B. Weinstein, "Echo cancellation in the telephone network," *IEEE Communications Magazine*, **15**, No. 1, Jan. 1977, pp. 9–15.

[Wer84]   J.J. Werner, "An echo-cancellation-based 4800 bit/s full-duplex DDD modem," *IEEE J. Selected Areas in Communications*, **SAC-2**, No. 5, Sept. 1984, pp. 722–730.

[Wer85]   J.J. Werner, "Effects of channel impairments on the performance of an in-band data driven echo canceler," *ATT Technical J.*, **64**, No. 1, Jan. 1985, pp. 91–113.

[Wid70]   B. Widrow, "Adaptive filters," from R.E. Kalman and N. DeClaris, eds. *Aspects of Network and System Theory*, Holt, Rinehart and Winston, 1970, pp. 563–587.

[Wid75]   B. Widrow et al., "The complex LMS algorithm," *IEEE Proceedings*, April 1975, pp. 719–720.

[YRL94]   J. Yang, S. Roy, and N.H. Lewis, "Data-driven echo cancellation for a multitone modulation system," *IEEE Trans. Communications*, **COM-42**, No. 5, May 1994, pp. 2134–2144.

INDEX